# Advanced
# Human Biology

# Advanced
# HUMAN BIOLOGY

**J. Simpkins**     Head of School: Biology and Health Studies
People's College of Further Education
Nottingham

**J. I. Williams**     College Coordinator: Curriculum
People's College of Further Education
Nottingham

Series Editor:     **M. K. Sands**

CollinsEducational
*An imprint of* HarperCollins*Publishers*

Published by
CollinsEducational
77–85 Fulham Palace Road
Hammersmith
London W6 8JB

© J. Simpkins and J. I. Williams 1987
First published 1987 by Unwin Hyman, London
Reprinted 1988 (twice), 1989

Published 1992 by CollinsEducational
Reprinted 1993 (twice)

British Library Cataloguing in Publication Data

Simpkins, J.
  Advanced human biology.
  I. Human biology
  I. Title   II. Williams, J. I.
  599.9 QP34.5

ISBN 000 322290 X

Cover illustration by Ian Hands

The test card on the back cover has been reproduced from *Ishihara's Tests
for Colour-Blindness*, published by KANEHARA & CO. Ltd., Tokyo,
Japan, but colour blindness testing cannot be conducted with this
material. For accurate testing, the original plate should be used.

Typeset by MS Filmsetting Ltd, Frome, Somerset
Printed and bound in Great Britain by
Butler & Tanner Ltd, Frome and London

# Acknowledgements

We gratefully acknowledge the assistance of:

Our editor, Margaret Sands, for constant encouragement and helpful guidance.

Christopher Blake and Pat Winter of Unwin Hyman for their endless patience and thoughtful advice in ensuring the completion of the project.

Professor P. H. Fentem of the Medical School, University of Nottingham, and Dr K. Howard of Liverpool Polytechnic, who devoted much time to reading the typescript. Their comments were immensely helpful.

Dr D. Brindley of the Queen's Medical Centre, University Hospital, Nottingham for his useful advice on biochemical nomenclature.

The Associated Examining Board and the Schools Examination Department of the University of London for their permission to use recent examination questions.

Mr J. Kugler of the Queen's Medical Centre, University of Nottingham and Mr D. Wark of King's Mill Hospital, Sutton-in-Ashfield, without whose expertise and commitment the task of obtaining most of the photographs would have been immensely more demanding. Nearly all of the photographs not acknowledged below or in the captions were provided by Mr Kugler and Mr Wark.

Mrs M. Hollingsworth of the Department of Anatomy, University of Sheffield, Mrs A. Tomlinson of the Queen's Medical Centre, University of Nottingham, and Mr B. Case of the Department of Botany, University of Nottingham, who kindly gave their time and skills in preparing some of the photographs.

Mr P. Bailey and Mr A. Bezear of the Queen's Medical Centre, University of Nottingham who supplied the photographs and drawings for Figures 20.19(c), 10.23(b), 13.25(a), 14.38, 16.8, 16.27 and 16.28.

Philip Harris Biological Ltd., for contributing photographs for Figures 6.37(b) and 6.39. Also for supplying microscopic preparations for which many other figures were produced.

Dr M. Davey of the Department of Botany, University of Nottingham for his generosity in providing Figures 6.7(b), 6.19, 6.21(a), 6.23(a) and (b) and 23.11(a).

Dr J. Freer of the University of Glasgow for providing Figure 25.6(b).

Dr M. A. Tribe of the School of Biological Sciences, University of Sussex for his help in acquiring Figure 5.11.

Mr D. Gould, of the Department of Dental Surgery, General Hospital, Nottingham for providing Figure 9.3.

Mr A. Pawley, Public Health Laboratory, University Hospital, Nottingham for providing Figures 25.1(b) (i) and (ii) and 25.3.

Mr A. J. Monk, Cytogenetics Department, City Hospital, Nottingham for providing Figures 20.19(a) and (b), 20.31 and 20.32.

Gene Cox for Figures 7.3(a), 10.1, 12.3(a) and (b) and 14.28.

St Bartholomew's Hospital Department of Medical Photography, Figures 10.31, 25.17, 25.18 and 25.20.

Table 21.2 was redrawn from Marshall, *Biology – Advanced Level* published by Macdonald and Evans as was Table 21.4 and Figure 21.14. Figure 20.25 was redrawn from Carter, *Human Heredity*, page 105, and was reprinted by permission of Penguin Books Ltd. Figure 24.14 was redrawn from *Evolution by Natural Selection*, Open University Science Foundation Course (S100) Unit 19.

The many other individual and institutions who provided or who gave permission to use illustrative material. They have been credited in the captions to the appropriate figures.

Any errors in the book are entirely the responsibility of the authors and publishers.

# Preface

Following the success of our first major textbook, *Advanced Biology*, we decided to write a companion volume for students preparing for GCE Advanced Level in Human Biology. In planning the Contents we have been guided to a large extent by the GCE Advanced Level syllabus in Human Biology of the Associated Examining Board. At the time of writing no other Examining Board offered an examination in Human Biology at Advanced Level. There is an overlap of content between syllabuses for GCE Advanced Level in Biology and Advanced Human Biology. Biochemistry, energy transformation, cell ultrastructure and division, mammalian anatomy and physiology, evolution and genetics, are core topics in both. Our experience in preparing students to sit the external examinations in Human Biology has enabled us to emphasise the human aspects of such topics in this book. Other topics such as demography, human diseases and parasites are clearly subjects of interest to those studying human biology.

In deciding the sequence in which the topics should appear we asked some fundamental questions such as, 'What is the human body made of?' Our bodies contain more water than any other substance, so we began by looking at those properties of water which are fundamental to human life. It is then a logical step to examine other vital molecules before turning to their organisation at the cellular level. After a study of functional histology the text moves on to organ systems and their integration as a whole body unit. Genetics follows naturally after chapters on reproduction, growth and development. At this stage students should be suitably prepared to appreciate evolutionary concepts. Population growth logically ensues from a study of human evolution. Despite many unique attributes, humans depend on the activities of plants and animals. Man's place in the biosphere, his impact on it and the threat posed to man by disease-causing organisms and parasites are dealt with in the final chapters.

Although the text is directed mainly at students preparing for GCE Advanced Level in Human Biology, it should also prove invaluable to those studying BTEC Standard Units in Mammalian Physiology II and III, Cell Biology II, and Biochemistry III as part of National Diploma and Certificate programmes. The stress placed on the human applications of biological principles makes it especially suitable for those studying courses appropriate to Medical Laboratory Sciences, Medical Physics, Physiological Measurement, Nursing and other related para-medical careers.

We have followed the recommendations of the Institute of Biology on biological nomenclature, units and symbols (1989).

J. SIMPKINS
J. I. WILLIAMS                                                   Nottingham 1989

# CONTENTS

# Section 2. Tissues and Body Systems

## Section 3. Heredity and Evolution

# Section 4. The Environment

# SECTION I
# Cell Structure, Function and Biochemistry

# 1 Water and aqueous solutions

**Table 1.1 Distribution of water in a human (after Armstrong and Bennett 1979)**

| Site | Water in body/% |
|------|------|
| intracellular | 55.0 |
| extracellular: | |
|   blood plasma | 7.5 |
|   tissue fluid and lymph | 22.5 |
|   connective tissue, | |
|     cartilage and bone | 15.0 |

**Table 1.2 Water content of a variety of organisms**

| Specimen | Water content/ % fresh mass |
|------|------|
| lettuce leaf | 93–95 |
| carrot taproot | 89–91 |
| strawberry fruit | 88–90 |
| jellyfish | 95–98 |
| earthworm | 82–84 |

Life is thought to have originated in water, and many species of plants and animals living today have the sea or freshwater as their habitat. Whether a creature lives in water, or lives on land as humans do, water is vital for its life processes. No less than two thirds of the fresh body mass of a human is water, about half occurring inside our cells and the remainder in our body fluids (Table 1.1). The bodies of all organisms contain a high percentage of water (Table 1.2).

It is in aqueous solution that all metabolic reactions take place. Water is a reactant in hydrolytic reactions (Chapter 3) and a raw material in photosynthesis (Chapter 5). However, there are many other ways in which water is indispensable to all living creatures. To appreciate why water is so vital it is first necessary to understand some of the properties of water.

## 1.1 The water molecule

A water molecule consists of two atoms of hydrogen and one atom of oxygen. The atoms are joined in such a way that the single electron of each of the hydrogen atoms is shared with electrons in the outer shell of the oxygen atom (Fig 1.1). The unshared pairs of electrons repel the shared pairs so that the hydrogen nuclei are pushed towards one another causing the molecule to be bent (Fig 1.2). In the region of the hydrogen nuclei the water molecule is positively charged, whilst the part of the oxygen atom away from the hydrogen nuclei is negatively charged.

**Fig 1.1** Formation of a molecule of water

**Fig 1.2** The non-linear water molecule

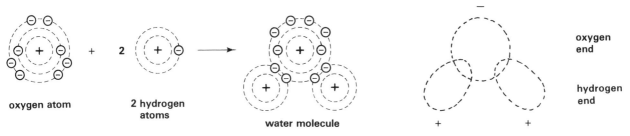

oxygen atom     2 hydrogen atoms     water molecule

oxygen end

hydrogen end

⊖    electrons

(+)    atomic nuclei

Such a molecule is **polar** because the positively charged part, or pole, differs from the negatively charged pole in the way it reacts with ions and charged molecules. The polarity of water molecules accounts for many of the properties of water which are important to living organisms.

## 1.2 Properties of water

### 1.2.1 Water as a solvent

A wide range of inorganic and organic substances readily dissolve in water. The reason why water acts as a **solvent** for so many inorganic compounds can be understood by considering the way a simple salt such as sodium chloride reacts with water. In a crystal of sodium chloride the sodium and chloride ions are held together by electrovalent bonds. When placed in

**Fig 1.3** Reaction between sodium chloride and water

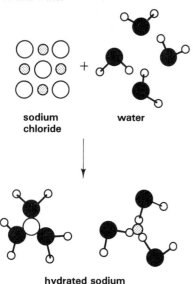

sodium
chloride

water

hydrated sodium
and chloride ions

water the positively charged sodium ions ($Na^+$) are attracted to the negatively charged oxygen poles of water molecules. Conversely, the negatively charged chloride ions ($Cl^-$) are pulled towards the positively charged hydrogen poles of water molecules. Consequently the sodium and chloride ions become separated by clusters of water molecules and an **aqueous solution** of sodium chloride is formed (Fig 1.3). The polarity of water molecules is clearly a key factor in the ability of water to dissolve inorganic substances.

Many biologically important organic substances do not ionise and yet they also dissolve in water. This is because they form hydrogen bonds with water. For example, hydroxyl (—OH) groups of sugar molecules, imino ( $>$NH) groups of proteins, and carbonyl ( $>$C$=$O) groups of organic acids react with water in this way. Molecules of such organic compounds thus become surrounded by water molecules and go into solution or suspension.

## 1 True solutions

For the reason outlined above many inorganic and organic particles smaller than $10^{-5}$ cm in diameter dissolve quickly in water to form a **true solution**. The dissolving power of water is particularly important in the uptake and transportation of substances inside the bodies of all living organisms. Metabolic reactions catalysed by enzymes also take place in aqueous solution (Chapter 3). Nevertheless, we should not overlook the fact that the inability of some substances to dissolve readily in water sometimes poses serious problems. For example, if we compare the volumes of some common gases in air and in water in direct contact with air, we see some startling differences (Table 1.3). Whereas carbon dioxide dissolves in water with ease, oxygen and nitrogen do not.

**Table 1.3 Amounts of some common gases in air and in water in direct contact with air (at 10 °C)**

| | Volume ($cm^3$) of gas in 100 $cm^3$ of air or water | | |
| | Oxygen | Nitrogen | Carbon dioxide |
| --- | --- | --- | --- |
| air | 20.95 | 78.0 | 0.03 |
| water | 0.79 | 1.2 | 0.03 |

The difference in the oxygen content of water and air has several consequences. Aquatic organisms, such as fish, have relatively less oxygen available to them for respiration in water than terrestrial organisms have in air; the lower concentration of oxygen in water is not sufficient to support respiration in humans, for example. Another consequence is that oxygen transport in our blood would be inefficient if it did not contain haemoglobin (Chapter 12).

The rate at which gases diffuse through water is much slower than through air. For this reason, the rate at which carbon dioxide diffuses through the protoplasm of photosynthetic cells before reaching the chloroplasts is a factor which limits the productivity of plants (Chapter 5).

## 2 The colloidal state

Particles between $10^{-3}$ and $10^{-5}$ cm in diameter, such as polysaccharides and proteins of high relative molar mass, do not form true solutions with water. They attract water to form a **colloidal state**, with water acting as a **dispersion medium** in which the particles, known as the **disperse phase**, are permanently suspended. Substances which react with water in this way are called hydrophilic colloids. Cytoplasm is an example of a colloidal suspension. Blood plasma is another. The blood proteins which form the disperse phase play important roles in blood clotting, combating infections and in the formation of tissue fluid (Chapters 10 and 11). Hydrophilic colloids cling to water and thus help minimise evaporation of water from organisms which live on land.

### 3 Suspensions

Particles larger than $10^{-3}$ cm in diameter can be temporarily dispersed in water to form a **suspension**. On standing, the disperse phase of a suspension will gradually separate from the dispersion medium unless an emulsifying agent is added. For example, bile emulsifies fats and oils in the mammalian gut (Chapter 9). The disperse phase of fats and oils then has a larger surface area on which fat-splitting enzymes can work.

### 1.2.2 Thermal properties

Compared with other compounds of about the same relative molar mass, water has higher melting and boiling points and is liquid over a wider temperature range (Table 1.4). This can be understood by looking at the way in which water molecules react with each other. Because particles of

**Fig 1.4** Hydrogen bonding between water molecules

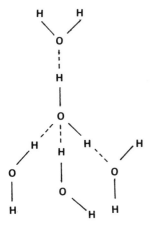

--- H = hydrogen bond

**Table 1.4 Some thermal properties of water and compounds of comparable relative molar mass**

| Substance | Formula | Relative molar mass | Melting point/°C | Boiling point/°C |
|---|---|---|---|---|
| water | $H_2O$ | 18 | 0 | 100 |
| ammonia | $NH_3$ | 17 | −78 | −33 |
| methane | $CH_4$ | 16 | −184 | −161 |

opposite charge attract each other, the negative pole of a water molecule is drawn towards the positive pole of another. Weak **hydrogen bonds** are formed between linked water molecules. In this way a water molecule can bind with up to four others, two attracted to the oxygen atom and one to each of the hydrogen atoms (Fig 1.4). Water therefore exists as **molecular clusters** rather than as individual molecules.

Temperature has an effect on the extent of bonding between water molecules. At 0 °C, the freezing point of water, the molecules are arranged in a regular, hexagonal, crystalline network in which each water molecule is hydrogen-bonded to four others (Fig 1.5(a)). At temperatures just above 0 °C some of the bonds are broken, producing clusters of molecules which fit together more compactly than the evenly spaced water molecules in ice (Fig 1.5(b)). This is why liquid water occupies a smaller volume and is therefore denser than ice. Water is most dense at 4 °C.

**Fig 1.5** (a) Crystalline structure of ice

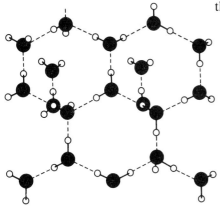

- oxygen atoms
- ○ hydrogen atoms
- --- hydrogen bonds

**Fig 1.5** (b) Molecules in liquid water

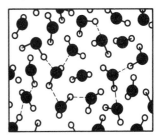

Changes in the density of water are important for the survival of many aquatic organisms. The temperature of a body of open water such as a pond or a lake is affected by the temperature of the air above the water. Very cold air causes the body of water to cool from the surface downwards. When the temperature of the upper layers falls to 4 °C the dense water sinks bringing warmer water to the surface. Convection currents of this sort delay the freezing of a body of water. Ultimately a point may be reached when the water near the surface freezes. The ice, being less dense than the water beneath, floats on the surface where it insulates the water from further heat loss. Consequently liquid water, in which aquatic life can survive, remains beneath the ice.

Heat energy can break the bonds between water molecules. Because of the extensive hydrogen-bonding in ice, a relatively large amount of heat $(333.6 \, J \, g^{-1}$ at 0 °C), called the **latent heat of melting**, is required to convert ice to liquid water. This is why water has a relatively high melting point. The formation of water vapour involves the breaking of hydrogen bonds between molecules of liquid water. Evaporation uses a lot of heat energy called the **latent heat of vaporisation** $(2411 \, J \, g^{-1}$ at 25 °C). This is why water has a relatively high boiling point. It also explains why a considerable cooling effect occurs when water evaporates from the bodies of living organisms. The evaporation of water from terrestrial organisms plays an important role in preventing them from becoming overheated (Chapter 17).

The amount of heat energy required to raise the temperature of 1 gram of a compound by 1 °C is called the **specific heat capacity**. Water has a much higher specific heat capacity $(4.2 \, J \, g^{-1}$ between 0 and 50 °C) than other compounds of similar relative molar mass. The significance of this to living organisms is that water does not quickly heat up or cool down as air does. It is for this reason that large bodies of water such as oceans and lakes have a steady low temperature which is suitable for cold-blooded organisms such as fish. The ability of water to act as a **thermal buffer** also helps to prevent rapid changes in body temperature of terrestrial organisms. Remember that land-dwelling creatures, like all forms of life, contain a large proportion of water.

### 1.2.3 Density and viscosity

A given mass of water is heavier than a similar mass of many other substances. The high **relative density** (specific gravity) of water explains why many living organisms, humans included, readily float in water. The buoyancy of water also helps the swimming of motile gametes and in the dispersal of fruits, seeds and spores.

**Viscosity** is a measure of the difficulty with which molecules slide over one another. The hydrogen bonds between water molecules are continually broken and re-formed so that the molecules slide over each other with relative ease. The viscosity of water is thus relatively low. Aqueous solutions and suspensions on the other hand can be very viscous. The high viscosity of blood plasma and lymph is of importance in the smooth flow of blood and lymph (Chapter 10).

To what extent are the properties of water reflected in the functions of human blood?

### 1.2.4 Surface tension

Another property of water attributable to the polarity of its molecules is its high **surface tension**. At the surface of an aqueous solution, water molecules are pulled downwards and inwards by the hydrogen bonds which join them to water molecules just below the surface. The result is that the water molecules in the surface layer draw together and form a skin. The surface tension in the layer of water lining our alveoli adds to the effort we need to make to stretch our lungs when breathing in. Phospholipid molecules in the alveolar membrane lower the surface tension, thus reducing the work required (Chapters 2 and 12).

### 1.2.5 Penetration by light

Light rays penetrate water with relative ease because water is transparent. The depth to which light can penetrate is of course reduced if the water is turbid due to suspended particles. In clear water, red and yellow light can reach to a depth of 50 m while blue and violet rays can go down to 200 m.

The ability of light to penetrate water enables photosynthetic organisms to inhabit the vast surface volumes of lakes and seas. It also means that light can easily penetrate the transparent water-filled epidermis of leaves and reach the underlying cells which contain the light-absorbing pigments (Chapter 5).

### 1.2.6 Dissociation of water

The nucleus of a hydrogen atom consists of a single positively charged particle called a **hydrogen ion** ($H^+$), also called a proton. In an aqueous solution small numbers of water molecules lose their hydrogen ions. The remainder of the water molecule is called a **hydroxyl ion** ($OH^-$). Such molecules are said to be **dissociated**. The process is usually written as:

$$H_2O \rightleftharpoons \quad H^+ \quad + \quad OH^-$$

water     hydrogen ion    hydroxyl ion

In pure water the concentration of hydrogen ions $[H^+]$ is $10^{-7}$ mol dm$^{-3}$ at 298 K. The product of the concentrations of hydrogen and hydroxyl ions, which is called the **ionic-product of water $K_w$**, is always $10^{-14}$ mol dm$^{-3}$ in any aqueous solution at 298 K:

$$K_w \quad = \quad [H^+] \quad \times \quad [OH^-]$$

ionic-product    concentration    concentration
of water       of $H^+$        of $OH^-$

Thus for pure water at 298 K: $K_w = 10^{-7} \times 10^{-7} = 10^{-14}$ mol dm$^{-3}$.

Remember: add indices when multiplying numbers expressed in this way.

$K_w$ is the basis of the **pH scale**, a means of indicating the concentration of hydrogen ions in an aqueous solution. The pH of a solution is defined as:

**the negative logarithm to the base 10 of the hydrogen ion concentration of the solution: $pH = -\log_{10}[H^+]$**

A solution of pH 7.0 is a **neutral** solution. If the pH is less than 7.0 the solution is **acidic** but if the pH is greater than 7.0 the solution is **alkaline**. The pH of pure water at 298 K is:

$$-\log_{10} 10^{-7} = 7.0, \quad \text{so pure water is neutral.}$$

Use the equation on the right to calculate the pH of a solution containing $1.5 \times 10^{-7}$ mol dm$^{-3}$ of hydrogen ions.

Table 1.5 gives the concentrations of hydrogen and hydroxyl ions in solutions ranging from pH 0 to pH 14, the full range of the pH scale. Note that the $[H^+]$ of a solution of pH 6.0 is ten times more than one of pH 7.0 and a hundred times more than a solution of pH 8.0.

**Table 1.5 Concentrations (mol dm$^{-3}$) of $H^+$ and $OH^-$ in solutions of pH ranging from 0 to 14**

| pH | 0.0 | 1.0 | 2.0 | 3.0 | 4.0 | 5.0 | 6.0 | 7.0 | 8.0 | 9.0 | 10 | 11 | 12 | 13 | 14 |
|---|---|---|---|---|---|---|---|---|---|---|---|---|---|---|---|
| $H^+$ | 1.0 | $10^{-1}$ | $10^{-2}$ | $10^{-3}$ | $10^{-4}$ | $10^{-5}$ | $10^{-6}$ | $10^{-7}$ | $10^{-8}$ | $10^{-9}$ | $10^{-10}$ | $10^{-11}$ | $10^{-12}$ | $10^{-13}$ | $10^{-14}$ |
| $OH^-$ | $10^{-14}$ | $10^{-13}$ | $10^{-12}$ | $10^{-11}$ | $10^{-10}$ | $10^{-9}$ | $10^{-8}$ | $10^{-7}$ | $10^{-6}$ | $10^{-5}$ | $10^{-4}$ | $10^{-3}$ | $10^{-2}$ | $10^{-1}$ | 1.0 |

The pH of intracellular fluid is normally between 6.5 and 8.0. The body fluids also have a pH within this range. Even more significant, the hydrogen ion concentration in living organisms is kept fairly constant. Examples of the way in which this is achieved are described later in the chapter. One of the advantages of maintaining a relatively fixed pH inside living organisms is that enzymes can then catalyse metabolic reactions at their optimum efficiency. Should the pH of body fluids go outside the pH range of 6.5–8.0, most enzymes will not work (Chapter 3).

## 1.3 Some properties of aqueous solutions

Some of the properties of aqueous solutions have already been touched on in this chapter. It is worthwhile exploring a number of these properties more fully. Two properties in particular are important to living organisms. Firstly, there are those properties governed by the nature of the solute which dissolves in water; of special importance in this context are **acids, bases** and **buffers**. Then there are the **osmotic properties** which are governed by the number of solute particles per unit volume of water, that is the concentration of an aqueous solution.

### 1.3.1 Acids, bases and buffers

An **acid** is a substance which dissociates in solution to release hydrogen ions or protons ($H^+$). Most organic acids are called **weak acids** because few of their molecules dissociate. Conversely, many inorganic acids are **strong acids** because most of their molecules dissociate. Both weak and strong acids are found in living cells. Carboxylic acids are weak acids produced by all living organisms. Hydrochloric acid, produced by cells in the stomach lining, is a strong acid.

Substances which accept $H^+$ are called **bases**. A reaction between an acid and a base involves a **conjugate acid–base pair** which consists of a $H^+$ donor and a $H^+$ acceptor. Acid–base pairs are found in solutions containing weak acids and their salts. Let us consider a mixture of carbonic acid and sodium hydrogencarbonate. The acid, being a weak acid, dissociates to yield a small number of hydrogen ions:

$$H_2CO_3 \rightleftharpoons H^+ + HCO_3^-$$

carbonic acid     hydrogen ion     hydrogencarbonate ion

However, the salt dissociates to produce a large number of hydrogencarbonate ions:

$$NaHCO_3 \rightleftharpoons HCO_3^- + Na^+$$

sodium salt     hydrogencarbonate ion     sodium ion

Because the hydrogencarbonate ions have a high affinity for $H^+$, they remove a large proportion of any further $H^+$ added to the mixture, so the pH of the solution remains almost constant. This effect is called the **buffering capacity** of the acid–base pair.

Blood and tissue fluid are buffered at about pH 7.2–7.4, by the carbonic acid–hydrogencarbonate acid–base pair (Chapters 10 and 12). The main buffer which prevents large fluctuations of pH in cells is the dihydrogenphosphate and hydrogenphosphate acid–base pair. This acid–base pair ionises as follows at pH 7.2:

$$H_2PO_4^- \rightleftharpoons H^+ + HPO_4^-$$

The importance of buffers in living organisms cannot be overstressed. They help to stabilize the pH of cellular and body fluids, so helping to maintain suitable conditions for enzymes to efficiently catalyse metabolic reactions (Chapter 3).

Fig 1.6 A simple osmometer

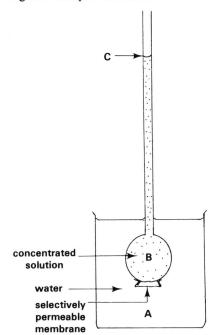

concentrated
solution

water

selectively
permeable
membrane

## 1.3.2 Osmotic properties

When pure water is separated from an aqueous solution by a **selectively permeable membrane**, water passes into the solution by **osmosis** (Fig 1.6). Water molecules on either side of the membrane move about at random. Those near it may hit the solid parts of the membrane or they may slip through the pores. In this way water moves either way across the membrane. However, in **B** the movement of water molecules is impeded by solute molecules to which water is bonded. Thus, in a given time, more water passes from **A** to **B** than vice versa. Osmosis can be stopped by applying a pressure at **C**. The pressure required to prevent a net movement of water is the **osmotic pressure** of the solution. The more concentrated a solution, the higher its osmotic pressure. Osmosis also occurs if a solution of osmotic pressure lower than at **B** is placed in **A** instead of water.

One way of defining **osmosis** is therefore:

> **the net movement of water molecules from an aqueous solution of low osmotic pressure to one of higher osmotic pressure through a selectively permeable membrane.**

In this context the term selectively permeable means that water can pass freely through the membrane, whereas solute molecules cannot.

Osmosis also occurs between aqueous suspensions separated by selectively permeable membranes. Thus plasma proteins cause blood to have a **colloidal osmotic pressure** which plays an important role in regulating the osmotic flow of water between our blood and body tissues (Chapter 10).

Another way of explaining osmosis is based on the amount of kinetic energy of water molecules. The total amount of kinetic energy of water molecules in a given system such as a cell is called its **water potential**. Pure water at STP has a water potential of 0. Aqueous solutions or suspensions have negative water potentials. Thus the water molecules at A have a greater water potential than the solution at B. Because diffusible molecules move from areas of high potential to areas of low potential, water diffuses from A to B. Osmosis may thus also be defined as:

> **the net diffusion of water through a selectively permeable membrane from a place of higher water potential to a place of lower water potential.**

# 1.4 Gain and loss of water by cells

## 1.4.1 Osmosis in human cells

An appropriate type of cell for studying osmosis is the red blood cell. On its outside there is a selectively permeable plasma membrane. Inside are substances dissolved and suspended in water, giving the cell an internal osmotic pressure. When immersed in an aqueous solution of lower osmotic pressure, such as weak saline, there is a net osmotic gain of water by the cell. Solutions having this effect are described as **hypotonic**. If sufficient water is absorbed the cell will burst. The bursting of red cells in this way is called **haemolysis**. The osmotic pressure of the solution required to cause haemolysis depends to some extent on the fragility of the plasma membrane of the red cells. This is the basis of the **osmotic fragility test** (Chapter 10).

When immersed in a **hypertonic** solution, whose osmotic pressure is higher than the cell contents, there is a net osmotic loss of water by the cell which therefore shrinks. Red cells appear **crenated** when this happens. In an **isotonic** (iso-osmotic) solution, whose osmotic pressure is identical to

**Fig 1.7** Osmotic behaviour of red blood cells

(a) Haemolysis

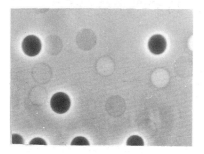

(b) Red blood cells, × 800

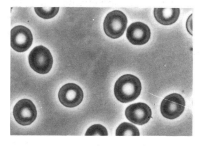

(c) Crenation

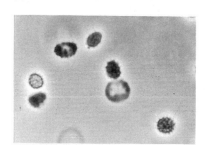

that of the cell, there is no net gain or loss of water by the cell. The cell's water content remains steady. These events are summarised in Fig 1.7.

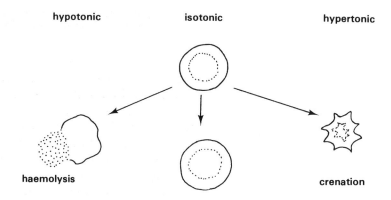

Cells cannot function efficiently if there are continuous fluctuations in their water content. A variety of **osmoregulatory** mechanisms have evolved in animals which maintain a relatively constant water content in their cells. In humans the kidneys help to keep the tissue fluid isotonic with cells (Chapter 13).

## 1.4.2 Non-osmotic movement of water

Osmosis is not the only mechanism which causes water to move into or out of cells. Water is taken in when pockets are formed in the plasma membrane during **pinocytosis** and **phagocytosis** (Chapters 6 and 10). Cells also contain substances such as polysaccharides and proteins which chemically attract water by a process called **imbibition. Evaporation** and subsequent **diffusion** of water vapour into the air is one of the main causes of water loss from the cells of our bodies (Chapters 12 and 17).

## 1.4.3 Water balance

Water is of such fundamental importance to us that it is essential we maintain a balance between gain and loss (Table 1.6). An adult human consumes on average about $1700 \, cm^3$ of water each day, about half in liquid drinks and the remainder in solid food. We make about $300 \, cm^3$ of water daily as a product of respiration (oxidation water). Water loss, mostly in urine but also in faeces and by evaporation from the skin and lungs, balances water intake and production.

**Table 1.6 Daily water balance of an adult human (from Armstrong and Bennett 1979)**

| | Gain ($cm^3$) | | Loss ($cm^3$) |
|---|---|---|---|
| in liquid drinks | 900 | in urine | 1050 |
| in solid food | 800 | from skin and lungs | 850 |
| oxidation water | 300 | in faeces | 100 |
| Total | $2 \, dm^3$ | Total | $2 \, dm^3$ |

The wide-scale famines in Africa in the 1980s illustrate that humans cannot survive without a constant supply of water. The cycling of water in our environment and the treatment of water for human consumption are described in Chapters 23 and 25 respectively.

# 2 Carbohydrates, lipoids and proteins

Many of the compounds which occur in living organisms are **organic substances**. They include carbohydrates, lipoids and proteins. Their molecules contain carbon atoms. Carbon is **tetravalent**: its atoms have four sites to which other atoms such as hydrogen, oxygen, nitrogen, phosphorus and sulphur can bond. Consequently a variety of functional groups such as amino, carboxyl and phosphate are found in naturally occurring organic compounds.

Carbon atoms also bond to each other, forming carbon chains as in fatty acid molecules and rings as in sugars. Atoms which bond to carbon are arranged spatially in three dimensions. An enormous range of organic molecules of various shapes and sizes thus exists.

## 2.1 Carbohydrates

**Carbohydrates** have the elements of a molecule of water for each carbon atom. The term carbohydrate is thus derived from 'hydrated carbon'. It reflects the fact that all carbohydrates are made up of carbon, hydrogen and oxygen, usually in the proportion 1:2:1 respectively.

### 2.1.1 Monosaccharides

Monosaccharides are the simplest of carbohydrates, having the empirical formula $(CH_2O)_n$ where $n = 3-7$. They are sweet-tasting and dissolve in water. One way of classifying monosaccharides is according to the number of carbon atoms in each of their molecules (Table 2.1).

**Fig 2.1** Reducing groups of monosaccharides

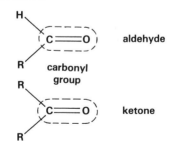

**Table 2.1 Some monosaccharide sugars**

|  | Molecular formula | Examples |
|---|---|---|
| trioses | $C_3H_6O_3$ | glyceraldehyde |
| pentoses | $C_5H_{10}O_5$ | ribose |
| hexoses | $C_6H_{12}O_6$ | glucose, fructose, galactose |

The molecules of all monosaccharides contain a **carbonyl group**, either as part of an **aldehyde** group or as part of a **ketone** group (Fig 2.1). The carbonyl group readily donates electrons. It is this property which makes monosaccharides **reducing sugars**. When heated with Fehling's or with Benedict's reagent they reduce copper(II) ions to copper(I), forming a brick-red precipitate of copper(I) oxide:

$$2Cu(OH)_2 + R-\underset{\underset{H}{|}}{C}{=}O \rightarrow Cu_2O + R.COOH + 2H_2O$$

copper(II) hydroxide     copper(I) oxide     sugar acid

**Aldoses** are monosaccharides whose molecules contain an aldehyde group, whereas **ketoses** have a ketone group instead. An alternative way of classifying monosaccharides is based on the nature of the reducing group and the number of carbon atoms per molecule (Table 2.2).

**Table 2.2 Classification of monosaccharides**

| Category | Example |
|---|---|
| aldotriose | glyceraldehyde |
| aldopentose | ribose, deoxyribose |
| aldohexose | glucose, galactose |
| ketohexose | fructose |

9

The molecules of all monosaccharides contain one or more **asymmetric carbon atoms**. They are identifiable because they have four different functional groups bonded to them. **Glyceraldehyde** has one asymmetric carbon atom to which the functional groups can be attached in two ways (Fig 2.2). Note that the hydroxyl group is on the right of the asymmetric carbon atom in D-glyceraldehyde and on the left in L-glyceraldehyde. The difference can be detected by passing a beam of polarised light through an aqueous solution of glyceraldehyde. The D-form rotates polarised light to the right (dextro-rotatory), the L-form to the left (laevo-rotatory). The two forms, called **stereoisomers**, cannot be superimposed. Imagine placing a molecule of D-glyceraldehyde in front of a mirror. The image you would see in the mirror is L-glyceraldehyde.

Fig 2.2 Stereoisomers of glyceraldehyde

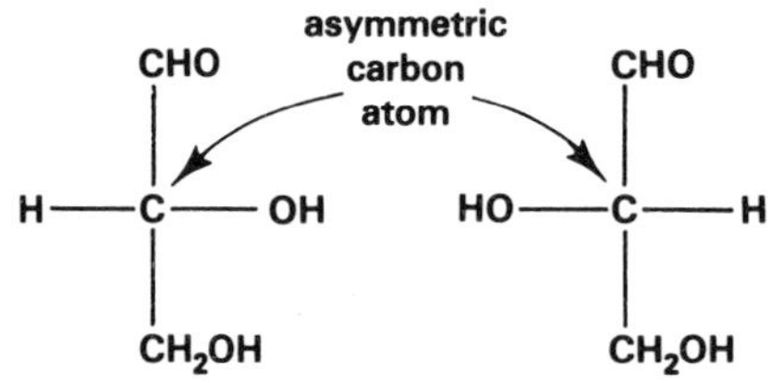

D-glyceraldehyde    L-glyceraldehyde

Pentoses and hexoses have several asymmetric carbon atoms per molecule. Of particular importance is the one furthest from the reducing group which is also the last-but-one carbon atom in the carbon chain. In the structural formulae of D-sugars, the hydroxyl group is on the right of this atom, as in D-glyceraldehyde, whereas in L-sugars it is on the left as in L-glyceraldehyde. Only D-isomers of sugars are commonly found in living organisms (Fig. 2.3).

Fig 2.3 Molecular structure of some D-sugars

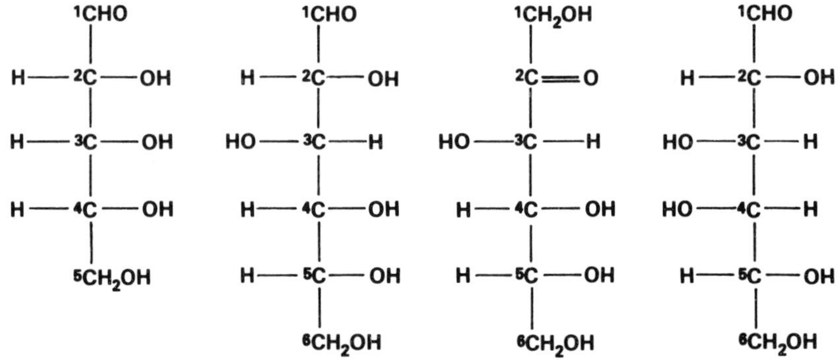

D-ribose    D-glucose    D-fructose    D-galactose

**Which of the numbered carbon atoms in each molecule is asymmetrical?**

The relatively long carbon chains of pentose and hexose sugars can bend, bringing the carbonyl group close enough to reduce one of the hydroxyl groups in the same molecule. In this way ring-shaped molecules are formed. **Glucose** forms six-sided rings where the hydroxyl group attached to carbon atom 5 is reduced. In the ring form, carbon atom 1 is asymmetric, having four different functional groups bonded to it. This is not so in the straight-chain form. The additional asymmetric carbon atom enables α-

and $\beta$-forms of D-glucose rings to exist. In **α-D-glucose** the hydroxyl group is below carbon atom 1, whereas it is above it in **β-D-glucose** (Fig 2.4).

**Fig 2.4** Ring forms of D-glucose

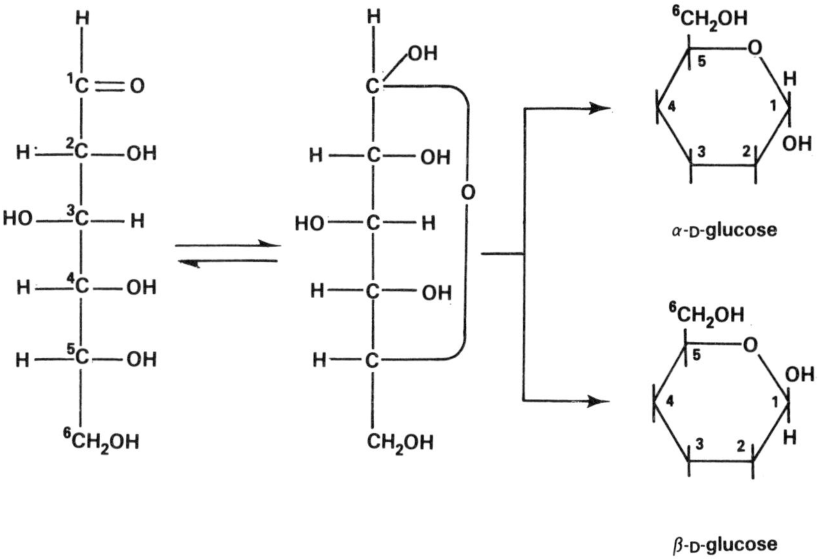

Fructose molecules can also exist as six-sided rings, but more often they occur as more stable five-sided rings in which the carbonyl group at carbon atom 2 reduces the hydroxyl group at carbon atom 5 (Fig 2.5).

**Fig 2.5** Ring forms of D-fructose

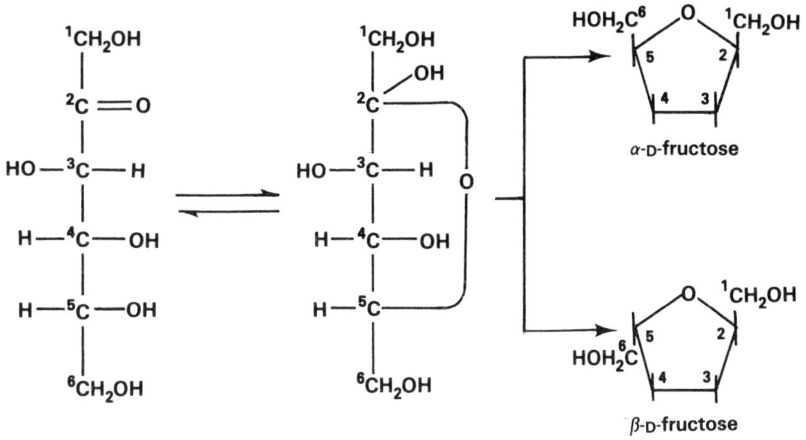

## 2.1.2 Disaccharides

Sucrose, lactose and maltose are important **disaccharides**. They are formed when two hexose molecules bond to each other in a condensation reaction:

$$2C_6H_{12}O_6 \rightarrow C_{12}H_{22}O_{11} + H_2O$$

**Sucrose** is consumed in large amounts to sweeten many of the foods we eat. It is stored in some plants such as sugar cane and sugar beet which are cultivated on a large scale. A molecule of sucrose is formed when one of α-D-glucose and one of $\beta$-D-fructose join in condensation. A bond is created between carbon atom 1 of the α-glucose ring and carbon atom 2 of the

11

How can Fehling's or Benedict's test be modified to show the presence of sucrose in a solution containing sucrose and glucose?

$\beta$-fructose. For this reason it is called an $\alpha(1\rightarrow2)$ **glycosidic linkage** (Fig 2.6(a)). It is at these carbon atoms that the carbonyl group occurs in each of the rings. It means that there are no free carbonyl groups in sucrose which is therefore a **non-reducing sugar**. However, sucrose can be easily hydrolysed into the monosaccharides of which it is made, by boiling sucrose with dilute hydrochloric acid or by incubating it with the enzyme sucrase (invertase).

**Lactose** is often called milk sugar because it is the main carbohydrate in milk. Its molecules consist of a ring of $\alpha$-D-glucose bonded by a $\beta(1\rightarrow4)$ glycosidic linkage to a ring of $\beta$-D-galactose (Fig 2.6(b)). The linkage uses the carbonyl group of the galactose ring, leaving that of glucose free. It is therefore a **reducing sugar**.

**Maltose** is an intermediate product of the hydrolysis of starch by the enzyme amylase (Chapter 9). A maltose molecule consists of two $\alpha$-D-glucose rings joined together by an $\alpha(1\rightarrow4)$ glycosidic linkage (Fig 2.6(c)). The linkage involves the carbonyl group of only one of the glucose rings. Maltose thus has a free carbonyl group and is a **reducing sugar**.

**Fig 2.6** Formation of some disaccharides     (a) Sucrose

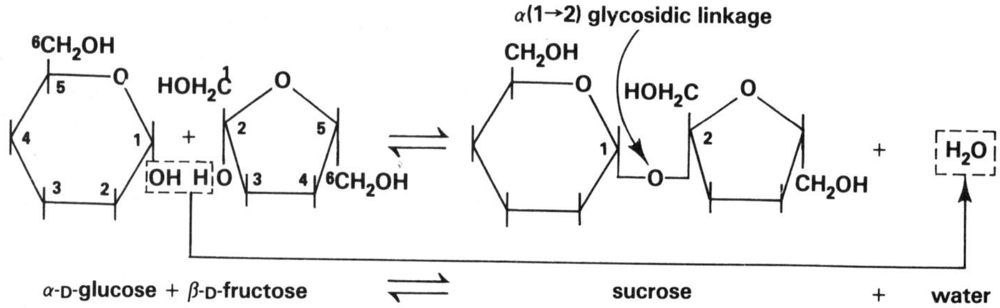

$\alpha$-D-glucose + $\beta$-D-fructose $\rightleftharpoons$ sucrose + water

(b) Lactose

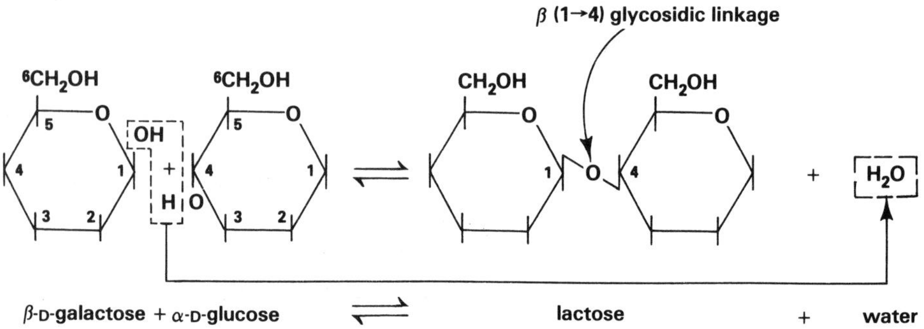

$\beta$-D-galactose + $\alpha$-D-glucose $\rightleftharpoons$ lactose + water

(c) Maltose

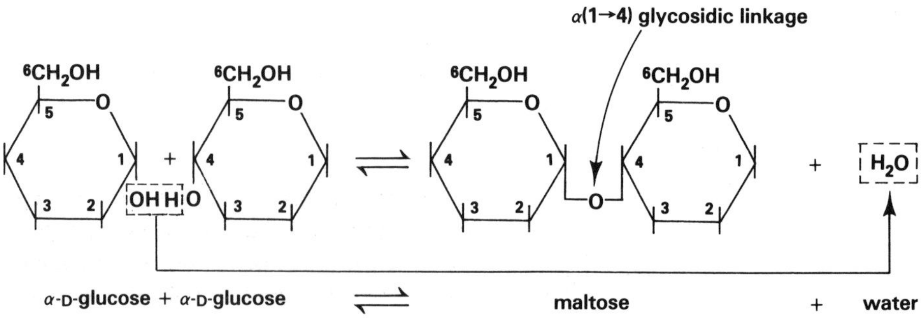

$\alpha$-D-glucose + $\alpha$-D-glucose $\rightleftharpoons$ maltose + water

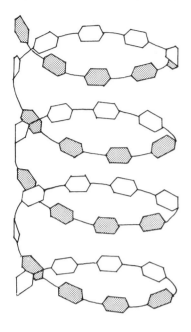

**Fig 2.7** Amylose helix

### 2.1.3 Polysaccharides

A polymer is a substance of large relative molar mass formed by the joining together of a large number of basically similar smaller molecules (monomers). Polysaccharides are **polymers** formed by the condensation of many monosaccharide molecules. D-glucose is the main sugar involved, and the products serve as storage and structural materials in plants and humans.

**Starch** is the main storage material of green plants. Its molecules have two components, **amylose** and **amylopectin**. The proportion of each varies from one type of starch to another.

'Soluble' starches consist mainly of amylose. Amylose does not truly dissolve in water but forms a colloidal suspension. Amylose is a polymer of several hundred to a few thousand $\alpha$-D-glucose rings joined by $\alpha(1\rightarrow4)$ glycosidic linkages. The resulting molecule is unbranched and wound into a helix (Fig 2.7). The bore of the helix is just big enough to trap iodine molecules. It is this reaction which produces the blue coloured complex when starch is mixed with iodine in potassium iodide solution. The iodine test is frequently used to investigate the presence of starch in materials of plant origin.

In contrast, amylopectin is a branched molecule, though again it is a polymer of $\alpha$-D-glucose. The backbone of the molecule is held together by $\alpha(1\rightarrow4)$ glycosidic linkages as in amylose. Branches arise about every twenty-fifth glucose ring where $\alpha(1\rightarrow6)$ glycosidic bonds occur (Fig 2.8). The branches which consist of 15–20 glucose rings joined by $\alpha(1\rightarrow4)$ linkages may themselves be branched. Each molecule of amylopectin contains several thousand glucose rings. Like amylose, amylopectin forms a colloidal suspension when mixed with water. The suspension reacts with iodine in potassium iodide solution giving a red-violet colour.

**Fig 2.8** (a) Part of amylopectin molecule

side branch

$\alpha$ (1 → 6)glycosidic linkage

$CH_2$

backbone

$\alpha$ (1 → 4)glycosidic linkages

**Fig 2.8** (b) Structure of amylopectin, simplified. Each circle represents an $\alpha$-D glucose ring

The two components of starch fit together to form a complex three-dimensional structure in which amylose helices are entangled in the branches of amylopectin molecules. Because starch is insoluble in water, it can be stored in large amounts without having any great effect on the water

13

potential of cells. It is usually found as grains in the tissues of storage organs such as potato tubers (Fig 2.9).

**Fig 2.9** Starch grains in potato tuber

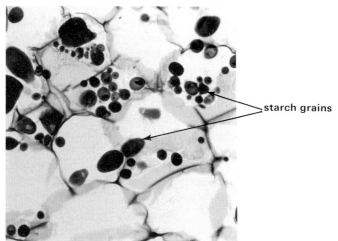

starch grains

**Fig 2.10** Structure of glycogen, simplified. Each circle represents an α-D glucose ring

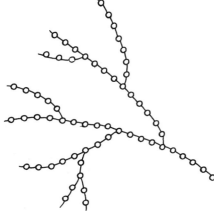

**Glycogen** is a storage polysaccharide found mainly in muscle and liver. In molecular structure it is very similar to amylopectin except that side branches occur more frequently, are somewhat longer and are more branched (Fig 2.10). Glycogen is insoluble in water and therefore is a useful storage material because it has little effect on the water potential of cellular fluid (Fig 2.11). Its reaction with iodine is similar to that of amylopectin.

**Fig 2.11** Glycogen granules in a liver cell

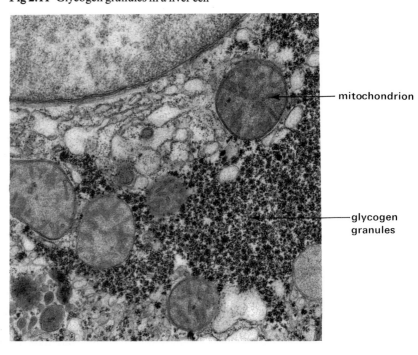

mitochondrion

glycogen granules

**Cellulose** is by far the most abundant structural polysaccharide in living organisms, although it is found only in plants. Unlike starch and glycogen, it is a polymer of $\beta$-D-glucose. Several hundred to a few thousand glucose rings are joined by $\beta(1{\rightarrow}4)$ glycosidic linkages in the long, unbranched molecules of cellulose (Fig 2.12). Attraction between hydroxyl (—OH) and hydrogen (—H) groups of the glucose rings of adjacent cellulose molecules results in the formation of **hydrogen bonds** which bind the molecules in a regular crystal-like lattice.

**Fig 2.12** (a) Part of a cellulose molecule

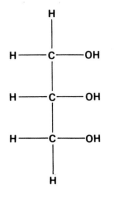

— $\beta(1\rightarrow4)$ **glycosidic linkages**

**Fig 2.12** (b) Cellulose lattice, simplified. Each circle represents a $\beta$-D glucose ring

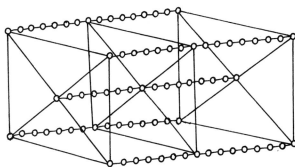

The crystalline property of cellulose has been known for a long time but the way in which cellulose is laid down in the walls of plant cells was not known until the electron microscope came into use. Electronmicrographs show that the plant cell walls consist of many fine **fibres** of cellulose which in some instances are laid down in parallel bundles but in others are distributed at random. Each fibre contains a large number of cellulose molecules. Other polysaccharides called hemicelluloses and pectic compounds cement the fibres together. In fibres and in xylem elements the walls are thickened with extra cellulose and stiffened with deposits of an alcohol polymer called lignin. This gives the tissues considerable rigidity, enabling them to help support the shoot system in the air.

## 2.2 Lipoids

The term lipoid includes a range of organic compounds which can be extracted from plant and animal tissues using non-polar organic solvents such as benzene and ether. **Lipoids** do not readily dissolve in polar solvents such as water. They include fats, oils, phospho-, sphingo- and glycolipids, waxes, steroids and sterols.

**Fig 2.13** Molecular structure of glycerol

### 2.2.1 Simple lipids

**Fats** and **oils** are sometimes called **simple lipids**. They are made of just two ingredients, **glycerol** and **fatty acids**. Glycerol is an alcohol derivative of glyceraldehyde, each of its molecules having three hydroxyl groups (Fig 2.13). Over seventy different fatty acids have been extracted from natural sources. All fatty acids have a long pleated hydrocarbon chain and a terminal carboxyl group. The length of the hydrocarbon chain differs from one fatty acid to another. Those most frequently found in living organisms have 15–17 carbon atoms, for example palmitic, stearic and oleic acids. In

15

saturated fatty acids, each carbon atom in the chain is joined to the next by single covalent bonds. In contrast, the hydrocarbon chains of **unsaturated fatty acids** have one or more double covalent bonds. Pronounced bends occur in the hydrocarbon chains where double bonds exist (Fig 2.14).

**Fig 2.14** Molecular structure of three fatty acids

(a) stearic and palmitic acids (saturated)

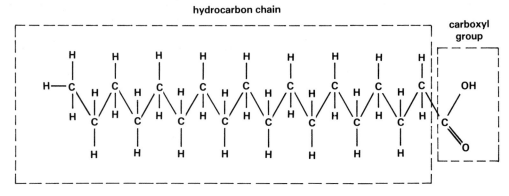

stearic acid: empirical formula $CH_3(CH_2)_{16}COOH$

palmitic acid: empirical formula $CH_3(CH_2)_{14}COOH$

(b) oleic acid (unsaturated)

double covalent bond

oleic acid: empirical formula $CH_3(CH_2)_7CH:CH(CH_2)_7COOH$

Carboxyl groups of fatty acids react with the hydroxyl groups of glycerol to form **acylglycerols** (glycerides) and water (Fig 2.15). The bonds which join the two components are called **ester linkages** and the process known as **esterification**. If only one hydroxyl group is esterified, the product is a monoacylglycerol (monoglyceride), if two a diacylglycerol (diglyceride), and if three a triacylglycerol (triglyceride). Where more than one hydroxyl group becomes esterified, the fatty acids involved may be similar or different. Many acylglycerols of animal origin are formed from saturated fatty acids. They are solid at room temperature and called fats. Butter and lard are products which consist almost entirely of fats. In contrast, acylglycerols formed from unsaturated fatty acids are liquid at 15–20 °C and are called oils. Most vegetable oils such as olive oil and corn oil are of this kind.

**Fig 2.15** Formation of an acylglycerol

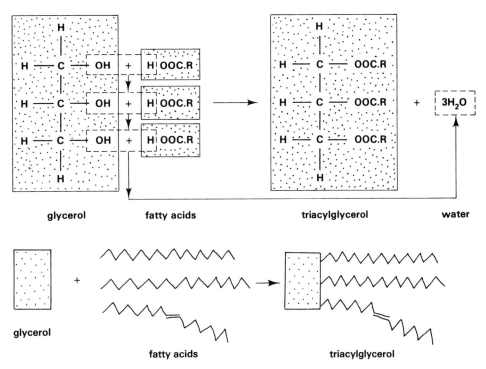

glycerol     fatty acids     triacylglycerol     water

glycerol     fatty acids     triacylglycerol

Fats and oils are virtually insoluble in water. For this reason they can be stored in the body without affecting the water potential of cells. In mammals, fats are stored in adipose tissue beneath the dermis of the skin and around internal organs (see Fig 7.2). Oil is often stored in the seeds of flowering plants (Fig 2.16).

**Fig 2.16** Oil stored in castor oil seed

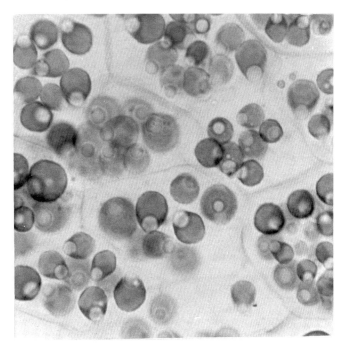

### 2.2.2 Complex lipids

**Phospholipids** are triacylglycerols in which one of the hydroxyl groups of glycerol is esterified by **phosphoric acid** which in turn is esterified to an **amino alcohol. Choline** is an amino alcohol commonly used for this

**Fig 2.17** Formation of a phospholipid

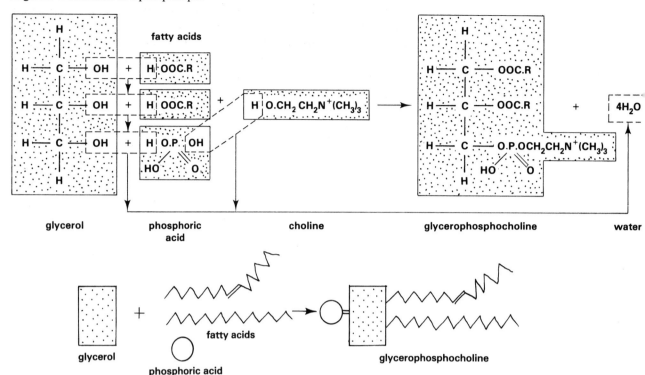

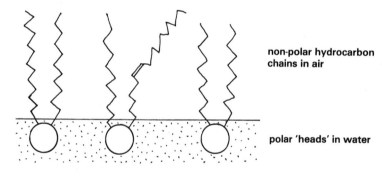

**Fig 2.18** (a) Phospholipid monolayer

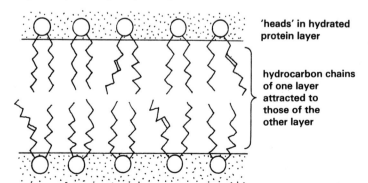

non-polar hydrocarbon chains in air

polar 'heads' in water

**Fig 2.18** (b) Phospholipid bilayer

'heads' in hydrated protein layer

hydrocarbon chains of one layer attracted to those of the other layer

purpose (Fig 2.17). **Lecithin** (glycerophosphocholine) is a phospholipid found in cell membranes (Chapter 6). The phosphate-containing 'head' of a phospholipid molecule attracts water whereas the hydrocarbon 'tails' of the fatty acids are hydrophobic. If lecithin is dissolved in ether and carefully poured on top of some water, the ether soon evaporates leaving a **monolayer** of the phospholipid on the surface of the water. The 'head' of each phospholipid molecule dissolves in the water whereas the 'tails' stand out of the water (Fig 2.18(a)).

The phospholipid monolayer greatly reduces the surface tension of the water, making it more easily stretched. Such an effect is important in the air sacs of our lungs because it enables the alveoli to become inflated with air and allows us to breathe without much effort. Some children are unable to produce phospholipids in their alveoli at birth. They suffer from respiratory distress syndrome. One of the symptoms of the condition is that a lot of effort is required when breathing to overcome the surface tension of the water layer on the alveolar membrane. Such children have to be kept in intensive care until they can synthesise lecithin.

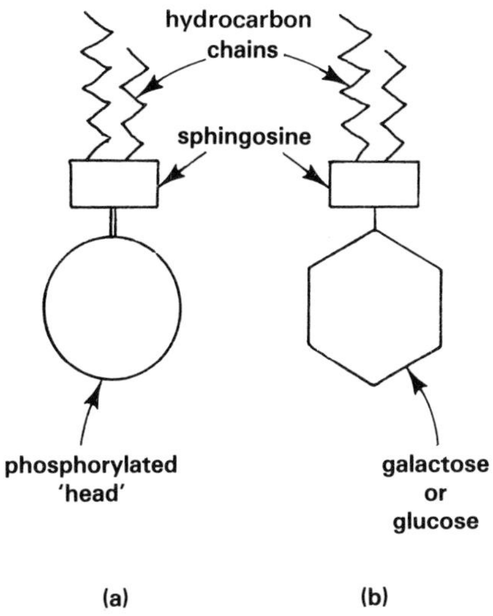

Another important effect of the phospholipid monolayer is that the hydrocarbon 'tails' pointing into the alveolar space reduce evaporation from the alveoli. If a second layer of phospholipid is carefully poured on top of the first, a **bilayer** is formed in which the 'tails' of the second layer are attracted to those of the first (Fig 2.18(b)). This arrangement is seen in the ultrastructure of cell membranes (Chapter 6).

**Sphingolipids** are comparable to phospholipids except that **sphingosine** is present instead of glycerol. Sphinogosine has a hydrocarbon chain of its own and two hydroxyl groups. One of these is esterified by phosphoric acid which in turn is esterified by an amino alcohol such as choline. Sphingosine also has an amino group which is esterified by a fatty acid with a long hydrocarbon chain. Each sphingolipid molecule thus has a water-soluble 'head', and one long and one shorter hydrocarbon 'tail' (Fig 2.19(a)). Sphingolipids are common in cell membranes.

**Glycolipids** are similar to sphingolipids except that an oligosaccharide made usually of glucose or galactose is bonded to sphingosine instead of the phosphoric acid esterified amino alcohol (Fig 2.19(b)). The oligosaccharide is water soluble, hence glycolipids have similar physical properties to sphingolipids. Galactocerebroside is the main glycolipid in the myelin sheath of nerve cells.

Fig 2.19 (a) A sphingolipid, simplified (b) A glycolipid, simplified

### 2.2.3 Waxes

**Waxes** are esters of fatty acids with long hydrocarbon chains and alcohols of high relative molar mass. The alcohols in waxes have only one hydroxyl group compared with three in glycerol.

A waxy coating called the **cuticle** covers the epidermis of leaves and stems of many kinds of terrestrial plants. It prevents excessive evaporation of water from the shoot system. The sebaceous glands in our skin release **sebum** onto the epidermis. Sebum is a mixture of waxes and triacylglycerols. It keeps the epidermis supple and inhibits some species of bacteria which would otherwise grow on the skin. Sebum also reduces the loss of water by evaporation from the surface of our bodies.

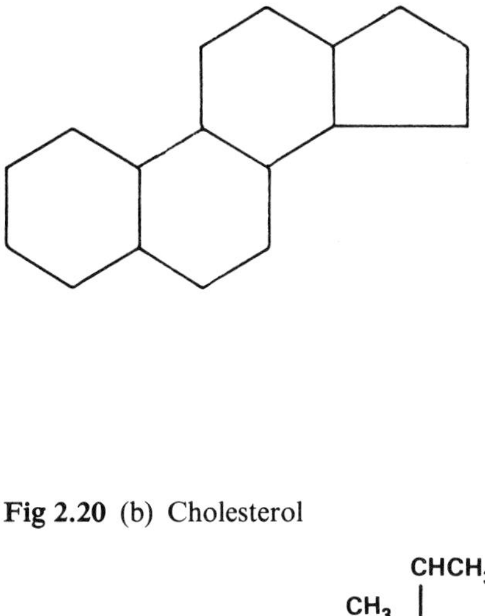

### 2.2.4 Steroids and sterols

The properties of these compounds have little in common with the other lipoids so far described, except that they are soluble in non-aqueous solvents. **Steroids** have a characteristic heterocyclic structure (Fig 2.20(a)). They include bile salts, hormones of the adrenal cortex and the sex hormones (Chapters 9, 16 and 18). **Sterols** are alcohol derivatives of steroids (Fig 2.20(b)). They include **cholesterol**, which is attracted to the hydrocarbon 'tails' of complex lipids in cell membranes (Chapter 6).

Fig 2.20 (a) A steroid, simplified

Fig 2.20 (b) Cholesterol

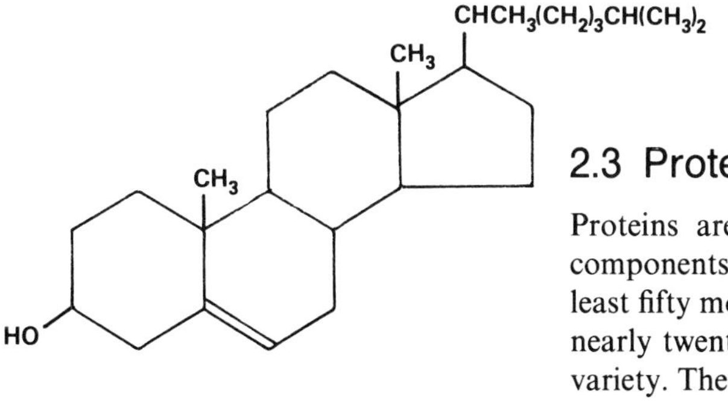

### 2.3 Proteins

Proteins are polymers of high relative molar mass. Their monomer components are **amino acids**. The smallest protein molecules contain at least fifty monomers, the largest over a thousand. A few proteins contain nearly twenty different amino acids; others are made of a much smaller variety. The number of ways in which amino acids can be put together in making proteins is infinite. It accounts for the enormous variety of proteins and their diverse functions.

**Fig 2.21** Structural formula of an amino acid

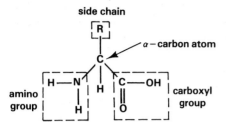

### 2.3.1 Amino acids

All amino acids contain the elements carbon, hydrogen, oxygen and nitrogen. Amino means nitrogen-containing. A few amino acids contain sulphur too. Every amino acid molecule has at least one **amino group** and one **carboxyl group** (Fig 2.21). The empirical formula for an amino acid is $NH_2.R.CH.COOH$. Chemists use Greek letters to denote the carbon atoms in molecules of this kind. The one next to the carboxyl group is the $\alpha$-carbon atom. It is to this carbon atom that the amino group also is bonded. For this reason such compounds are called $\alpha$-amino acids. With the exception of glycine, the $\alpha$-carbon atom of all amino acids is asymmetrical. It means that D- and L-stereoisomers of most amino acids can exist (Fig 2.22). In contrast to carbohydrates it is the L-isomers of amino acids which are mainly found in living organisms.

**Fig 2.22** Stereoisomers of alanine. Compare with Fig 2.2. Note that the amino group is on the right of the $\alpha$-carbon atom in D-alanine and on the left in L-alanine

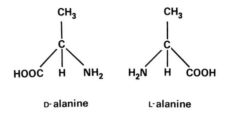

**Table 2.3** $\alpha$-amino acids commonly found in plant and animal proteins

| | | | |
|---|---|---|---|
| alanine | glutamine | leucine | serine |
| arginine | glutamic acid | lysine | threonine |
| asparagine | glycine | methionine | tryptophan |
| aspartic acid | histidine | phenylalanine | tyrosine |
| cysteine | isoleucine | proline | valine |

The twenty different amino acids isolated from plant and animal proteins are listed in Table 2.3. They differ according to the nature of the amino acid side-chain, R. In glycine, the simplest amino acid, R is a hydrogen atom. The amino acid side chain can be a short hydrocarbon chain as in alanine, a hydrocarbon ring as in tyrosine, or sulphur-containing as in cysteine. The R group of glutamic acid has a carboxyl group whereas in lysine it contains an amino group (Fig 2.23).

**Fig 2.23** Molecular structure of some amino acids

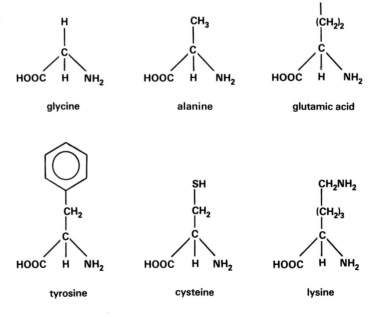

A knowledge of the nature of amino acid side chains is vital as a basis for understanding some of the techniques used to separate mixtures of amino acids and also proteins, and in appreciating some of the properties of proteins. Consider the simplest amino acid glycine. In aqueous solution its carboxyl group ionises to release hydrogen ions. It is this property which gives amino acids their acidic characteristic. However, the amino group is basic and so attracts hydrogen ions.

Amino acids thus have acidic and basic properties. For this reason they are described as **amphoteric**. In very acidic solutions ionisation of the carboxyl group is suppressed whilst the amino group attracts hydrogen ions and becomes positively charged. The effect is to give glycine an overall net electrostatic charge of $+1$. Conversely, in very alkaline solutions the carboxyl group ionises and the amino group is uncharged, giving glycine an overall net charge of $-1$. Between pH 6.5 and 7.5 both carboxyl and amino groups are charged. Such an ion with both positive and negative charges is called an **amphion** (Fig 2.24). The pH at which amphions predominate is called the **isoelectric point**.

**Fig 2.24** Effect of pH on the ionisation of amino acid: (a) glycine (b) lysine (c) glutamic acid

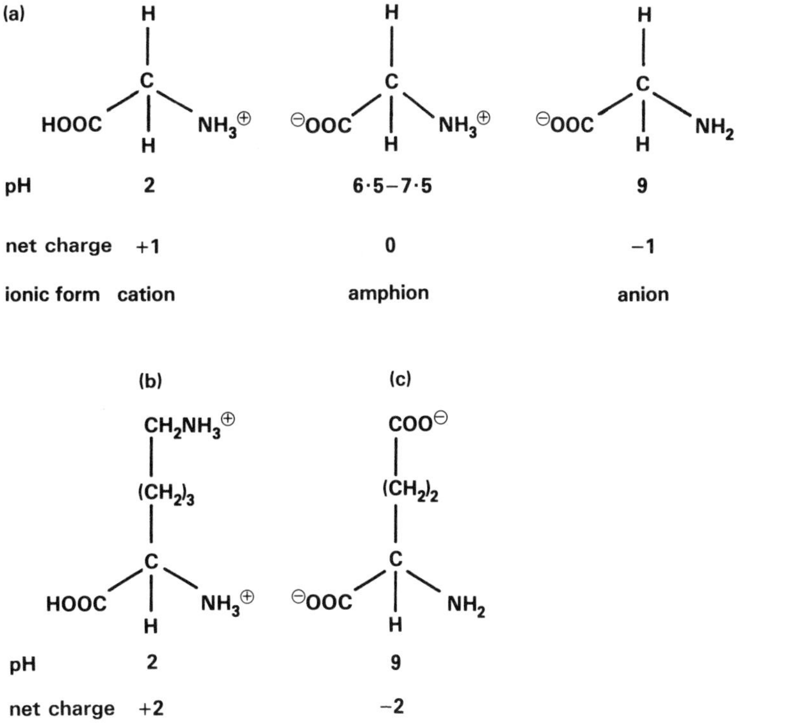

Where the amino acid side chain is also ionisable the situation is a little more complex. Nevertheless the same principle applies. If it contains an amino group this attracts hydrogen ions in acidic solutions giving an additional positive charge. Hence lysine has a net electrostatic charge of $+2$ at low pH. For similar reasons, glutamic acid has a net charge of $-2$ at high pH. Thus at a given pH some amino acids in a mixture may be cationic, others anionic and others amphionic. Furthermore, some may be more strongly cationic or anionic than others. This is the basis on which mixtures of amino acids or proteins can be separated by electrophoresis (section 2.4.2).

## 2.3.2 Polypeptides and proteins

In suitable conditions amino acids polymerise. The α-amino group of one molecule joins to the carboxyl group of another in a condensation reaction which results in the formation of **peptide linkages** (Fig 2.25). The amino and carboxyl groups at each end of the **dipeptide** can form peptide linkages with other amino acid molecules, so building up a chain of amino acid residues. Some short-chain peptides are of biological importance. Vasopressin (antidiuretic hormone) and ocytocin, two mammalian hormones, are each made from just nine amino acid molecules (Chapter 16). Long chains of amino acid residues are called **polypeptides**. One or more polypeptide chains occur in a protein molecule. The number of amino acid residues ranges from about fifty in the smallest proteins to several thousand in the largest. The relative molar mass of proteins ranges from 6000 to over 1 000 000.

**Fig 2.25** Formation of a dipeptide

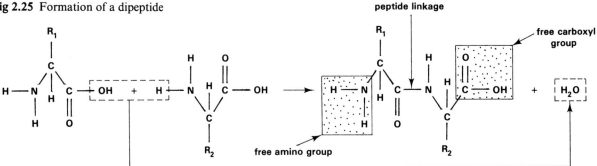

Chromatographic and other techniques have made it possible to identify the variety of amino acids in many proteins. However, the entire sequence of amino acids in any protein molecule was a mystery until the early 1950s when Frederick Sanger analysed the structure of the hormone insulin. Sanger showed that insulin consists of two polypeptide chains bonded to each other by disulphide bridges (Fig 2.26). One chain contains 21 amino acid molecules, the other 30. In the late 1950s and during the 1960s, the amino acid sequences of other small protein molecules were determined. Corticotropin (ACTH) was found to contain 39 amino acid molecules, ribonuclease has 124 while the $\alpha$-globin and $\beta$-globin chains of haemoglobin have 141 and 146 respectively. More recently proteins of much higher relative molar mass have been analysed and a great deal of information has been compiled on the composition and amino acid sequences of a range of proteins from many different sources.

**Fig 2.26** Structure of insulin

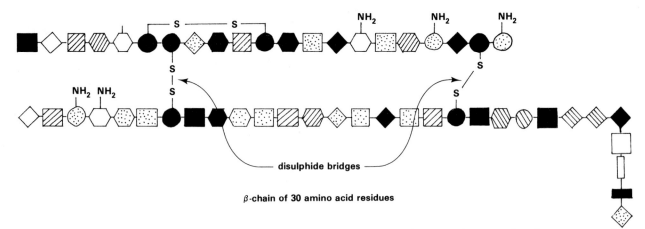

Such studies have shown that few proteins contain all twenty $\alpha$-amino acids. Insulin, for example, has seventeen different amino acids. Some proteins consist of a small range of amino acids; in others a much wider range is found. However, the results of the sequence analyses have revealed two general principles. Firstly, the sequence of amino acids differs from one type of protein to another. Secondly, the sequence in any one type of protein is precisely fixed.

Without question, amino acid composition and sequence studies have yielded much valuable information about proteins. Yet proteins display many properties which cannot be accounted for from the results of such studies. In order to understand some of these properties it is necessary to know something about protein structure.

## 1 Protein structure

The **primary structure** is the number and sequence of amino acids in a polypeptide chain. The only form of bonding in a protein's primary structure is the **peptide linkage**. Protein molecules have a very complex three-dimensional shape which cannot be explained in terms of the primary structure of their polypeptide chains. X-ray diffraction analysis has been a very important tool in unravelling the spatial arrangement of the atoms in protein molecules. In this technique molecules are bombarded with a beam of X-rays. Some of the rays are deflected by atoms in the molecules and are then passed through a photographic film to give a diffraction pattern. From the pattern, which is characteristic for a particular compound, the shape of the molecules can be deduced. It all sounds very simple but the interpretation of the pattern necessitates the use of a computer.

In 1939 Astbury obtained an X-ray diffraction pattern for keratin, a fibrous protein found in hair. The pattern indicated that the polypeptide chains in keratin were twisted or folded in a regular manner. Pauling and Corey in 1951 showed that the chains were twisted into a right handed helix. In this **α-helix** the peptide linkages form the backbone from which the R groups of the amino acids jut out in all directions. The helical structure is stabilised by **hydrogen bonds** which occur between the carbonyl ($>C=O$) and imino ($>N-H$) groups of every fourth peptide link (Fig 2.27). It is now generally agreed that many fibrous proteins have a **secondary structure** of this kind.

**Fig 2.27** The α-helix, simplified

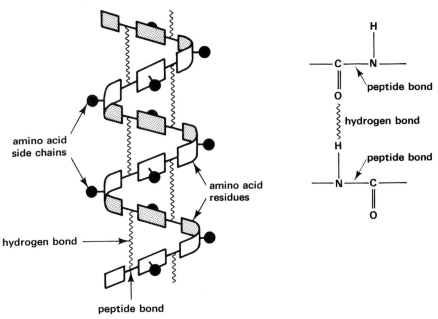

Another form of secondary structure also identified by Pauling and Corey is the **pleated sheet**. Here two or more pleated polypeptides are hydrogen bonded to each other. Once again there is hydrogen bonding

23

between adjacent carbonyl and imino groups. When the terminal amino groups are at the same end of each polypeptide a parallel pleated sheet is formed. If at opposite ends, the pleated sheet is **antiparallel** (Fig 2.28(a)). In some polypeptides a mixture of α-helix alternates with a pleated sheet secondary structure. Other parts of the same polypeptide have no regular arrangement and exist as a **random coil**.

**Fig 2.28** (a) Pleated sheets

**Fig 2.28** (b) Elastin

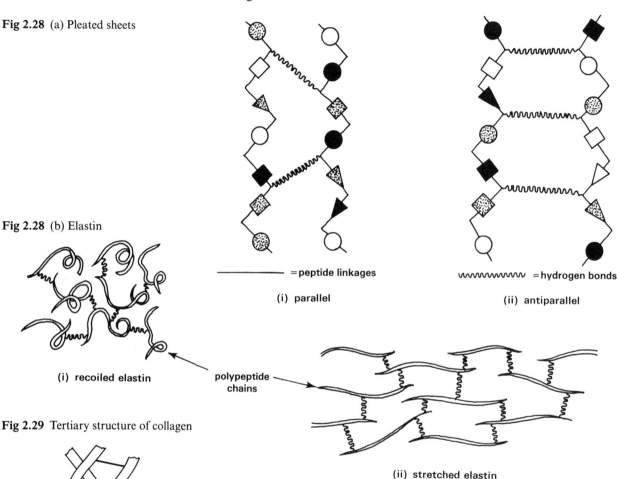

——————— = peptide linkages          ⟋⟍⟋⟍⟋⟍ = hydrogen bonds

(i) parallel                    (ii) antiparallel

(i) recoiled elastin

polypeptide chains

(ii) stretched elastin

**Fig 2.29** Tertiary structure of collagen

extended polypeptide chains

hydrogen bonds

Elastin (Fig 2.28(b)) in elastic connective tissue consists of many cross-linked polypeptides. When relaxed they are randomly coiled but when extended they take on a regular shape. It is this property which gives elastin its considerable elasticity. In unstretched keratin, three to seven α-helices are coiled around each other like the strands of a rope. The strands are held together by **disulphide ($-S-S-$) cross linkages** between the thiol ($-SH$) groups of cysteine molecules. The rope-like arrangement has considerable strength. In collagen of white fibrous tissue, the polypeptide chains are permanently extended. Three such chains are wound around each other and are bound to one another by hydrogen bonds (Fig 2.29). This arrangement has less elasticity but gives considerable tensile strength to fibrous connective tissue as in tendons.

Globular proteins such as myoglobin, haemoglobin and enzyme proteins have a **tertiary structure**. Again using X-ray diffraction analysis, J. C. Kendrew in the late 1950s showed that myoglobin, the oxygen-carrying pigment in muscles, consists of a single polypeptide chain folded asymmetrically. The chain has eight relatively straight segments in which the polypeptide is coiled in an α-helix. The segments are joined by bent peptide bonds and the chain bonded to an iron-containing haem group (Fig 2.30).

**Fig 2.30** Tertiary structure of myoglobin

**Disulphide, hydrogen, hydrophobic** and **electrovalent bonds** help to stabilise the three-dimensional shape of such molecules (Fig 2.31).

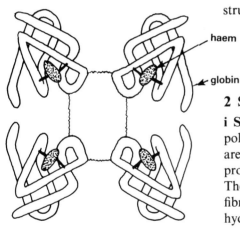

**Fig 2.31** Some of the bonds which maintain tertiary structure

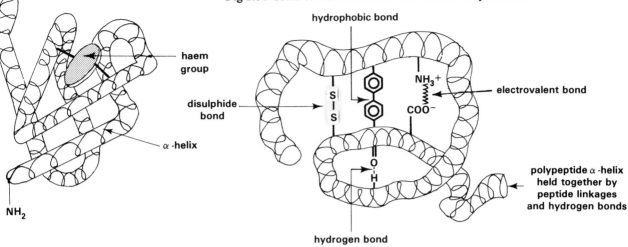

Most globular proteins with a relative molar mass of more than 50 000 consist of two or more polypeptide chains. The way in which they fit together is called the **quaternary structure** of the protein. Haemoglobin for example consists of four chains, two of α-globin and two of β-globin, each bonded to a haem group (Fig 2.32). The α- and β-chains are very similar in structure to myoglobin. Many enzymes have a quaternary structure, their molecules consisting of two to four globular units.

**Fig 2.32** Quarternary structure of haemoglobin

The sub-units are held more tightly together than illustrated here.

### 2 Some properties of proteins in aqueous solutions

**i Solubility** Globular proteins are generally **soluble in water** because the polar R groups jutting outwards from the molecule attract water. Enzymes are globular proteins which function in aqueous solutions. Some globular proteins of high relative molar mass form **colloidal suspensions** in water. They include some of the proteins in blood plasma. The **insolubility** of fibrous proteins in water is accounted for by the large number of non-polar hydrophobic R groups on the exterior of their molecules.

**ii Buffering capacity** The ability of a protein to accept or donate hydrogen ions ($H^+$) depends on the number and charge of ionisable R groups in its molecules. The most important R groups in this respect are the amino (—$NH_2$) and carboxyl (—COOH) groups. Amino groups bind $H^+$ in acidic solutions (pH < 7) whilst carboxyl groups donate $H^+$ in alkaline solutions (pH > 7). In this way proteins act as **buffers** in stabilising the pH of their surroundings (Fig 2.33).

**Fig 2.33** Buffering by a protein

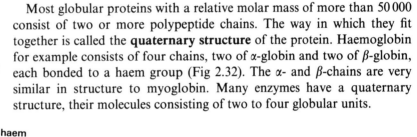

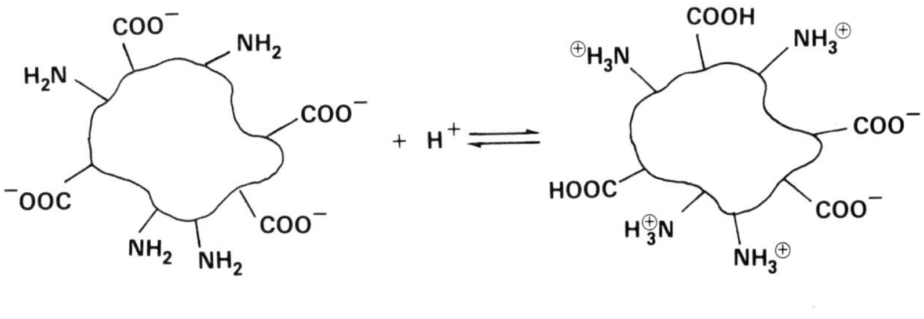

**iii Isoelectric point** At a precise pH in aqueous solution each protein exists as an amphion which carries no net electrostatic charge. This pH is called the **isoelectric point** of the protein. If the protein has a large number of amino groups in its side chains the isoelectric point is at a pH above 7, whilst proteins with a large number of carboxyl groups form amphions at a pH below 7 (Fig 2.34). A knowledge of the isoelectric points of proteins is exploited in separating mixtures of proteins by electrophoresis (section 2.2).

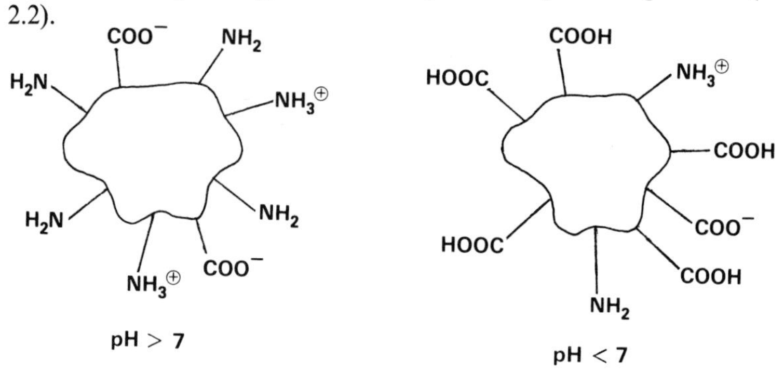

**Fig 2.34** Isoelectric points of two proteins

pH > 7

pH < 7

## 3 Protein denaturation

Most proteins function effectively only within a limited range of temperature and pH. At extremes of temperature and pH proteins undergo **denaturation**. Denaturation is a change in the physical shape of a protein molecule. It happens because some of the bonds which normally maintain the protein's three-dimensional structure are broken. Excessive heat can cause atoms in the molecule to vibrate so strongly that bonds between them are destroyed. Proteins also become denatured in very acidic or very alkaline solutions. At such extremes of pH electrovalent bonds do not develop. Denatured proteins are unable to carry out their normal functions.

Provided denaturation is not too extreme, protein molecules often spontaneously return to their original form when placed in ideal conditions of temperature and pH.

## 4 Functions of proteins

Proteins are used for an extraordinary range of activities in living organisms. Some of their more important functions are summarised as follows.

| | | |
|---|---|---|
| i | Enzymes | eg pepsin, amylase, lipase |
| ii | Contraction | eg actin and myosin in muscle |
| iii | Protection | eg antibodies, fibrinogen, prothrombin |
| iv | Hormones | eg insulin, somatotropin, corticotropin |
| v | Transport | eg haemoglobin, transferrin (carries iron) |
| iv | Support | eg collagen, elastin |
| vii | Storage | eg myoglobin, ferritin (stores iron) |

## 5 Biochemical tests for proteins and amino acids

One of the most common biochemical tests used to demonstrate the presence of amino acids in extracts from living organisms is the **ninhydrin reaction**. When heated with ninhydrin the α-amino group forms a blue coloured complex. Proline gives a yellow colour in this reaction.

The **biuret test** is frequently used to investigate the presence of compounds with peptide links in materials of living origin. Proteins and polypeptides form a violet complex when their peptide linkages react with alkaline copper(II) sulphate solution.

What is the main limitation of Millon's test in identifying the presence of proteins in biological fluids?

**Millon's reagent** is a solution of mercury(II) nitrate in nitric acid. Mercury reacts with thiol groups of amino acids, such as cysteine which is found in many proteins. At the same time electrostatic forces in the protein molecules break due to the high concentration of hydrogen ions from the acid. The protein becomes denatured and a white precipitate is formed. When heated the precipitate takes on a red coloration if the protein contains the amino acid tyrosine.

### 2.3.3 Glycoproteins and lipoproteins

These are conjugated proteins, the molecules of which contain an organic **prosthetic group** in addition to protein. In **glycoproteins** the additional component is usually a monosaccharide sugar such as galactose or mannose. Glycoproteins are part of the structure of cell membranes, assisting cell adhesion (Chapter 6). The α- and γ-immunoglobulins, prothrombin and fibrinogen are also glycoproteins (Chapter 10). In **lipoproteins** the prosthetic group is a simple lipid, phospholipid or cholesterol. The β-immunoglobulins are lipoproteins.

## 2.4 Biochemical techniques

Some technological developments used in biochemistry such as X-ray crystallography have already been mentioned earlier in this chapter. Let us now look at other techniques which have been used even more widely during the past twenty-five years or so to provide a wealth of information about the organic compounds found in living organisms. Some of the techniques are simple and you can use them yourself. Scientific progress does not rely exclusively on sophisticated equipment and complicated technology. The techniques described here are mainly concerned with the separation and identification of biological compounds in mixtures and the analysis of complex molecules.

### 2.4.1 Chromatography

Chromatography is a technique used to separate the components of mixtures. It is based on the partitioning of compounds between a stationary phase and a moving phase. In **partition chromatography** the stationary phase is a liquid whilst in **adsorption chromatography** the stationary phase is a solid. Partition chromatography depends mainly on the relative solubility of substances in two or more solvents. When a solute is added to a mixture of equal volumes of two immiscible solvents, it may dissolve entirely in one or other of the solvents. If the solute dissolves in both solvents, it may not dissolve to the same extent in each solvent. The ratio:

$$\frac{\text{concentration of solute in solvent 1}}{\text{concentration of solute in solvent 2}}$$

at equilibrium is called the **partition coefficient**. In partition chromatography, use is made of differences in the partition coefficient to separate components of mixtures.

In contrast, adsorption chromatography relies mainly on **electrostatic interaction** between solutes and the stationary phase. Some adsorption also occurs in partition chromatography. Paper, column and thin-layer are examples of the various forms of chromatography commonly used by biochemists.

## 1 Paper chromatography

**Paper chromatography** is used extensively in biological research yet it is a simple technique and is often practised in school laboratories (Fig 2.35). A **starting line** is drawn in pencil about 1.0 cm from one edge of a sheet of absorbent paper which is the **support medium**. On the line are marked a number of **origins** about 1.5 cm apart. A drop of the mixture to be analysed is spotted on to one of the origins. On each of the remaining origins is spotted a drop of a pure solution of a substance suspected to be in the mixture. The technique is extremely sensitive and can be used to detect minute quantities of unknown substances. For this reason a micropipette is used to spot the origins.

**Fig 2.35** Paper chromatography

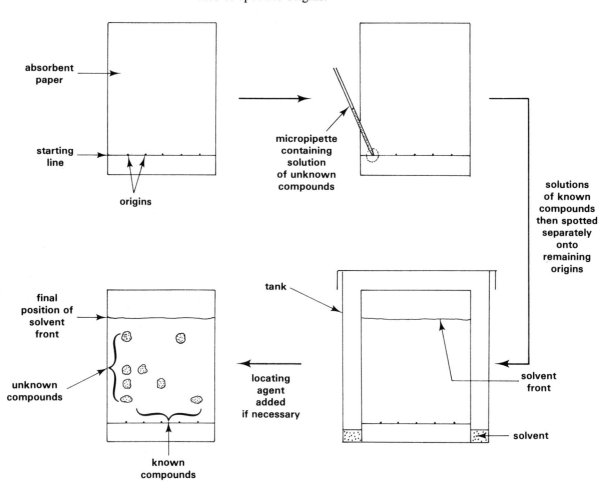

Before the paper is prepared, the **developing solvent** is placed in a chromatography tank and the lid replaced. Some of the solvent evaporates to saturate the air in the tank with solvent vapour. The paper is then placed vertically in the tank so that the starting line is just above the solvent which runs up (in **ascending chromatography**) or down (in **descending chromatography**) the support medium by capillarity. The known compounds and the components of the unknown mixture are carried different distances according to their partition coefficients between the solvent and water present in the paper. Adsorption to the support medium also affects movement of the solutes as the chromatogram develops.

When the solvent has travelled most of the length of the paper, the distance it has moved, the **solvent front**, is marked and the paper is dried. If the solutes are coloured, as in the case of leaf pigments, the positions of the components can be seen without further steps. The positions of

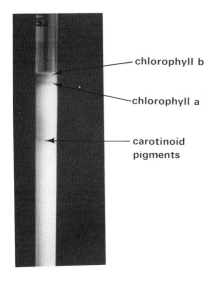

Fig 2.36 (a) A column chromatogram showing separated photosynthetic pigments

— chlorophyll b

— chlorophyll a

— carotinoid pigments

colourless substances have to be detected using **locating agents** which react with the compounds under investigation to form coloured end products. Ninhydrin, for example, is used to locate amino acids (section 2.3.2). The ratio:

$$\frac{\text{distance travelled by a compound}}{\text{distance travelled by solvent front}}$$

is called the **retardation factor ($R_f$ value)**. For a given solvent at a given temperature every compound has a characteristic $R_f$ value. Thus by comparing the $R_f$ values of the components of a mixture with those of the known compounds it is usually possible to identify some if not all of the substances in the mixture (Fig 2.35). Should the first solvent not produce a satisfactory separation of all the compounds in the mixture, the chromatogram may then be turned through 90° and a second solvent run along it. This is **two-dimensional chromatography**.

The amounts of substances which have been separated can also be determined. The first step is to cut out the coloured spots and place them separately in a standard volume of a solvent. This procedure is called **elution**. After a period of time the paper discs are removed and the intensity of colour in the solvent is measured using a colorimeter. The amount of substance in solution is determined by comparing its absorbance value read on the scale of the colorimeter with the absorbance of a standard solution containing a known concentration of the same substance.

## 2 Column chromatography

A long glass tube is packed with a support medium such as hydrated starch, silica gel or a powder of diatom shells called kieselguhr. A solution of the mixture to be analysed is poured into the top of the column. The components of the mixture move down the column at different rates. Differences in partition coefficients between the solvent and water in the support medium mainly account for the rates at which the solutes travel. Again, the separation of coloured compounds such as leaf pigments can be seen without further steps. When the solvent begins to run out at the bottom of the column, fractions collected at staggered intervals of time will contain different components of the mixture. The amounts of the separated components can be measured using a colorimeter (Fig 2.36).

Fig 2.36 (b) Stages in the separation of the components of a mixture by column chromatography

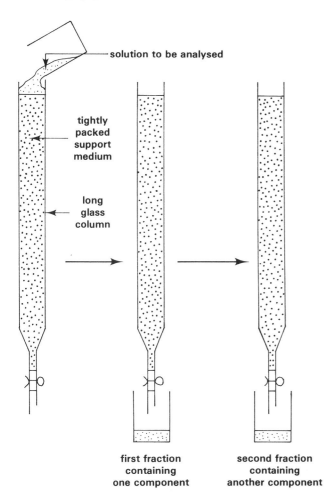

— solution to be analysed

tightly packed support medium

long glass column

first fraction containing one component

second fraction containing another component

further fractions may be collected

List the factors affecting the separation of a mixture of substances in paper chromatography. What measures should be taken to ensure that the results of such a technique can be reliably reproduced?

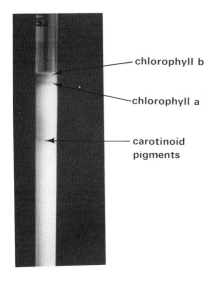

29

### 3 Thin-layer chromatography

A thin glass or aluminium plate is covered with an aqueous slurry of a support medium, for example, silica gel or cellulose powder. The plate is dried and the mixture to be analysed, together with known compounds, spotted on to separate origins on the starting line. The plate is placed vertically in a tank so that the solvent almost touches the starting line. As in ascending paper chromatography the solvent rises by capillarity and the components of the mixture become separated according to their partition coefficients and adsorption to the support medium. When the solvent front has moved a sufficient distance to separate the components of the mixture, the plate is removed from the tank, dried, and if necessary a locating agent sprayed on to the support medium. The procedure can be made two-dimensional if the first solvent gives an incomplete separation. Once more the amounts of substances in the mixture can be determined. After scraping off the coloured spots and eluting the coloured complex, the intensity of colour can be determined with a colorimeter.

Sugars, amino acids, lipids and pigments in biological extracts can be separated using any of the methods of chromatography described above. Thin-layer chromatography is especially useful in analysing the acyl-glycerols in natural fats and oils.

### 2.4.2 Electrophoresis

An amino acid or protein molecule may be cationic, amphionic or anionic in aqueous solution depending on the nature of the amino acid side chains R and the pH of the solution. At a given pH some molecules may be more strongly cationic or anionic than others (section 2.3.1). When an electric current is passed through such a mixture, cations will move towards the cathode, anions to the anode, while amphions will fail to move either way. It is this difference which is exploited in separating mixtures of amino acids and of proteins by **electrophoresis**.

The pH of the mixture is first adjusted with a buffer so that each of the expected components has a different charge or a different strength of the same charge. In paper electrophoresis the mixture is applied to a starting line marked across the centre of a piece of paper or cellulose acetate support medium which has been soaked in the same buffer. Each end of the paper is then immersed in separated troughs of the buffer solution (see Fig 2.37). An

**Fig 2.37** Diagram of electrophoresis tank

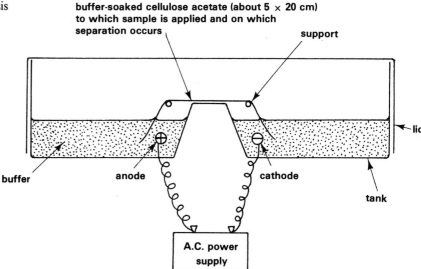

buffer-soaked cellulose acetate (about 5 × 20 cm) to which sample is applied and on which separation occurs

support

lid

buffer

anode

cathode

tank

A.C. power supply

electric current is then passed through the buffer. The current must be **direct** not alternate as on mains electricity, otherwise the components will migrate towards one pole them immediately back towards the other as the current alternates. A **power pack transformer** converts alternating mains electricity to the direct current required and also lowers the voltage (Fig 2.38). If the voltage is too high the support medium becomes overheated and may cause denaturation of proteins as well as evaporation of the buffer. After a period of time the power supply is disconnected and the

**Fig 2.38** Equipment for electrophoresis

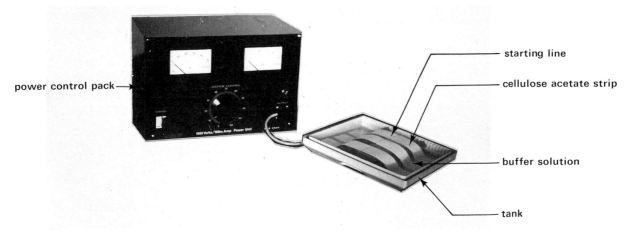

power control pack

starting line

cellulose acetate strip

buffer solution

tank

support medium treated with a locating agent to find the positions of the separated components. Ninhydrin can be used to locate amino acids and a dye such as light green for proteins. The pattern of separation is compared with that of known mixtures run under identical conditions. Unknowns which migrate similar distances to knowns are probably identical.

Electrophoresis is used extensively to analyse plasma proteins such as immunoglobulins (Chapter 11).

# 3 Enzymes

In living cells hundreds of different biochemical reactions take place rapidly and simultaneously. The reactions go on at relatively low temperatures and are normally controlled in such a way that useful products are made and wastes removed at rates which satisfy the metabolic needs of cells. How is it possible for there to be such orderliness in what must be a potentially chaotic situation? How can reactions take place so rapidly at such modest temperatures? The answers to these questions come from a study of enzymes.

The first enzyme to be discovered was amylase which catalyses the conversion of starch to maltose. Its presence in malt extract was detected in 1833 by two French chemists Payen and Persoz. However, it was not until 1876 that the term **enzyme** was proposed by Wilhelm Kühne, the distinguished German biochemist.

## 3.1 The structure of enzymes

To appreciate the structure of enzymes it is necessary to have a knowledge of protein structure (Chapter 2). Over 90 % of enzymes are simple **globular proteins**. The remainder are **conjugated proteins** which have a non-protein fraction called the **prosthetic group**. Many enzymes have relative molar masses of between 10 000 and 500 000.

Amino acid analysis, X-ray diffraction studies and other techniques have in recent years provided a lot of information about the structure of proteins (Chapter 2). The three-dimensional structure of a number of enzymes is now known in detail. Chymotrypsinogen, an inactive precursor of the pancreatic enzyme chymotrypsin, consists of three polypeptide chains containing 245 amino acid molecules (Fig 3.1).

**Fig 3.1** Three-dimensional structure of chymotrypsinogen (after P B Singler et al *J. Mol. Biol.* **35,** 143, 1968)

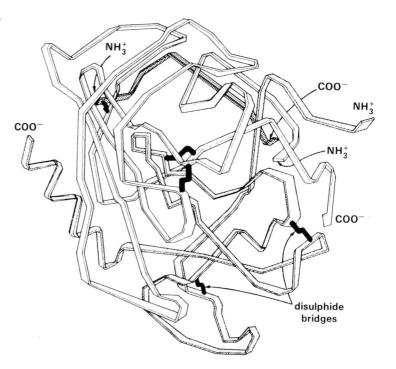

## 3.2 Enzyme catalysis

Enzymes are soluble in water and work in aqueous solution in living cells. They are sometimes described as **organic catalysts**. A catalyst is a substance which affects the rate of chemical reactions. Many of the biochemical reactions catalysed by enzymes are reversible reactions. An example of a reversible reaction is as follows:

$$A + B \rightleftharpoons C$$
reactants     product

At equilibrium the rate at which A and B are converted to C is equal to the rate at which C is converted back to A and B. The position of the equilibrium depends on the energy difference between the reactants and the product. Enzymes do not alter the direction of a reaction; they speed up the rate at which equilibrium is reached. In so doing they can catalyse reversible reactions in either direction provided it is energetically feasible.

Biochemical reactions involve the formation or the destruction of chemical bonds. When two or more reactants are joined, chemical bonds are formed. When a complex molecule is split into simpler components, chemical bonds are destroyed. In either case energy is required to bring about the changes. The energy required for a chemical reaction to proceed is called **activation energy** (Fig 3.2(a)). Heat can be used as a source of activation energy. Indeed, many chemical reactions do not proceed quickly unless the reactants are raised to relatively high temperatures. In living cells however, reactions take place rapidly at relatively low temperatures. Enzymes lower the amount of activation energy needed, making it possible for reactions to occur at temperatures which are otherwise energetically unfavourable (Fig 3.2(b)).

**Fig 3.2** Energetics of biochemical reactions

(a) Uncatalysed

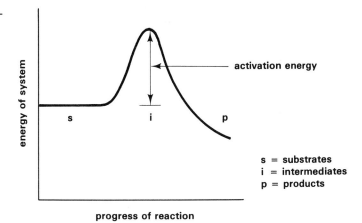

s = substrates
i = intermediates
p = products

(b) Catalysed

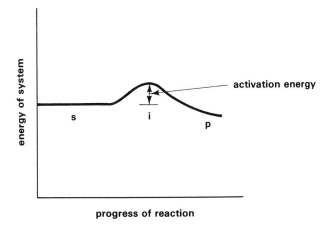

### 3.2.1 The lock and key mechanism

Emil Fischer in 1894 proposed the **lock and key mechanism** to explain how enzymes and **substrates** interact when mixed. Fischer suggested that enzyme and substrate molecules combine to form an **enzyme–substrate complex** before the products of the reaction are released (Fig 3.3).

**Fig 3.3** The lock and key mechanism of enzyme action

(a) Bond formation

(b) Bond destruction

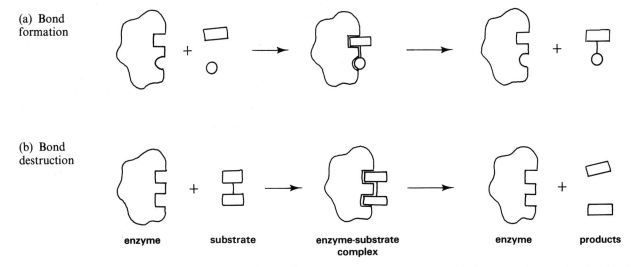

enzyme    substrate    enzyme-substrate complex    enzyme    products

One of the systems used to name enzymes is based on the substrates with which they combine. For example, lipases react with lipids. What types of substrates do you think carbohydrases and peptidases react with?

The sites on enzyme molecules where substrates fit are called **active centres**. Bringing substrate molecules close to each other at the active centre increases the probability that they will react. Because enzyme and substrate molecules are three-dimensional structures, the formation of an enzyme–substrate complex requires that the shapes of the reactants and the active centres of enzymes are **complementary**. Otherwise the enzyme and substrate cannot unite (Fig 3.4(a)). It is comparable to the way in which a lock can only be opened by a key of a specific shape. The phenomenon whereby most enzymes work with only one or with a limited range of substrates is called **enzyme specificity**. The specificity of enzymes supports the lock and key hypothesis.

**Fig 3.4** (a) Enzyme specificity

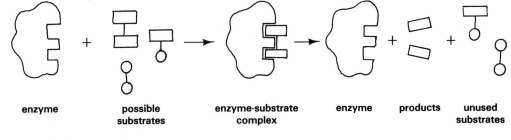

enzyme    possible substrates    enzyme-substrate complex    enzyme    products    unused substrates

**Fig 3.4** (b) Induced fit mechanism

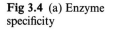

enzyme    substrate    induced fit enzyme-substrate complex    enzyme    products

Koshland has more recently proposed the **induced-fit mechanism** to explain the action of some enzymes. According to this theory, the shape of the active centre is changed when a substrate molecule binds to the enzyme (Fig 3.4(b)).

### 3.2.2 Control of enzyme action

Biochemical pathways usually involve a number of linked reactions. Each reaction is catalysed by a specific enzyme:

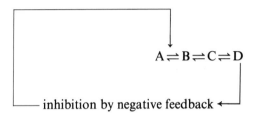

$$\text{enzyme X} \quad \text{enzyme Y} \quad \text{enzyme Z}$$
$$A \rightleftharpoons B \rightleftharpoons C \rightleftharpoons D$$

In some instances accumulation of one of the products formed near the end of the chain inhibits the action of an enzyme used in one of the earlier reactions. In the hypothetical example shown above, the presence of a high concentration of product D for example may slow down the rate at which enzyme X converts A to B. This is an example of **negative feedback inhibition,** an important device in the co-ordination of the metabolism of cells.

$$A \rightleftharpoons B \rightleftharpoons C \rightleftharpoons D$$
inhibition by negative feedback

Negative feedback ensures that reactants are used efficiently and prevents the excess manufacture of end products. Like other homeostatic devices described elsewhere in this book, the control of enzyme action helps to maintain a stable environment in living organisms.

## 3.3 Factors affecting enzyme action

To function at all, some enzymes need the presence of **co-factors**. To function efficiently, enzymes need a suitable **temperature** and **pH** and there must be enough **substrate** available. Enzyme action can be stopped partially or completely if **inhibitors** are present in the reaction mixture.

### 3.3.1 Co-factors

A **prosthetic group** is an essential co-factor attached to the protein part of a conjugated enzyme. Some prosthetic groups are organic, others are atoms of metals such as copper. If the prosthetic group is removed the enzyme fails to function. Mineral ions must be mixed with the reactants before some enzymes will work. Zinc, iron and magnesium act as **enzyme activators** in this way. It is partly for these reasons that minerals are essential to living organisms (Chapters 9 and 24).

Other enzymes have **co-enzymes** as co-factors. Co-enzymes are organic, non-protein substances which are not bonded to enzyme molecules like prosthetic groups. Several important co-enzymes are vitamin derivatives. **Nicotinamide-adenine dinucleotide ($NAD^+$)** and its phosphate ester **nicotinamide-adenine dinucleotide phosphate ($NADP^+$)** for example, are derived from nicotinamide, a B-group vitamin. Riboflavin (vitamin $B_2$) forms part of **flavin-adenine dinucleotide (FAD)**. $NAD^+$, $NADP^+$ and FAD function as hydrogen acceptors in reactions catalysed by dehydrogenase enzymes (section 3.4).

### 3.3.2 Inhibitors

Many substances inhibit the activity of enzymes. **Inhibitors** fall into two categories, **reversible** and **non-reversible**.

### 1 Reversible inhibitors

Reversible inhibitors are substances which prevent enzymes from combining with substrates. Activity of the enzyme is restored when the inhibitor is removed.

**Competitive inhibitors** affect enzyme action by becoming attached to active centres, so stopping the substrate from binding to the enzyme. A well-known example of this behaviour is inhibition of the enzyme succinic dehydrogenase by malonic acid. Succinic dehydrogenase catalyses the oxidation of succinic acid to fumaric acid in Krebs' tricarboxylic acid cycle (Chapter 5). In the presence of malonic acid the reaction rate is slowed down. Malonic acid has a molecular structure which is very similar to that of succinic acid. Inhibition happens because the active centre of some of the dehydrogenase enzyme molecules becomes occupied by malonic acid rather than by the normal substrate succinic acid. In effect, malonic acid and succinic acid are competing for the active centre of the enzyme (Fig 3.5).

**Fig 3.5** (a) Competitive inhibition

**Fig 3.5** (b) Molecular structures of succinic and malonic acids

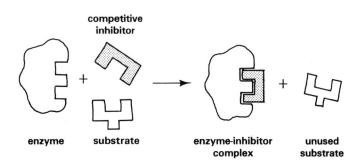

enzyme  substrate    enzyme-inhibitor complex    unused substrate

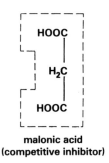

**Fig 3.5** (c) Molecular structures of PABA and prontosil

$H_2N$—〇—COOH      $H_2N$—〇—$SO_2NH.R$

p-aminobenzoic acid (PABA)      prontosil

Inhibitors of this kind can be used as drugs to reduce the rate at which undesirable reactions occur in the human body. Inhibiting such reactions is one of the ways of treating some forms of cancer. Another application of competitive inhibition is the use of sulphonamide drugs such as prontosil to combat bacterial infections. Prontosil is similar in molecular structure to *p*-aminobenzoic acid (see Fig 3.5) which bacteria use to synthesise folic acid. When unable to make folic acid, they cannot grow and multiply in the human body (Chapter 25).

The degree of inhibition by a competitive inhibitor is less if the ratio of substrate-to-inhibitor molecules is high (Fig 3.6).

**Fig 3.6** Effect of concentration of a competitive inhibitor on the rate of an enzyme catalysed reaction

(a) Low concentration

(b) High concentration

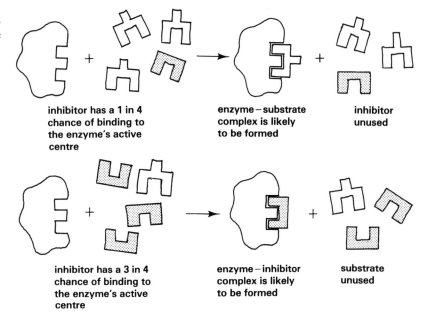

inhibitor has a 1 in 4 chance of binding to the enzyme's active centre

enzyme – substrate complex is likely to be formed

inhibitor unused

inhibitor has a 3 in 4 chance of binding to the enzyme's active centre

enzyme – inhibitor complex is likely to be formed

substrate unused

The degree of inhibition by **non-competitive inhibitors** cannot be reduced by increasing the number of substrate molecules (Fig 3.7). Here the inhibitor becomes attached to the enzyme at a position other than the active centre. Nevertheless the enzyme, the substrate, or possibly both, become changed so that enzyme activity stops. Disulphide bridges are important in maintaining the tertiary structure of enzyme molecules (Chapter 2). If the disulphide bridges are broken the three-dimensional shape of the enzyme changes. The ions of heavy metals such as mercury (Hg), silver (Ag) and copper (Cu) affect enzymes in this way. $Hg^{2+}$, $Ag^+$ and $Cu^{2+}$ ions combine with thiol (—SH) groups in enzymes, so denaturing enzyme molecules and inhibiting enzyme activity. Cyanide is another non-competitive inhibitor. It blocks the action of some enzymes by combining with iron which may be present in a prosthetic group or which may be required as an enzyme activator. It is not surprising therefore that the salts of heavy metals and cyanide are potent poisons to living organisms. Nevertheless, non-competitive inhibitors do not bind strongly to enzymes and can be removed by dialysis. Enzyme activity is then restored.

**Fig 3.7** Comparative effects of a non-competitive and competitive inhibitor on the rate of an enzyme-catalysed reaction

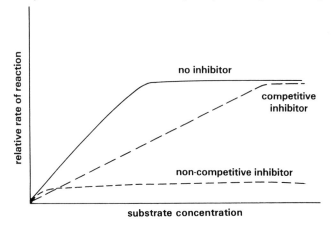

relative rate of reaction

no inhibitor

competitive inhibitor

non-competitive inhibitor

substrate concentration

### 2 Non-reversible inhibitors

Organophosphorus insecticides such as malathion are good examples of **non-reversible inhibitors**. They become firmly bound to active centres so that substrate molecules cannot bind to enzymes and activity of the enzyme is permanently stopped. Insecticides of this type inactivate the enzyme cholinesterase which is essential for the functioning of the nervous system (Chapter 14). Nerve gases such as sarin do the same.

### 3.3.3 Temperature

Heat supplies kinetic energy to reacting molecules, causing them to move more rapidly. The chances of molecular collision taking place are thus increased at higher temperatures so it is more likely that enzyme–substrate complexes will be formed. However, heat energy also increases the vibration of the atoms which make up enzyme molecules. If the vibrations become too violent, chemical bonds in the enzyme break and the precise three-dimensional structure, so essential for enzyme activity, is lost. At high temperatures therefore enzymes become **denatured**.

When the effect of temperature on enzyme activity is investigated experimentally, a temperature usually called the **optimum temperature** is observed at which the reaction proceeds most rapidly. This temperature is not necessarily that at which the enzyme is most stable. It is the resultant of the contrary effects of temperature on the movement of reactants and of enzyme denaturation (Fig 3.8).

**Fig 3.8** Effect of temperature on the rate of an enzyme-catalysed reaction

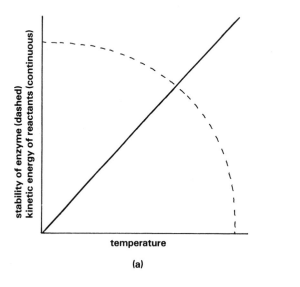

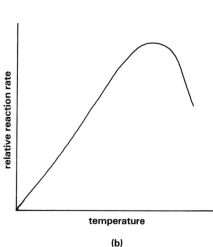

(a)

(b)

The term **temperature coefficient ($Q_{10}$)** is used to express the effect of a 10 °C rise in temperature on the rate of a chemical reaction.

$$Q_{10} = \frac{\text{rate of reaction at } t + 10\,°\text{C}}{\text{rate of reaction at } t\,°\text{C}}$$

Between 4 °C and 37 °C, the optimum temperature for enzymes in the human body, the $Q_{10}$ for enzyme-catalysed reactions is 2.

---

At 35°C it took an enzyme-catalysed reaction 5 minutes to be completed compared with 10 minutes at 25°C. Calculate the $Q_{10}$ for the reaction.

---

### 3.3.4 pH

The symbol pH refers to the concentration of hydrogen ions in solution (Chapter 1). The concentration of hydrogen ions $[H^+]$ affects the stability of the electrovalent bonds which help to maintain the tertiary structure of protein molecules (Chapter 2). Extremes of pH cause the bonds to break resulting in enzyme **denaturation**.

For every enzyme there is an **optimum pH** at which the reaction it catalyses proceeds most rapidly (Fig 3.9). Many enzymes work within a pH range of 5–9 and catalyse reactions most efficiently at pH 7. There are exceptions. For example, pepsin and rennin secreted in the mammalian stomach work best at pH 1.5–2.5 (Chapter 14). Alkaline phosphatase in the kidneys has an optimum pH of 10.

**Fig 3.9** Effect of pH on the rates of enzyme-catalysed reactions

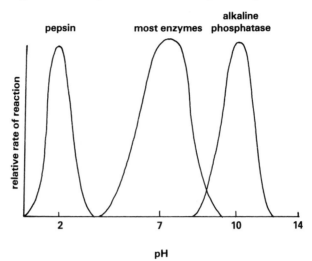

It must be remembered that even small changes in pH mean relatively large changes in $[H^+]$. A change of 1 on the pH scale involves a ten-fold increase or decrease in $[H^+]$ whilst a change in pH of 2 represents a hundred-fold change in $[H^+]$ (Chapter 1). Thus even small changes in pH can have a great effect on enzyme activity.

Apart from the effect in denaturing enzymes, changes in $[H^+]$ can alter the ionisation of the amino acid side chains at the active centres of enzymes. Ionisation of substrate molecules can also be affected. The formation of enzyme–substrate complexes sometimes depends on the active centres and substrate molecules having opposite electrostatic charges. If the charges are altered by changes in pH, some enzymes fail to function (Fig 3.10).

**Fig 3.10** One way in which pH may affect an enzyme

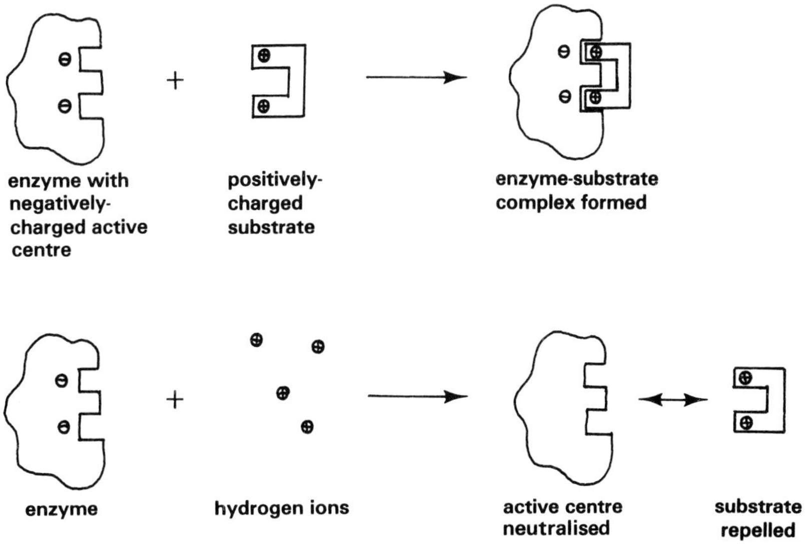

enzyme with negatively-charged active centre

positively-charged substrate

enzyme-substrate complex formed

enzyme

hydrogen ions

active centre neutralised

substrate repelled

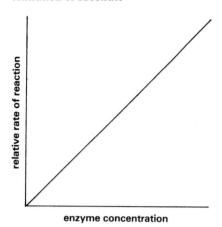

### 3.3.5 Enzyme concentration

Enzymes catalyse reactions rapidly at very low enzyme concentrations. This is because enzyme molecules form complexes with substrates only very briefly. The products of the reaction are quickly released and the enzyme is then available for further activity.

The rate at which enzymes use substrates is described as the **turnover number**. For some enzymes the turnover number is very high. A molecule of catalase for example can break down 40 000 molecules of hydrogen peroxide into water and oxygen every second! Even the slowest of enzymes have turnover numbers of about $100\,s^{-1}$. The larger the number of enzyme molecules present the greater is the amount of substrate used in a given period of time, provided that there is an excess of substrate available (Figs 3.11 and 3.12).

**Fig 3.12** Formation of enzyme-substrate complexes at three concentrations of enzyme

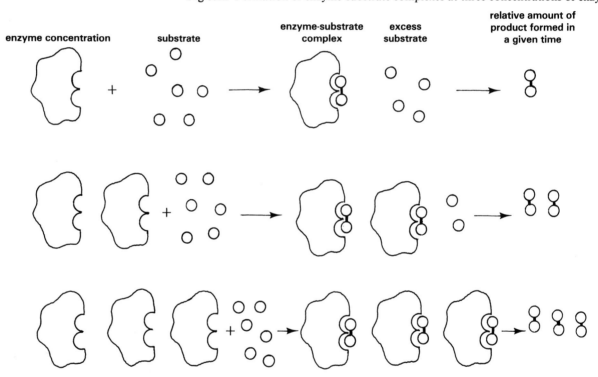

### 3.3.6 Substrate concentration

Fig 3.13 shows the effect of substrate concentration on the rate of a biochemical reaction when the amount of enzyme is limited. At low concentrations of substrate there is a linear relationship between the reaction rate and substrate concentration. In these conditions the ratio of enzyme to substrate molecules is high. Consequently some active centres are always free for substrate molecules to bind with the enzyme. However, a point is reached when a further increase in substrate concentration does not cause the reaction to go any faster. The enzyme-to-substrate ratio is then lower, and there are more substrate molecules present than there are free

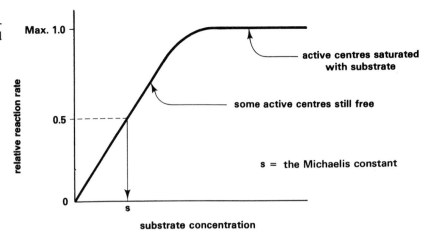

**Fig 3.13** Effect of substrate concentration on the rate of an enzyme-catalysed reaction

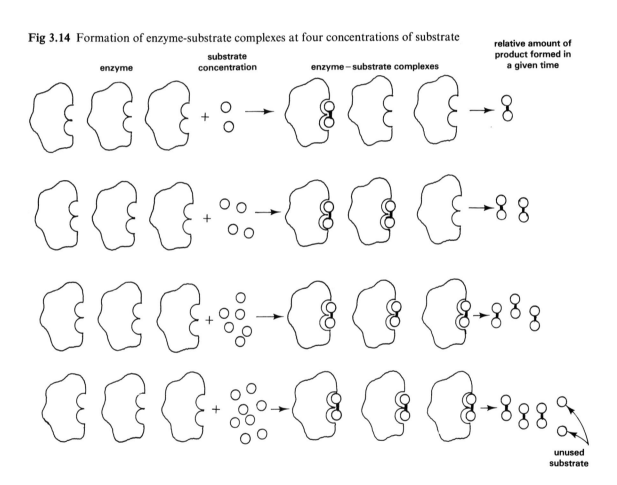

**Fig 3.14** Formation of enzyme-substrate complexes at four concentrations of substrate

active centres with which to bind. Adding more substrate will not make the reaction go more quickly (Fig 3.14).

**The concentration of substrate at which the reaction rate is half of its maximum is characteristic for each combination of enzyme and substrate.**

It is called the **Michaelis constant ($K_M$)**. A high $K_M$ value indicates that the enzyme has a low affinity for its substrate. Conversely a low $K_M$ value indicates a high affinity.

## 3.4 The classification of enzymes

In the past twenty years or so, many new enzymes have been discovered. Older methods of classifying enzymes became impractical and in the early 1960s the International Union of Biochemistry (IUB) recommended a new system for the classification and naming of enzymes. According to the IUB system there are six groups of enzymes.

### 3.4.1 Hydrolases

These enzymes catalyse reactions in which a substrate is hydrolysed into two simpler products. During **hydrolysis**, hydrogen atoms from water enter one of the products whilst the hydroxyl groups end up in the other product. The reaction is shown in a simplified form as follows:

$$AB + H.OH \rightleftharpoons AH + B.OH$$

substrate     water     products

Fig 3.15 shows how some important substrates found in living organisms are hydrolysed. Hydrolases bring about the breakdown of materials in lysosomes (Chapter 6) and the digestion of food in the gut (Chapter 9).

**Fig 3.15** Hydrolysis of some important biological compounds

(a) Glycosidic linkage in a disaccharide

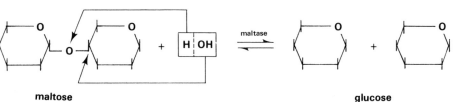

maltose        glucose

(b) Peptide linkage in a dipeptide

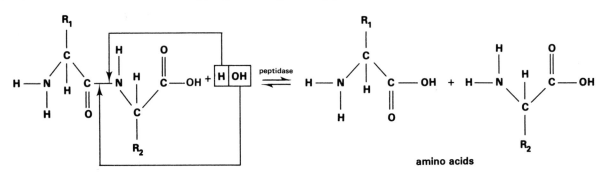

amino acids

(c) Ester linkage in a lipid

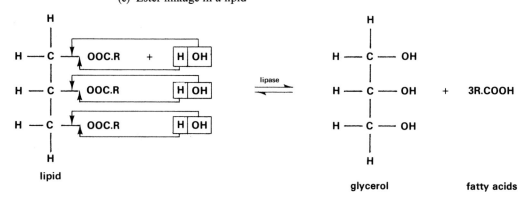

lipid        glycerol        fatty acids

### 3.4.2 Oxido-reductases

The oxidation of substrates is catalysed by this group of enzymes of which there are two kinds:

#### 1 Oxidases

Oxidases catalyse the transfer of hydrogen to molecules of oxygen. A simplified way of expressing the reaction is as follows:

$$AH_2 \;+\; \tfrac{1}{2}O_2 \;\rightleftharpoons\; A \;+\; H_2O$$

substrate    oxygen    oxidised    water
            substrate    (reduced oxygen)

An example is **cytochrome oxidase** which catalyses the oxidation of reduced cytochrome:

$$Cyt.H_2 \;+\; \tfrac{1}{2}O_2 \rightleftharpoons cytochrome + H_2O$$

reduced
cytochrome

Hydrogen peroxide is sometimes formed instead of water.

#### 2 Dehydrogenases

Dehydrogenases catalyse the oxidation of substrates by transferring hydrogen to co-enzymes such as $NAD^+$ and $NADP^+$.

$$AH_2 \;+\; 2NAD^+ \;\rightleftharpoons\; A \;+\; 2NADH$$

substrate    co-enzyme    oxidised    reduced
                        substrate    co-enzyme

For example, **alcohol dehydrogenase** controls the rate at which ethanol is oxidised to ethanal:

$$CH_3CH_2OH \;+\; 2NAD^+ \rightleftharpoons CH_3CHO \;+\; 2NADH$$

ethanol                    ethanal

In reactions catalysed by **oxido-reductases**, substrates are oxidised whilst oxygen or co-enzymes are reduced. Oxidation–reduction reactions usually release energy. The importance of these reactions is described more fully in Chapter 5.

### 3.4.3 Transferases

Transferases catalyse the transfer of functional groups from one substrate to another:

$$AB + C \rightleftharpoons A + BC$$

**Phosphotransferases** control the transfer of phosphate groups in respiration (Chapter 5):

$$glucose + ATP \rightleftharpoons glucose\text{-}6\text{-}phosphate + ADP$$

**Fig 3.16** Action of an aminotransferase enzyme

**Aminotransferases** regulate the transfer of amino groups (Fig 3.16 and Chapter 9).

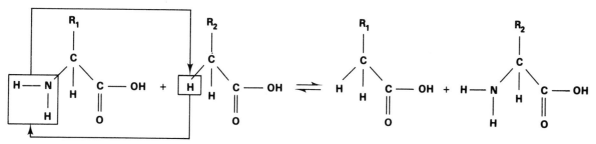

### 3.4.4 Isomerases

Isomerases control the conversion of one isomer of a compound to another isomer of the same compound:

$$ABC \rightleftharpoons ACB$$

The interconversion of sugar isomers in glycolysis is catalysed by isomerase enzymes such as **hexosephosphate isomerase** (Chapter 5):

$$\text{glucose-6-phosphate} \rightleftharpoons \text{fructose-6-phosphate}$$

### 3.4.5 Ligases

This group of enzymes catalyses reactions in which new chemical bonds are formed. ATP provides the energy to make the new chemical bonds. Ligases control the synthesis of macromolecules such as proteins and nucleic acids. An example is **DNA ligase** (Chapter 4).

### 3.4.6 Lyases

Lyases catalyse the breakdown of complex substrates into simpler products but, unlike the case of hydrolytic reactions, water is not used:

$$AB \rightleftharpoons A + B$$

**Decarboxylases** which regulate the release of carbon dioxide from respiratory substrates are examples of lyase enzymes (Chapter 5). **Deaminases** catalyse the release of ammonia from amino acids. Their role in the metabolism of unwanted amino acids absorbed from the gut is outlined in Chapter 9.

## 3.5 Commercial production of enzymes

Large-scale cultures of bacteria and fungi are grown these days for commercial production of enzymes. Live cells of such micro-organisms are homogenised and a variety of enzymes are present in the cell extract. Biochemists have perfected techniques, such as iso-electric precipitation and molecular gel filtration, which make it possible to separate mixtures of enzymes. A wide range of pure enzymes can thus be prepared for domestic, industrial and medical uses.

### 3.5.1 Domestic and industrial uses

A well-known application of enzymes is in **'biological' detergents.** They contain **peptidase** enzymes extracted from cultures of the bacterium *Bacillus subtilis*. Peptidases catalyse the digestion of proteins in such things as egg yolk and blood which are difficult to remove from fabrics washed with conventional detergents or with soap. However, some people involved in the manufacture of 'biological' detergents, as well as housewives using them, sometimes develop respiratory allergies or dermatitis. For this reason they have become less popular in recent years.

**Pectinase** enzymes, mostly from the moulds *Aspergillus* and *Penicillium*, are used to clarify fruit juices. The cloudiness of fresh fruit juice is often caused by the presence of pectin. It is hydrolysed by pectinase, thus clearing the juice. Wine can be clarified in the same way.

**Amylase** from *Aspergillus niger* is produced on a large scale. An important application of amylase is in **brewing**. In brewing, the first step is to hydrolyse cereal starch. The traditional way is to germinate grains of barley, when amylase in the seeds converts stored starch into maltose. The process is called malting. These days many maltsters make an aqueous suspension of ground grain to which commercially prepared amylase is added. The new procedure of malting is very efficient and thus more economical.

### 3.5.2 Medical uses

Some forms of cancer are dependent on a supply of the amino acid asparagine. It can be broken down to aspartic acid and ammonia by the enzyme **asparaginase**. When the enzyme is injected into patients with such cancers the growth becomes starved of asparagine and fails to develop further.

One of the potentially dangerous side-effects of surgery is the development of blood clots which can be carried in the bloodstream and block vessels supplying vital organs (Chapter 10). **Streptokinase**, produced from cultures of the bacterium *Streptococcus*, is often injected into the bloodstream of patients who have had major surgery. The enzyme dissolves fibrin in blood clots causing them to disperse.

The cells of human tissues are stuck to each other by a cell coat containing hyaluronic acid (Chapter 6). The acid can be destroyed by injecting the enzyme **hyaluronidase** into the tissue. In this way spaces can be made under the skin for sub-cutaneous administration of large doses of drugs such as insulin. It avoids the necessity for daily injections.

Much of what is known about enzymes has come from investigations carried out *in vitro* on purified enzyme extracts. *In vitro* means in glass, indicating that it is easier to obtain information from test-tube studies than it is *in vivo*, in living cells. The way in which enzymes control vast numbers of different reactions taking place simultaneously in microscopic packets of protoplasm is therefore all the more remarkable. Compartmentalisation of cells (Chapter 6) undoubtedly helps: each of the organelles of a living cell has a specific range of enzymes so that different kinds of metabolic reactions are separated. Even so, the way in which enzymes regulate the many, rapid and complex metabolic reactions in living cells surely gives cause for wonder.

# 4 Nucleic acids and protein synthesis

In 1869 a German physician named Meischer reported on a chemical analysis he had made of the nuclei of human pus cells collected from bandages. He found the nuclei to be rich in nitrogen, sulphur and phosphorus. Meischer set a pattern for investigating the composition of the nuclei of cells using the methods of the analytical chemist.

During the next eighty years analytical chemists revealed small but significant facts about the composition of nuclei. Crude extracts of nuclei were found to be acidic in reaction, hence the term **nucleic acids** to describe the most important substances present in nuclei. Gradually the building blocks which make up the nucleic acids were identified. But the way in which the building blocks were put together and the roles of the nucleic acids were still poorly understood until the middle of this century. It was then that giant strides in our knowledge and understanding of the nucleic acids were made.

The turning point came in the early 1950s when James Watson and Francis Crick proposed a structure for **deoxyribonucleic acid (DNA)**. The discovery of how exact copies of DNA are made in living cells soon followed and its biological significance was quickly realised. An even greater leap forward was the unravelling of the genetic code, the 'secret of life' as it has been called. We shall see in this chapter that important steps continue to be made in unlocking the secrets of nucleic acids, one of which has recently confronted biologists with a moral dilemma.

## 4.1 Nucleic acids

**Deoxyribonucleic acid (DNA)** and **ribonucleic acid (RNA)** are the two main kinds of nucleic acid. DNA is found mainly in the nuclei of cells, with small amounts in mitochondria and chloroplasts. RNA occurs mainly in the cytoplasm, particularly at the ribosomes.

Both DNA and RNA are polymers, and their building blocks are called nucleotides. For this reason the nucleic acids are described as **polynucleotides**.

### 4.1.1 Deoxyribonucleic acid

A **nucleotide** is made up of three parts, a **pentose sugar**, a **nitrogenous base** and **phosphate**. In DNA the sugar is **deoxyribose** (Fig 4.1). Four different nitrogenous bases occur in DNA (Fig 4.2). **Adenine (A)** and

**Fig 4.1** Molecular structure of deoxyribose

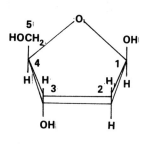

deoxyribose

**Fig 4.2** Nitrogenous bases of DNA

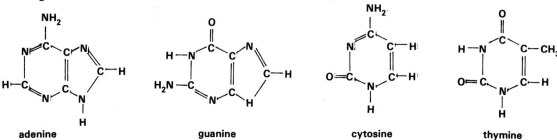

adenine　　　　guanine　　　　cytosine　　　　thymine

**guanine (G)** belong to a group of compounds callled **purines**. **Cytosine (C)** and **thymine (T)** are **pyrimidines**. The sugar, nitrogenous base and phosphate are bonded as shown in Fig 4.3 to form a nucleotide. In a DNA molecule the phosphate groups join nucleotides together by their sugar molecules (Fig 4.4).

**Fig 4.3** A nucleotide

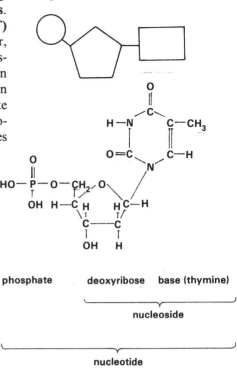

phosphate     deoxyribose    base (thymine)

nucleoside

nucleotide

**Fig 4.4** Part of a polynucleotide

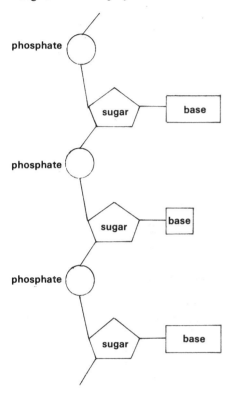

During the early 1950s biochemists established that the number of adenine molecules in DNA from different organisms is the same as the number of thymine molecules. The ratio of cytosine and guanine molecules is also 1:1. However, the ratio of $(A+T):(C+G)$ varies according to the source of the DNA. At this time Rosalind Franklin was working on the structure of DNA using X-ray crystallographic techniques (Chapter 2). Her results were part of the evidence which, in 1953, led James Watson and Francis Crick to propose that a DNA molecule consists of two polynucleotide strands each coiled in a **right-handed helix**. The two strands of the double helix are held together by **hydrogen bonding** between the nitrogenous bases of adjacent nucleotides. For each complete twist of the double helix there are 10 pairs of nucleotides, each twist measuring 3.4 nm in length (Fig 4.5).

**Fig 4.5** Watson-Crick model of DNA

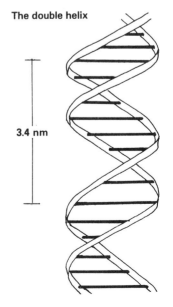

The double helix

3.4 nm

It is important to appreciate that the two polynucleotide strands of a DNA molecule are not identical. They are **complementary**. Where adenine occurs in one strand, thymine is found in the other. Where cytosine appears in a strand, guanine is present in the complementary strand (Fig 4.6(a)). Watson and Crick were aware that such a structure could explain how exact copies of a DNA molecule can be made (section 4.1.4). Altogether there can be several thousand pairs of nucleotides in a single molecule of DNA. The exact number depends on the origin of the DNA. The sequence in which the pairs of bases occurs also differs in DNA from different species (Fig 4.6(b)). In theory the number of permutations of the bases is infinite.

**Fig 4.6** (a) Pairing of the bases in the polynucleotides of DNA

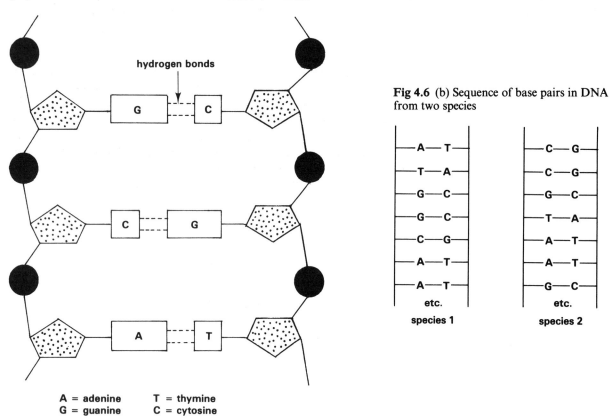

A = adenine    T = thymine
G = guanine    C = cytosine

**Fig 4.6** (b) Sequence of base pairs in DNA from two species

### 4.1.2 Ribonucleic acid

There are three different kinds of ribonucleic acid, **messenger RNA (mRNA), transfer RNA (tRNA)** and **ribosomal RNA (rRNA)**. The nucleotides of which they are made contain the sugar **ribose**, not deoxyribose as in DNA. Another difference is that thymine is not found in RNA. The nitrogenous base **uracil (U)** is present instead (Fig 4.7).

**Fig 4.7** Molecular structure of ribose and uracil

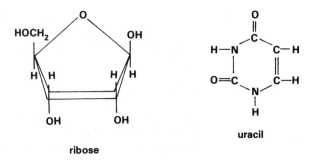

ribose

uracil

## 1 Messenger RNA

Between 3 and 5 % of the RNA in a cell is **mRNA**. The molecules of mRNA are single, helical strands made of up to several thousand nucleotides (Fig 4.8). Messenger RNA is made in the nucleus. Its sequence of bases is complementary to part of one of the helices of DNA. The mRNA passes through the nuclear membrane to the ribosomes where triplets of bases in the mRNA act as **codons** in the synthesis of proteins (section 4.2.1).

**Fig 4.8** Messenger RNA. There are as many different kinds of mRNA as there are proteins. They differ in the sequence of codons in their molecules

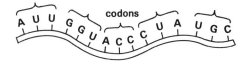

## 2 Transfer RNA

Transfer RNA makes up between 10 and 15 % of a cell's RNA content. The single strand of 75–90 nucleotides which make up a **tRNA** molecule is wound into a double helix which usually has three prominent bulges (Fig 4.9). One of the free ends of every tRNA molecule ends with nucleotides containing the following order of bases **A←C←C←**, with **A** at the very end. There are at least twenty different kinds of tRNA. They differ in the sequence of base triplets making up the **anticodons** by which tRNA binds to the codons of mRNA during the synthesis of proteins.

**Fig 4.9** Transfer RNA

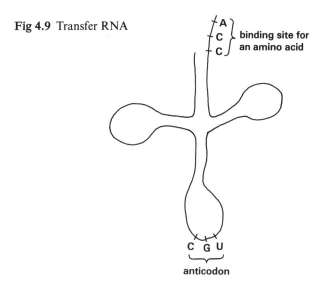

## 3 Ribosomal RNA

The many thousands of nucleotides which make up a molecule of **rRNA** are wound into a complex structure consisting partly of single and partly of double helices. Ribosomal RNA is made in the nucleus under the control of the nucleoli. It enters the cytoplasm and binds with protein molecules to become ribosomes. Over half the mass of a ribosome consists of rRNA and more than 80 % of the total RNA in a cell is rRNA. Even so, the precise function of rRNA is still not known.

### 4.1.3 DNA as the hereditary substance

In 1928 Fred Griffith, an English bacteriologist, noted two distinct strains of the bacterium *Pneumococcus*. Each bacterium of the **S-strain** is enclosed in a capsule, and when inoculated on to an agar-based medium it grows into a smooth, glistening colony. The other, called the **R-strain**, lacks a capsule and its colonies have a rough, dull appearance. Griffith also observed that mice injected with the S-strain soon developed pneumonia and died. The R-strain is non-pathogenic and mice injected with it show no symptoms of illness. However, when heat-killed S-pneumococcus was mixed with the R-strain and then injected into the mice, the animals died of pneumonia. Moreover Griffith was able to isolate live S-pneumococcus from the mice subjected to this treatment (Fig 4.10). He was, however, unable to explain his findings.

**Fig 4.10** Griffith's experiment

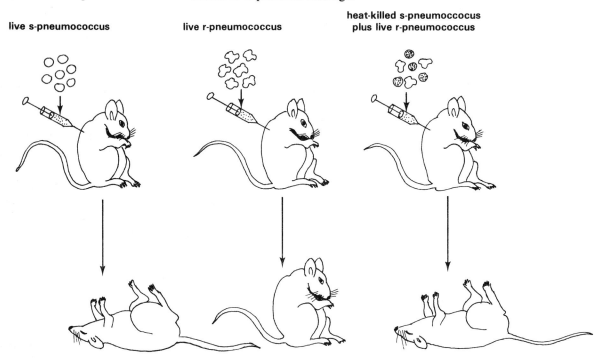

live s-pneumococcus    live r-pneumococcus    heat-killed s-pneumoccocus plus live r-pneumococcus

The reason for the peculiar behaviour of Griffith's *Pneumococcus* was given by Avery in 1943. Avery reported that the non-pathogenic R-strain could be converted to the S-strain by simply adding DNA from the S-strain to the medium in which the R-strain was growing. Furthermore, the newly-acquired characteristic was permanent and was passed to subsequent generations when the bacterium reproduced. Avery gave the name **transformation** to the process whereby the R-strain was converted to the S-strain. Here was a convincing piece of experimental evidence to suggest that DNA is the substance responsible for the characteristics of living organisms.

### 4.1.4 Replication of DNA

Watson and Crick's proposed double helix provided a strong hint as to how exact copies of DNA are normally produced, but it was a few years later before there was any experimental proof.

The hydrogen bonds between the base pairs break and the two polynucleotide strands unwind. Each of the strands acts as a molecular mould or **template** on which a new polynucleotide strand is made. The new strands are complementary to the original strands (Fig 4.11). Synthesis of DNA is controlled by an enzyme called **DNA ligase**. The process gives rise to two molecules of DNA identical to each other and which are **replicas** of the original molecule. This mode of DNA **replication** is described as **semi-conservative**. Compare it with conservative replication as in Fig. 4.12.

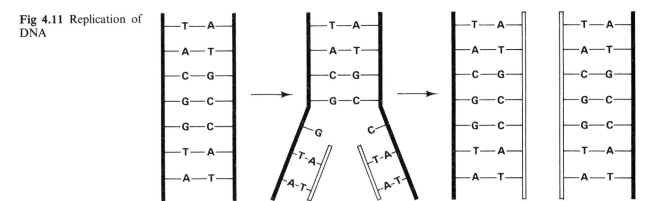

**Fig 4.11** Replication of DNA

**Fig 4.12** Semi-conservative and conservative replication

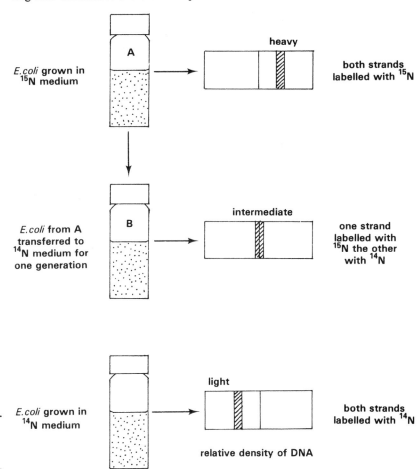

semi-conservative

conservative

Refer to Fig 4.12 and make drawings to show the result of replication of the DNA molecules shown in stage 2 in both cases.

Experimental evidence that DNA replication is semi-conservative was obtained in 1957 by Meselson and Stahl. They grew the bacterium *Escherichia coli* for several generations in a medium containing $^{15}NH_4Cl$ as the only source of nitrogen. The DNA of the bacterium was thus **labelled** with $^{15}N$, the heavy isotope of nitrogen. The bacteria were then transferred to a medium in which $^{14}NH_4Cl$ was the only source of nitrogen. As *E. coli* multiplied the density of its DNA was measured. After one generation the density of the DNA was intermediate between DNA of the bacterium made with the heavy isotope $^{15}N$ and DNA made using $^{14}N$, the normal isotope of nitrogen (Fig 4.13). The result was possible only if the DNA of *E. coli* had replicated in a semi-conservative way.

**Fig 4.13** Meselson's and Stahl's experiment

*E.coli* grown in $^{15}N$ medium

A

heavy

both strands labelled with $^{15}N$

*E.coli* from A transferred to $^{14}N$ medium for one generation

B

intermediate

one strand labelled with $^{15}N$ the other with $^{14}N$

*E.coli* grown in $^{14}N$ medium

light

both strands labelled with $^{14}N$

relative density of DNA

## 4.2 Protein synthesis

If amino acids, the building blocks of proteins, are labelled with a radioactive isotope and are fed to rats, the ribosomes of the rats' liver cells soon become radioactive. Furthermore, peptide links are formed between the amino and carboxyl groups of amino acid molecules incubated with a suspension of ribosomes, RNA and ATP. Evidence of this sort indicates that the ribosomes are the sites of protein synthesis. But what parts do RNA and ATP play in the synthesis of proteins?

Explain how amino acids which are fed to rats soon appear in their livers.

### 4.2.1 Role of the nucleic acids

Proteins are polymers consisting of up to twenty different amino acids joined together in a specific linear sequence and folded in a particular three-dimensional way (Chapter 2). Many different kinds of protein can be made inside a living cell. The accuracy with which ribosomes repeatedly assemble such complex molecules suggests that some sort of control system is at work inside cells coding the production of proteins. The code, called the **genetic code**, is the sequence of nitrogenous bases in DNA. How is information contained in DNA conveyed to and interpreted by the ribosomes?

In working out the mechanism of protein synthesis, molecular biologists were faced with the problem of explaining how the nucleotides found in DNA could provide a code for the synthesis of many different kinds of protein. Single nucleotides can hardly be the basis of the code as there are just four different nucleotides but at least twenty different amino acids. Neither does it seem logical that pairs of nucleotides are responsible as this would give only $4^2 = 16$ possible combinations of two nucleotides. It was therefore deduced that combinations of at least three nucleotides are required. The **triplet code** provides $4^3 = 64$ different permutations from the four nucleotides (Fig 4.14).

**Fig 4.14** The triplet code

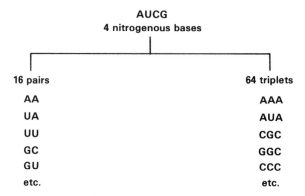

In the early 1960s Nirenberg and his colleagues began to experiment with the production of artificial polypeptides using ribosomes extracted from *Escherichia coli*. When incubated with a mixture of amino acids, ATP and synthetically-made mRNA containing only the nitrogenous base uracil, the ribosomes synthesised a polypeptide made solely from the amino acid called phenylalanine. The triplet code for phenylalanine is therefore **UUU**. Nirenberg also discovered that **AAA** is the triplet code for lysine and **CCC** for proline. Since then the triplet codes for all twenty amino acids have been worked out. Methionine and tryptophan each have a single triplet code, the other eighteen amino acids have more than one triplet permutation. Some have just two alternative triplets, others have up to six. For this reason the genetic code is described as **degenerate**. A few of the sixty-four triplets, called **non-sense triplets**, do not code for any amino acid.

Protein synthesis takes place in three distinct stages.

### 1 Transcription

Before proteins are synthesised, mRNA is made inside the nucleus. The double helix of DNA unwinds and single-stranded mRNA molecules are produced, part of one of the DNA strands acting as a template (Fig 4.15). Remember that the mRNA is complementary to its DNA template. In this way the genetic code is **transcribed** as the sequence of bases in mRNA. Synthesis of mRNA is controlled by the enzyme **RNA ligase**. The mRNA now passes through the pores of the nuclear membrane into the cytoplasm. Here one end of the mRNA molecule becomes attached to a ribosome.

**Fig 4.15** Transcription

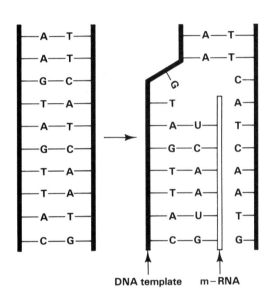

DNA template   m – RNA

### 2 Activation

During **activation**, energy from ATP (Chapter 5) is used to combine tRNA molecules with amino acid molecules. There are at least twenty different kinds of tRNA, the important difference between them being the sequence of nitrogenous bases in their anticodons (section 4.1.2). Each type of tRNA binds with a specific amino acid. The amino acid molecules join to the free ends of the tRNA molecules where bases A←C←C← are found (Fig 4.16). This rules out the possibility that the anticodon determines the amino acid with which the tRNA combines. The tRNA–amino acid complexes now move to the ribosomes.

**Fig 4.16** Activation

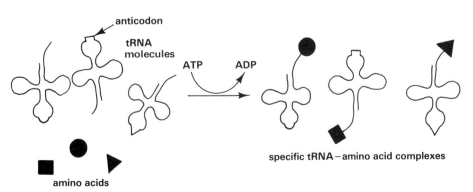

53

### 3 Translation

It is now necessary for the transcribed genetic code to be **translated**. A knowledge of protein structure (Chapter 2) is assumed in the following description of **translation**. Starting at one end of an mRNA molecule, a ribosome works its way along, positioning the **anticodon** of each tRNA molecule on to a complementary **codon** of the mRNA strand (Fig 4.17). For example, the triplet code for methionine, the first amino acid in a polypeptide chain, is AUG. The tRNA–methionine complex having the anticodon

**Fig 4.17** Translation

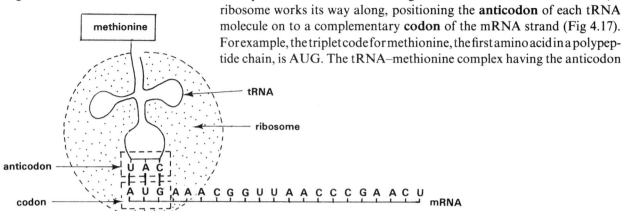

UAC is brought to the end of the mRNA molecule where the codon AUG is located. The ribosome binds the codon and anticodon, then moves on to the next triplet of the mRNA strand. Here **codon–anticodon binding** again takes place and the next amino acid molecule is brought into position. A peptide link is formed with methionine and the polypeptide chain begins to take shape. As soon as an amino acid is linked to its neighbour its tRNA partner is released back into the cytoplasm to pick up another molecule of the same amino acid (Fig 4.18). The ribosome continues to work its way along the mRNA strand until it reaches a triplet for which there is no anticodon. This is the signal that the polypeptide chain is complete. The part of a DNA molecule which codes the synthesis of a polypeptide is usually called a **gene**. Protein molecules are made of one or

**Fig 4.18** Polypeptide synthesis

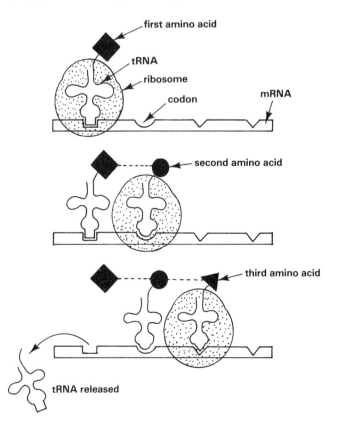

more polypeptide chains (Chapter 2). For example, a molecule of haemoglobin contains four, each called a globin unit. Normal adults synthesise haemoglobin A (HbA), the molecules of which contain two α- and two β-globins. The α-globins have 141 amino acids in each chain, whereas there are 146 in each β-chain (Chapter 12).

**Polysomes**, groups of ribosomes connected by a common strand of mRNA, are often seen in cells (Chapter 6). This arrangement may mean that several polypeptides are made at the same time on one mRNA molecule (Fig 4.19). When synthesis is completed, the polypeptides are moved from the ribosomes to the cytoplasm and constructed into proteins for internal use by the cell or for secretion.

**Fig 4.19** Polypeptide synthesis by a polysome

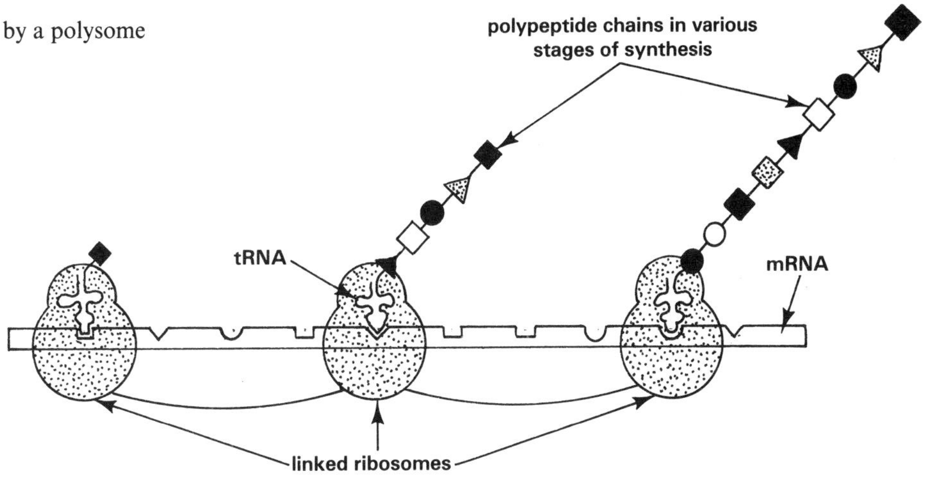

### 4.2.2 Gene mutations

Sometimes mistakes arise when DNA replicates. There are several ways in which errors can occur. One pair of bases may be replaced by another, for example C:G may be replaced by A:T. A pair of nucleotides may be lost or an extra pair added (Fig 4.20). Whatever the cause, the new DNA is not an exact copy of the original. Such changes are called **gene mutations**. When a gene mutates the changes in sequence of the bases in DNA causes a

**Fig 4.20** Gene mutations

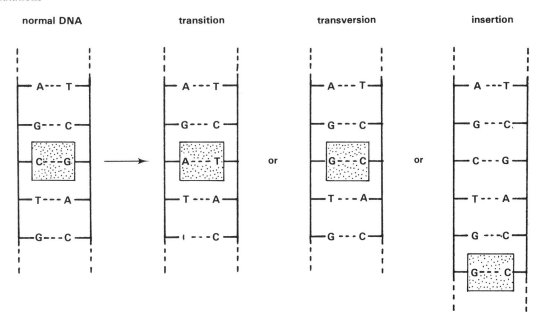

55

List some of the factors which may result in gene mutations.

complementary change in base sequence in a codon of mRNA. The altered codon is translated either as **non-sense**, causing synthesis of a polypeptide with one or more amino acids missing. Alternatively the codon is translated as **mis-sense** in which case another amino acid is substituted in or added to the polypeptide chain (Fig 4.21). Protein molecules built from such polypeptides are usually defective and cannot carry out their normal functions. **Abnormal haemoglobins** and **inborn errors of metabolism** are examples of the consequences of gene mutations.

**Fig 4.21** Some consequences of gene mutations

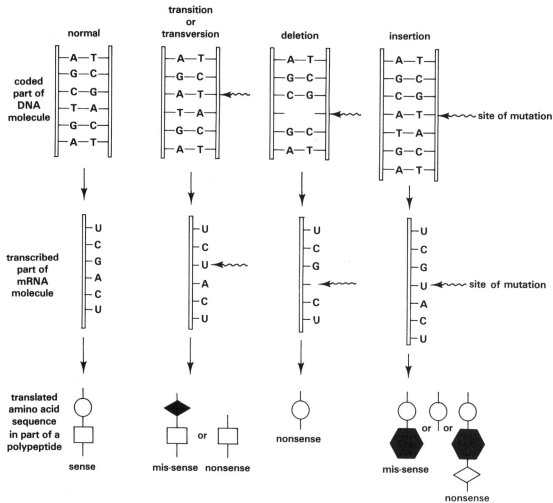

## 1 Abnormal haemoglobins

People who have **sickle cell disease** produce haemoglobin S (HbS) instead of HbA. In tissues where there is a low oxygen tension, HbS molecules become rigid chains, causing the red blood cells to become sickle-shaped. Such cells can block vital blood vessels. Haemoglobin S also has a lower affinity for oxygen than HbA (Chapter 12). In each of its $\beta$-globin chains HbS has the amino acid valine instead of glutamic acid (Fig 4.22). Our understanding of the genetic code now makes it possible to explain the biochemical difference between HbS and HbA. The codon for glutamic acid is GAG which corresponds to the triplet CTC in the DNA template. If the triplet is changed to CAC, the corresponding codon in mRNA becomes GUG which is the codon for valine. The kind of mutation seen here is **transversion**. The effect is that HbS is synthesised instead of HbA.

**Fig 4.22** Two-dimensional chromatographic analysis of haemoglobins

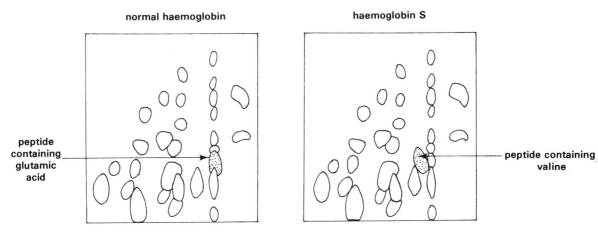

normal haemoglobin          haemoglobin S

peptide containing glutamic acid

peptide containing valine

Gene mutations are also the cause of people with **thalassaemia syndrome** producing an abnormal type of haemoglobin. In α-thalassaemia synthesis of α-globin polypeptide chains fails to occur. In β-thalassaemia β-globin synthesis is suppressed (Chapter 12). It is presumed that a **deletion, transition** or **transversion** in the DNA template is the cause of the failure. As with sickle cell disease, people who suffer from thalassaemia are extremely anaemic.

### 2 Inborn errors of metabolism

A number of human diseases can be attributed to the inability to synthesise specific enzymes or to the synthesis of enzymes which have reduced catalytic powers.

### A. Phenylketonuria

In the liver of a normal person the essential amino acid phenylalanine is converted to another amino acid tyrosine by the enzyme **phenylalanine hydroxylase**. People with **phenylketonuria (PKU)** cannot bring about the conversion. It is assumed that this is because they cannot synthesise an active form of the enzyme. During foetal life it is not a problem because the mother's liver carries out the reaction, but after birth phenylalanine begins to accumulate in the child's blood. Some phenylalanine is converted to phenylpyruvic acid which is excreted in the urine. However, the remaining phenylalanine prevents the child's brain from absorbing sufficient amounts of other essential amino acids from its blood. As a result the brain and other organs and tissues such as muscles and cartilage fail to grow and develop normally.

Unless PKU is detected soon after birth, the child usually becomes mentally retarded and cannot walk properly. At one time children with PKU were identified by testing their urine for phenylpyruvic acid with iron(III) chloride, $FeCl_3$. However, the acid is not in sufficient quantities for diagnosis until the child is several weeks old. By that time brain damage may have already begun. These days a blood test makes it possible to identify children with PKU at birth. They can then be fed a special diet containing just enough phenylalanine to allow normal growth without it accumulating in the body. The diet can be stopped at about nine years of age when the brain should be fully developed. Females with PKU must return to the diet when they become pregnant. If they do not, the high concentration of phenylalanine in their blood would damage the brain of the developing foetus.

## B. Alkaptonuria

The urine of people with **alkaptonuria** turns black when exposed to light. Chemical analysis shows the urine to contain a high concentration of **homogentisic acid**, also called **alkapton**. The acid is produced in the liver from excess amounts of tyrosine. In normal people the breakdown of alkapton to carbon dioxide and water is catalysed by enzymes. In people with alkaptonuria, failure to produce one of the enzymes causes alkapton to accumulate. In early life there are no apparent ill-effects. Later on, alkapton is deposited in cartilage causing the tip of the nose and the pinnae of the ears to turn black. A painful form of arthritis often accompanies these changes for which there is no known cure.

## C. Galactosaemia

Children with **galactosaemia** are often apparently normal when they are born, but within a few weeks they begin to vomit much of the milk they drink and fail to thrive. If the condition is not diagnosed early the child may become blind and mentally retarded. Galactosaemia occurs when a child's liver fails to produce the enzyme which catalyses the conversion of **galactose** to glucose. In the small intestine, lactose (milk sugar) is hydrolysed by the enzyme lactase to form galactose (Chapter 9). If this cannot be converted to glucose, it accumulates in the blood and causes the damage described earlier. The symptoms disappear if the condition is diagnosed early and the child is given a lactose-free diet. It is necessary to continue with the diet throughout life.

## D. Acatalasia

The blood of most people contains the enzyme **catalase** which is synthesised in the liver. If a drop of blood or a small piece of fresh liver is added to some hydrogen peroxide, catalase very rapidly breaks down the peroxide to water and oxygen. The release of oxygen causes the solution to froth vigorously. Some people cannot synthesise catalase. The condition is called **acatalasia**. About half of those with acatalasia are prone to infection of the tissues of the mouth. The remainder are as healthy as normal people.

## 4.2.3 Regulating protein synthesis

It is necessary for cells to regulate the amounts and types of proteins they make because enzymes and metabolites need to be produced only as and when required. The efficiency of metabolism is thus optimised and a steady state is maintained. One device which helps in **cellular homeostasis**, the maintenance of a steady state in living cells, is the temporary inhibition of an enzyme when the concentration of a substrate becomes too high (Chapter 3). Another is to regulate the synthesis of enzymes.

In the late 1950s two French scientists, François Jacob and Jacques Monod, reported that the bacterium *Escherichia coli* was able to produce several enzymes 'on demand'. *E. coli* can synthesise all the metabolites and enzymes it requires from a medium consisting of glucose, mineral salts and water. Jacob and Monod observed that if lactose was substituted for glucose, the *E. coli* makes an additional enzyme called $\beta$-D-galactosidase which hydrolyses lactose to glucose and galactose. When returned to the glucose medium the bacterium stops making $\beta$-D-galactosidase. The presence of lactose in the medium induces *E. coli* to make an enzyme capable of hydrolysing lactose. This is an example of **enzyme induction**.

*E. coli* can make all the amino acids it requires from a medium containing $NH_4^+$ as the only source of nitrogen. However, if the amino acid arginine is added to the medium the bacterium stops making the enzymes needed for the synthesis of arginine. In this case the presence of arginine in the medium inhibits the production of an enzyme, an example of **enzyme repression**.

Jacob and Monod produced a hypothesis to explain enzyme induction and repression. They proposed that the sequence of amino acids in an enzyme is determined by a **structural gene** which can be switched on or off by a **regulator gene**. The regulator gene transcribes the production of mRNA which has the codons to make a **repressor protein**. The repressor moves from the ribosomes where it is made to the nucleus. Here it binds with the structural gene, stopping it from transcribing the manufacture of mRNA used in the synthesis of an enzyme. As a consequence, production of the enzyme is repressed. However, if an inducer substance is present it combines with the repressor. The inducer–repressor complex is unable to prevent transcription of the structural gene. Synthesis of the enzyme can then proceed. The formation of the inducer–repressor complex must be reversible because enzyme repression occurs when the inducer is removed. Induction and repression are thus linked. Induction is freedom from repression. The Jacob–Monod hypothesis is summarised in Fig 4.23.

**Fig 4.23** The Jacob-Monod hypothesis

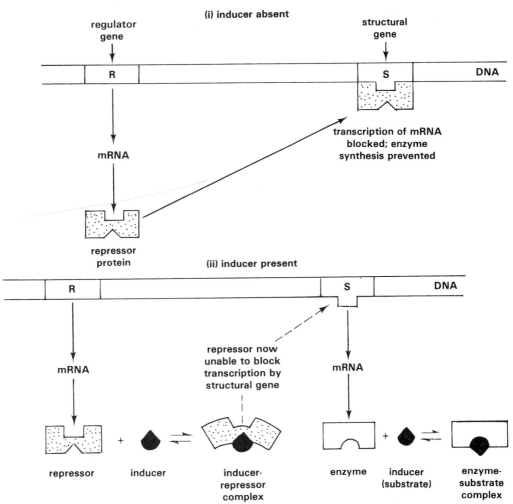

Recently it has been shown that induction and repression of enzymes also occurs in the cells of humans. Liver cells are particularly efficient at regulating the types and amounts of enzymes they produce. The liver can therefore adjust to the fluctuating levels of nutrients it receives from the gut (Chapter 9).

Enzyme induction and repression are also important in the differentiation of tissues in living organisms. How a one-celled fertilised egg grows into a multicellular body made up of a number of different types of tissue is a process about which a lot has yet to be learned. However, induction and repression of enzymes probably enables the various types of tissue to differentiate by allowing different combinations of genes to be expressed at different stages during growth and development. A good example of the differential action of genes is seen in the synthesis of haemoglobin. In the foetal stage of development we produce **foetal haemoglobin (HbF)**. Each molecule contains two α-globin and two γ-globin polypeptide chains. The haemoglobin we produce after birth has two α-globin and two β-globin chains in each molecule. This is haemoglobin A (HbA). Foetal haemoglobin has a higher affinity for oxygen than adult haemoglobin and thus absorbs oxygen across the placenta from haemoglobin in the mother's blood (Chapters 12 and 19).

A better understanding of gene regulation may make it possible to switch off defective genes and to switch on genes which are not functioning. In this way it may be possible to prevent production of abnormal haemoglobins and cure inborn errors of metabolism.

## 4.3 Genetic engineering

Avery in 1943 showed that it was possible to alter the genetic make-up (genome) of the R-strain of *Pneumococcus* by adding to its culture medium DNA from the S-strain (section 4.1.3). Since then, biologists have devised sophisticated procedures whereby the genomes of experimental organisms such as bacteria can be altered with great precision. The experimental manipulation of the genomes of organisms is called **genetic engineering**. One of the methods used involves the transfer of genes from the cells of one organism to those of another. The new DNA formed in the recipient is called **recombinant DNA**.

The technique involves isolating strands of DNA from one species and breaking them at specific points using **nuclease enzymes**. The same enzymes are then used to fragment strands of DNA from another species. Because of the specificity of the enzymes, the base sequence at the ends of some of the pieces of DNA from both sources are the same. They are said to have **sticky ends**. Recombinant DNA is formed when the two samples of DNA are incubated in the presence of **DNA ligase**, an enzyme which catalyses binding of the pieces of DNA at their sticky ends (Fig 4.24). If the recombinant DNA is then put back into one of the cells from which it came, it replicates and the daughter cells have characteristics of both species. So far, experiments in genetic engineering have been mainly with microbes. However there is no reason in principle why it should not be extended to other groups of living organisms.

**Fig 4.24** Genetic engineering

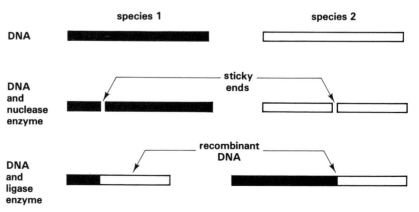

60

Advantages of immense importance to the human race could arise from genetic engineering. For example, the roots of leguminous plants form a symbiotic relationship with the bacterium *Rhizobium leguminosarum*. The bacterium can fix nitrogen gas and some of the organic nitrogenous substances it makes are used by the leguminous host plants. The hosts as a result grow more profusely. Because legumes, such as peas, beans and clover, are extensively grown crop plants, the symbiotic relationship is of great importance to food production. But what of the many other types of crop plant which do not have this facility? The soil in which they are grown has to be treated with expensive artificial fertilisers or with manure (Chapter 24). Clearly it would be advantageous if non-legumes too could fix gaseous nitrogen which is plentiful in the atmosphere. For this reason attempts have been made to transfer the genes for nitrogen fixation into the cells of cereals and other non-leguminous plants.

Genetic engineering is also of interest to the medical world. When attacked by viruses, mammalian tissues produce interferon, an antiviral agent (section 11.2.4). Until recently it has not been possible to manufacture adequate quantities of interferon for treating humans suffering from viral diseases. The very small amounts available came mainly from leucocytes taken from blood given by donors. However, within the past few years molecular geneticists have succeeded in incorporating the gene for making interferon from human leucocytes into the DNA of the bacterium *Escherichia coli*. When *E. coli* containing the gene is grown in laboratory conditions it secretes interferon into the medium. Mass culture of *E. coli* with recombinant DNA has thus substantially increased the manufacture of an important life-saving drug. Comparable procedures have been used to make anti-haemophilic factor, a variety of antibodies, and hormones.

There is, however, another side to genetic engineering. Concern has been expressed about the danger of producing organisms whose new genomes pose a threat to the well-being of humans. One of the organisms widely used in such experiments is the gut bacterium *Escherichia coli* (Fig 4.25) which lives commensally in the bowels of humans and other vertebrates. What would be the consequence if a strain of genetically-engineered *E. coli* which was highly pathogenic became resident in a human population? As Dr Erwin Chargaff, an eminent molecular geneticist has put it:

'You cannot recall a new form of life. Once you have constructed a viable *E. coli* cell into which a piece of eukaryotic DNA has been spliced, it will survive you and your children and your children's children. An irreversible attack on the biosphere is something so unheard of, so unthinkable to previous generations, that I could only wish that mine had not been guilty of it.'

The potential hazards of such experiments have led to a call for strict precautions in certain types of genetic engineering. In Britain the Genetic Manipulation Advisory Committee was set up to identify the precautions required to prevent the escape of genetically engineered microbes and to suggest the kinds of organisms to be used. These days most genetic engineers agree that the benefits which may accrue from such experiments greatly outweigh the potential risks which are generally thought to be negligible.

What are the potential implications of engineering the genetic make-up of humans?

**Fig 4.25** *E. coli* which has been osmotically shocked to cause extrusion of its DNA (Science Photo Library)

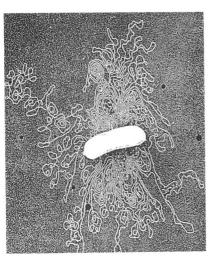

# 5 Energy conversion in living cells

The composition of our bodies is very different from our surroundings. Whilst the non-living environment consists of relatively simple substances such as gases, water and minerals, we are made up of very complex molecules (Chapter 2). Most of the atoms in the environment are distributed in energy-poor compounds. In contrast, the atoms which make up our bodies are mainly in energy-rich compounds.

The sun is the main source of energy available to the living world. Energy in sunlight is changed to chemical energy by green plants in a process called **photosynthesis**. The products of photosynthesis are complex energy-rich molecules from which energy can be released. Green plants are called the **producer organisms** of ecosystems. Organisms which cannot photosynthesise are called **consumers**. Primary consumers acquire energy-rich compounds by feeding on green plants. Secondary consumers feed on primary consumers. Whatever the feeding level of an organism, energy is made available when energy-rich substances are oxidised in living cells during **respiration**. Thus photosynthesis and respiration bring about the conversion of energy in sunlight into chemical energy which can be used to do work in living organisms.

## 5.1 Photosynthesis

Photosynthesis consists of two main phases, the **light-dependent** and **light-independent** (dark) reactions.

### 5.1.1 Light-dependent reactions

White light is a mixture of light of different wavelengths (Fig 5.1). The energy of light is inversely proportional to its wavelength (measured in nanometres, nm), and so light of short wavelength has more energy than light of long wavelength. Green plants convert the energy of some wavelengths of visible light into the energy which bonds together the atoms of complex organic molecules.

**Fig 5.1** The visible spectrum

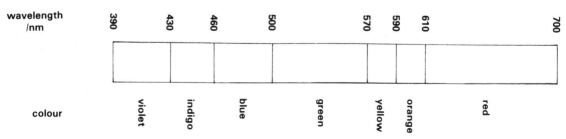

### 1 Absorption and action spectra

Green plants do not use energy from all of the components of white light for photosynthesis. This was neatly demonstrated by the German plant physiologist H. T. Engelmann in 1882. He placed filaments of the green alga *Cladophora* in a drop of water on a microscope slide. The filaments were illuminated with light of different wavelengths. He then watched the distribution of aerobic bacteria in the water. Engelmann noted that the bacteria clustered near to the filaments when blue light (450 nm) or red light (650 nm) was used (Fig 5.2).

**Fig 5.2** Engelmann's experiment

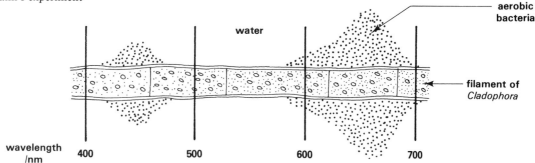

Knowing that the alga gave off oxygen as it photosynthesised, Engelmann deduced that blue and red light are the most effective for photosynthesis. It has since been shown that photosynthesis in nearly all green plants occurs most rapidly when they are illuminated by blue and red light.

An **action spectrum** is produced when the rate of photosynthesis is plotted against wavelength of light (Fig 5.3). The reason why light from the blue and red parts of the spectrum is effective is that it is absorbed efficiently by the pigments contained in the chloroplasts of green plants. The photosynthetic pigments most commonly present are the **chlorophylls** and **carotinoids** (Table 5.1). They can be readily extracted by grinding up chopped leaves in an organic solvent such as propanone. The pigments in the extract can then be separated by paper chromatography (Fig 5.4).

**Fig 5.3** An action spectrum for photosynthesis

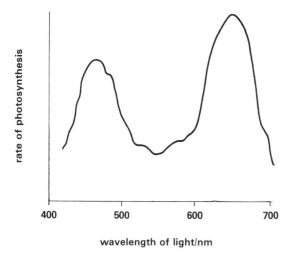

**Table 5.1 Photosynthetic pigments commonly found in leaves of green plants**

| Group | Pigment | Colour |
|---|---|---|
| *chlorophylls* | chlorophyll *a* | green |
| | chlorophyll *b* | green |
| *carotinoids* | xanthophyll | yellow |
| | carotene | orange |

**Fig 5.4** A chromatogram of photosynthetic pigments

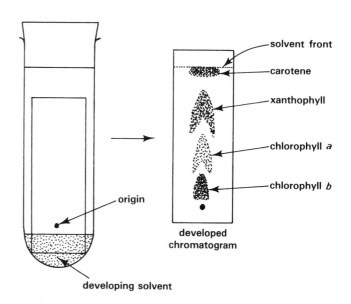

Chlorophylls belong to a group of chemicals called **porphyrins**, which includes cytochromes and haemoglobin. They all have four nitrogen-containing pyrrole rings (Fig 5.5). In the chlorophylls a magnesium atom is held between the rings. In cytochromes and haemoglobin there is an iron atom instead. Chlorophylls are green because they reflect green light. The wavelengths they absorb most strongly are in the blue and red parts of the visible spectrum.

**Fig 5.5** Structure of a chlorophyll molecule. Compare with Fig 12.14 (b)

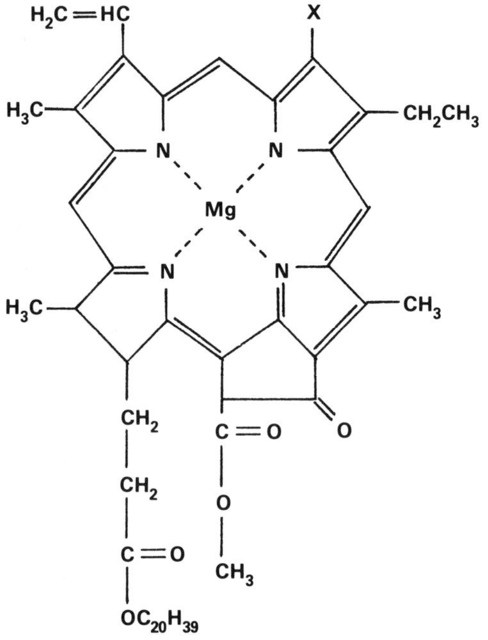

In chlorophyll $a$ X = $CH_3$; in chlorophyll $b$ X = CHO

Formula:  $C_{55}H_{72}O_5N_4Mg$        $C_{55}H_{70}O_6N_4Mg$

Carotene and xanthophyll are long-chain **hydrocarbons** (Fig 5.6). Their colours indicate that they reflect orange and yellow light respectively. If the percentage light absorption is plotted against wavelength of light, an **absorption spectrum** is obtained for each pigment (Fig 5.7). Notice the close correlation between the absorption and action spectra. Chlorophylls absorb both red and blue light, whereas the carotinoids absorb mainly blue light.

**Fig 5.6** Structure of a $\beta$-carotene molecule

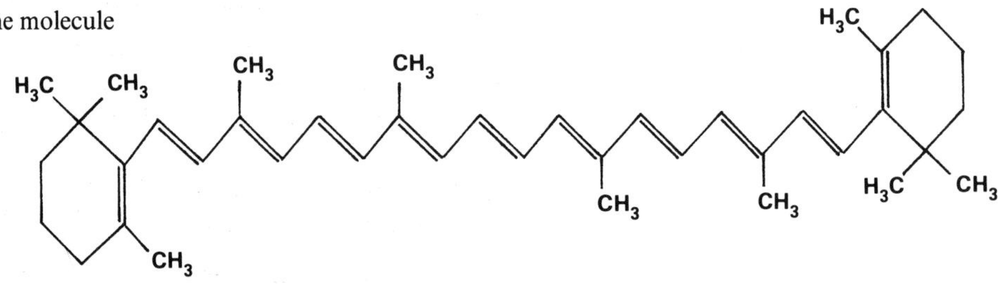

Formula of carotene:  $C_{40}H_{56}$

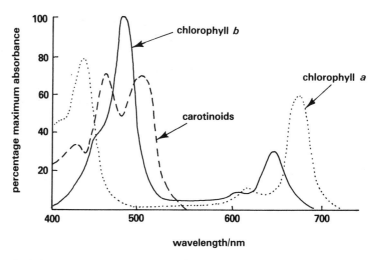

**Fig 5.7** Absorption spectra of chloroplast pigments

## 2 Electron transfer in chloroplasts

What happens when chloroplast pigments absorb light? One clue comes from the behaviour of the pigments extracted from leaves in organic solvents such as ethanol or propanone. When the solution is placed in white light the pigments give off red light. This is called fluorescence. Before illumination, the electrons in the outer shell of magnesium atoms in the chlorophyll molecules are in the ground-state energy level. Light energy temporarily raises the energy level, causing the electrons to become displaced. In their high-energy state the electrons are said to be **excited**. Red light appears during fluorescence because, shortly after being displaced, the electrons fall back into the ground state. The extra energy they held is emitted as red light.

What has just been described is of course an experimental observation. In living tissues the solvent is water and not an organic solvent. In nature, we do not see green plants emitting red light when they photosynthesise. Also, green plants give off oxygen during photosynthesis. These facts strongly suggest that in living plants the energy of the excited electrons in chlorophyll molecules is linked to other reactions that involve the release of oxygen. Where does the oxygen come from, and what happens to the excited electrons?

Some idea of what happens to the excited electrons was established in 1937 by the English biochemist Robin Hill. He observed an illuminated suspension of chloroplasts in water evolving oxygen, and at the same time transferring electrons to iron(III) ions, reducing them to iron(II) ions. This phenomenon, called the **Hill reaction**, suggests that chloroplasts contain acceptors of the excited electrons from chlorophyll molecules. In the 1950s it was discovered that the main acceptor of these excited electrons was **nicotinamide-adenine dinucleotide phosphate (NADP$^+$)**.

## 3 Photolysis

Having given electrons to NADP$^+$, chlorophyll molecules are in an electron-deficient (oxidised) state. How is the pigment reduced to its former, stable condition? One way it could be reduced is to receive electrons from another source. Water is the most likely source of the electrons. Water is a weak electrolyte and a small proportion of its molecules dissociate into hydrogen ions (H$^+$) and hydroxyl ions (OH$^-$). When electrons are removed from hydroxyl ions by oxidised chlorophyll, oxygen is given off:

$$2OH^- + \text{oxidised chlorophyll} \rightarrow \tfrac{1}{2}O_2 + H_2O + \text{reduced chlorophyll}$$

Equilibrium is re-established as more water molecules dissociate. Thus one

65

result of light absorption by chlorophyll is a rapid splitting of water molecules into hydrogen ions and oxygen. This is called **photolysis** (photo = light; lysis = to split). Using isotopes it is possible to prove that the oxygen given off in photosynthesis comes from water. When heavy water containing the heavy isotope of oxygen $^{18}O$ is supplied to photosynthesising plants instead of normal water (containing the isotope $^{16}O$), heavy oxygen is given off:

$$CO_2 + 2H_2{}^{18}O \rightarrow {}^{18}O_2 + (CH_2O) + H_2O$$

If a control experiment is also run in which the $^{18}O$ is supplied in carbon dioxide, while normal water is provided, the evolved oxygen is of the $^{16}O$ type.

What happens to the hydrogen ions from photolysed water is explained in the next section.

### 4 Photophosphorylation

Excited electrons displaced from chlorophyll *b* pass through an electron transport chain in which cytochromes act as electron carriers (Fig 5.8). As excited electrons pass through the chain, some of their energy is used to make **adenosine 5′-triphosphate (ATP)** (section 5.2.1). When they leave the chain the electrons have a normal, ground-state amount of energy. The ground-state electrons reduce chlorophyll *a* which had previously lost excited electrons. The excited electrons from chlorophyll *a* pass through a second electron transport chain in which ferredoxin is the main electron carrier. Once again ATP is made. The production of ATP in the two electron transport chains is called **non-cyclic photophosphorylation**.

**Fig 5.8** Non-cyclic photophosphorylation

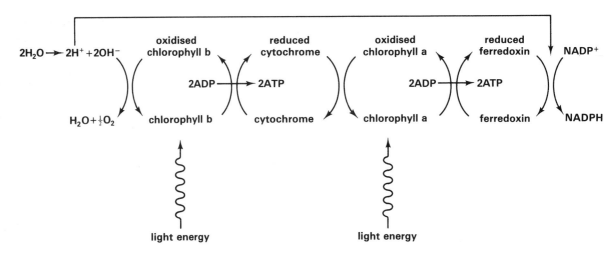

**Reduced NADP⁺(NADPH)** is another important product of the second electron transport chain. Hydrogen ions from photolysed water combine with ground-state electrons coming out of the chain. The hydrogen atoms so formed reduce $NADP^+$ to NADPH. Oxygen is given off when hydroxyl ions from photolysed water donate electrons to oxidised chlorophyll *b*.

ATP is also made in a third electron transport chain called **cyclic photophosphorylation**. Excited electrons from chlorophyll *a* pass first to ferredoxin, then to cytochromes. The reduced cytochromes then pass the electrons back to oxidised chlorophyll *a* (Fig 5.9).

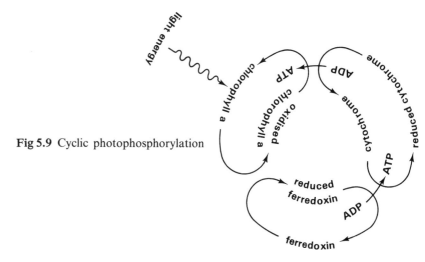

Fig 5.9 Cyclic photophosphorylation

The important products of the light-dependent reactions of photosynthesis are thus ATP, NADPH and oxygen. ATP is a source of energy required for the dark reactions. NADPH is also used in the dark reactions to reduce 3-phosphoglyceric acid (section 5.1.2) Oxygen given off in photosynthesis can be used by living organisms for respiration.

No mention has yet been made of the role of the carotinoid pigments. They are thought to shield the chlorophylls from excessive oxidation in intense light. The carotinoids are not involved in electron transfer, but they can absorb light energy which they transfer as excitation energy to chlorophyll molecules. This is why photosynthesis occurs only in the green parts of variegated leaves.

## 5.1.2 Light-independent reactions

### 1 Fixation of carbon dioxide

For a long time, plant scientists had little idea of what happens to carbon dioxide in photosynthesis. In the late 1940s Professor Melvin Calvin of the University of California decided that one way to find out was to allow plants to photosynthesise using carbon dioxide labelled with the radio-nuclide $^{14}C$. Calvin experimented with unicellular green algae such as *Chlorella* and *Scenedesmus* which he grew in a mineral solution held in flat glass containers he called 'lollipops' (Fig 5.10). The isotope was added as sodium hydrogencarbonate solution which breaks down to form $^{14}CO_2$.

Fig 5.10 Calvin's experiment

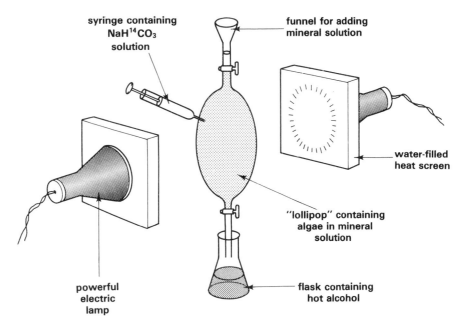

syringe containing
NaH$^{14}$CO$_3$
solution

funnel for adding
mineral solution

water-filled
heat screen

"lollipop" containing
algae in mineral
solution

powerful
electric
lamp

flask containing
hot alcohol

After a fixed period of illumination the algae were rapidly killed by running the mineral solution into a flask of hot ethanol. The cell contents were then analysed by autoradiography to find out which substances were labelled.

First the cell extracts were separated by paper chromatography. A variety of known organic compounds was run on the chromatogram at the same time in order to identify the components of the cell extracts. The chromatogram was then placed on a sheet of X-ray film. Radiations emitted by radionuclides cause black spots called **fogging** to appear on the film (Fig 5.11). When the fog marks on the autoradiogram are checked against the positions of the known compounds on the chromatogram it is possible to identify which of the components of the cell extracts are probably labelled with the C[14].

**Fig 5.11** Autoradiographs of photosynthetic products containing $^{14}$C in *Scenedesmus* (courtesy Dr M Tribe)

(a) After 5 seconds

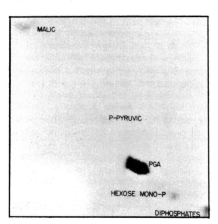

(b) After 15 seconds

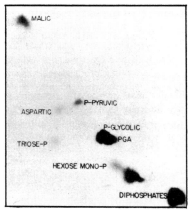

(c) After 60 seconds

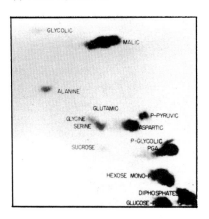

Calvin argued that after only a short period of illumination the radionuclide should appear in the first products of carbon dioxide fixation. With longer exposures to light, the intermediate and end products of photosynthesis should be labelled. Table 5.2 shows a summary of his findings. These reactions take place in the stroma of chloroplasts (Chapter 6). The first product of carbon dioxide fixation is 3-phosphoglyceric acid (PGA). But what does $CO_2$ combine with to form PGA? Because each PGA molecule contains three carbon atoms Calvin thought it was logical to search for a $CO_2$-acceptor molecule containing two carbon atoms. Despite much effort, no such compound was found in green plants. Later a pentose sugar, ribulose 1,5-bisphosphate (RBP) was shown to be the main $CO_2$ acceptor in many plant species. Each molecule of RBP combines with one of carbon dioxide to form two molecules of PGA. But how is PGA built up into carbohydrates? Furthermore, how is a constant supply of RBP maintained so that $CO_2$ fixation can continue indefinitely?

**Table 5.2 Components of algal cell extract labelled with radioactive carbon after different periods of photosynthesis in the presence of $^{14}CO_2$**

| Time of exposure to light after isotope added/s | Main substances containing $^{14}$C |
| --- | --- |
| 5 | 3-phosphoglyceric acid (PGA) |
| 15 | PGA, hexose phosphates |
| 60 | PGA, hexose phosphates, sucrose, amino acids |
| 300 | PGA, hexose phosphates, sucrose, starch, amino acids, proteins, lipids |

Fig 5.12 (a) The Calvin cycle

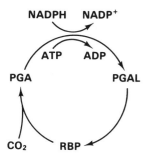

Fig 5.12 (a) The Calvin cycle

## 2 Link between light and dark stages

So far there seems to be no connection between the light-dependent reactions, and carbon dioxide fixation which is light-independent. However NADPH and ATP, the products of the light stage, are used in the dark stage. PGA is reduced by NADPH to form the triose sugar 3-phospho-glyceraldehyde (PGAL). The reaction uses energy provided by ATP. One sixth of the PGAL is built into hexose sugars, the remainder is used to resynthesise the carbon dioxide acceptor, RBP. These events are sometimes called the **Calvin cycle** (Fig 5.12(a)). The link between the light and dark stages of photosynthesis is summarised in Fig 5.12(b). Hexoses are condensed into disaccharides and polysaccharides. Amino acids and lipids are formed in other linked pathways.

Fig 5.12 (b) Link between the light and dark stages of photosynthesis

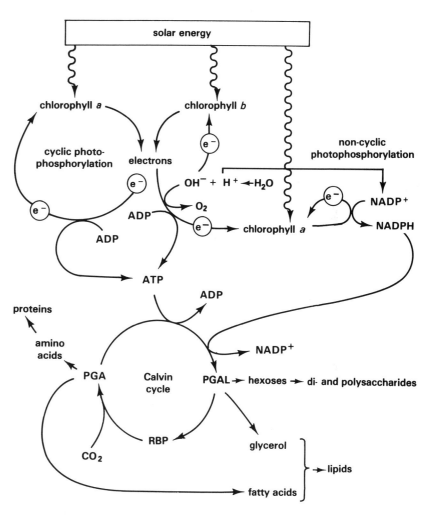

Fig 5.13 The Hatch–Slack (C₄) pathway

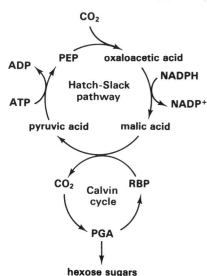

## 3 Other pathways for carbon dioxide fixation

The fixation of carbon dioxide in some green plants occurs in a different way from that described above. The plants include some important tropical crops such as sugar cane, maize and sorghum. When given $^{14}CO_2$ the carbon radionuclide first appears in oxaloacetic acid. Oxaloacetic acid contains four carbon atoms in each of its molecules compared with three in PGA. This pathway is sometimes called the **Hatch–Slack** or **C4 pathway**, and its link to the Calvin cycle is outlined in Fig 5.13.

The source of energy for photosynthesis in C4 plants is sunlight, and ATP and NADPH are again the products of the light-dependent reactions.

Some ATP is used to phosphorylate pyruvic acid, with the formation of phosphoenolpyruvic acid (PEP). On losing its phosphate group, PEP releases energy which is used to fix $CO_2$. Oxaloacetic acid is the first product of carbon dioxide fixation in C4 plants. NADPH reduces the oxaloacetic acid to malic acid which reacts with RBP to form pyruvic acid and PGA. The fate of PGA is the same as that described for C3 plants.

Among the products of $CO_2$ fixation in all green plants is hydroxyethanoic acid. The acid is immediately oxidised in C3 plants in a process called **photorespiration**, and carbon dioxide is given off. Up to 30 % of carbon dioxide fixed in photosynthesis can be recycled in this way. Consequently a considerable amount of the solar energy used to fix carbon dioxide is wasted. C4 plants photorespire less, so their photosynthesis is more efficient in making raw materials for growth. Furthermore, with a rise in temperature the rate of photorespiration increases more rapidly than carbon dioxide fixation. C4 plants are therefore much more productive than C3 plants, especially in tropical climates.

### 5.1.3 Factors affecting the rate of photosynthesis

The rate at which oxygen is given off by green plants can be used to measure the rate of photosynthesis. Aquatic plants such as *Elodea* are particularly suitable for this purpose. Bubbles of oxygen are given off from the cut end of the stem of *Elodea* when it is immersed in illuminated pond water or a dilute solution of sodium hydrogencarbonate (Fig 5.14)

**Fig 5.14** Measuring the rate of photosynthesis in the pondweed *Elodea*

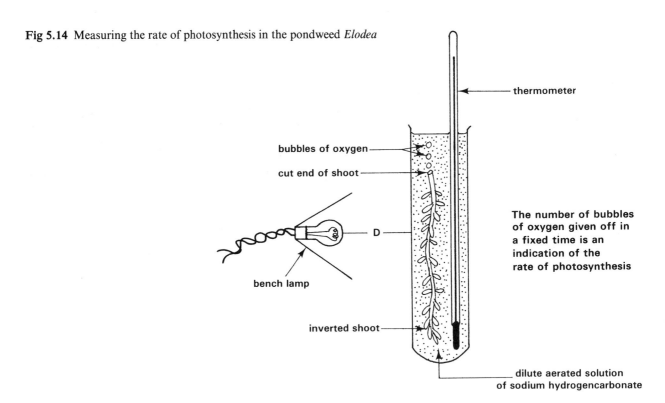

thermometer

bubbles of oxygen

cut end of shoot

D

The number of bubbles of oxygen given off in a fixed time is an indication of the rate of photosynthesis

bench lamp

inverted shoot

dilute aerated solution of sodium hydrogencarbonate

### 1 Light intensity

Using this method in the 1930s, the British plant physiologist F. F. Blackmann investigated the effect of light intensity on the rate of photosynthesis. The intensity of light was varied by increasing or decreasing the distance $D$ between the lamp and the plant. Light intensity $I$ is inversely proportional to the square of the distance: $I \propto 1/D^2$. The results of the experiment are shown in Fig 5.15.

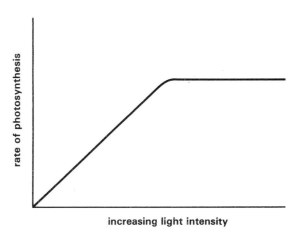

**Fig 5.15** Effect of light intensity on the rate of photosynthesis

Over a range of low intensities of light, Blackmann observed an increase in the rate of photosynthesis as the light intensity was increased. However, once a critical light intensity was reached, the rate of photosynthesis remained constant. The results surprised Blackmann who reasoned that if photosynthesis is driven by light energy, then it should go faster as more intense light is provided. He concluded that factors other than the amount of energy from sunlight must limit the rate of photosynthesis at high light intensities.

## 2 Concentration of carbon dioxide

Knowing that carbon dioxide is an essential raw ingredient for photosynthesis, and that the process was probably catalysed at least in part by enzymes, Blackmann went on to investigate the effects of carbon dioxide concentration and temperature on the rate of photosynthesis. The results of these investigations are shown in Figs 5.16 and 5.17. Clearly, photosynthesis is controlled by a combination of factors and it is apparent that the

**Fig 5.16** Effect of $CO_2$ concentration on the rate of photosynthesis

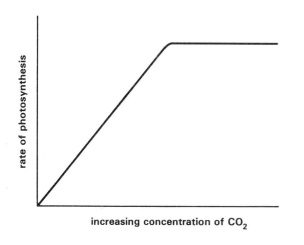

**Fig 5.17** Effect of temperature on the rate of photosynthesis

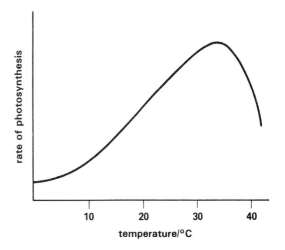

71

photosynthetic rate is limited by whichever factor is nearest its minimum value. It is an illustration of the **law of limiting factors** (Fig 5.18).

**Fig 5.18** Effect of temperature on the rate of photosynthesis at high and low intensities of light

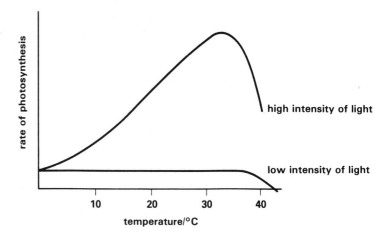

On a warm sunny day the concentration of carbon dioxide in the air is probably the factor which limits photosynthesis more than any other. Enriching air with carbon dioxide has a significant effect on the photosynthetic rate of crop plants grown in greenhouses (Table 5.3). It is now common practice for commercial growers of salad plants to raise the concentration of carbon dioxide of the air in greenhouses during daylight. The gas is either pumped in directly or is released by the burning of fuels such as paraffin, propane and natural gas which are used to heat greenhouses. Concentrations of up to 0.2% carbon dioxide produce increases in crop yield which are economically worthwhile.

**Table 5.3 Yield of lettuces and tomatoes grown in normal air and in air enriched with carbon dioxide**

| Crop | Without added $CO_2$ | With added $CO_2$ | Yield |
|---|---|---|---|
| lettuces | 0.9 | 1.1 | Fresh mass/kg per 10 heads |
| tomatoes | 4.4 | 6.4 | Fresh mass/kg per plant |

Of course, increasing the concentration of atmospheric carbon dioxide for crops grown in the open air is impracticable. On average, the amount of carbon dioxide in the atmosphere outdoors is 0.03% (300 ppm). It is controlled mainly by the rate at which living organisms give off carbon dioxide from respiration compared with the rate at which it is used by green plants for photosynthesis. An old-fashioned way of improving crop production in greenhouses was to add large quantities of manure to the soil. Carbon dioxide, released by soil microorganisms which decomposed the manure, accumulated in the air and stimulated photosynthesis, resulting in bigger crops. Plants which use the C3 pathway of photosynthesis continue fixing carbon dioxide until its concentration in the air falls to about 50 ppm. In contrast C4 species carry on photosynthesising until the concentration of carbon dioxide is as little as 1 ppm.

## 5.2 Respiration

**Respiration** is the oxidation of energy-rich substrates. Carbohydrates, sugars especially, are the respiratory substrates most used, although lipids and proteins can also be oxidised. Polysaccharides are first hydrolysed to the hexose sugar glucose, lipids to glycerol and fatty acids, proteins to amino acids. Most forms of life are **aerobes**. They respire using oxygen.

Many aerobic organisms are **facultative anaerobes**. They can respire for short periods of time in the absence of oxygen. A few species of bacteria are **strict anaerobes** and respire only in the absence of oxygen. Whatever the kind of respiration, it is important to understand what energy is provided and how this energy is used in processes which could not occur without respiratory energy. In this context it is useful to know more about adenosine 5′-triphosphate.

### 5.2.1 Adenosine 5′-triphosphate

The energy which binds the structure of molecules is called **free energy** (the energy potentially available to do work). Products of a biochemical reaction may contain less free energy than the substrate molecules. This is because energy is *released* in the reaction. In other instances, the products have more free energy. This happens when energy is *used* in the reaction which forms them. Many important reactions occur in our cells which require an input of energy. They include the synthesis of nucleic acids and proteins, active transport and muscle contraction. Energy for such processes is provided by **adenosine 5′-triphosphate (ATP)** which has the structure shown in Fig 5.19.

**Fig 5.19** (a) ATP simplified

**phosphate groups**

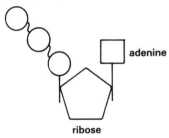

ribose

**Fig 5.19** (b) Structure of a molecule of ATP

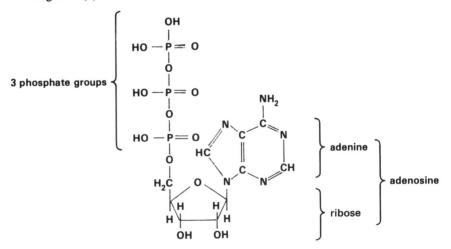

It has been calculated that when the bond holding the terminal phosphate group of ATP is broken by hydrolysis, $30.66\,\mathrm{kJ\,mol^{-1}}$ of energy is released. The same amount of energy is released when the terminal phosphate bond of adenosine 5′-diphosphate (ADP) is hydrolysed. However, when the phosphate bond of adenosine 5′-monophosphate (AMP) is hydrolysed, only $14.28\,\mathrm{kJ\,mol^{-1}}$ of energy is released. The amounts of energy released in such reactions in living cells is probably higher than the figures quoted here.

Similar amounts of energy are required to make the phosphate bonds. For example, when ADP and inorganic phosphate are converted to ATP, $30.66\,\mathrm{kJ\,mol^{-1}}$ of energy is needed. The two bonds between the first, second and third phosphate groups of ATP are called **high-energy bonds**, and adenosine 5′-triphosphate is sometimes written as:

$$A—P \sim P \sim P \quad \text{where}$$

A = adenosine
P = phosphate group
$\sim$ = high-energy bond
— = low-energy bond

The bonds in other substances found in living cells yield even more energy than the high-energy bonds of ATP. Phosphocreatine in striped muscle is an example of such a compound (Chapter 8). However, ATP is the only substance present in sufficient quantities in most living cells to be of general use as a provider of energy. ATP is therefore indispensable in the transfer of energy from energy-rich substrates to reactions in which energy is used.

But how is ATP made in living cells, and where does the energy for its synthesis come from?

### 5.2.2 Oxidation of respiratory substrates

In anaerobic conditions, energy-rich substrates such as glucose are oxidised to lactic acid or to ethanol and carbon dioxide. Lactic acid is a product of anaerobic respiration in some bacteria and in animal cells. Ethanol and carbon dioxide are made by yeasts and higher plant cells when oxygen is absent. The overall processes can be summarised as follows:

Lactic fermentation:

$$\underset{\text{glucose}}{C_6H_{12}O_6} \rightarrow \underset{\text{lactic acid}}{2CH_3CHOHCOOH}$$

Alcoholic fermentation:

$$\underset{\text{glucose}}{C_6H_{12}O_6} \rightarrow \underset{\text{ethanol}}{2CH_3CH_2OH} + \underset{\substack{\text{carbon} \\ \text{dioxide}}}{2CO_2}$$

The equations give us no idea of the amount of energy released, nor do they tell us anything about the way in which the substrate is oxidised. What is more, they disguise the fact that the two processes are very similar, each sharing a number of common steps. **Glycolysis** (glyco = sugar; lysis = to split) is the name given to the common steps. Glycolysis also takes place when sugars are respired aerobically.

### 1 Glycolysis

Fig 5.20(a) shows the more important common steps in the oxidation of glucose, whether in anaerobic or aerobic conditions. Far from producing energy, some of the earlier reactions need an input of energy. The energy comes from the hydrolysis of high-energy bonds of ATP. Without a supply of energy from ATP, the reactions could not proceed. The splitting of each fructose 1,6-bisphosphate molecule yields two molecules of the triosephosphate called 3-phosphoglyceraldehyde (PGAL), which is then oxidised to pyruvic acid. Oxidation of PGAL involves the removal of hydrogen and is catalysed by dehydrogenase enzymes (Chapter 3). At the same time the coenzyme nicotinamide-adenine dinucleotide (NAD⁺) becomes reduced to NADH. Oxidation of each PGAL molecule produces enough energy for two ATP molecules to be synthesised from ADP and inorganic phosphate.

The fate of the pyruvic acid depends on the organism in which it was produced and on whether oxygen is available or not. If lactic acid is a product of anaerobic respiration, the NADH formed in glycolysis transfers its hydrogen to pyruvic acid, and lactic acid is formed. This is what happens in mammalian muscle cells when they have an oxygen debt (Chapter 12).

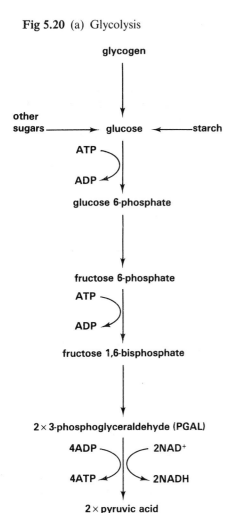

**Fig 5.20** (a) Glycolysis

glycogen

other sugars → glucose ← starch

ATP → ADP

glucose 6-phosphate

fructose 6-phosphate

ATP → ADP

fructose 1,6-bisphosphate

2 × 3-phosphoglyceraldehyde (PGAL)

4ADP → 4ATP    2NAD⁺ → 2NADH

2 × pyruvic acid

Bacteria which sour milk also make lactic acid when oxygen is not available.

If yeasts and higher plant cells are kept without oxygen, pyruvic acid is split into carbon dioxide and ethanal. The ethanal is then reduced to ethanol by hydrogen from NADH (Fig 5.20(b)). Either way, $NAD^+$ is remade and once again acts as a hydrogen acceptor when more PGAL is oxidised. The enzymes which control glycolysis are found in the cytoplasm.

**Fig 5.20** (b) Fate of pyruvic acid in the absence of oxygen

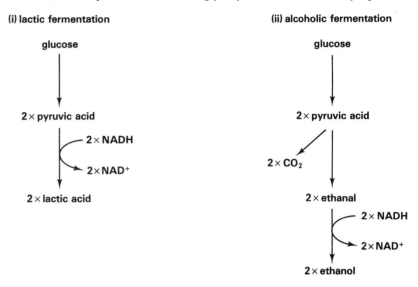

In aerobic conditions the pyruvic acid is taken instead into the tri-carboxylic acid cycle which occurs in mitochondria.

## 2  The tricarboxylic acid cycle

At the beginning of this century, Thunberg devised a special tube; with it he discovered that animal tissues contain dehydrogenase enzymes which catalyse the transfer of hydrogen from carboxylic acids (Fig 5.21). In 1935 Szent-Gyorgyi noticed that dehydrogenation of succinic acid in muscle is blocked by the competitive inhibitor malonic acid. Furthermore, the inhibition stopped respiration in the muscle. Two years later Hans Krebs showed that respiration in pigeon breast muscle was stimulated by a specific variety of carboxylic acids. Krebs went on to carry out a series of brilliant experiments from which he deduced that, in aerobic conditions, pyruvic acid combines with oxaloacetic acid to form citric acid. The citric acid is then dehydrogenated via a number of intermediate compounds back to oxaloacetic acid. One of the intermediates is succinic acid. Carbon

**Fig 5.21** A Thunberg tube. The top of the tube is rotated so that the hole coincides with the outlet and the air in the tube is sucked out using a vacuum pump. The top is then turned to seal the contents from the external air. Finally the tube is inverted so that the blue dye mixes with the respiring cells. The dye is gradually reduced to colourless leuco-methylene blue. If air is left in the tube gaseous oxygen is reduced instead and the dye remains in its oxidized blue state

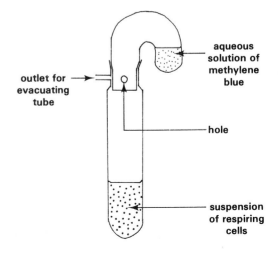

dioxide is released in these reactions. It was some time later before it was found that pyruvic acid is converted to acetyl coenzyme A before it enters the cycle (Fig 5.22). Today the citric acid cycle is called **Krebs' tricarboxylic acid (TCA) cycle**.

**Fig 5.22** Krebs' TCA cycle and some important biochemical pathways to which it is linked

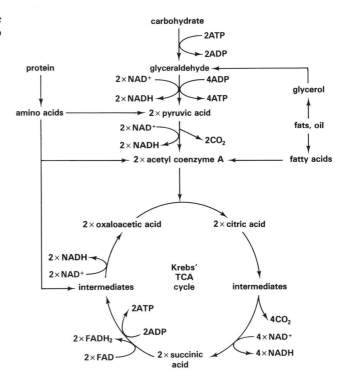

In the late 1940s mitochondria, separated from other cell components by differential centrifugation (Chapter 6), were found to break down pyruvic acid to carbon dioxide and water. The dehydrogenase enzymes which catalyse the oxidations in the TCA cycle are thus in the mitochondria. Biochemists are now fairly sure that the enzymes are in the matrix enclosed by the inner membrane of mitochondria.

In the TCA cycle oxidation of carboxylic acid molecules is catalysed by dehydrogenase enzymes. Dehydrogenation of the acids is accompanied by reduction of $NAD^+$ to NADH. The fate of NADH in aerobic conditions is, however, different from NADH formed in anaerobic conditions. When oxygen is available the hydrogen from NADH is passed via a succession of hydrogen acceptors until it finally reduces oxygen to water. The succession is called a **respiratory** or **electron transport chain** (Fig 5.23). After NADH the hydrogen is passed to flavin-adenine dinucleotide (FAD). The hydrogen atoms now split into electrons and protons. Reduced FAD ($FADH_2$) transfers the electrons to the next acceptors called cytochromes. Reduced cytochromes then pass the electrons to cytochrome oxidase. Finally reduced cytochrome oxidase transfers the electrons back to the protons. The hydrogen atoms so produced, reduce oxygen to form water. Each of the reactions in the chain is an oxidation–reduction reaction. As

**Fig 5.23** A respiratory chain

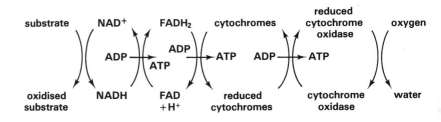

the hydrogen atoms and later the electrons move along the chain, the reduced acceptors are oxidised and more of them can be taken into the chain.

What is particularly important is that three of the oxidation–reduction reactions in an electron transport chain yield enough energy to make ATP from inorganic phosphate ions and ADP. This type of reaction is called **oxidative phosphorylation**. It takes place in the oxysomes projecting from the cristae of mitochondria (Chapter 6). For each molecule of NADH taken into a respiratory chain, three molecules of ATP are formed. In one of the oxidations in the TCA cycle, FAD is reduced. Only two molecules of ATP are produced for each molecule of $FADH_2$ entering a respiratory chain.

Note that the aerobic oxidation of a molecule of glucose causes the reduction of ten molecules of $NAD^+$ and two of FAD. The twelve molecules of reduced coenzymes reduce twelve atoms (six molecules) of oxygen to produce twelve molecules of water. The summary equation for aerobic respiration of a mole of glucose is thus:

$$C_6H_{12}O_6 + 6O_2 + 6H_2O \rightarrow 6CO_2 + 12H_2O$$

### 5.2.3 Energy yields from respiratory substrates

How many ATP moles are formed from each mole of glucose oxidised aerobically? How does this figure compare with the number obtained when glucose is oxidised anaerobically? The breakdown of one mole of glucose in glycolysis uses energy from the terminal high-energy bonds of two moles of ATP. On the other hand, enough energy is released in glycolysis to produce four moles of ATP from ADP and inorganic phosphate ($P_i$):

The net release of energy from one mole of glucose broken down to pyruvic acid, therefore, is the energy released when the terminal high-energy bonds of two moles of ATP are later hydrolysed. Thus living cells respiring anaerobically release $2 \times 30.66$ kJ of energy from every mole of glucose oxidised.

In aerobic conditions, ATP is also made mainly in the respiratory chains into which NADH and $FADH_2$ from the TCA cycle are fed. Altogether, 38 ATP moles are made from each mole of glucose oxidised aerobically, 8 from glycolysis and 30 from the TCA cycle (Table 5.4). Living cells respiring aerobically thus release $38 \times 30.66$ kJ of energy from a mole of glucose. Aerobic respiration is clearly much more efficient than anaerobic respiration in releasing energy from respiratory substrates.

**Table 5.4 Origin of ATP formed in aerobic respiration of glucose**

| Source | No. of ATP molecules |
| --- | --- |
| direct synthesis in glycolysis | 2 (net gain) |
| 2NADH formed in glycolysis | 6 (2 × 3) |
| 8NADH formed in Krebs' TCA cycle | 24 (8 × 3) |
| 2FADH$_2$ formed in Krebs' TCA cycle | 4 (2 × 2) |
| direct synthesis in TCA cycle | 2 |
| **Total** | **38** |

For a triglyceride consisting of glycerol and palmitic acid, the net gain of ATP is shown in Table 5.5. However, a molecule of tripalmitin ($C_{51}H_{95}O_6$) is nearly five times heavier than a glucose molecule ($C_6H_{12}O_6$). On a weight-for-weight basis, lipids yield a little more than twice as much energy as carbohydrates (Chapter 9). The energy yield from protein is about the

**Table 5.5 Net gain of ATP for respiration of a molecule of tripalmitin**

| | |
|---|---|
| glycerol$\rightarrow CO_2 + H_2O$ | 20 ATP |
| $3 \times$ palmitic acid$\rightarrow CO_2 + H_2O$ | 390 ATP |
| **Total** | **410 ATP** |

same as for a similar weight of carbohydrate. When burned to carbon dioxide and water in a **calorimeter** (Fig 5.24), a mole of glucose releases 2880 kJ of heat energy. This is the amount of energy potentially available to do work in a living cell. We can now calculate the efficiency of energy change when glucose is respired aerobically:

$$\text{Efficiency of energy change} = \frac{38 \times 30.66}{2880} \times 100 = 44\,\%.$$

This means that less than half of the potential energy in glucose is used to do work.

You may wonder what has happened to the rest of the energy in the respiratory substrate. The Second Law of Thermodynamics states that whenever one form of energy is changed into another, some of the energy is converted into heat. In aerobic respiration, over half of the energy in the substrate is released as heat. Heat energy helps to maintain the constant body temperature of warm-blooded animals (Chapter 17). However, remember that heat energy cannot be used to do work in living organisms.

**Fig 5.24** A simple calorimeter. When the supply of electricity is switched on the heating element becomes red hot and the sample ignites in the stream of oxygen. The heat given off by the burned sample causes the temperature of the water to rise. The increase in temperature, volume of water and mass of sample are used to work out its calorific value.

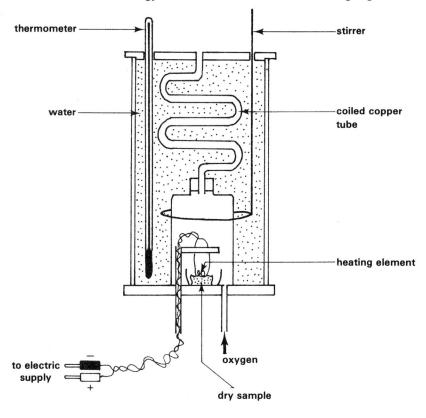

Many of the biochemical reactions in glycolysis and Krebs' cycle are reversible. They constitute a metabolic 'hub' which meets the cell's immediate energy needs. Excess input of energy-rich substrates such as glucose leads to their storage as fat. At times of shortage, reserves of energy can be drawn on by converting fat to sugar.

### 5.2.4 Measuring the rate of respiration

In the 1930s Otto Warburg devised a constant-volume manometer to measure the rate of oxygen uptake by slices of living tissue. The apparatus is often called a **Warburg manometer** (Fig 5.25). To begin with, the index fluid in both arms is brought to the same height at the reference point (P) by turning the adjusting screws.

**Fig 5.25** A Warburg manometer

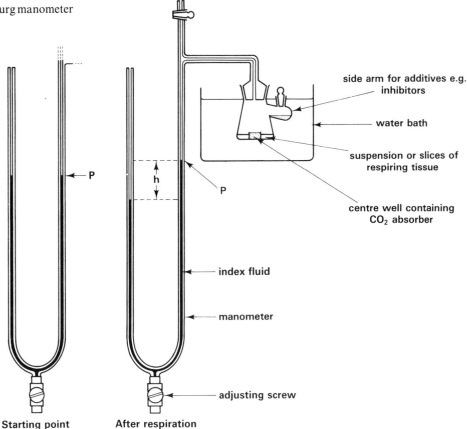

side arm for additives e.g. inhibitors

water bath

suspension or slices of respiring tissue

centre well containing $CO_2$ absorber

P

h

P

index fluid

manometer

adjusting screw

Starting point

After respiration

As the tissue respires the carbon dioxide it evolves is removed by the potassium hydroxide in the centre well. The volume of air enclosed in the manometer is thus reduced. After a given time, the adjusting screw is used to return the index fluid in the right arm to P. The reduction in pressure of the air in the manometer causes the height of the fluid in the left arm to fall. The change in height ($h$) is multiplied by the flask constant ($K$) to determine the volume of oxygen consumed. Oxygen consumption is usually expressed as $mm^3$ oxygen absorbed $h^{-1} g^{-1}$ of tissue. Note that the volume of air in the manometer is always the same on each occasion a reading is taken. Hence the name constant-volume manometer. A control manometer, called a **thermobarometer** lacking the tissue, is set up at the same time. It is necessary because changes in air pressure during the experiment would alter the volume of air in the manometer. Such changes have to be accounted for in the calculations.

**Simple respirometers** (Fig 5.26) are often used in school and college laboratories to measure respiratory rates. A known mass of germinating seeds or small invertebrate animals such as woodlice, is placed in the **respiration chamber**. The organisms are given time to adjust to the chosen temperature and the valves are then closed. Once again the potassium hydroxide absorbs evolved carbon dioxide, so the volume of air in the respiration chamber decreases as oxygen is consumed. Consequently the manometer fluid moves towards the respiration chamber. Knowing the distance moved by the meniscus of the fluid and the bore of the manometer tube, the rate of oxygen uptake can be calculated. Alternatively the manometer fluid can be returned to its original height by pushing the plunger in the syringe. The volume of air required to do this can then be measured directly on the syringe. The **compensation tube** functions as a thermobarometer. It contains the same volume of air as the respiration chamber. Thus changes in air pressure or temperature during the experiment have the same effect on the volume of air in both tubes. In this way the changes cancel out each other.

**Fig 5.26** A simple respirometer

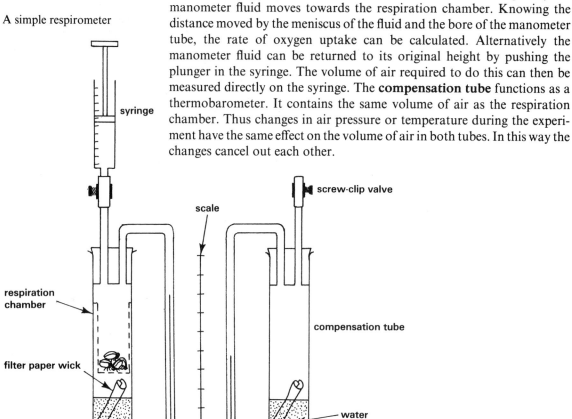

### 5.2.5 Respiratory quotient

The volume of carbon dioxide given off divided by the volume of oxygen taken up in a fixed period is called the **respiratory quotient, RQ**:

$$RQ = \frac{\text{volume } CO_2 \text{ given off}}{\text{volume } O_2 \text{ taken up}}$$

RQ values indicate the type of respiration, aerobic, anaerobic or both. They also tell us which substrates or combination of substrates are oxidised.

In aerobic conditions an RQ of 1.0 indicates that the respiratory substrate is a carbohydrate:

$$\underset{\text{carbohydrate}}{C_6H_{12}O_6} + 6O_2 \rightarrow 6CO_2 + 6H_2O$$

For each volume of oxygen taken up, a similar volume of $CO_2$ is given off:

$$RQ = \frac{6}{6} = 1.0$$

When a lipid is oxidised aerobically, the RQ is 0.7:

$$C_{57}H_{110}O_6 + 81.5O_2 \rightarrow 57CO_2 + 55H_2O$$
tristearoylglycerol

$$RQ = \frac{57}{81.5} = 0.7$$

For the aerobic oxidation of protein, an RQ of 0.99 is obtained. The aerobic breakdown of a mixture of carbohydrates, lipids and proteins gives an RQ of 0.8–0.9. Animals fed a balanced diet use mainly carbohydrates and lipids as respiratory substrates and have an RQ of about 0.85. The RQ of starving animals is between 0.9 and 1.0. Can you explain why this is so?

A mixture of aerobic and anaerobic respiration takes place when there is a shortage of oxygen. Here the RQ is usually greater than 1.0. The exact value depends on the respiratory substrates used and on the relative rates of the two types of respiration. For example, when anaerobic and aerobic respiration of glucose occur at similar rates, an RQ of 1.33 is obtained:

What might the RQ be if aerobic respiration takes place more rapidly, or more slowly than anaerobic respiration of glucose?

Anaerobic respriation:   $C_6H_{12}O_6 \rightarrow 2CO_2 + 2CH_3CH_2OH$

Aerobic respiration:   $C_6H_{12}O_6 + 6O_2 \rightarrow 6CO_2 + 6H_2O$

*Total*   $6O_2 \rightarrow 8CO_2$

$$RQ = \frac{8}{6} = 1.33$$

Simple respirometers can be used to determine the RQ of living organisms. One is set up in the way described earlier to measure oxygen consumption, say $a$ mm$^3$ h$^{-1}$. The other lacks the KOH solution and thus measures the difference in volumes of oxygen consumed and carbon dioxide evolved, say $b$ mm$^3$ h$^{-1}$. The differences in readings make it possible to calculate the RQ as follows:

$$RQ = \frac{a-b}{a}$$

Use the formula to calculate RQ values from the following readings:

(i)   $a = 10$ mm$^3$ h$^{-1}$; $b = 0$ mm$^3$ h$^{-1}$
(ii)  $a = 10$ mm$^3$ h$^{-1}$; $b = +3$ mm$^3$ h$^{-1}$
(iii) $a = 10$ mm$^3$ h$^{-1}$; $b = -3$ mm$^3$ h$^{-1}$

### 5.2.6 Basal metabolic rate

Recent studies have shown that about two thirds of the energy content of the diet consumed by a resting person is used to maintain the **basal metabolic rate, BMR.** This is the basic rate at which energy must be released to perform vital functions such as beating of the heart, breathing, peristalsis, and biosynthesis of proteins and other important molecules. In infancy and childhood the BMR is relatively high as much of the energy is required for biosynthesis of cellular components necessary for growth. When we mature the BMR levels off until middle age. In old age there is a gradual fall in BMR as metabolism beings to slow down (Fig 5.27).

**Fig 5.27** The effect of age on BMR

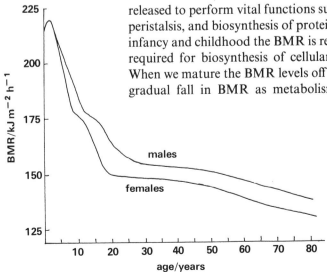

81

Throughout life, men usually have a higher BMR than women. This is because men generally have less fat per unit body mass and surface area. Calculated on the basis of unit lean body mass, the BMR is similar in both sexes.

The expenditure of energy in an active person includes the amount used in body movement. Hence the total energy requirement of any individual depends on a variety of factors including age, sex, body mass and the extent of physical activity engaged in (Tables 5.6 and 5.7). The energy requirements of women increase in pregnancy to support foetal growth and, after birth, to meet the demands of lactation. This is why the World Health Organisation recommends an additional energy intake of 1500 kJ per day for pregnant women, especially during the second half of the gestation period.

**Table 5.6 Daily energy expenditure for average men and women (from Taylor 1978)**

| Occupation | Sleep 8 h (basal)/kJ | Work 8 h/kJ | Leisure 8 h/kJ | Total 24 h/kJ |
|---|---|---|---|---|
| *Men* | | | | |
| sedentary | 2000 | 3500 | 5500 | 11 000 |
| moderately active | 2000 | 5000 | 5500 | 12 500 |
| very active | 2000 | 7500 | 5500 | 15 000 |
| *Women* | | | | |
| home or office | 1750 | 3500 | 3750 | 9 000 |

**Table 5.7 Recommended daily energy intakes for people of different ages (from Taylor 1978)**

| Age | Energy intake/ kJ kg$^{-1}$ body mass |
|---|---|
| 0–3 months | 500 |
| 6–9 months | 460 |
| 1–2 years | 430 |
| 3–4 years | 410 |
| 4–5 years | 400 |
| adult man | 180 |

If the intake of energy-producing foods is more than our immediate requirements, the energy is stored, mainly as fat. Consequently body mass increases. Some people can respire the excess intake in brown fat (Chapter 17), so do not put on weight. The present UK Government recommendations are that the energy intake for women should be between 7030 and 10 510 and for men between 10 080 and 14 070 kJ per day depending on how active they are. Research carried out on women in recent years at the Dunn Nutrition Unit, Cambridge has shown that physical activities take up less energy than was previously thought. The study indicates that the Government figures are between 1260 and 1680 kJ a day too high. It is for this reason that slimmers' diets based on Government recommendations may not result in a decrease in body mass. Finding an appropriate intake of energy is a useful contribution to maintaining health. Many cardiovascular and respiratory ailments of modern societies are thought to be caused by excess body mass.

# 6 Cell structure and division

What is known about cell structure has largely depended on the development of microsopes and microscopical techniques. A **simple microscope** was invented by Galileo in 1610, but there are no reports that he used it to examine living organisms. In 1676 a Dutch draper, van Leeuwenhoek, whose hobby was the grinding of lenses, used one of his simple microscopes (Fig 6.1) to examine rainwater in which grains of pepper had been soaked. He observed a variety of unicellular organisms which he called 'animalcules'. It is now known that among the organisms he saw were bacteria. A decade earlier Robert Hooke in England had made a **compound microscope** with which he observed, among other things, thin slices of cork tissue. He saw that the cork was porous, rather like a honeycomb, consisting of a great many small compartments which he called **cells**.

**Fig 6.1** Van Leeuwenhoek's simple microscope

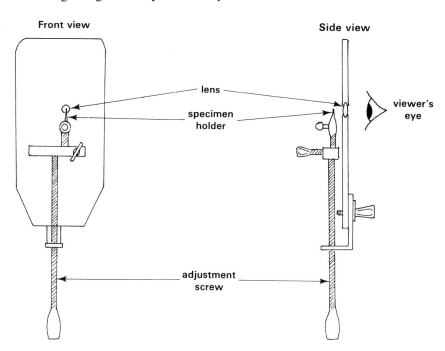

During the next 150 years, with further improvement of the compound microscope, a great deal was learned of the structure of cells and tissues from many plants and animals. In the 1840s the **cell theory** of Schleiden and Schwann became generally accepted by biologists. It stated that cells are the basic structural units of all living organisms. For some time after the theory was first proposed there was a tendency for biologists to emphasise the importance of individual cells; their structure and activity were thought to mirror that of the whole organism. Recently there has been more interest in the ways in which different types of cell interact in the functioning and development of multicellular organisms.

## 6.1 Cell structure as seen with the light microscope

The compound microscope is used by most biologists to examine cells and tissues (Fig 6.2). With this instrument it is possible to observe living material. Good images showing much detail can be obtained especially if the microscope is of the **phase-contrast** type (Fig 6.3). A phase-contrast microscope exaggerates small differences in the refractive index of cell components to create an image in which the components are clearly distinguished by the human eye.

**Fig 6.2** A modern compound microscope (courtesy Vickers Instruments)

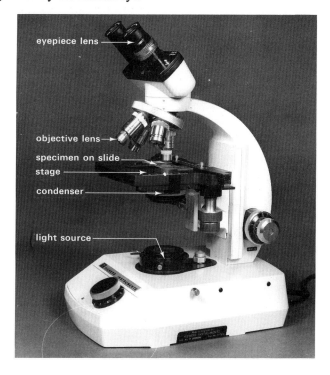

**Fig 6.3** An epithelial cell from the mouth lining as seen using (a) a conventional light microscope (b) a phase contrast microscope, ×900

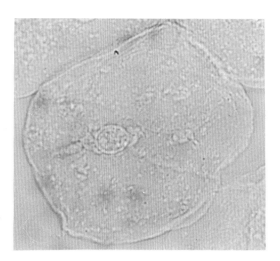
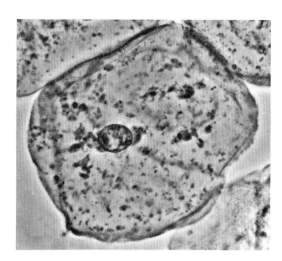

Nevertheless, most of what was known about the structure of cells up to the 1940s was obtained from observations on dead tissue which had been treated with preservatives. In the technique, which had been developed for over a hundred years, the tissue is first immersed in a fixative such as methanol or ethanol in order to prevent deterioration and to keep the structure as life-like as possible. Following **fixation**, the tissue is **dehydrated** with ethanol and then **cleared** with an organic solvent such as dimethylbenzene which is miscible with paraffin wax. The next stage is to

**embed** the tissue in molten wax which, on hardening, supports the tissue while thin sections, 2–10 μm thick, are cut using a **microtome**. After attaching the sections to microscope slides, the wax is removed before the tissue is **stained**.

Staining enables cell components to be differentiated when the material is examined microscopically. Any of the above steps can lead to distortion of the specimen, so it is necessary to guard against artificial structures, called **artefacts**, which appear in the specimen during treatment but are not present in living cells. Using such methods, coupled with observations on living cells, it is possible to build up a fairly detailed picture of cell structure (Fig 6.4).

**Fig 6.4** Structure of animal and plant cells revealed by a compound microscope

(a) (i) Section of a liver cell, × 1200

(a) (ii) Cell from gastric gland

(b) (i) Section of a leaf mesophyll cell, × 1200

(b) (ii) Structure of leaf mesophyll cell

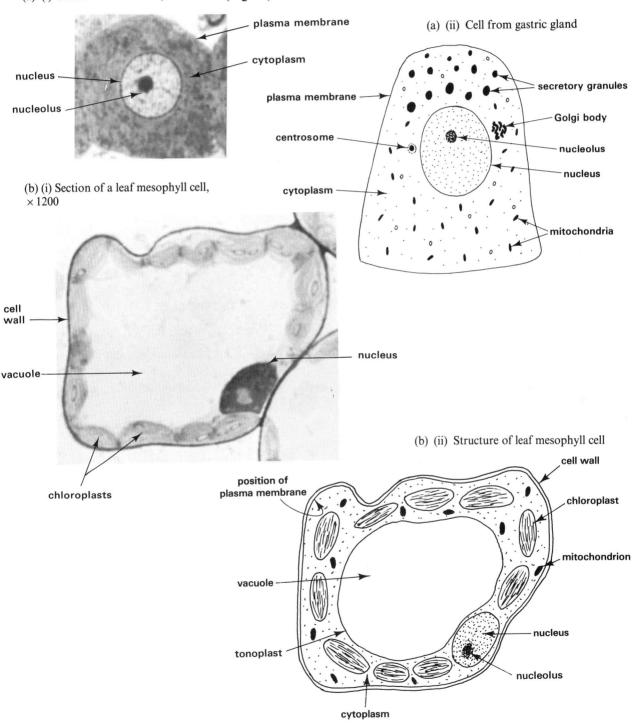

The living material of cells is called **protoplasm** and is enclosed in a **plasma membrane**. In plant cells a **cell wall**, mainly of cellulose, surrounds the plasma membrane. Adjacent plant cells are held together by a thin layer composed mainly of calcium pectate and known as the **middle lamella**. Other distinctive features of plant cells are pigment-containing bodies called plastids, the most common of which are the green **chloroplasts**, and also large sap-filled **vacuoles**.

Both plant and animal cells contain a **nucleus** at some stage in their development. The nucleus is surrounded by a **nuclear membrane** and contains granular chromatin in which one or more dense areas known as **nucleoli** are suspended. During nuclear division the nucleoli disappear and the chromatin appears as thread-like **chromosomes** (section 6.5).

The protoplasm outside the nucleus is called **cytoplasm**. At the end of the nineteenth century, Camillo Golgi stained brain cells with silver salts, and observed a cytoplasmic structure which looked like a tiny net. It was subsequently called the **Golgi body**.

**Mitochondria** are also structures of the cytoplasm common to most cells, and are just visible as tiny granules with a compound microscope. **Storage materials** are often seen in the cytoplasm. In plant cells **starch** is the main storage substance, while in animal cells granules of **glycogen** are commonly found.

Although all living cells have many common features, there is no such thing as a typical or generalised cell. Attention has already been drawn to the main differences between plant and animal cells. It is important to realise that multicellular plants and animals consist of a variety of cell types. The precise structure of a cell is suited to the functions it performs (Chapter 7).

## 6.2 Cell ultrastructure as revealed by the electron microscope

Since the 1950s biologists using the electron microscope (Fig 6.5) have made tremendous strides in our knowledge of the detailed structure of cells.

**Fig 6.5** A modern electron microscope (courtesy Kratos Ltd)

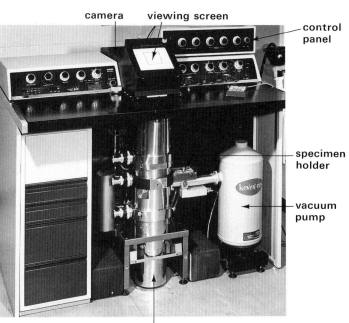

camera  viewing screen

control panel

specimen holder

vacuum pump

housing for condenser objective and projector lenses

The preparation of specimens for sectioning prior to examination with an electron microscope is in some ways similar to the method used for light microscopy, but there are important differences.

To create an image, the electron microscope directs a beam of electrons at and through the material. The image results from the way the material scatters the electrons. Atmospheric atoms and molecules would interfere by also scattering the electron beam, so the material has to be held in a vacuum. Therefore it has always to be prepared in a way which resembles the technique used by light microscopists for preserved specimens. For fixation, osmium(VIII) oxide is often used. It binds to lipids and proteins, making them electron-dense. Such cell components scatter electrons strongly and appear as dark areas in the image. Otherwise, glutaraldehyde is used which fixes the material without rendering it electron-dense.

The sample is next **dehydrated** with ethanol as described for light microscopy, then **cleared** ready for **embedding**. Paraffin wax breaks if cut into very thin sections, so it is unsuitable as an embedding substance for electron microscopy. Instead, clear plastic epoxy resins such as Araldite are used. The material is cleared in liquid resin which is hardened by gentle heat.

The sample, now embedded in a tough, clear supporting substance, can be cut into very thin slices. Sections of cells and tissues must be extremely thin (0.01–0.5 $\mu$m thick), to allow some electrons to pass through them. Sectioning is carried out with an **ultra-microtome**, with the blade usually made by breaking a thick piece of glass to give a hard cutting edge.

The sections are then transferred to tiny circular copper grids on which they may be **stained**. Solutions of lead salts are taken up by lipid components in the specimen, whilst uranyl salts react with proteins and nucleic acids. The effect is to make the lipids, proteins and nucleic acids electron-dense so that they contrast with other cell components in the final image. Such staining is not necessary if osmium oxide is used as a fixative. A special holder is used to place the sections, still supported on their grids, in the electron microscope which is evacuated of air before the electron beam is switched on (Fig 6.6).

When electron microscope techniques were first used, both plant and animal cells were shown to have a detailed structure previously unsuspected. Not only were some components discovered for the first time;

**Fig 6.6** Stages in the preparation of thin sections for electronmicroscopy: *left* cutting sections with an ultramicrotome, *right* mounting sections on to a copper grid

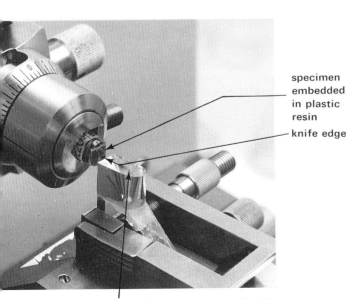

specimen embedded in plastic resin

knife edge

water bath for sections to float in

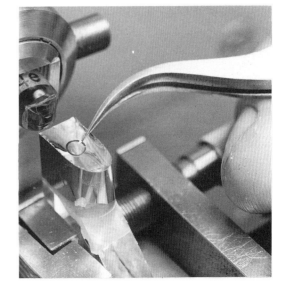

**Fig 6.7** (a) Electronmicrograph of a section through a pancreatic cell, × 10 000

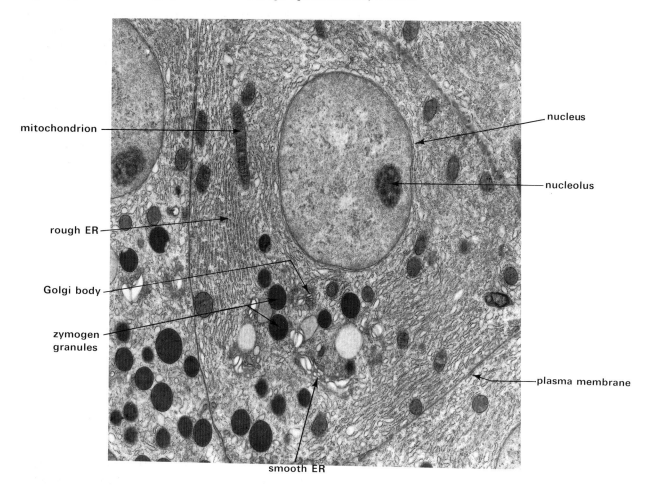

mitochondrion

nucleus

nucleolus

rough ER

Golgi body

zymogen granules

plasma membrane

smooth ER

**Fig 6.7** (b) Electronmicrograph of a section through parts of two adjacent cells from a tobacco leaf, × 15 000

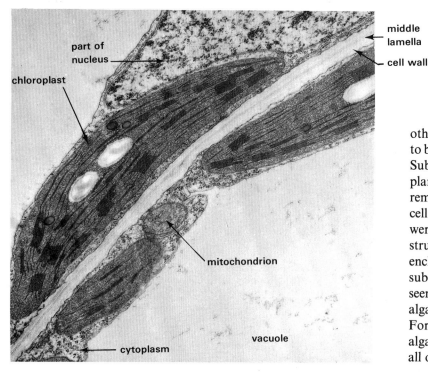

part of nucleus

middle lamella

cell wall

chloroplast

mitochondrion

cytoplasm

vacuole

others already well known were found to be very complex in structure (Fig 6.7). Subcellular components common to plant and animal cells were seen to be remarkably similar in appearance. The cells of bacteria and blue-green algae were found to be relatively simple in structure. They lack a membrane-enclosed nucleus and have few of the subcellular structures called organelles seen in fungi, protozoa, most kinds of algae, multicellular plants and animals. For this reason bacteria and blue-green algae are termed **procaryotic** whereas all other organisms are **eucaryotic**.

### 6.2.1 Cell membranes

The outer boundary of the protoplast, the **plasma membrane**, is invisible with the light microscope. Even so, its presence can be inferred because protoplasm leaks out of animal cells when the cell surface is punctured. Overton in 1895 suggested that the membrane was made of fatty substances. Other workers later deduced that two layers of lipid were present in the plasma membrane. In 1935 Danielli and Davson proposed a model for membrane structure in which a **lipid bilayer** was coated on either side with **protein** (Fig 6.8). Mutual attraction between the hydrocarbon chains of the lipids, and electrostatic forces between the protein and the 'heads' of the lipid molecules, were thought to maintain the stability of the membrane.

**Fig 6.8** The Danielli–Davson model of the cell membrane

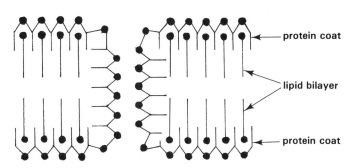

Using evidence from electronmicrographs Robertson, in 1960, proposed a **unit membrane hypothesis** based on the observation that all membranes of cells have a comparable appearance when viewed with the electron microscope. The two outer layers of protein are each about 2 nm thick and appear densely granular. They enclose a clear central area about 3.5 nm wide consisting of lipid (Fig 6.9). The term **unit membrane** is now used to denote all cell membranes which are similar to this in structure.

**Fig 6.9** Electronmicrograph of a section through microvilli from intestinal epithelium, × 180 000. A unit membrane surrounds each microvillus

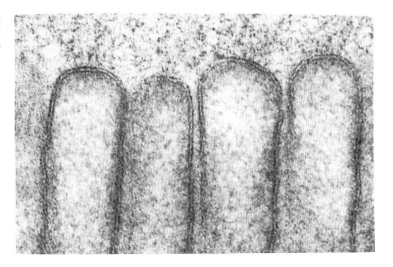

Much has since been learned about the composition and probable organisation of cell membranes. The lipids are mainly **phospholipid** molecules which are polar at the phosphate group end (Chapter 2). There is considerable variation in the fatty acid content of membrane lipids from the cells of different species. The saturated fatty acids of some lipids attract molecules of **cholesterol**. The amount of protein relative to the quantity of lipid also varies from one cell type to another.

Proteins in the outer parts of the membrane may even differ from those of the inner part. Many are **carrier proteins** which transport substances across the membrane. The current view of membrane structure which is generally held is the **fluid-mosaic model** proposed in 1972 by Singer and Nicholson. It suggests that the protein component of the membrane is patchy rather like a mosaic. Constant movement of the phospholipid molecules gives the membrane fluidity. The plasma membrane is therefore a dynamic rather than a static structure. Some proteins on the exterior of cell membranes are **antigenic**. Lymphocytes have antibodies attached to the cell membrane. Antigens and **antibodies** provide a means whereby our cells can distinguish self from non-self (Chapter 11). The way in which the components of cell membranes are now thought to be arranged is shown in Fig 6.10.

**Fig 6.10** The fluid-mosaic model of the cell membrane

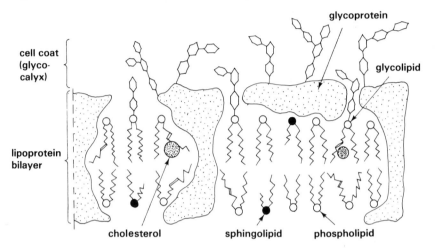

The exterior of the cell membrane of animal cells is called the **cell coat**. It is made of **mucopolysaccharides, glycolipids, glycoproteins** and **hyaluronic acid**. The cell coat is sticky, enabling animal cells to adhere to each other.

The plasma membrane is much more than just a protoplasmic boundary. It provides a means of controlling the passage of materials both into and out of the cell. Some materials are taken in by **phagocytosis** and **pinocytosis**. Phagocytosis is a mechanism that enables large suspended particles to be taken wholesale into cells. Pinocytosis is the intake of droplets of liquid by the formation of tiny pockets in the membrane. The plasma membrane is a **selectively permeable membrane** controlling the passage of water and dissolved substances into or out of the cell. Water passes through the membrane by **osmosis** (Chapter 1). Water-soluble substances cross the membrane by **diffusion**, by **facilitated transport** or by **active transport**. It is now generally agreed that many water-soluble solutes are transported through the membrane by carrier proteins. Lipid-soluble compounds can pass more quickly through membranes by dissolving in the phospholipid layer.

Diffusion is the random movement of ions, atoms or molecules. It results in such particles moving from places where they are highly concentrated to where their concentration is less. For example, oxygen diffuses through the plasma membrane of the epithelial cells lining our alveoli and into blood circulating in nearby capillaries. Carbon dioxide diffuses in the opposite direction (Chapter 12). Carrier proteins may bind with and facilitate the diffusion of some kinds of particles across the plasma membrane (Fig 6.11). The absorption of most nutrients from the gut occurs by facilitated transport (Chapter 9). Active transport requires the use of energy from

ATP to move particles across the plasma membrane against a concentration gradient. Reabsorption of sodium ions from the renal filtrate, and the sodium pump in nerve impulse transmission, are examples of active transport (Chapters 13 and 14 respectively).

**Fig 6.11** (a) Relative permeability of plasma membrane.

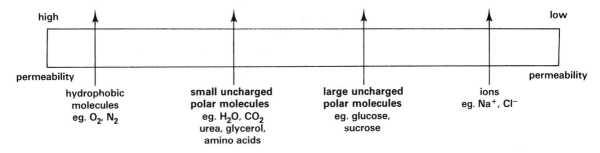

(b) Transmembrane transport mechanisms

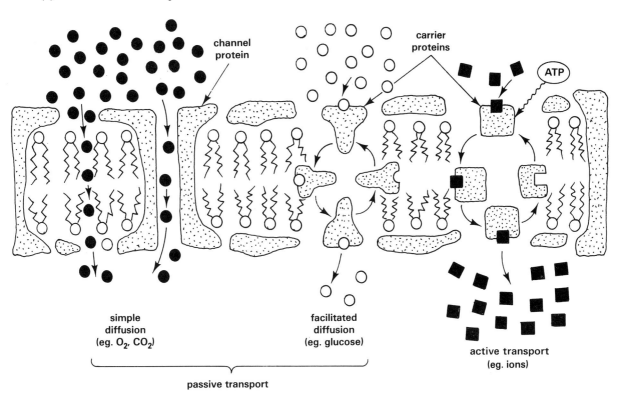

Passive mechanisms move substances along a concentration gradient using the kinetic energy of the transported atoms, molecules or ions.
Active transport uses energy from ATP to move substances against a concentration gradient.

### 6.2.2 Endoplasmic reticulum

Biologists once regarded the cytoplasm as a homogeneous jelly. This view was radically changed when sections of cells were examined with electron microscopes. Extending throughout the cytoplasm is a three-dimensional network of sac-like and tubular cavities called **cisternae** bounded by a unit membrane. These structures are collectively called the **endoplasmic reticulum (ER)** (Fig 6.12). In places the membranes are covered on the cytoplasmic side with ribosomes, the **rough ER**. Ribosomes are small bodies about 15 nm in diameter. They consist of protein and ribosomal RNA. Elsewhere the ribosomes are lacking, the **smooth ER**. The total area of the ER membranes in a cell of volume $5000 \, \mu m^3$ can be as much as $40\,000 \, \mu m^2$.

**Fig 6.12** (a) Electronmicrograph of a section through rough ER, $\times 100\,000$

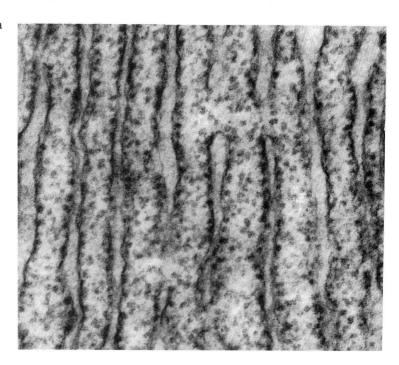

**Fig 6.12** (b) Diagram showing three-dimensional structure of ER

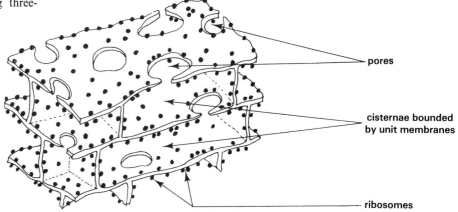

pores

cisternae bounded by unit membranes

ribosomes

If labelled amino acids are introduced into live cells, radioactivity first appears in the ribosomes. Within a few minutes it is found in the membrane-enclosed sacs of the rough ER. The reason for this is that proteins are produced at the ribosomes, threaded through the membrane, and are stored temporarily in the sacs of the rough ER before they are used inside the cell or are secreted to the exterior. It is thus not surprising that the rough ER is very prominent in **enzyme-secreting** cells such as those of the pancreas. The enzymes enter vesicles formed by the Golgi body before they are moved through the cell membrane by reverse pinocytosis.

The smooth ER is prominent in **steroid-secreting** cells such as the interstitial cells of the testes and the adrenal cortex, and also in cells concerned with lipid metabolism such as the epithelial cells of the intestine. The smooth ER also gives rise to the Golgi body (section 6.2.3). Both types of ER are continuous with the nuclear membrane.

The main functions of the ER are to provide a relatively large surface area for protein synthesis, and to permit the rapid transport of such molecules within the cell, and from the inside to the outside of the cell.

### 6.2.3 The Golgi body

The **Golgi body** was discovered in brain cells. Since then it has been seen in cells from almost every group of living organisms. In transverse section the Golgi body usually appears as closely packed, parallel curved pockets (Fig 6.13). The pockets are bounded by unit membranes and are called

**Fig 6.13** Electronmicrograph of a section through a Golgi body, × 24 000

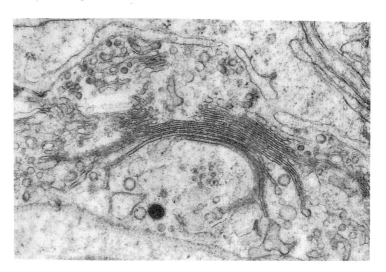

cisternae. From the edges of the cisternae tiny **vesicles** arise. Some of the vesicles become lysosomes (section 6.2.4), some fuse with and enlarge the plasma membrane, others carry secretions to the plasma membrane for release to the exterior. Recently it has been shown that the cisternae are net-like (Fig 6.14). Like the endoplasmic reticulum, Golgi bodies are well developed in cells whose secretions include glycoproteins (Chapter 2). In the Golgi body carbohydrate is added to protein coming from the ER, and the glycoprotein product is secreted at the cell surface. **Mucus** is a typical glycoprotein secreted by goblet cells which abound in the respiratory and gastro-intestinal tracts of mammals.

**Fig 6.14** Diagram showing three-dimensional structure of a Golgi body

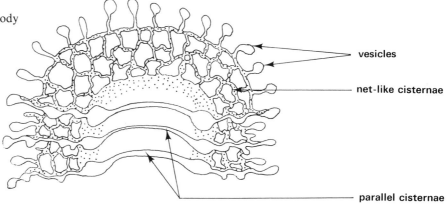

vesicles

net-like cisternae

parallel cisternae

### 6.2.4 Lysosomes

The term **lysosome** was given by Christian de Duve in 1955 to tiny organelles containing hydrolytic enzymes lying near the nucleus of liver cells (Fig 6.15). Electronmicrographs show **primary lysosomes** as small vesicles bounded by a double unit membrane arising from the edge of the Golgi body. Larger **secondary lysosomes** are formed by fusion of primary lysosomes with small vacuoles. The vacuoles arise by infolding of the plasma membrane. Primary lysosomes may also fuse with **autophagosomes** which are internally-formed membranous pockets enclosing worn-out organelles such as mitochondria and ribosomes (Fig 6.16).

**Fig 6.15** Electronmicrograph of a section through lysosomes, × 28 000

lysosome

**Fig 6.16** Summary of the functions of lysosomes

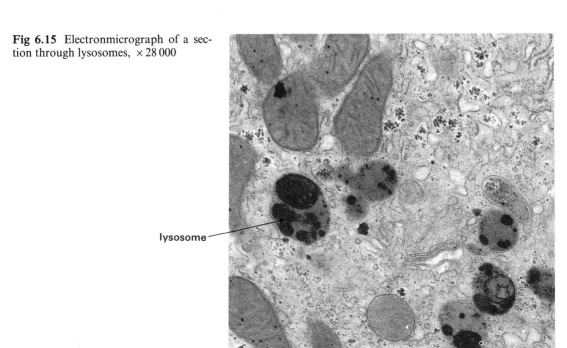

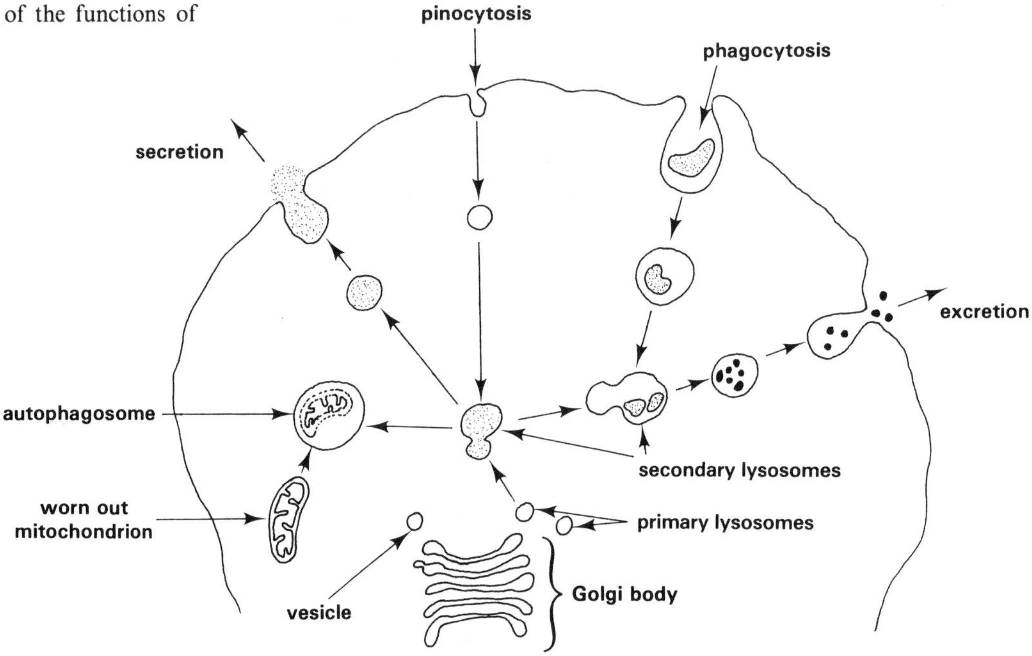

pinocytosis

phagocytosis

secretion

excretion

autophagosome

secondary lysosomes

worn out mitochondrion

primary lysosomes

vesicle

**Golgi body**

The main function of lysosomes is the digestion of materials made in the cell or taken into the cell from outside. A large variety of enzymes has been demonstrated in lysosomes from different sources. Many are lipases, carbohydrases and peptidases which hydrolyse lipids, carbohydrates and proteins respectively. Lysosomal enzymes cause the destruction of foreign particles such as bacteria engulfed by phagocytes, and the breakdown of ageing organelles in all cells.

In old and diseased cells, enzymes released internally by lysosomes bring about self-destruction, **autolysis**, of the protoplast. Erosion of cartilage in rheumatoid arthritis is an example of such activity.

### 6.2.5 Mitochondria

The presence of small elongated bodies in the cytoplasm of plant and animal cells was first reported in the 1850s. The name **mitochondria**, meaning thread-granules, was later given to these structures which are just visible with a light microscope. It was not until a hundred years later that thin sections of mitochondria examined under the electron microscope showed that they had a structural complexity previously unsuspected. They have a smooth outer unit membrane and a much folded inner unit membrane of relatively large surface area (Fig 6.17). Many stalked spherical bodies called **oxysomes** are attached to the folds which are known as **cristae**. The inner membrane encloses a space called the **matrix** containing enzymes and DNA. The DNA codes the synthesis of proteins in mitochondrial membranes. In this way mitochondria replicate when a cell divides.

**Fig 6.17** (a) Electronmicrograph of a section through a mitochondrion, × 81 000

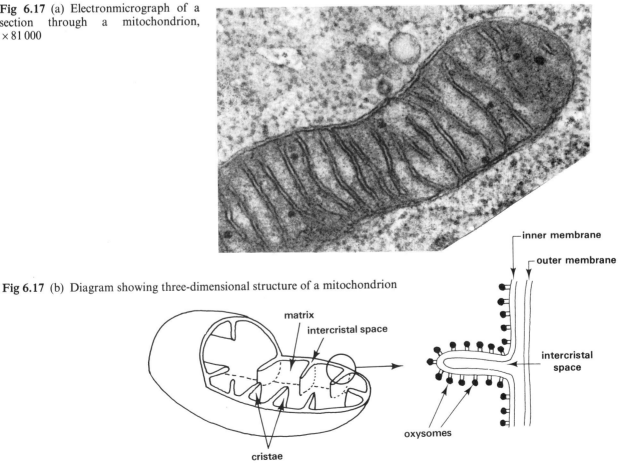

**Fig 6.17** (b) Diagram showing three-dimensional structure of a mitochondrion

The shape of mitochondria is variable, from rod-shaped to spherical, spiral and even cup-shaped. They also vary in size depending on their source. The mitochondria from liver cells measure $1-2\,\mu m$ long and $0.3-0.7\,\mu m$ wide, while those from pancreatic cells, although of the same width, are up to $10\,\mu m$ long. Cells which are metabolically very active contain large numbers of mitochondria. The tubule cells of kidney nephrons, muscle fibres and axon terminals, for example, are packed with mitochondria.

Mitochondria provide sites isolated from the cytoplasm on which the enzyme-catalysed reactions of **aerobic respiration** occur. For instance, mainly at the oxysomes, **adenosine 5′-triphosphate (ATP)** is produced from adenosine 5′-diphosphate (ADP) and inorganic phosphate ions. Conversion of ADP to ATP is an energy-consuming reaction. Energy for the conversion comes from the oxidation of energy-rich substrates such as pyruvic acid derived from glycolysis of sugars in the cytoplasm. ATP provides the cell with energy for energy-consuming processes such as active transport and muscle contraction. The biochemistry of respiration is described in detail in Chapter 5.

### 6.2.6 Peroxisomes

**Peroxisomes** are also called **microbodies**. They are tiny vesicles about the size of lysosomes but bounded with a single unit membrane. The vesicles contain the enzyme **catalase** which catalyses the breakdown of hydrogen peroxide, a by-product of aerobic respiration. Hydrogen peroxide is toxic if allowed to accumulate, so is best disposed of as quickly as it forms.

### 6.2.7 Nucleus

The **nucleus** is usually the largest of the cell's organelles and can thus be seen with a light microscope. Van Leeuwenhoek is accredited with the discovery of the nucleus in the red cells of salmon blood towards the end of the seventeenth century. A distinct nucleus is present at some stage in the cells of all forms of life apart from bacteria, blue-green algae and viruses. Although usually more or less spherical, the nucleus can be more complex in shape. For example, the nucleus of a neutrophil white blood cell is lobed (Chapter 10).

In electronmicrographs the nucleus is seen to be bounded by a double-layered **nuclear membrane** (Fig 6.18). Each layer is a unit membrane, the outer one often covered with ribosomes and continuous with the endoplasmic reticulum. Between the two layers is a **perinuclear space** about 20 nm wide. A prominent feature of the nuclear membrane is the presence of numerous **pores**. They may occupy up to 15 % of the membrane's surface area, each pore being approximately 50 nm in diameter (Fig 6.19). The pores provide routes for the passage of large molecules, such as messenger RNA, from the nucleus to the cytoplasm and vice versa. Inside the nuclear membrane are two main ingredients, **nucleic acids** and protein. Both RNA and DNA are present, the nucleus being the main store of the cell's DNA. Nuclear DNA is bonded to a number of proteins collectively called **histone**. When the nucleus is not dividing the nucleic acid–protein complex appears as tiny granules of **chromatin**. Among the chromatin granules are one or more densely granular bodies called **nucleoli** composed mainly of DNA.

During nuclear division, the nuclear membrane and nucleoli disappear and the chromatin becomes visible as thread-like bodies called **chromosomes**. At telophase the nucleoli reappear. It is thought that nucleoli are the sites where ribosomal RNA is made.

**Fig 6.18** Electronmicrograph of a section through a nucleus, × 14 000

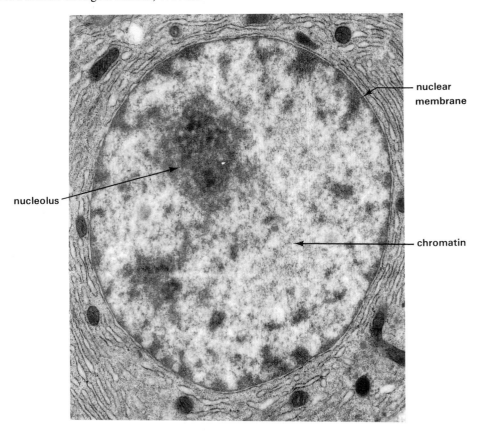

nuclear membrane

nucleolus

chromatin

**Fig 6.19** Electronmicrograph of a freeze-etched nuclear membrane showing nuclear pores, × 45 000

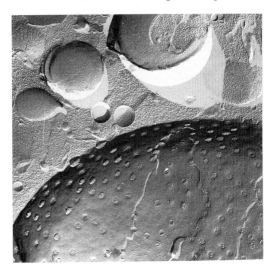

The way in which the nucleus transmits hereditary materials is described more fully in section 6.5. In Chapter 4 you can read how the nucleic acids provide the genetic code for protein synthesis at the ribosomes.

### 6.2.8 Centrioles, cilia and flagella

**Centrioles** are characteristic of animal and fungal cells but are not found in plants. A pair of **centrioles**, each with its long axis at 90° to the other,

97

usually lies near the nucleus. A centriole is a cylinder made of nine tubular filaments about $0.2\,\mu m$ in length. At very high resolution, each filament can be seen to consist of three fused hollow fibrils (Fig 6.20(a)).

During nuclear division the centrioles divide and a pair of them moves to each pole of the cell. They produce a system of **microtubules** called **spindle fibres** radiating towards the equator of the cell. Chromosomes become attached to the spindle equator before migrating to the poles of the cell, seemingly connected to the microtubules (section 6.5).

In some cells, centrioles divide to produce **basal bodies** from which flagella and cilia develop. **Cilia** contain longitudinal fibrils, two in the centre surrounded by nine outer pairs, all enclosed in a unit membrane (Fig 6.20(b)). Alternate contraction and relaxation of the fibrils causes rhythmical bending of the cilia. Ciliary movement wafts liquids and particles in suspension over the surface of the cell. **Flagella** are whip-like organelles, relatively long compared with cilia, but with the same ultrastructure. They are used in locomotion by human sperm (Chapter 18).

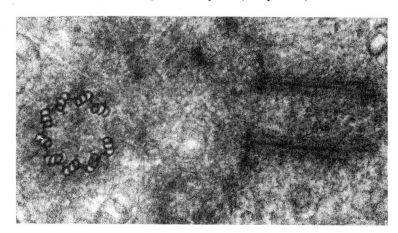

**Fig 6.20** (a) Electronmicrograph of a section through a pair of centrioles, $\times 65\,000$. The left centriole has been cut transversely, the right longitudinally.

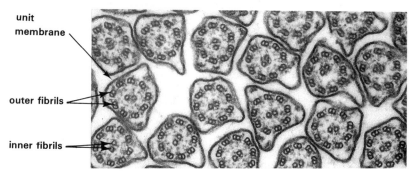

**Fig 6.20** (b) Electronmicrograph of transverse section through cilia, $\times 5000$ (courtesy Biophoto Associates)

unit membrane

outer fibrils

inner fibrils

## 6.2.9 Chloroplasts

In photosynthetic plants the light-absorbing pigments are housed in complex organelles called **chloroplasts**. The chloroplasts of flowering plants are shaped like biconvex lenses $4$–$10\,\mu m$ in diameter and $2$–$3\,\mu m$ thick. They are found chiefly in the mesophyll cells of leaves. Being so large, chloroplasts can be seen with a light microscope and they have been studied from the mid-seventeenth century.

Thin sections of chloroplasts viewed under an electron microscope show them to be bounded by a double unit membrane. The outer membrane is smooth while the inner is extended inwards as a system of layers called **lamellae** in which the photosynthetic pigments are located. In places the lamellae appear as flat discs known as **grana** piled on top of each other.

These are connected by intergrana lamellae. The entire system of internal membranes is suspended in an aqueous matrix called the **stroma** which contains protein and DNA. Following a period of illumination, photosynthetic end-products such as starch grains and lipid globules appear in the stroma (Fig 6.21).

**Fig 6.21** (a) Electronmicrograph of a section through chloroplasts, × 30 000

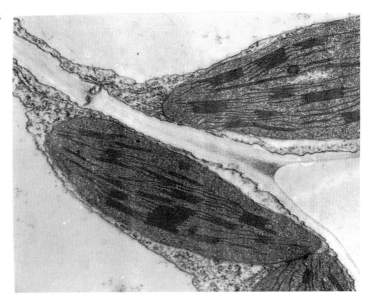

**Fig 6.21** (b) Drawing of a thin section through a chloroplast

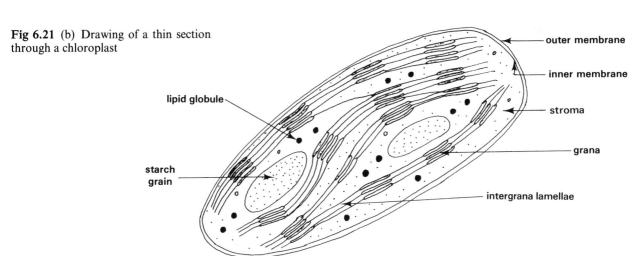

**Chlorophyll** and other photosynthetic pigments in the membranes are concentrated in the grana which are just visible with a light microscope as darker green spots inside chloroplasts. Chloroplasts provide sites on which the biochemical and photochemical reactions of **photosynthesis** can proceed independent of those going on in the rest of the cytoplasm. Details of the reactions are given in Chapter 5. In the grana especially, solar energy is used to produce **reduced nicotinamide-adenine dinucleotide phosphate (NADPH)** and **ATP** in the light-dependent reactions of photosynthesis. The stroma contains the enzymes necessary for the light-independent reactions in which NADPH, carbon dioxide and energy from ATP are used to synthesise energy-rich organic compounds such as sugars and starch. Chloroplasts are thus the organelles in which energy from sunlight is converted to chemical bond energy.

### 6.2.10 The plant cell wall

A characteristic of plant cells is that they have a **cell wall**. Structurally it is like fibreglass, consisting of fibres embedded in an amorphous matrix. The fibres are of **cellulose**, an unbranched polymer of $\beta$-D-glucose (Chapter 2). Each fibre is made of several hundred microfibrils in which about 2000 cellulose molecules are held together by hydrogen bonds (Fig 6.22). The matrix consists of **pectic acid** and its salts calcium and magnesium pectate, and **hemicelluloses** which are polymers of various pentose and hexose sugars. Pectic substances also make up most of the middle lamella which binds adjacent plant cells to one another.

**Fig 6.22** Diagrammatic representation of the structure of a cellulose fibre

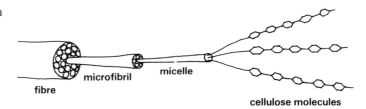

In young cells the wall is thin and called the **primary wall**. The cellulose fibres are orientated at random (Fig 6.23). As a cell grows, more cellulose fibres are laid down on the inside of the primary wall, forming a thicker **secondary wall**. The cellulose fibres of the secondary wall are closely packed and laid down in an orderly way. The wall of fibrous cells becomes impregnated with an alcohol polymer called **lignin** which gives great strength. In cork tissues, **suberin** impregnates the cell wall, while in the outer walls of epidermal cells of leaves and young stems **cutin** appears instead. Both suberin and cutin are waxy materials which provide an effective waterproof covering to the aerial surface of plants.

**Fig 6.23** Electronmicrographs of (a) the primary, and (b) the secondary wall of a plant cell

(a)                                    (b)

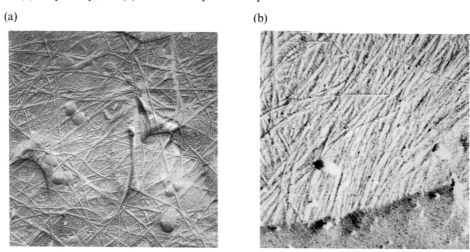

## 6.3 Comparison of light and electron microscopes

Why is it that so much more detail can be made out when specimens are examined with an electron microscope as opposed to a light microscope? Before we can answer this question it is necessary to compare the principles on which light and electron microscopes work.

Fig 6.24 shows the paths of radiation in the two instruments. In both microscopes a tungsten filament lamp is used as a **source of radiation**, but whereas the light microscope uses **visible light** to create the final image, the electron microscope uses a beam of **electrons**. The radiation is focused on to the specimen by a **condenser**, which in the light microscope consists of thick glass lenses mounted beneath the stage. In electron microscopes the condenser is a vertical magnetic field produced by a large cylindrical electromagnet which straightens and intensifies the electron beam. The light rays or electrons, as the case may be, now pass through the specimen, after which the radiation is focused by an **objective lens**. An **eyepiece lens** in the light microscope further enlarges the image. The human eye cannot see electrons, so in the electron microscope the final image is focused by a **projector lens** on to a viewing screen coated with a fluorescent compound such as zinc sulphide. When irradiated with electrons, the fluorescent substance emits light visible to the human eye. A comparable process is used to create a picture on a television screen. With both microscopes it is possible to photograph the final image to produce **photomicrographs** and **electronmicrographs**.

**Fig 6.24** Comparison of the components and pathways of radiation:

(a) of an electronmicroscope

(b) of a conventional light microscope

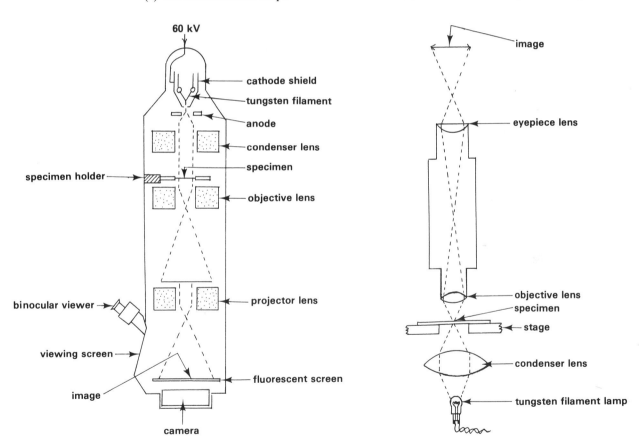

Perhaps the most obvious difference in the final images is the extent to which the specimens are magnified. With a high quality compound microscope fitted with an oil-immersion objective, a magnification of about 1500 times is possible. An electron microscope can magnify up to 500 000 times. However, it is the **resolving power** of the electron microscope rather than its magnifying power which enables it to produce images containing so much more detail. Resolving power or **resolution** is the ability to make out bodies which lie close to one another as separate entities.

The power of resolution $R$ of a microscope depends mainly on the wavelength of the radiation used and on the numerical aperture of the objective lens.

$$R = \frac{0.5\,\lambda}{n \cdot \sin\theta}$$

where $\lambda$ = wavelength of radiation

$n$ = refractive index (RI) of medium between specimen and objective lens

$\theta = \frac{1}{2}$ angle of aperture (Fig 6.25(a))

The expression $n \cdot \sin\theta$ is called the **numerical aperture (NA)** of the objective. The smaller the value of $R$, the better is the resolving power of the microscope. Using a compound microscope fitted with an oil-immersion objective, $R = 210\,\text{nm}$ (Fig 6.25(b)). In other words this instrument is theoretically capable of resolving particles lying as close as 210 nm to one another. In practice it is difficult to resolve as well as this when viewing unstained preparations with a conventional compound microscope.

**Fig 6.25** Angle of aperture:

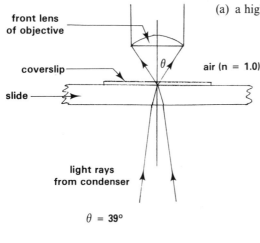

(a) a high power dry objective

front lens of objective

coverslip

slide

air (n = 1.0)

light rays from condenser

$\theta = 39°$

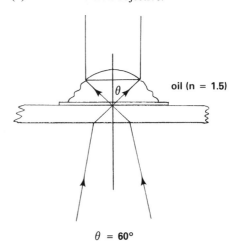

(b) an oil immersion objective.

oil (n = 1.5)

$\theta = 60°$

Study Fig 6.25(a) and (b) and calculate the resolution of each objective assuming $\lambda = 500$ nm.

The **phase-contrast microscope** provides better powers of contrast but does not improve resolution. In this instrument a circular **backstop** is fitted into the substage to create a hollow beam of light which is then focused onto the specimen. Light rays not refracted by the specimen are then put out of phase of the refracted rays by an objective lens which has a circular groove cut in it (Fig 6.26). The effect is that structures can be seen which are difficult to make out with ordinary bright-field illumination. It does not however show any further detail of structure.

**Fig 6.26** Light path through a phase-contrast microscope

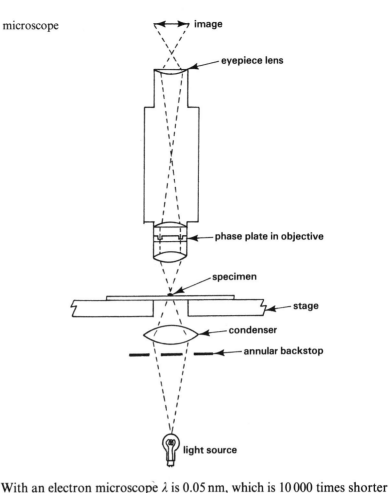

With an electron microscope $\lambda$ is 0.05 nm, which is 10 000 times shorter than the average wavelength of white light. Other things being equal, this should produce a corresponding increase in resolution, but there are technical difficulties preventing this from being realised in practice. Present-day electron microscopes have resolving powers of about 0.5 nm, about 400 times greater than the light microscope. Compare these figures with the resolving power of the human eye which is about $1.0 \times 10^8$ nm. The major disadvantage of the electron microscope is that it cannot be used to examine live specimens.

Compare the advantages and limitation of light microscopes and electron microscopes.

## 6.4 Cell fractionation

The wealth of information which now exists on cell structure has clearly depended on technological improvements in microscopy. The electron microscope is the most sophisticated of instruments available to date for probing the fine structure of cells. Useful as it is to know the detailed structure of cells, electronmicrographs tell us little about the functions of the various cellular components. Cell biologists therefore make use of a variety of techniques in investigating the relationship between structure and function.

How is so much known about the functions of the various components of cells? One technique which has been of enormous help involves separating or **fractionating** the organelles. The activities of the organelles can then be studied without interference from all of the other reactions which take place in whole cells.

Live tissue is first chopped up in a cold **isotonic buffer** solution. The isotonic solution prevents distortion of the organelles. The chopped tissue is then ground up in an **homogeniser**. A domestic blender can be used for this purpose but it usually breaks most of the organelles. Sophisticated work employs a motor-driven ground glass pestle which fits into a tube. This type of homogeniser develops shearing forces just sufficient to rupture the cells. Cells can also be ruptured using **ultrasonic waves**. The homogenate is then transferred to a **centrifuge** in which the mixture is spun at specific speeds at which organelles are known to sediment separately. The main factors governing sedimentation are the magnitude of the centrifugal force, which depends on the spinning speed, and the size and density of the suspended organelles relative to the medium in which they are suspended. Exact times and speeds of centrifugation vary from one tissue to another and are determined by trial and error.

Centrifugal forces of up to 1000 times the force of gravity ($1000 \times g$) can be attained with simple bench centrifuges used in school laboratories. Large organelles such as nuclei and chloroplasts can be sedimented by spinning at $500–600\,g$ for 5–10 minutes. If the supernatant liquid is spun at $10\,000–20\,000\,g$ for 15–20 minutes, mitochondria and lysosomes are sedimented. Fragmented endoplasmic reticulum with attached ribosomes, collectively called microsomes, can be sedimented from the supernatant liquid by spinning it for 60 minutes in an **ultracentrifuge** (Fig 6.27(b))

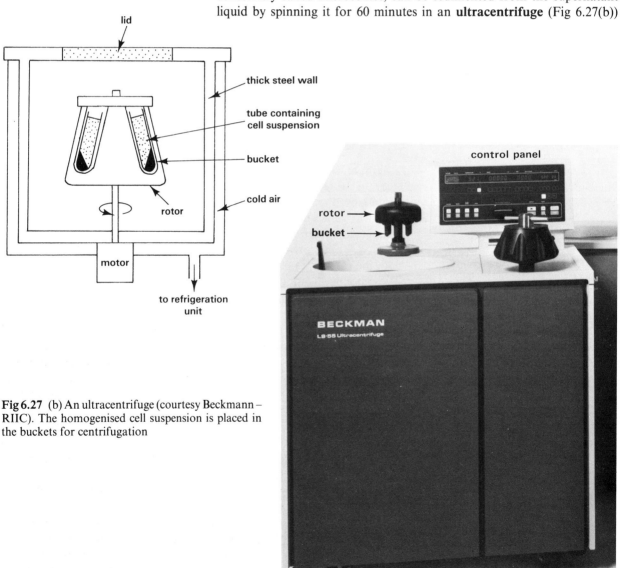

**Fig 6.27** (a) Diagram of the interior of an ultracentrifuge

**Fig 6.27** (b) An ultracentrifuge (courtesy Beckmann – RIIC). The homogenised cell suspension is placed in the buckets for centrifugation

in which forces of 100 000 $g$ and above are developed. Fig 6.28 summarises the various steps in fractioning cell organelles by **differential centrifugation**. Of course, separating organelles in this way can lead to their physical and chemical damage. They may not then function as they normally do in the intact cell where their activities are closely co-ordinated. Nevertheless, by suspending the separated components in a medium which closely resembles the intracellular environment, some of the functions or organelles can be investigated.

**Fig 6.28** Major stages in the fractionation of organelles by differential centrifugation

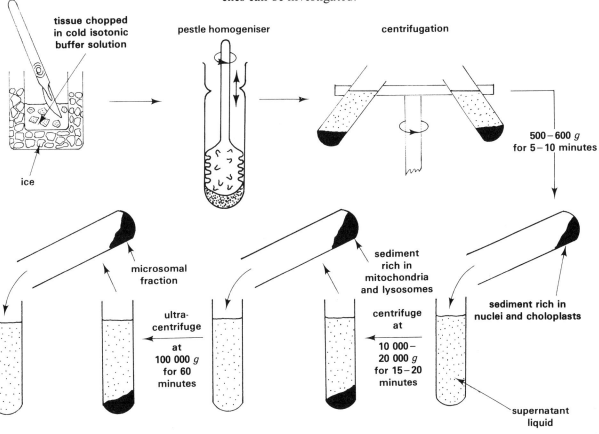

**Fig 6.29** Human chromosomes, × 1000

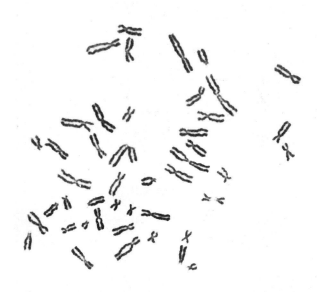

## 6.5 Cell division

There are two types of cell division, **mitosis** and **meiosis**. Mitosis takes place when new cells are added to multicellular organisms as they grow and when tissues are repaired or replaced. Meiosis occurs in the production of gametes by organisms which reproduce sexually.

During both types of division the DNA and histone proteins of the cell nucleus can be seen as threads called **chromosomes** (Fig 6.29). Chromosomes can be stained and are visible using a conventional compound microscope (chromo = coloured; soma = body). The number of chromosomes in the nucleus is fixed for each species of living organism. What is more, the chromosomes can be arranged in **homologous pairs**. In humans for example, the body cells contain 46 chromosomes, 23 homologous pairs (Fig 20.31).

Mitosis normally ensures that the cells produced contain exactly the same number of chromosomes as the cells from which they were formed. The cells of the bone marrow of humans, for example, constantly give rise to cells which have 46 chromosomes. On the other hand, cells produced by meiosis have half the chromosome number. For example, human sperm and eggs each have 23 chromosomes. However, there is much more to it than this. Let us take a close look at the two types of cell division to see in what other ways they differ.

### 6.5.1 Mitosis

Before starting to divide, a cell is at the **interphase** stage. The nucleus appears as a granular body. Inside the nuclear membrane are one or more dense nucleoli (Fig 6.30). The absence of any visible signs of activity disguises the fact that intense metabolism is taking place. It is during interphase that replication of DNA and synthesis of nuclear proteins occurs, new ribosomes are made, and mitochondria and centrioles divide. The proteins which later make up the microtubules of the spindle

**Fig 6.30** A cell at interphase

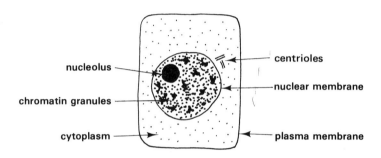

are also made at interphase, although the spindle is not yet constructed. The energy for the various forms of activity comes from respiration. It is therefore not surprising that the respiratory rate of cells at interphase is very rapid.

During both mitosis and meiosis the nucleus divides first, followed by cleavage of the cytoplasm.

### 1 Mitotic nuclear division

For convenience of description it is usual to separate mitotic division of the nucleus into four main stages. It is possible to see the various stages by microscopic examination of stained preparations of bone marrow cells or of growing embryos or other suitable material (Fig 6.31). In this way what appears to be a series of static events can be observed. However, mitosis is an active process, the different stages merging into each other. Time-lapse

**Fig 6.31** Mitotic division in a fish embryo × 1000

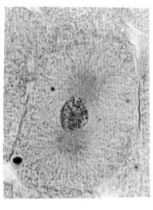

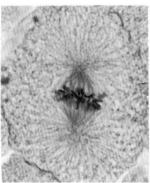
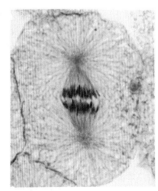
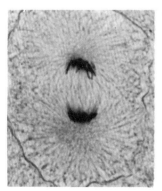
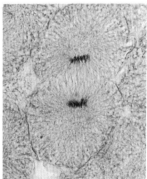

photomicrography provides a means of seeing the dynamic nature of mitosis. Dividing cells are placed in a nutrient solution on a microscope slide and photographed through the microscope every 30 seconds or so, over a period of several hours or even days. The exposed film is developed and run through a ciné projector. It then becomes apparent that during division cells are particularly active.

## A. Prophase

At the start of **prophase**, chromosomes become visible inside the nuclear membrane. At first they are long, thin, entangled threads. As time goes on the threads become shorter and thicker. The chromosomes disentangle and can be seen as separate structures. As the chromosomes become visible the nucleoli gradually disappear. The centrioles which duplicated at interphase begin to migrate to opposite ends (poles) of the cell. As they move apart the centrioles lay down microtubules which extend from one pole of the cell to the other. The microtubules are the **spindle**, a fibrous structure which is widest at the centre (equator) of the cell. A mass of microtubules called an **aster** also radiates from the centrioles at each of the poles (Fig 6.32(a)).

**Fig 6.32** (a) Stages of mitotic nuclear division (only four chromosomes shown for clarity)

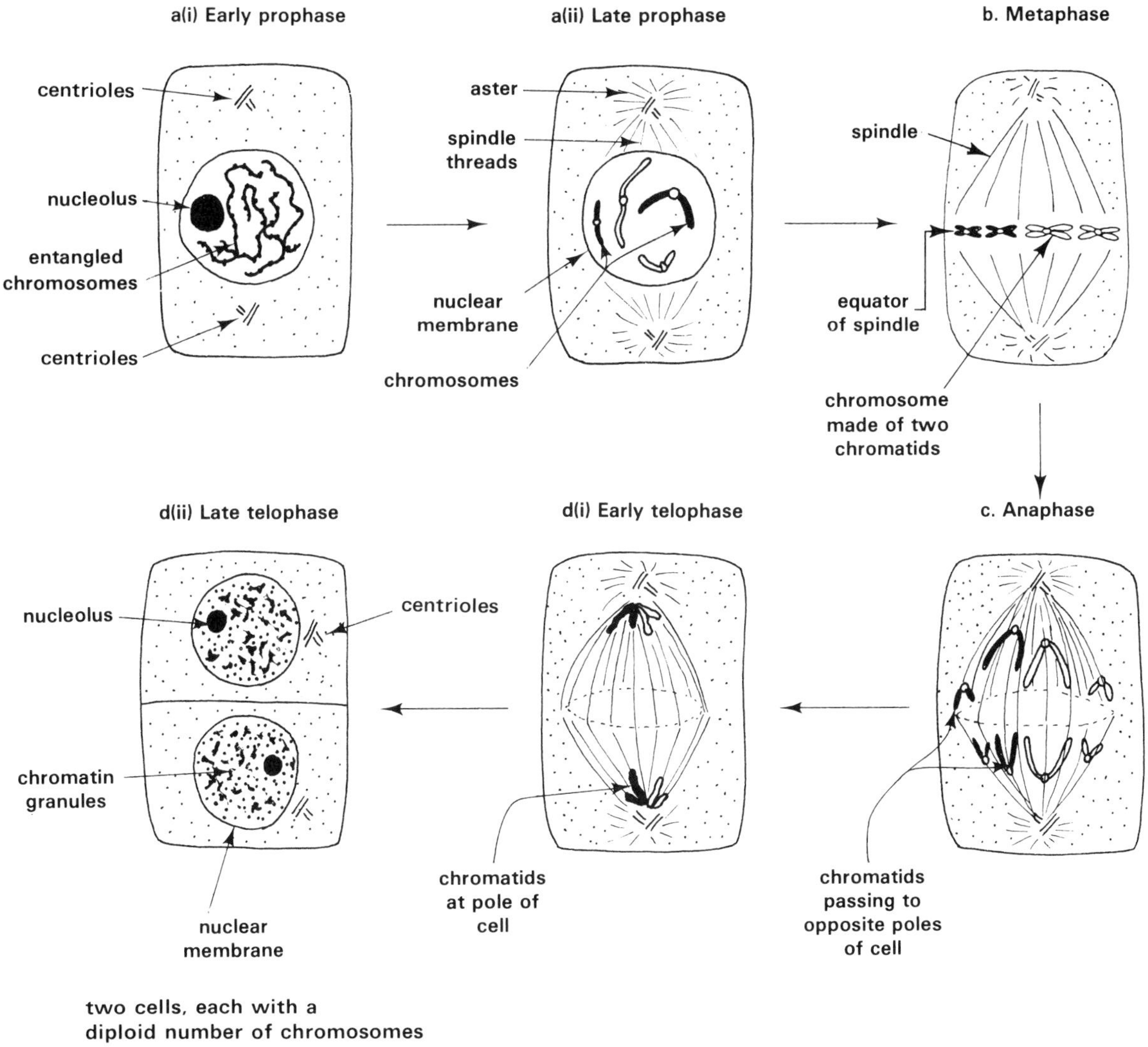

a(i) Early prophase

centrioles

nucleolus

entangled chromosomes

centrioles

a(ii) Late prophase

aster

spindle threads

nuclear membrane

chromosomes

b. Metaphase

spindle

equator of spindle

chromosome made of two chromatids

d(ii) Late telophase

nucleolus

chromatin granules

nuclear membrane

two cells, each with a diploid number of chromosomes

centrioles

d(i) Early telophase

chromatids at pole of cell

c. Anaphase

chromatids passing to opposite poles of cell

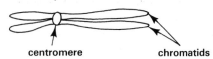

**Fig 6.32** (b) A chromosome

centromere          chromatids

### B. Metaphase

The nuclear membrane breaks down. How and why this occurs is still a matter for debate. Mitochondria often gather near the nuclear membrane at this stage. They may provide energy for some of the spindle microtubules to pull the nuclear membrane apart.

Each chromosome can now be seen to consist of two threads called **chromatids** joined at a **centromere** (Fig 6.32(b)). Unlike the rest of the chromosome, the centromere is not easily stained. Its position differs from one chromosome to another. Independently of each other the chromosomes become attached by their centromeres to the equator of the spindle.

### C. Anaphase

The centromere of each chromosome splits and the chromatids move to opposite poles of the cell. Separation of the chromatids appears to be caused by a shortening of the spindle microtubules to which the centromeres are attached. Consequently the chromatids are dragged, centromere first, away from the equator of the spindle. During their passage the chromatids slide over other spindle microtubules which extent from pole to pole.

### D. Telophase

The two groups of chromatids come together at opposite poles. Each group becomes surrounded by a newly formed nuclear membrane. It is not clear whether the new membranes are put together from fragments of the nuclear membrane destroyed in prophase or are made anew. Whatever their origin, they are assembled from pieces of membrane. Inside the nuclear membranes the chromatids become uncoiled, nucleoli reform and the nucleus takes on the granular appearance it had at interphase.

### 2 Cytoplasmic cleavage

Soon after nuclear division the cytoplasm is separated into two more or less equal parts, each part enclosing one of the newly formed nuclei.

During **cytoplasmic cleavage**, a ring of **microfilaments** appears around the middle of the cell just inside the plasma membrane. A shallow **furrow** develops in the membrane, possibly caused by contraction of the filaments. Further shortening of the filaments ultimately pinches the cytoplasm into two more or less equal parts, each part surrounding a nucleus (Fig 6.33).

**Fig 6.33** Cytoplasmic cleavage in a dividing bone marrow cell, × 1800

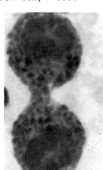

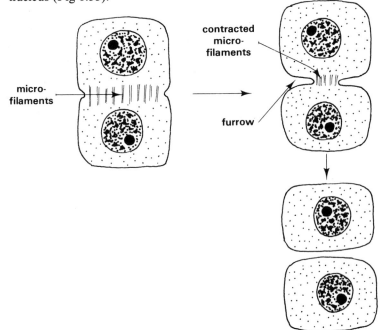

contracted micro-filaments

micro-filaments

furrow

108

### 3 The cell cycle

When incubated at their optimum temperature and supplied with ample nutrients and space, some kinds of human cell grow and divide by mitosis every 24 hours. Interphase occupies by far the longest part of the **cell cycle** (Fig 6.34) and can be divided into the following stages:

i. The **G₁ stage**, in which synthesis of RNA and proteins occurs. It takes about 8 hours to complete.
ii. The **S stage**, in which replication of DNA is completed and duplication of histone proteins occurs. It lasts about 6 hours.
iii. The **G₂ stage**, in which synthesis and replication of organelles such as mitochondria occurs. About 4 hours are normally required for this stage to be completed.

Fig 6.34 The cell cycle

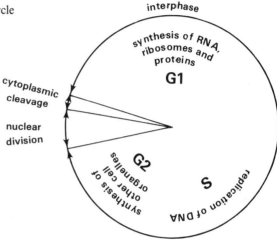

Mitotic nuclear division and cytoplasmic cleavage occupy the remainder of the cycle. In the body of an adult human, few kinds of cells cycle at this rate. Skin cells divide on average once a week, whereas brain cells never multiply. A better understanding of the factors which regulate the cell cycle would be extremely valuable. Restoring growth to vital, damaged organs such as the brain and heart and checking growth of tumours may then be possible.

### 4 The significance of mitosis

The essential feature of mitosis is that it provides a means of distributing the hereditary material DNA equally between two cells. This does not mean that the DNA content is halved at each mitotic division. The hereditary material is normally reproduced in each cell exactly as it was in the parent cell.

One way of finding out when DNA replicates is to measure the amount of DNA in a cell at different times during mitosis. Fig 6.35 shows the results of such an investigation. The amount of DNA exactly doubles during interphase and is halved at anaphase. In this way each of the new cells receives the same amount of DNA as in the parent cell. It does not, however, prove that an exact copy of DNA is distributed to the new cells.

**Fig 6.35** Changes in the amount of DNA in a nucleus during mitosis

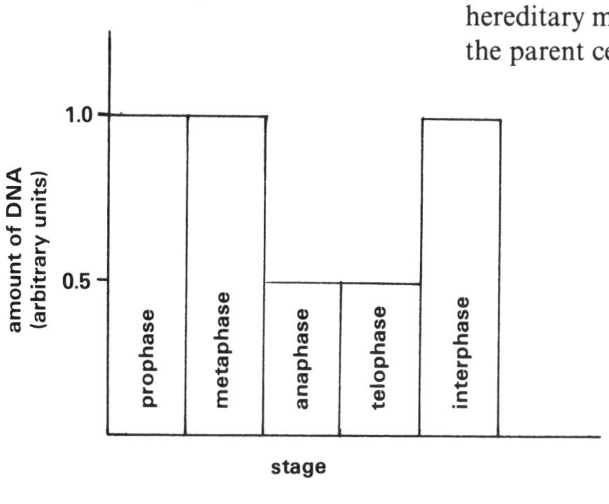

Another approach is to use radionuclides to label the DNA. Cells are grown in a nutrient solution containing thymine labelled with tritium ($^3H$), a radionuclide of hydrogen. As the cells divide, the labelled thymine is incorporated into their DNA, and after a while nearly all the DNA is radioactive. The cells are then placed in a non-radioactive medium. At each subsequent division some of the cells are removed and placed in photographic emulsion. Radiations given off by the labelled DNA cause the emulsion to develop, giving labelled chromosomes a dark colour when viewed under a microscope. The results of such an experiment are shown in Fig 6.36. They prove conclusively that replication of DNA does occur during the cell cycle.

**Fig 6.36** Chromosomes of a chinese hamster after labelling with $^3H$ thymidine (based on autoradiographs)

(a) Metaphase immediately after labelling

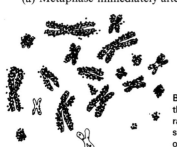

Both chromatids of most of the chromosomes are very radioactive. A few of the smaller chromosomes are only lightly labelled.

(b) Metaphase of next division

Only one chromatid of most of the chromosomes is very radioactive.

The significance of what has just been described is that mitosis normally gives rise to cells with the same combination of genes, the **genome**. Successful genomes can be perpetuated generation after generation in organisms which multiply asexually. In this way breeders can maintain pure strains of many useful plants which can be propagated vegetatively. Some of these are important crop plants such as the potato.

Even so, variations do arise in organisms which do not reproduce sexually. The variation is due in some instances to gene mutations (Chapter 4). In others it is caused by chromosome mutations (Chapter 20).

### 6.5.2 Meiosis

**Meiosis** occurs in the formation of gametes in organisms which reproduce sexually. In meiosis, nuclear division is followed by cytoplasmic cleavage, but in contrast with mitosis there are two nuclear divisions not one. Thus four nuclei are formed from a cell which undergoes meiosis, not two as in mitosis. However, there are other differences which are just as important. The interphase stage of meiosis is the same as in mitosis.

### 1 The first meiotic division

Once more, for convenience of description, division of the nucleus is separated into a number of stages. The Roman numeral I is placed after each of the stages in the first meiotic division of the nucleus to distinguish them from the stages of the second meiotic division.

#### A. Prophase I

The events of **prophase I** are much more complex than those which occur in prophase of mitosis. At the start of prophase I the chromosomes appear as long, thin entangled threads inside the nuclear membrane (Fig 6.37(a)). They then come together in pairs called **bivalents**. Each bivalent is a **homologous pair** of chromosomes. The homologous chromosomes are positioned so that their centromeres are adjacent. Gradually the chromosomes shorten and thicken. At this stage the two chromatids making up each

**Fig 6.37** (a) Stages of the first meiotic division (only four chromosomes shown for clarity)

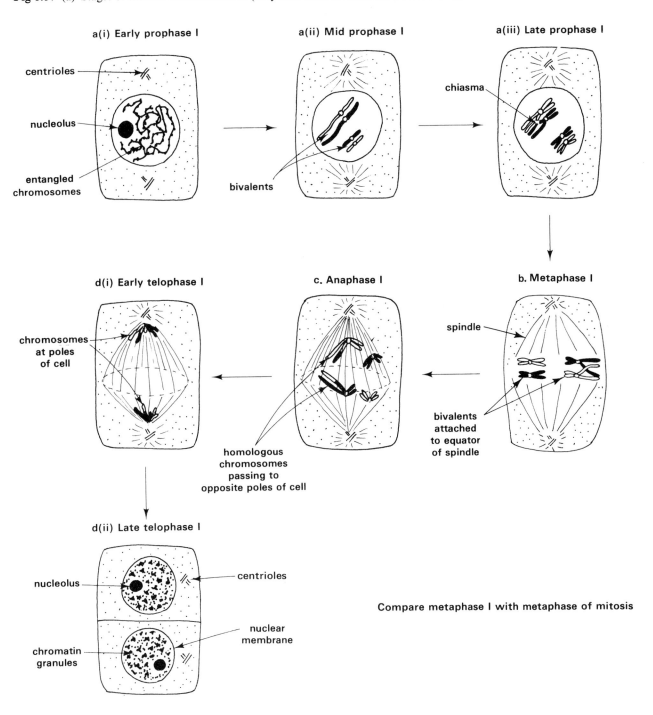

a(i) Early prophase I

centrioles

nucleolus

entangled chromosomes

a(ii) Mid prophase I

bivalents

a(iii) Late prophase I

chiasma

d(i) Early telophase I

chromosomes at poles of cell

c. Anaphase I

homologous chromosomes passing to opposite poles of cell

b. Metaphase I

spindle

bivalents attached to equator of spindle

d(ii) Late telophase I

nucleolus

chromatin granules

centrioles

nuclear membrane

Compare metaphase I with metaphase of mitosis

**Fig 6.37** (b) Chiasmata in grasshopper chromosomes (courtesy Philip Harris Biological Ltd)

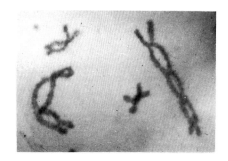

chromosome can be clearly seen. Cross-links called **chiasmata** frequently develop between chromosomes and their homologous partners. Bivalents of long chromosomes often display several chiasmata (Fig 6.37(b)).

Pairing of homologous chromosomes and the formation of chiasmata do not occur during prophase of mitosis. However, two other events which take place in mitotic prophase also occur in prophase I. The nucleoli disappear and a spindle is laid down in the cytoplasm by the centrioles which divided at interphase.

What are the similarities and differences between the events of prophase I of meiosis and prophase of mitosis?

### B. Metaphase I

The nuclear membrane breaks down and the bivalents become attached by their centromeres to the microtubules at the equator of the spindle. One chromosome of each bivalent is directed towards one of the poles of the cell, its homologous partner towards the other pole. It is important to realise that each bivalent is orientated at random with respect to each of the other bivalents.

### C. Anaphase I

Shortening of the spindle microtubules drags the homologous chromosomes of each bivalent apart, pulling them to opposite poles of the cell.

### D. Telophase I

The two groups of chromosomes come together at opposite poles. Each group becomes surrounded by a new nuclear membrane. The chromosomes uncoil, the nucleoli reappear and the nuclei take on a granular appearance. Cleavage of the cytoplasm may occur as in mitosis.

The two nuclei have half the number of chromosomes as the nucleus from which they were derived. For this reason the first division of meiosis is sometimes called **reduction division**. A short interphase may follow, but there is no replication of DNA as happens during interphase of mitosis. Often there is no cleavage of the cytoplasm and the two nuclei proceed directly to the second division of meiosis.

## 2 The second meiotic division

The Roman numeral II is used after each of the stages to distinguish them from the stages of the first meiotic division.

### A. Prophase II

Chromosomes appear in both of the nuclei formed in the first division of meiosis. The centrioles move to opposite poles of the cells, laying down the microtubules of the spindle. There is no pairing of chromosomes and chiasmata do not develop as in prophase I (Fig 6.38).

### B. Metaphase II

The nuclear membranes disappear and the chromosomes become attached by their centromeres to the microtubules at the equators of the spindles. The two chromatids of each chromosome are now easily seen. What is not so obvious is that the chromosomes are orientated at random with respect to one another.

### C. Anaphase II

In what important way does anaphase II of meiosis differ from anaphase of mitosis?

The centromeres of the chromosomes break in two and the chromatids are pulled, centromere first, towards opposite poles of the cell.

**Fig 6.38** Stages of second meiotic division

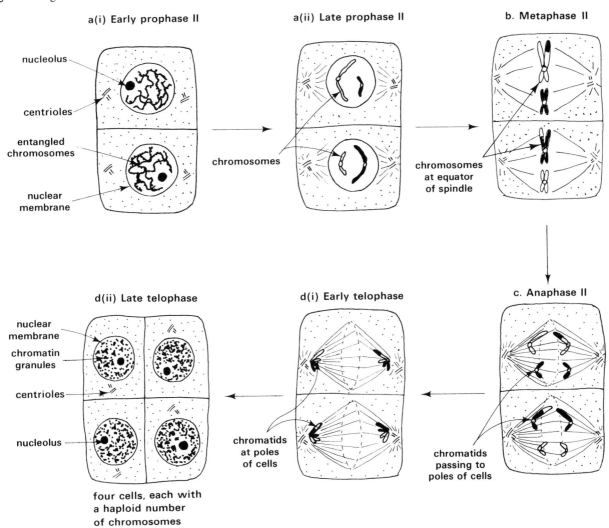

a(i) Early prophase II

nucleolus

centrioles

entangled chromosomes

nuclear membrane

a(ii) Late prophase II

chromosomes

b. Metaphase II

chromosomes at equator of spindle

d(ii) Late telophase

nuclear membrane

chromatin granules

centrioles

nucleolus

four cells, each with a haploid number of chromosomes

d(i) Early telophase

chromatids at poles of cells

c. Anaphase II

chromatids passing to poles of cells

## D. Telophase II

The chromatids come together at opposite poles of the cells. Here they become surrounded by nuclear membranes, uncoil, and the nucleoli appear. The spindle disappears and cleavage of the cytoplasm follows. Altogether four cells, each with half the number of chromosomes, are produced from each cell which divides by meiosis. The events of the second meiotic division are similar to those of mitosis, except that there are half the number of chromosomes.

Examine the photographs in Fig 6.39. Which stage of meiosis is shown in each picture?

**Fig 6.39** Meiotic division in grasshopper testis (courtesy Philip Harris Biological Ltd)

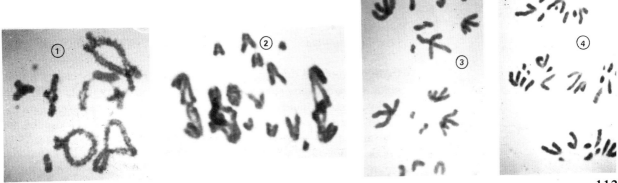

①  ②  ③  ④

113

## 3 The significance of meiosis

Whilst mitosis normally produces cells with an exact replica of the genetic material found in the parent cell, meiosis gives rise to cells with half the amount of genetic material. Our body cells have a **diploid** number of chromosomes, gametes have a **haploid** number. At fertilisation the diploid number is restored. Meiosis thus ensures that the chromosome number is normally kept constant in each generation (Fig 6.40).

**Fig 6.40** Mammalian life cycle

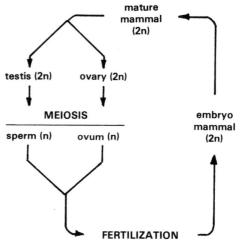

Another significant feature of meiosis is that it produces gametes with varied combinations of genes. There are two important events in meiosis which create new genomes.

### A. Crossing over

**Crossing over** takes place when bivalents appear in prophase I. Chiasmata are formed and homologous chromosomes exchange genes (Fig 6.41). The homologous chromosomes later separate and end up in different gametes. As a result of crossing over, linked genes are parted and gametes with new genomes are thus produced (Chapter 20).

**Fig 6.41** Crossing over. Because of crossing over, genes carried on the same chromosome (linked genes) are separated.

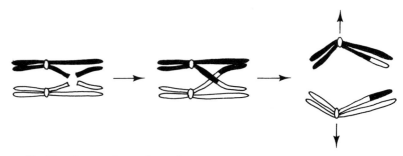

### B. Random orientation of chromosomes

The separation of a pair of homologous chromosomes at anaphase I is independent of the separation of other pairs. As the chromosomes are orientated at random, the alleles on one pair of homologous chromosomes separate independently of the alleles on other (Fig 6.42). Because of **random assortment** a vast permutation of genes is possible in the gametes.

Crossing over and random assortment ensure that progeny resulting from sexual reproduction are genetically different from their parents. This is why no two persons, unless they are identical twins, have the same genome. Clearly meiosis is important in producing **genetic variation**. Variation can also arise from gene and chromosome mutations (Chapters 4 and 20 respectively).

**Fig 6.42** Random orientation of chromosomes. The genes on one pair of homologous chromosomes segregate independently of those on the other pair. For clarity crossing over has not been shown

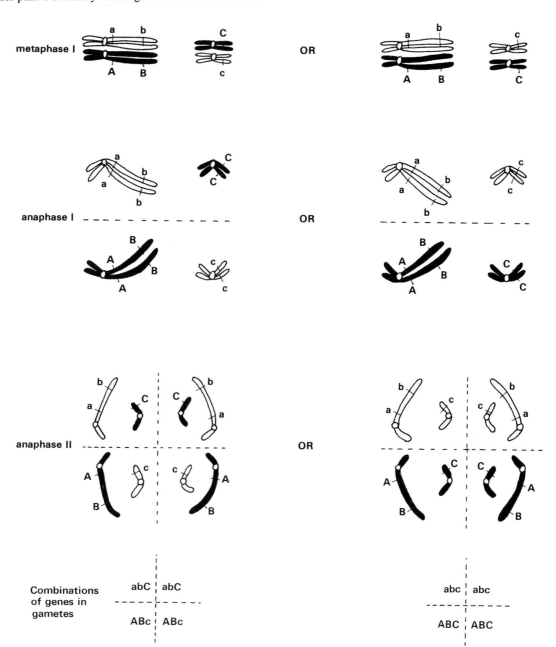

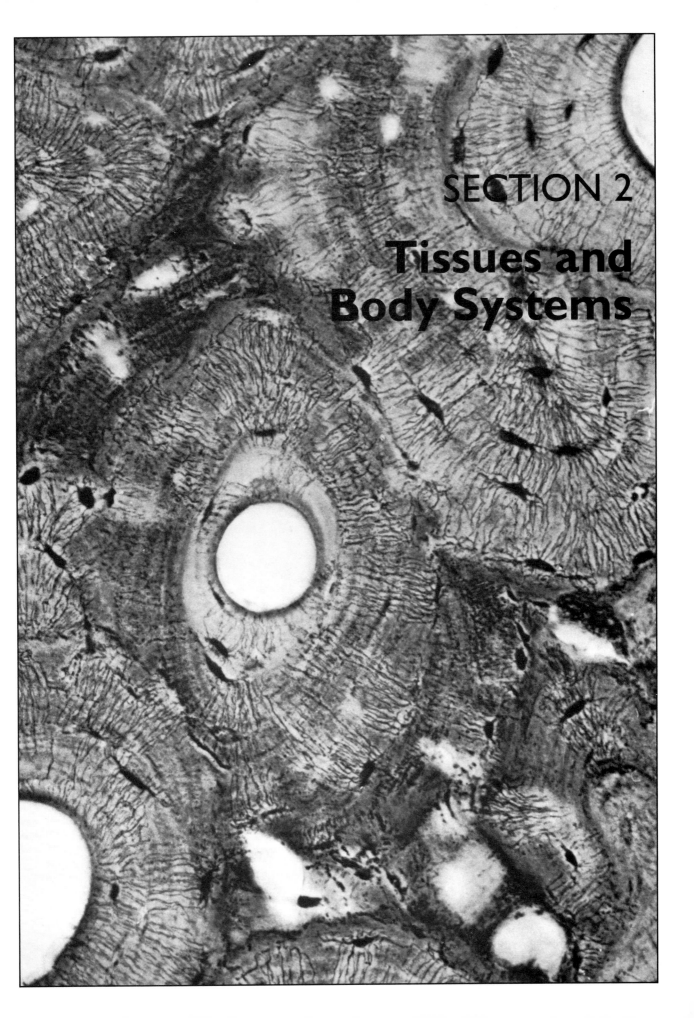

# SECTION 2
# Tissues and Body Systems

# 7 Tissues

The structures of the body can be described at four different levels. They are the **molecular** level (Chapters 2, 3, 4 and 5), the **cellular** level (Chapter 6), the **tissue** level and the level of **organ systems** (Chapters 8 to 19). A tissue is defined as:

> **a group of cells of common origin with some specialised structure and common function.**

An organ consists of:

> **a collection of different tissues which combine to effect one or more physiological functions.**

In the embryo, tissues arise from three **germ layers** called **ectoderm** on the outside, **endoderm** lining most of the gut, and **mesoderm** in between the other two layers (Chapter 19). Tissues are classified into four groups: **connective, epithelial, nervous** and **muscular**.

---

**In what ways do different cells cooperate in the body to perform one or more physiological functions?**

---

## 7.1 Connective tissues

Connective tissues originate in the mesoderm and consist of several types of cells within an **extracellular matrix** (Fig 7.1). Blood and lymph are examples of fluid connective tissues. They are fully described in Chapter 10. Here we shall look at non-fluid connective tissues.

**Fig 7.1** Main components of connective tissue

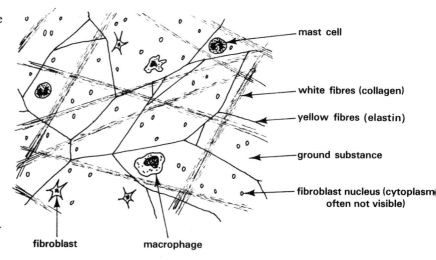

mast cell

white fibres (collagen)

yellow fibres (elastin)

ground substance

fibroblast nucleus (cytoplasm often not visible)

fibroblast          macrophage

**Fig 7.2** Photomicrograph of a thin section of adipose connective tissue, × 400

### 7.1.1 Cells

The several kinds of cell found in connective tissue include the following:

**i. Fibroblasts** secrete the extracellular matrix including a variety of protein **fibres** (section 7.1.2).

**ii. Adipocytes** store **fat. Adipose tissue**, such as that found under the skin, is a connective tissue made of masses of adipocytes (Figs 7.2 and 17.7).

**iii. Defensive cells** include **mast cells** and **macrophages**. Mast cells are similar to the basophils in blood (section 10.1.2). They contain histamine granules. Histamine is released from mast cells following injury to the tissues or in an allergic response, and may trigger the process of **inflammation** (section 11.1.5). Vasodilation occurs, increasing the rate at which phagocytes enter the affected tissues from blood. Macrophages are similar to monocytes in blood. They phagocytose particulate debris and infecting microbes (section 10.1.2).

**iv. Chondrocytes and osteocytes** produce cartilage and bone respectively (Chapter 8).

Fig 7.3 Photomicrograph of thin sections of connective tissue

(a) areolar connective tissue, × 450

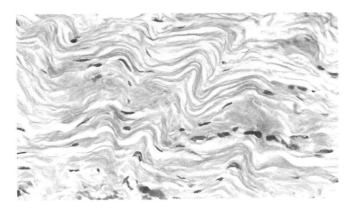

(b) white fibrous connective tissue, × 100

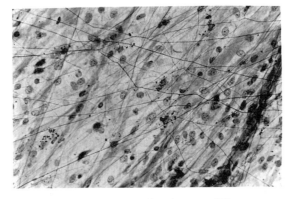

Fig 7.4 Photomicrograph of a thin section of reticular connective tissue, × 100

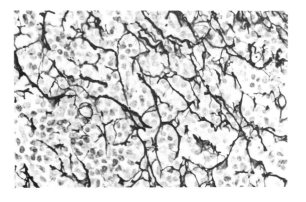

## 7.1.2 Extracellular matrix

The extracellular matrix of connective tissues contains **ground substance**. In connective tissues other than blood, bone and cartilage, ground substance consists of polysaccharides and proteins. In it are suspended protein fibres secreted by fibroblasts. Some pathogenic bacteria secrete the enzyme **hyaluronidase** which hydrolyses hyaluronic acid, one of the constituents of ground substance. However, ground substance is a mechanical barrier not easily penetrated by most kinds of microbes.

The fibres of the matrix are of three main kinds and are present in all connective tissues apart from blood. Their relative proportions vary and reflect the specific functions of different connective tissues.

**i. Collagen fibres** are the most common and are made of the protein **collagen**. It has limited elasticity, providing strength and structural support for connective tissues. In **areolar** tissue, collagen fibres are loosely packed. Conversely in **white fibrous** tissue, which makes up tendons and ligaments, they are closely packed (Fig 7.3(b)).

**ii. Reticular fibres** are composed of a protein called **reticulin**. It is delicately branched and supports tissues containing many cells in organs such as the liver, lymph nodes and endocrine glands (Fig 7.4).

**iii. Elastic fibres** are made of **elastin**. It may be found as fibres or discontinuous sheets. Elastic fibres can be stretched but will recoil to resume their original length. Elastic recoil is vital to the normal functioning of arteries (section 10.2.5) and lungs (section 12.1).

**Cartilage** and **bone**, **blood** and **lymph** are also connective tissues. They are described in Chapters 8 and 10.

**Fig 7.5** Diagram of the main kinds of epithelia

(a) Squamous

(b) Cuboidal

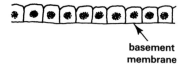

basement
membrane

(c) Columnar ciliated

cilia

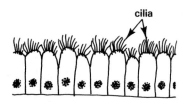

(d) Pseudostratified

(e) Transitional

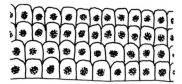

(f) Stratified

flattened
outer cells

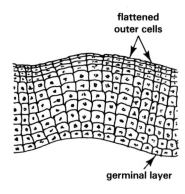

germinal layer

# 7.2 Epithelia

Epithelia arise from all three embryonic germ layers. They form the **boundaries** at the surfaces of other tissues and organs. They line the inner surfaces of structures such as blood vessels where they are called **endothelia**. Such boundary layers protect the underlying tissues and allow selective interchange of metabolites between the underlying tissues and their immediate surroundings. Epithelia may also develop secretory **glands**.

The cells of epithelia are bonded to a **basement membrane** consisting of reticular fibres and glycoprotein.

## 7.2.1 Boundary epithelia

Boundary epithelia are classified according to the shapes and number of layers of cells of which they are composed. **Simple** epithelia are one cell thick, whereas **compound** epithelia are several cells thick (Fig 7.5).

**1 Simple epithelia**

**i. Squamous epithelium** is sometimes called **pavement** epithelium since its cells are flat like paving slabs. The walls of blood capillaries, alveoli and Bowman's capsules of nephrons are made of squamous epithelium. Transport of metabolites across squamous epithelium is relatively fast because the distance across flat cells is relatively short (Fig. 7.5(a)).

**ii. Cuboidal epithelium** consists of cells that appear cube-shaped when seen in section (Fig 7.5(b)). They are found lining the tubular part of kidney nephrons and collecting ducts (Fig. 13.4(a)). Secretory surfaces such as thyroid follicles (Fig 16.2) are also made of cuboidal epithelium. Lining the ventricles of the brain and the spinal canal are ependymal cells, a layer of ciliated cuboidal epithelium. Here, the flickering cilia set up currents in the cerebro-spinal fluid, thus assisting the circulation of metabolites.

**iii. Columnar epithelium** consists of relatively tall cells. They may possess **microvilli** which greatly increase the area on the epithelial surface. Such an arrangement helps the process of absorption in the small intestine (Chapter 9). Microvilli are only clearly visible with the aid of an electron microsocpe. At lower magnification, obtained with the light microscope, microvilli often appear like the bristles of a brush. Hence the term **brush border** has been used to describe this appearance. A brush border is also present on the cuboidal epithelium of nephrons where it facilitates reabsorption from the renal filtrate.

**Ciliated** columnar epithelium is found lining the oviducts. Beating of the cilia sucks ova into the oviducts where they may be fertilised (Figs 7.5(c) and 7.6 and Chapter 18). Ciliated columnar epithelium is also found in the terminal bronchioles (Chapter 12).

**iv. Pseudostratified epithelium** is so called because the cells appear to be arranged in several layers. However, the appearance is due to the cells being of different sizes. All the cells contact the basement membrane (Fig 7.5(d)). Pseudostratified ciliated columnar epithelium is found in the trachea, bronchi and bronchioles. The back and forth movement of the cilia pushes mucus towards the pharynx. The mucus traps inhaled dust and microbes (Fig 12.3).

**Fig 7.6** Photomicrograph of a thin section of ciliated columnar epithelium in an oviduct, × 1000

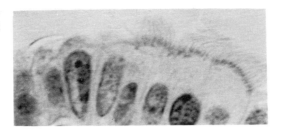

## 2 Compound epithelia

**i. Transitional epithelium** consists of several layers of cells of similar shape and size. It lines the urinary bladder (Figs 7.5(e) and 7.7). When the bladder is full of urine the epithelium stretches and becomes a single layer of cells. After urine has been passed, the bladder wall relaxes and the epithelial lining is then of several layers.

**Fig 7.7** Photomicrograph of a thin section of transitional epithelium in urinary bladder, × 75

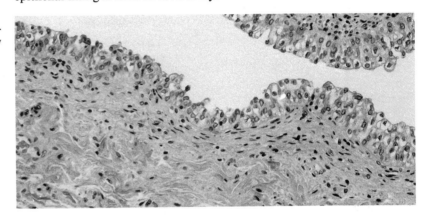

**ii. Stratified epithelium** also consists of several layers of cells. The cells lying next to the basement membrane divide constantly and comprise the **germinal layer**. As the products of cell division are pushed upwards towards the epithelial surface, they become flattened. Eventually the cells degenerate and are sloughed off into the surroundings to be replaced by more cells from underneath. Stratified epithelia are found lining the mouth, pharynx, oesophagus, rectum and vagina. In the epidermis of the skin, the degenerated surface cells become filled with a protein called **keratin**. It enables the epithelium to resist mechanical abrasion and protects the underlying tissues from desiccation (Figs 7.5(f), 7.8 and 17.7).

**Fig 7.8** Photomicrograph of a thin section of stratified epithelium in skin, × 150

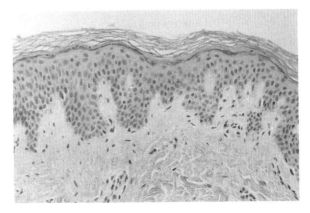

## 7.2.2 Glandular epithelia

There are two main kinds of **gland** formed from epithelia, **endocrine** and **exocrine**. Endocrine glands become separated from the epithelia from which they develop (Fig 7.9(b)). They are **ductless** and secrete **hormones** directly into the body fluids, usually the blood. Hormones are distributed quickly throughout the body in this way. The main endocrine glands and their functions are described in Chapter 16.

Exocrine glands remain attached to the epithelium from which they develop (Fig. 7.9(a)). Ducts channel their secretions onto the epithelial surface or into the cavity, **lumen**, above it. Exocrine glands are classified according to the shape of the secretory structure and the complexity of the ducts (Fig 7.10).

**Fig 7.9** Development of (a) exocrine and (b) endocrine glands

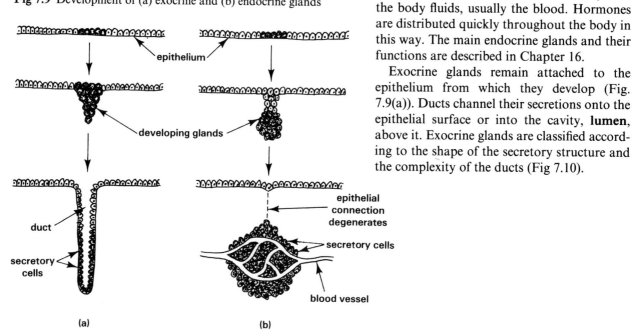

epithelium

developing glands

duct

secretory cells

epithelial connection degenerates

secretory cells

blood vessel

(a)    (b)

**Fig 7.10** Diagram of the main kinds of exocrine glands

simple tubular
eg. gastric pit

simple coiled tubular
eg. sudorific gland

simple branched tubular
eg. fundic gland in stomach

simple alveolar
eg. mucus gland

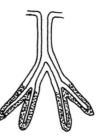
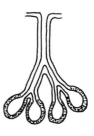
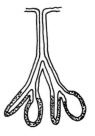

simple branched alveolar
eg. meibomian gland in eyelids

compound tubular
eg. Brunner's gland

compound alveolar
eg. lactating mammary gland

compound tubular-alveolar
eg. sub-maxillary gland

## 7.3 Nervous tissue

Nervous tissue consists of a complex network of excitable cells called **neurons** together with a variety of non-neuronal cells. The functioning of the nervous system and the mechanisms of impulse conduction are described in Chapter 14.

### 7.3.1 Neurons

All neurons possess three main parts: **cell body, dendrites** and **axon** (Fig 7.11).

**Fig 7.11** Main parts of a neuron

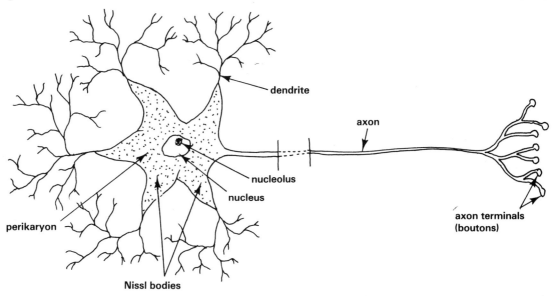

**Fig 7.12** Main categories of neurons: (a) multipolar (the commonest) (b) bipolar and (c) pseudo-unipolar

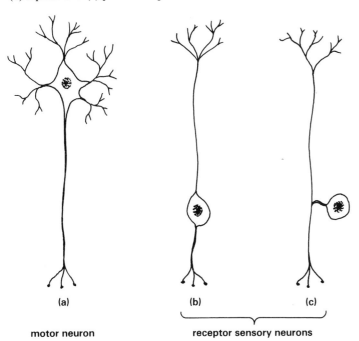

(a)

motor neuron

(b)          (c)

receptor sensory neurons

**i. The cell body** or **soma** contains the nucleus and a mass of cytoplasm called the **perikaryon**. A mass of granules called **Nissl bodies** is usually distributed in the perikaryon. They are aggregates of ribosomes on rough endoplasmic reticulum. The proteins they make include enzymes active in the synthesis of synaptic transmitters (section 7.3.2).

**ii. Dendrites** project from the cell body as a mass of branched threads containing cytoplasm. They receive impulses from other neurons.

Several kinds of neurons can be distinguished. They are classified according to the nature of their dendrites (Fig. 7.12).

**iii. Axon** Projecting from the cell body and with a similar basic structure to the dendrites is a single axon. It conveys impulses away from the cell body and on to other neurons (section 14.2.1). Dendrites may be thought of as the *aerials* and the axon as the *transmitter* for the neuron.

Some axons run the length of our limbs. They are up to a metre long. Other axons, such as those in the cerebral cortex of the brain, are indistinguishable from the dendrites on the same neuron (Fig 7.13).

**Fig 7.13** (a) Pyramid cell from the cerebral cortex

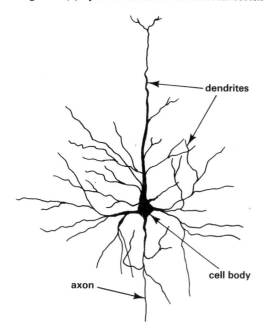

**Fig 7.13** (b) Photomicrograph of a thin section of cerebral cortex of a rat, × 75

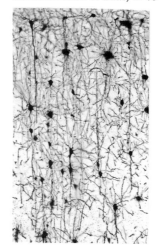

Axons are branched at their distal endings and terminate in small, swollen structures called **boutons**. Boutons form the synaptic junctions with other neurons by which complex nerve pathways are built up (sections 7.3.2 and 14.1.5).

Electronmicroscopy reveals a system of very fine **microtubules** inside the cytoplasm (axoplasm) running from the cell body to the boutons. They transport various chemical substances along the axon. The substances include **synaptic transmitters** that are synthesised in the cell body, but are required for secretion in the axon terminals.

### 7.3.2 Synapses and motor end-plates

Synapses are the junctions between successive neurons. They occur where axons terminate on the dendrites and cell bodies of other neurons. We can distinguish **axodendritic** and **axosomatic** synapses depending on their sites (Fig 7.14). Electronmicrographs show a narrow gap about 10 to 20 nm across, called the **synaptic cleft**. Inside the boutons, mitochondria and vesicles are found (Figs 7.15 and 14.13). The vesicles contain neuro-transmitters which enable impulses to cross the synaptic cleft to the postsynaptic neuron.

At the end of each nerve pathway **neuroeffector junctions** are found between the last neuron and an effector, such as a muscle. These junctions are similar to synapses in structure and mode of functioning (Figs 7.16 and 14.17).

The mechanisms of synaptic and neuroeffector transmission are described in section 14.1.5.

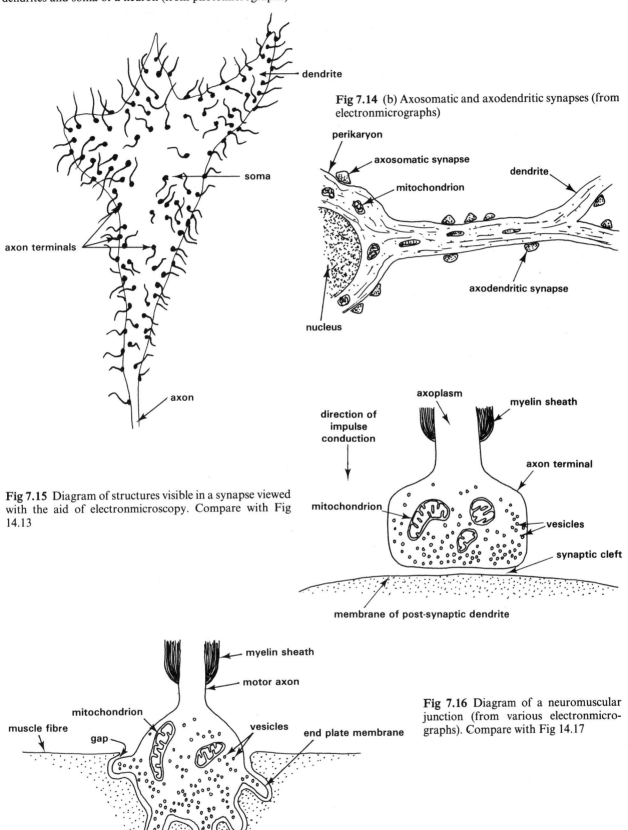

**Fig 7.14** (a) Many axons terminate as synapses on the dendrites and soma of a neuron (from photomicrographs)

dendrite

soma

axon terminals

axon

**Fig 7.14** (b) Axosomatic and axodendritic synapses (from electronmicrographs)

perikaryon

axosomatic synapse

mitochondrion

dendrite

axodendritic synapse

nucleus

**Fig 7.15** Diagram of structures visible in a synapse viewed with the aid of electronmicroscopy. Compare with Fig 14.13

axoplasm

myelin sheath

direction of impulse conduction

axon terminal

mitochondrion

vesicles

synaptic cleft

membrane of post-synaptic dendrite

myelin sheath

motor axon

mitochondrion

muscle fibre

gap

vesicles

end plate membrane

**Fig 7.16** Diagram of a neuromuscular junction (from various electronmicrographs). Compare with Fig 14.17

123

### 7.3.3 Myelination

Neurons in the peripheral nervous system are enveloped by **Schwann cells**. The Schwann cells may grow around the axons to form a **myelin sheath** made up of many layers of cell membrane (Fig 7.17). Such axons are **myelinated**. The myelin sheath of any one axon is provided by many Schwann cells, each covering only a small portion of the axon up to 1 mm long. Between adjacent Schwann cells along the sheath there are short

**Fig 7.17** (a) Development of myelin sheath from Schwann cell seen in transverse section

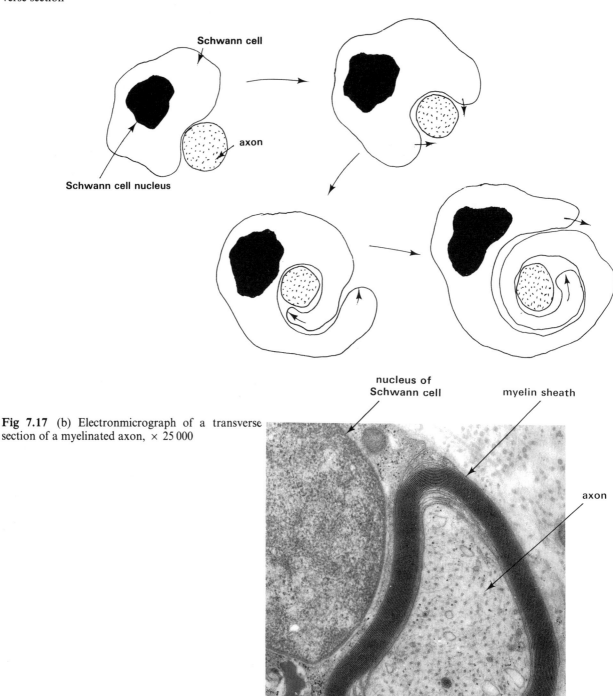

**Fig 7.17** (b) Electronmicrograph of a transverse section of a myelinated axon, × 25 000

intervals where the axon is uncovered. The spaces are called **nodes of Ranvier** (Fig 7.18). It is only at the nodes that the movements of ions necessary for impulse conduction occur. The insulating effect of the myelin sheath in speeding up impulse conduction is described in section 14.1.3. Schwann cells are absent from the central nervous system where myelination is provided by **oligodendrocytes** (section 7.3.4). Some neurons do not become myelinated. Axons of small diameter, such as those of the autonomic system, simply lie in shallow grooves in the Schwann cell cytoplasm. Many **non-myelinated** axons may be found in the same Schwann cell (Fig 7.19).

**Fig 7.18** Photomicrograph showing some myelinated neurons and nodes of Ranvier, × 400

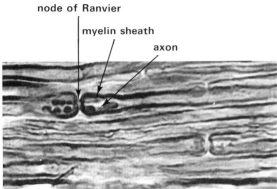

node of Ranvier

myelin sheath

axon

**Fig 7.19** Diagram of non-myelinated neurons as seen in transverse section (from various electronmicrographs). Compare with Fig 7.17(b)

Schwann cell cytoplasm

Schwann cell nucleus

axons

### 7.3.4 Neuroglia

In the central nervous system where there are no Schwann cells, the non-neuronal cells are called **neuroglia**. They account for about 50% of the mass of the central nervous system and provide mechanical and metabolic support for the neurons. There are four main kinds of neuroglia:

**i. Oligodendrocytes** are relatively large cells from which project a few short, branched processes (Fig 7.20(a)). The processes envelop nearby axons to form a myelin sheath in much the same way that Schwann cells do in the peripheral system. A single oligodendrocyte may contribute to the myelin sheaths of up to 50 different neurons. Oligodendrocytes are the most abundant neuroglia in white matter (section 14.2.2).

**ii. Astrocytes** are the most abundant neuroglia in grey matter. They have many branched projections. Some of the projections become **perivascular feet** wrapped around nearby blood capillaries (Fig 7.20(b)). Other projections of the same astrocytes connect with neurons. Astrocytes may thus help metabolic exchanges between blood and neurons.

**iii. Microglia** are relatively small with many fine, branched projections (Fig 7.20(c)). Microglia can transform into mobile phagocytes similar to macrophages (section7.1.1), usually in response to tissue damage.

**iv. Ependymal cells** form a layer of ciliated cuboidal epithelium (Fig. 7.20(d)) which lines the brain's ventricles and the spinal cord. The beating of the cilia sets up currents in the cerebrospinal fluid (CSF), thus helping the circulation of metabolites. Astrocytes are connected to the bases of ependymal cells which may assist exchange of metabolites between CSF and neurons in the central nervous system.

**Fig 7.20** Main kinds of neuroglia

(a) Oligodendrocytes

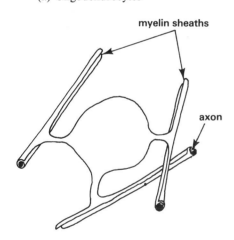

myelin sheaths

axon

(b) Astrocytes

perivascular feet

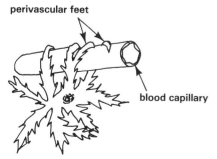

blood capillary

(c) Microglia

(d) Ependymal cells

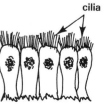

cilia

## 7.4 Muscle

There are three kinds of muscle in our bodies, **skeletal**, **visceral** and **cardiac**. The mechanism of muscle contraction is described in Chapter 14.

### 7.4.1 Skeletal muscle

Skeletal muscle may be controlled consciously to move the bones of the skeleton (Chapter 8). For this reason it is called **voluntary** muscle. The arrangement of the the protein filaments that bring about contraction gives skeletal muscle a banded appearance when viewed with the aid of a microscope (Figs 7.22 and 14.18). For this reason it is also called **striped** or **striated** muscle.

Skeletal muscle consists of **fibres** which are arranged in large groups called **fasciculi**. Each fasciculus is surrounded by a fibrous membrane called the **perimysium**. The entire muscle is surrounded by a membrane called the **epimysium**. Tendons are composed of dense collagenous connective tissue (section 7.1.2). They attach the tapered ends of muscles to the bones (Fig 7.21).

**Fig 7.21** Structure of striped muscle

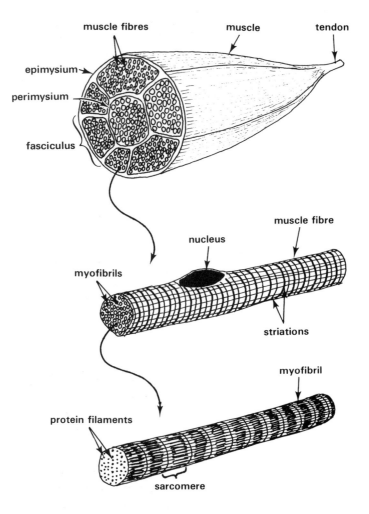

Skeletal muscle fibres are between 10 and 100 $\mu$m in diameter and may be up to 300 000 $\mu$m long. The fibres have an outer membrane called the **sarcolemma** which contains large numbers of **myofibrils**, each about 2 $\mu$m in diameter. Nuclei are scattered under the sarcolemma along the length of each fibre.

The myofibrils are divided into compartments called **sarcomeres** by internal membranous partitions called **Z-lines** (Figs 7.22 and 14.18). There are several bands across the myofibrils and hence the fibres. The **I-bands** are light in appearance and straddle each Z-line. They are regions where only thin filaments of the protein actin are found inside the sarcomeres. Between the I-bands and in the middle of each sarcomere are the **A-bands**. They are dark in appearance and contain the thick filaments of the protein myosin as well as overlapping thin filaments. In the middle of each A-band is a lighter **H-band** where there are thick filaments only (Figs 7.23(a) and 14.18).

**Fig 7.22** Photomicrograph of some skeletal muscle fibres teased apart, × 600

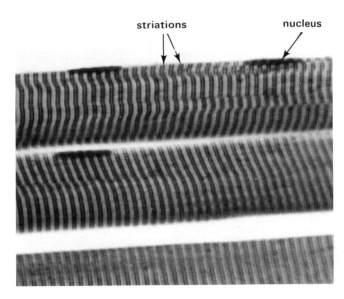

**Fig 7.23** Diagram of the microscope appearance of sarcomeres when (a) relaxed and (b) contracted

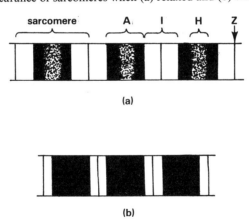

When the muscle fibres contract the thin filaments slide in between the thick filaments and may meet in the middle of each sarcomere (section 14.1.6). When this happens the H-bands become narrower and may disappear. Similarly, the I-bands become narrower as the Z-lines are drawn closer to each other. Since the filaments do not shorten during contraction the A-bands remain the same length (Fig 7.23(b)).

Skeletal muscle normally only contracts when stimulated by nerve impulses. It is **neurogenic**.

## 7.4.2 Cardiac and visceral muscle

The walls of the heart are mainly composed of cardiac muscle (Chapter 10). Cardiac muscle is striated like skeletal muscle (Fig 7.24). However, the fibres are branched and divided longitudinally by **intercalated discs** into cells each containing a central nucleus. The discs are composed of several layers of membranes which strengthen the fibres and help to conduct the cardiac impulse (section 10.2.2). Branching of the fibres enables contraction to spread rapidly and smoothly through the heart wall.

**Fig 7.24** Photomicrograph of some cardiac muscle fibres, × 450

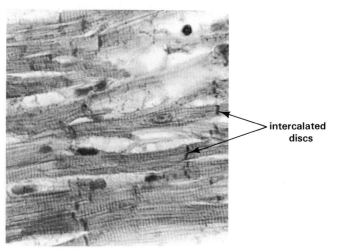

intercalated discs

Cardiac muscle is **involuntary** since its contraction cannot normally be controlled consciously.

Visceral muscle is found in the walls of many tubular structures in the body such as blood vessels (Chapter 10), the gut (Chapter 9) and the urinogenital system (Chapters 13 and 18). Like cardiac muscle, it is involuntary. Visceral muscle cells are not arranged into fibres. The cells are nucleate, and usually tapered with fine longitudinal striations in the cytoplasm (Fig 7.25).

**Fig 7.25** Photomicrograph of some visceral muscle cells, × 1000

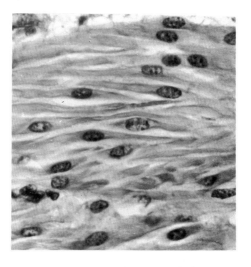

Contraction in visceral muscle involves sliding of actin and myosin filaments. However, they are not so highly organised as in skeletal and cardiac muscle.

Cardiac and visceral muscle can contract and relax rhythmically on their own, even without nervous stimulation. They are **myogenic**. The nerves connected to these muscles control activity rather than initiate it.

# 8 Skeletal and muscle systems

We move our bodies by coordinated contraction and relaxation of muscles attached to the bones of the **skeleton**. Bones articulate with each other at **joints**. Movable joints allow considerable manoeuvrability while retaining the necessary strength to support the body's mass during movement. An important feature of the human skeleton is that it has evolved to allow **bipedal** movement, that is, walking upright on the hind limbs. In this way its design differs from closely related species.

**Fig 8.1** The skeleton, ventral view

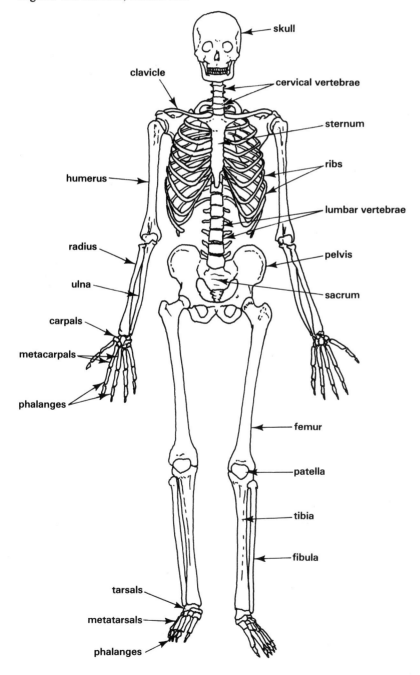

skull
clavicle
cervical vertebrae
sternum
ribs
humerus
lumbar vertebrae
radius
pelvis
ulna
sacrum
carpals
metacarpals
phalanges
femur
patella
tibia
fibula
tarsals
metatarsals
phalanges

## 8.1 The skeleton

The skeleton consists of bone and cartilage (section 8.2). They provide support and protection for the body's organ systems. Many of our blood cells are produced in the marrow of our bones (Chapter 10). Some bones perform specialised functions. For example, the middle ear ossicles conduct sound waves to the cochlea (Chapter 16). The skeleton (Fig 8.1) may be considered in two main parts. The **axial skeleton** consists of the skull, vertebrate and rib cage. The girdles and limbs constitute the **appendicular skeleton**.

## 8.1.1 The skull

The **skull** consists of 22 bones (Fig 8.2 and Table 8.1). Enclosing the brain, eyes and ears is the **cranium**. The human has a cranial capacity of about 1500 cm³ compared with 400–500 cm³ for apes such as chimpanzees and gorillas. The cranial bones include the large, upright **frontal** bones which

**Fig 8.2** Skull (a) frontal view (b) side view

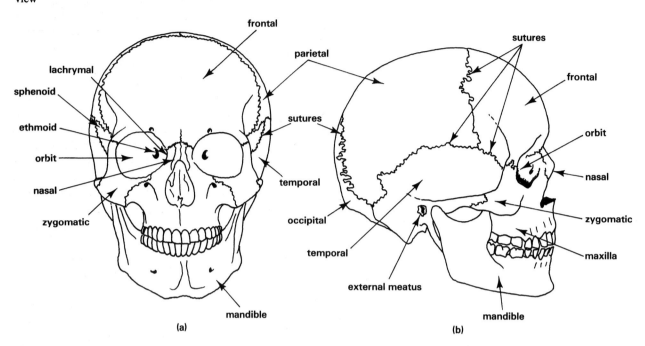

**Table 8.1 Main functions of the skull bones**

| Name | Main function(s) |
|---|---|
| frontal bone | Forehead, anterior cranium, roofs of orbits, anterior floor of cranium. Contains cavities (frontal sinuses) which act as sound chambers, giving resonance to the voice. |
| parietal bones | Sides and roof of cranium. |
| temporal bones | Lower sides and part of floor of cranium. Enclose the inner ears. External auditory meati lead to middle ears. On mastoid processes behind external meati are points of attachment for neck muscles. |
| occipital bone | Posterior and part of floor of cranium. Foramen magnum in base through which passes medulla oblongata. |
| sphenoid bone | Part of the floor of cranium. Part of the floors and sides of orbits. Houses the pituitary body. Part of the walls of the nasal cavities. |
| ethmoid bone | Part of the floor of cranium between the orbits. Main supporting structure of the nasal cavities. |
| nasal bones | Part of the bridge of the nose and upper part of face. |
| maxillae | Upper jaw bone; part of floors of orbits, part of roof of mouth, most of the hard palate, part of walls and floor of nasal cavities. Upper teeth set into maxillae. |
| malar (zygomatic) bones | Cheekbones; part of walls and floors of orbits. |
| mandible | Lower jaw; the only movable skull bone. Condylar processes articulate with temporal bones. Lower teeth set into mandible. Movements enable feeding, chewing and speech. |
| palate bones | Posterior part of hard palate. Part of floor and walls of nasal cavities. Separate nasal and oral cavities. |
| lachrymal bones | Part of medial wall of orbits. |
| inferior turbinated bones | Part of wall of nasal cavities. Like parts of the ethmoid bone, they allow circulation and filtration of inhaled air before it enters the respiratory passages. |
| vomer | Part of nasal septum; together with cartilage it divides nose into left and right. |
| hyoid bone | Found in neck between mandible and larynx. Supports tongue. |

**Fig 8.3** Skull of a newborn baby showing fontanelles

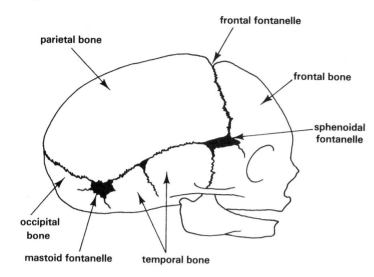

give humans the characteristic high forehead. At the back of the cranium the **occipital** bones lack the ridges for attachment of powerful neck and jaw muscles that occur in apes. The **foramen magnum** is on the underside of the skull, thus allowing considerable reduction of musculature at the nape of the neck. Inserted on the mastoid swellings at the base of the skull are the sternomastoid muscles which allow free rotation of the head. Many of the skull bones are held together by immovable joints called **sutures**.

The skull bones develop from connective tissue and are not fully developed at birth (section 8.2.2). Between the cranial bones of a new-born baby there are membrane-filled spaces called **fontanelles** (Fig 8.3). They gradually close as the cranial bones grow and usually disappear by about 12 months of age.

The shortened **facial** bones give humans a more or less vertical face, apart from the nose, which extends the nasal channel. Inhaled air can be filtered and warmed before entering the respiratory system. The forward pointing **orbits** position the eyes so that their fields of vision overlap, thus allowing binocular vision (Chapter 15). The **jaws** are relatively short compared with other primates and the teeth arranged in semicircular rows. In apes the dental arcade is U-shaped. The lower jaw in humans is not as shortened as the upper and is strengthened on the outside to make the **chin**. In contrast, apes have a strengthening **simian shelf** of bone on the inside of the **mandible**. Their mandibles are larger and more robust than man's, especially at the posterior ramus.

### 8.1.2 Vertebrae

The **vertebrae** make up the slightly S-shaped **spinal column** or backbone. In foetal life the thoracic curve of the backbone develops. After birth the cervical curve appears when a baby learns to hold up its head. When a child begins to walk the lumbar curve develops. There are 33 vertebrae in the five main regions of the spinal column: **cervical**, **thoracic**, **lumbar**, **sacral** and **coccygeal** (Fig 8.1). The vertebrae are modified to perform different functions in each region. However, each vertebra conforms to a common structural pattern (Figs 8.4 and 8.5).

**Fig 8.4** Basic structure of vertebrae

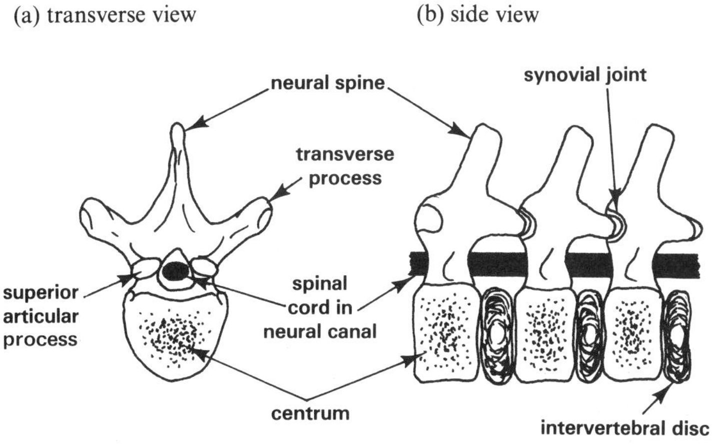

(a) transverse view          (b) side view

131

The main load-bearing portion of a vertebra is a solid disc of bone called the **centrum**. The centra are relatively broad compared with those of some other vertebrate animals, thus helping to support an upright body. The centra of adjacent vertebrae are separated by **intervertebral discs** of cartilage. The discs act as shock absorbers and allow the spinal column to flex and arch. Sometimes the intervertebral discs become distorted, possibly due to excessive pressure from the spinal column, causing a disc to burst. The deformed disc may push against nearby nerve tissue and cause pain. The condition is often called a **slipped disc**.

Projecting from the centrum dorsally is a **vertebral arch** which encloses the **neural canal**. The canal houses the spinal cord (Chapter 14).

Several bony projections arise from the vertebral arch. On each side there is a **transverse process**. These and a dorsal **neural spine** are points of attachment for muscles. Two **superior** (anterior) and two **inferior** (posterior) **articular processes** articulate with the vertebra above and below (Fig 8.4(b)).

The seven cervical vertebrae possess a relatively small centrum and a relatively large neural canal. The very short, forked neural spine provides sites for the insertion of posterior neck muscles which, in humans, are much less developed than in apes. There is a small canal in each transverse process which encloses the vertebral artery and vein and nerve fibres. The first cervical vertebra is called the **atlas** (Fig 8.5(a)). It articulates with the occipital bone of the skull and the head is thus balanced on the neck when

**Fig 8.5** Structures of vertebrae, not drawn to scale. See p.133 for (c)–(f)

(a) atlas  (b) axis

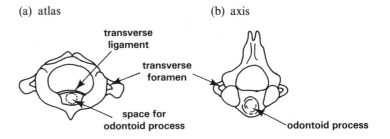

transverse ligament
transverse foramen
space for odontoid process
odontoid process

the body is upright. The second cervical vertebra is called the **axis** (Fig 8.5(b)). Its centrum has an anterior projection called the **odontoid process** which fits into a space in the atlas. The atlas has no centrum. In its place is a space which accomodates the odontoid process. The arrangement acts as a pivot allowing the head to rotate sideways (Figs 8.5(c) and 8.17(d)). The odontoid process is separated from the spinal cord in the neural canal by a **transverse ligament**. However, sudden jolting of the neck may push the odontoid process into the spinal cord or even the medulla oblongata of the hind brain (Fig 14.37). The result is usually sudden death. The use of head restrainers on car seats has reduced the number of deaths caused by this type of injury in road traffic accidents. The twelve thoracic vertebrae have relatively long, backward-pointing neural spines (Fig 8.5(d)). They also possess relatively long transverse processes with **facets** for the attachment of the ribs. Facets are also found on the centra.

The five lumbar vertebrae are relatively bulky (Fig 8.5(e)). They have relatively short but sturdy neural spines and transverse processes to which are attached the powerful muscles of the lower back.

The five sacral vertebrae are fused into a single triangular structure called the **sacrum** (Fig 8.5(f)). The fusion of its constituent bones provides the sacrum with strength and rigidity. It acts as a firm point of attachment for the pelvic girdle (section 8.1.4).

The **coccyx** consists of four fused coccygeal vertebrae attached to the posterior apex of the sacrum (Fig 8.5(f)). The coccyx performs no known function in the human skeleton and is a vestigial tail.

(c) atlas–axis articulation, side-view

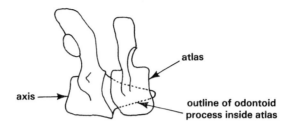

(d) thoracic vertebra    (e) lumbar vertebra

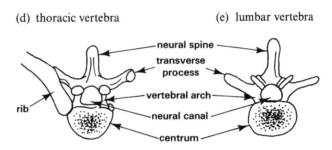

(f) sacrum

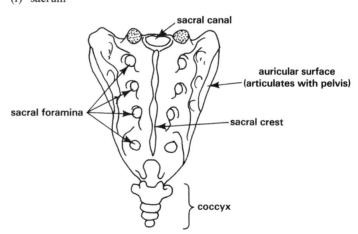

How is the basic plan of vertebrae modified in different regions of the vertebral column to perform different skeletal functions?

### 8.1.3 Rib cage

The rib cage of the **thorax** encloses the heart and the lungs. Breathing is partly brought about by rhythmical movements of the **ribs**, resulting from the actions of **intercostal muscles** (Fig 12.8). The rib cage also provides support for the pectoral girdle and hence the arms.

There are twelve pairs of ribs, each articulating with a thoracic vertebra (Fig 8.5(d)). The first ten pairs of ribs are attached to the **sternum** (breastbone) by strips of cartilage called the **costal cartilages** (Fig 8.6). The first seven pairs of ribs are called **true ribs** because their costal cartilages attach them directly to the sternum. The costal cartilages of the next three pairs of ribs attach to those of the seventh rib pair. For this reason the eighth, ninth and tenth rib pairs are called **false ribs**. The two most posterior pairs of ribs are not attached to the sternum at all. They are called **floating ribs**.

The sternum consists of three parts. Most anteriorly is the **manubrium**. The clavicles (collar bones) articulate with the manubrium and the scapula (section 8.1.4). The costal cartilages of the first rib pair also attach to the manubrium. The other costal cartilages are attached to the **body** of the sternum. The small **xiphoid process** projects from the posterior edge of the sternal body. Bone marrow samples are often taken from the sternum. The technique is called **sternal puncture** and involves pushing a broad needle into the sternal marrow.

**Fig 8.6** Rib cage, ventral view. Ribs are numbered 1 to 12

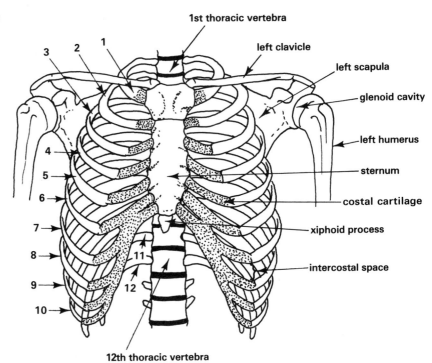

### 8.1.4 Girdles

The arms articulate with the **pectoral girdle**, the legs with the **pelvic girdle**.

The pectoral girdle consists of two ventral **clavicles** (collar bones) and two dorsal **scapulae** (shoulder blades). The only bony attachment to the rest of the skeleton is by the articulation of the clavicles with the sternum (Fig 8.6). The scapulae are attached to the dorsal rib cage by a complex of muscles. The freedom of movement thus allowed to the arms reflects the swinging habit of our tree-living ancestors.

The pelvic girdle is more firmly attached to the spinal column. This enables the mass of the body to be transmitted to the ground through the pelvic girdles and the legs. The pelvic girdle consists of two **pelvic bones** which attach to the sacrum dorsally (Fig 8.5(f)) and to one another ventrally at the **symphysis pubis**. The pelvic bones are each formed in the embryo by the fusion of three bones called the **ilium, ischium** and **pubis** (Fig 8.7). The pelvic girdle, sacrum and coccyx comprise the **pelvis**.

**Fig 8.7** Pelvic girdle, ventral view

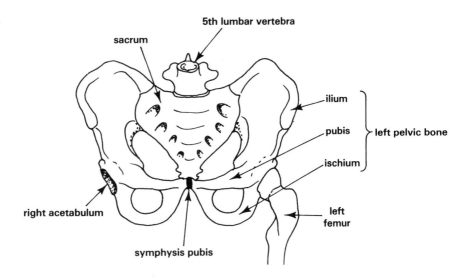

The broad ilium provides a large surface for the attachment of muscles which enable us to adopt a bipedal gait. Especially important are the relatively large gluteal muscles. They give us buttocks, another typical human feature. Also attached to the ilium are the abdominal muscles which carry much of the weight of the viscera of the abdomen and the iliacus muscle, a flexor of the hip.

### 8.1.5 Limbs

Our arms and legs are **pentadactyl limbs** (Fig 8.8). The upper part of each limb contains a single long bone, the **humerus** in the arm and the **femur** in the leg. The head of the humerus articulates with a depression in the scapula

**Fig 8.8** Pentadactyl limb, general plan

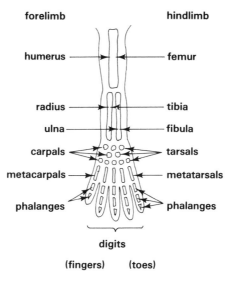

called the **glenoid cavity** (Figs 8.6 and 8.9). The distal end of the humerus articulates with the **ulna** and **radius** at the **elbow**. In the leg the femur (thigh bone) articulates with the pelvic girdle. The head of the femur is ball-shaped and fits into a cavity called the **acetabulum** in the pelvic bone (Figs 8.7 and 8.10). The distal end of each femur articulates with the tibia at the **knee**. The femur is straight, but its articular surface at the knee is at an angle. Hence, although the heads of the femurs are far apart, the knees can be held together. Our lower legs and feet are thus immediately under our centre of gravity when we are standing.

The two bones of the forearms are the ulna and radius. The ulna is the longer of the two and includes the **olecranon process** which makes the elbow. The possession of two freely-articulating bones in the forearm allows a twisting action at the wrist such as may occur when using a screwdriver. The human forearm is much shorter than that of apes. Its muscles are also less strong. For this reason we find it difficult to pull our bodies upwards when swinging from a bar. In contrast, apes can do so with ease.

The two bones of the lower leg are the **tibia** and **fibula**. The tibia (shin bone) is the longer of the two and is the main load-bearing bone of the lower leg. Its upper end articulates with the femur at the knee and its lower end with the bones of the ankle. The **patella** (kneecap) is a small bone that develops in the tendon of the **quadriceps extensor** muscle. This large muscle connects the pelvic girdle and femur to the patella and tibia. It serves to keep the knee extended and checks the forward momentum of the body when walking or running. Contraction of the calf muscles, especially the soleus which extends from the heel to the tibia, creates the propulsive thrust for bipedal movement.

Each wrist contains eight small bones called **carpals**, joined together by ligaments. The five bones in the palms of the hands are called **metacarpals**. The fingers each contain three bones called **phalanges**. The second finger is the largest and the **index** finger is about as long as or longer than the third. The **pollex** (thumb) contains two phalanges (Fig 8.9). The joint between the first carpal and metacarpal enables the thumb to be fully opposed to the fingers. No other primate can do so. This arrangement enables us to hold and manipulate objects with great dexterity. It has enabled humans to manufacture and use elaborate tools and instruments and to write and draw with great accuracy. Such activities distinguish human life from that of all other animals.

**Fig 8.9** Skeleton of the arm, dorsal view

The bones in our limbs are arranged in a pentadactyl pattern which is common to all classes of vertebrate animals. What adaptations of our limbs enable us to perform delicate manipulations with our hands and to walk upright?

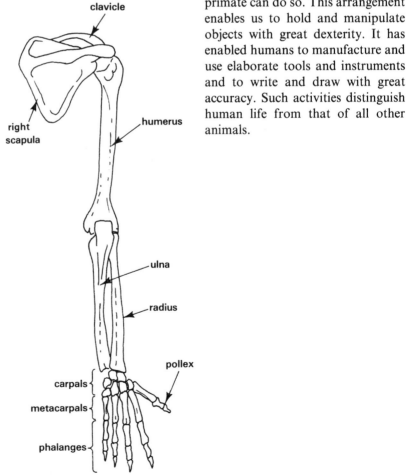

**Fig 8.10** Skeleton of the leg, ventral view

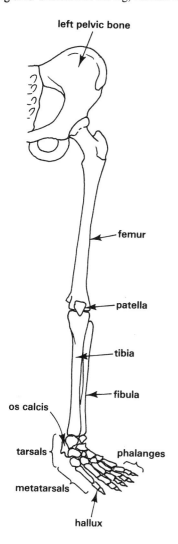

left pelvic bone

femur

patella

tibia

fibula

os calcis

tarsals

phalanges

metatarsals

hallux

Each ankle contains seven small bones called **tarsals**. The largest of these, called the **os calcis** (calcaneum) constitutes the **heel**. The main body of the foot contains five **metatarsals** from which project the toes at the extremities. Like the fingers, each toe contains three phalanges, except the **hallux** (big toe) which, like the thumb, has only two (Fig 8.10). Man is unique in having a big toe longer than his other toes. The articular surface at the ankle joint is at right angles to the tibia. Consequently when we stand, our weight is transferred from the tibia partly forwards to the metatarsals and partly backwards to the os calcis. The foot is arched and acts as an efficient weight-bearing organ. When we walk, our toes remain in contact with the ground, thus maintaining the forward momentum.

## 8.2 Cartilage and bone

The skeleton is mainly composed of **bone**. In the embryo, however, the first skeletal material to develop is **cartilage**. Embryonic cartilage ossifies during development to become bone. Some bones, called **membrane bones**, arise from embryonic connective tissue and do not begin as cartilage.

### 8.2.1 Cartilage

**Hyaline cartilage** consists of a matrix of **chondrin**, containing **collagen** (section 7.1.2). Chondrin is secreted by cells called **chondroblasts** which lie in the matrix in small clusters (Fig 8.11). Cartilage is bounded by a fibrous layer called the **perichondrium** in which there are blood vessels. It is here that new cartilage is laid down. In adults some hyaline cartilage remains in the skeleton. One of its main functions is to protect bones from abrasion where they articulate at joints (Figs 8.12 and 8.16). Hyaline cartilage also joins the ribs to the sternum (Fig 8.6) and allows limited flexing of the ribs during breathing. Rings of hyaline cartilage in the trachea and bronchi support the air passages and keep them open (Fig 12.3). **Elastic cartilage** contains yellow elastic fibres arranged in all directions in the matrix. The fibres allow greater flexibility compared with hyaline cartilage. Elastic cartilage is found in the epiglottis and in the pinna of the ear. **Fibro-cartilage** has many white fibres of collagen in the matrix. They resist stretching and give this type of cartilage considerable strength. Fibro-cartilage is found in the intervertebral discs of the spinal column (Fig 8.4(b)).

**Fig 8.11** Photomicrograph of hyaline cartilage, × 100

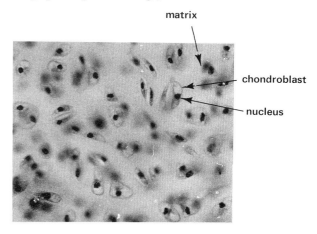

matrix

chondroblast

nucleus

Many parts of our skeleton contain cartilage. What special properties does cartilage have which enables those parts of the skeleton to perform their functions?

### 8.2.2 Bone

Fig 8.12 Diagram of the main parts of a bone

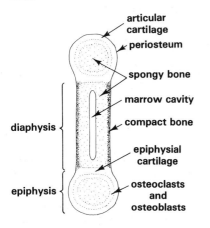

Fig 8.12 Diagram of the main parts of a bone

Most bones have a shaft called the **diaphysis**. It is usually hollow and contains the bone **marrow** which produces most of the various kinds of blood cells (Chapter 10). Bone is covered by a tough, fibrous membrane called the **periosteum**. At the ends, bones are usually expanded to form the **epiphyses**. These are sites for the attachment of tendons and articulation with other bones (Fig 8.12).

The substance of bones may be **spongy** or **compact** and is arranged to provide maximum strength with minimum mass. The matrix of the bone is a mesh of collagen fibres impregnated with concentric **lamellae** of calcium salts, especially phosphate (section 16.1.5). In the matrix is a system of **Haversian canals** arranged parallel to the longitudinal axis of the bone (Fig 8.13). They contain nerves, blood and lymphatic vessels and connect with the periosteum and the marrow. Bone cells called **osteocytes** are

**Fig 8.13** Photomicrograph of compact bone, × 100

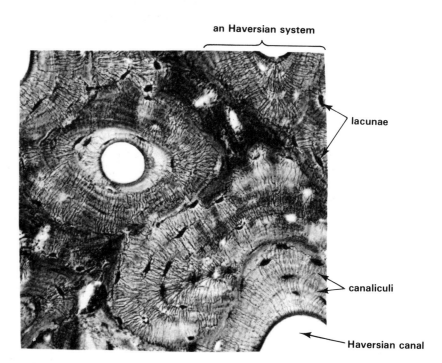

found in concentric rings in the lamellae. The cells lie in cavities called **lacunae** which are linked by a series of fine **canaliculi**. The osteocytes secrete the bone matrix. The production of bone from embryonic cartilage is called **endochondral ossification**. The limb bones are formed in this way. Chondroblasts multiply and become arranged in longitudinal columns. Then, starting from the centre, the chondroblasts deposit calcium salts in the matrix. The cartilage thus becomes ossified. Many of the chondroblasts degenerate and are replaced by large, amoeboid cells called **osteoclasts**. They cause erosion of much of the calcified cartilage which then becomes invaded by blood vessels. Cells called **osteoblasts** now secrete a series of bony columns called **trabeculae** in the centre of the bone shaft. Ossification gradually spreads outwards towards the periosteum. At the same time osteoblasts just beneath the periosteum produce dense bone. Some of the osteoblasts are enclosed in lacunae and become osteocytes (Fig 8.14).

**Fig 8.14** Diagram of the stages of ossification

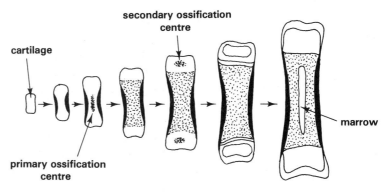

The production of bone from embryonic connective tissue is called **intramembranous ossification**. The bones of the skull are formed in this way (section 8.1.1). Fine bundles of white fibres appear in the connective tissue matrix and calcium salts are deposited around them. Osteoclasts then erode much of the calcified matrix. Osteoblasts now form bony trabeculae, the periosteum appears and new bone is laid down beneath it. Osteocytes are active throughout life. Their activities help the healing of bones which have been fractured.

## 8.3 Joints and movement

### 8.3.1 Joints

The bones of the skeleton articulate at **joints**. The actions of muscles bring about the movement of bones at a **synovial joint** (Fig 8.15). The articulating surfaces of the bone are covered with pads of articular cartilage. The joint is enclosed in a fibrous **synovial capsule** which, with the aid of ligaments, keeps the bones together. Lining the capsule is a **synovial membrane**. It secretes **synovial fluid** into the capsule's cavity. The synovial fluid is a lubricant allowing movement of the bones at the joint.

**Fig 8.15** Diagram of a joint

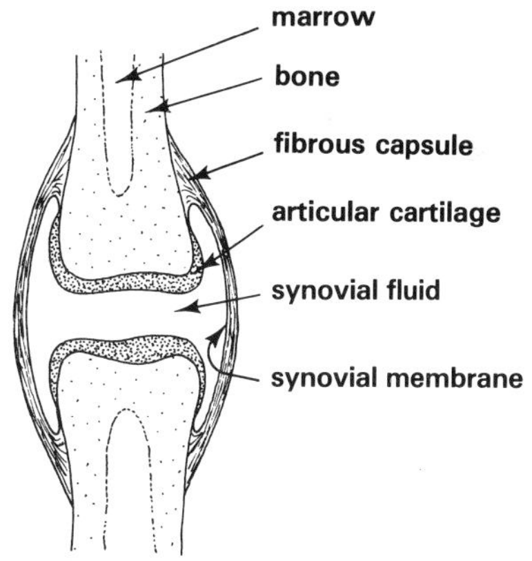

**Fig 8.16** Weeping lubrication (from McNeil-Alexander, after McCutchen)

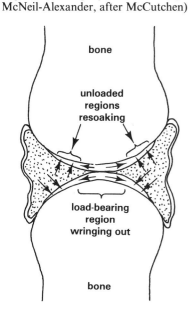

The exact mechanism of synovial joint action is not fully understood. Lewis and McCutchen have suggested the theory of **weeping lubrication**. The articular cartilage acts like a sponge, absorbing synovial fluid from the synovial cavity. When a load is applied to the joint the articular cartilages of the bones may be pushed together (Fig 8.16). Synovial fluid is squeezed out of the articular cartilage at the point of contact. Since the pores in the cartilage are very small, the fluid is wrung out only very slowly. Consequently the synovial fluid and not the cartilage bears most of the load applied to the joint. There are several different kinds of synovial joint depending on the ways in which the bones move against each other.

**i. Gliding movements** occur, for example, between the ribs and thoracic vertebrae (Fig 8.5(d) and 8.17(a)). Back and forth or side to side movements occur between the articulating bones.

**ii. Hinged joints** allow movements which alter the angle in one plane between two articulating bones. An example is the elbow (Figs 8.9 and 8.17(b)).

**iii. Ball and socket joints** involve the ball-shaped end of one bone moving within a cup-shaped cavity of another bone. Movement including rotation is possible in all directions. An example is the hip joint between the femur and pelvic girdle (Figs 8.7, 8.10 and 8.17(c)).

**iv. Pivot joints** allow rotation. An example is the articulation of the axis and atlas in the neck. It allows rotation of the head (Figs 8.5(c) and 8.17(d)). Another example is the articulation of the elbow between the radius and ulna. It allows the lower arm to be twisted (section 8.1.5).

**Fig 8.17** Main types of synovial joint    (a) gliding, such as between vertebrae    (b) hinge, such as elbow

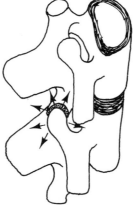

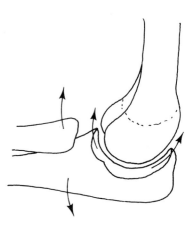

(c) ball and socket, such as hip    (d) pivot, such as atlas–axis

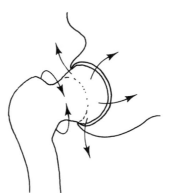

atlas    axis

odontoid process

How does the structure of joints allow a variety of different skeletal movements?

Where free movement does not occur, the bones are separated by connective tissue as in the sutures of the skull (Fig 8.2), or by pads of fibro-cartilage as in the joints between vertebrae (Fig 8.4(b)).

The bones of movable joints are held together by **ligaments**. They are made of white fibrous tissue containing non-stretchable collagen fibres (Fig 8.18).

**Fig 8.18** Elbow joint showing ligaments

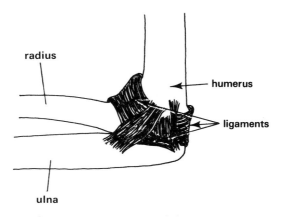

### 8.3.2 Movement

The movement of bones is brought about by the actions of muscles. Muscles are attached at each end to bones by collagenous fibres called **tendons**. Contraction of the muscle moves the bones in a manner dictated by the joint where they articulate. Most skeletal movement is controlled by **antagonistic** muscles. They are opposing sets of muscles which act in the opposite way at any one time. For example, the upper and lower arm articulate at the elbow. To raise the lower arm, **flexion** requires contraction of the **biceps** muscle. The biceps connects the scapula to the radius. At the same time the **triceps** muscle relaxes (Fig 8.19(a)). The triceps connects the scapula and humerus to the ulna. When the biceps relaxes and the triceps contracts, **extension** of the lower arm occurs (Fig 8.19(b)).

**Fig 8.19** Movement of the forearm showing (a) flexion and (b) extension

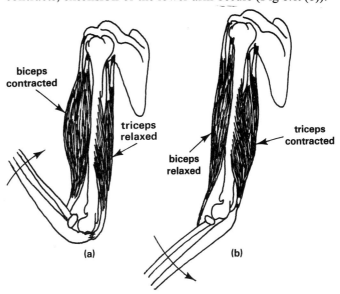

Similar antagonistic muscle activity occurs at other joints. The coordination of such activity often involves **reflexes** (section 14.2.2). Reflex pathways to antagonistic muscles include inhibitory and excitatory synapses so that contraction is inhibited in one muscle while it is stimulated in the other (Fig 14.33).

The movements of bones can be described in terms of **levers**. Each joint acts as a **fulcrum** or **pivot**. The force used to move a lever is called the **effort**. It is provided by muscle contraction. The force to be overcome by the effort is called **resistance**. It is provided by the action of gravity on the mass of the body and any additional masses being lifted by the skeleton. Friction at the joint also adds minimally to resistance. Levers are classified according to the relative positions of the effort, resistance and fulcrum.

**i. First class levers** have the fulcrum in the middle. The effort and resistance are applied at opposite ends. An example is the movement of the skull on the spinal column (Fig 8.20).

**ii. Second class levers** possess the fulcrum at one end, the resistance in the middle. The effort is applied at the other end. An example is seen when raising the body on the toes (Fig 8.21).

**iii. Third class levers** are the commonest type in the body. The fulcrum is at one end of the lever and the resistance at the other end. The effort is applied to the middle. An example is the flexion of the forearm (Figs 8.19(a) and 8.22).

**Fig 8.20** First class lever    **Fig 8.21** Second class lever

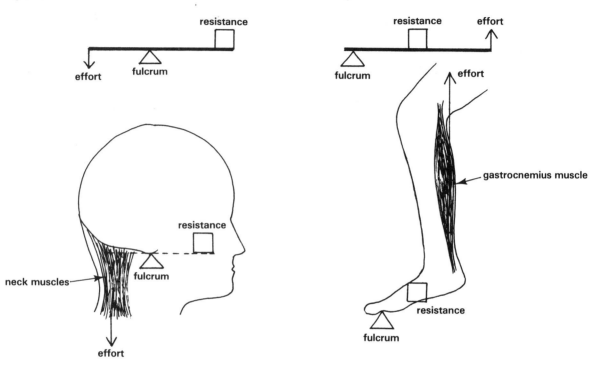

**Fig 8.22** Third class lever

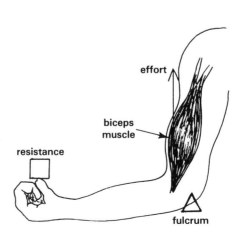

The advantage of using levers to move large masses has long been realised by engineers. Generally the further away from the fulcrum the effort is applied, the less is the force required to raise the mass; that is, the greater the **leverage**. This phenonemon can be illustrated by the **principle of moments** (Fig 8.23).

**Fig 8.23** Principle of moments.
A 0.5 kg load is applied to the lever 20 cm from the fulcrum, **F**

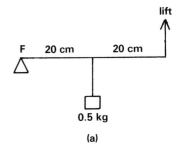

(a)

(a) A lifting force, $x$, is applied a further 20 cm away from the load at the other end of the lever from the fulcrum. The lifting force just balances the load when

$$x \times 40 = 0.5 \times 20$$

i.e. $\quad x = \dfrac{0.5 \times 20}{40} = 0.25 \text{ kg}$

(b) If a lifting force, $y$, is applied 15 cm from the load then it just balances the load when

$$y \times 35 = 0.5 \times 20$$

i.e. $\quad y = \dfrac{0.5 \times 20}{35} = 0.28 \text{ kg}$

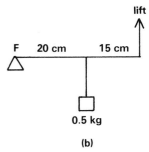

(b)

# 9 Digestive system

A constant supply of organic and inorganic nutrients is required to sustain the many and complex activities which take place in the body.

Organic nutrients such as proteins and lipids are the raw materials for the synthesis of cellular components in growth and repair. The energy needed to drive synthetic processes comes mainly from the oxidation of sugars in cellular respiration (Chapter 5). We also require energy for movement and to help maintain a constant body temperature (Chapter 17). Vitamins are organic nutrients which participate in a wide range of body functions.

Inorganic nutrients (minerals) are needed to create and maintain electrical potentials across the membranes of nerve and muscle cells (Chapter 14). The coagulation of blood (Chapter 10), activation of enzymes (Chapter 3), growth of teeth and bones and maintenance of the osmotic potential of the body fluids, are among the many other essential activities which depend on a supply of minerals.

We obtain nutrients from our diet which consists of food and water. Food is a mixture of organic and inorganic nutrients. Water contains dissolved minerals.

## 9.1 Nutritional requirements

Our nutritional requirements can be divided into three categories, **organic nutrients, vitamins** and **inorganic nutrients**.

### 9.1.1 Organic nutrients

The chemical nature of the main organic constituents of the diet is described in Chapter 2.

**1 Carbohydrates**

**Starch, sugars** and **cellulose** are the main carbohydrates we eat. About 50 % of the energy from the average human diet in Britain comes from starch and sugars. However, a larger proportion of energy may be derived from dietary carbohydrates in people who lead very active lives (Chapter 5). Reserves of carbohydrate are stored as **glycogen** in the liver and skeletal muscles. Cellulose is an important constituent of **dietary fibre** (section 9.4.4).

**2 Proteins**

Proteins are necessary in the diet as a source of **amino acids**. Amino acids may also be respired, but only about 5% of the human body's energy requirement normally comes from amino acids. Amino acids are used mainly for the synthesis of our proteins. A constant supply of amino acids is needed for the production of enzymes and the proteins in cell membranes. Growth and repair especially depend on the presence of proteins in the diet. Proteins are also made in secretory cells for export to other parts of the body. Plasma proteins in the blood, for example, are made in the liver. Antibodies are proteins made in the lymph nodes, and some hormones are short-chain polypeptides. Transamination in the liver enables **non-essential amino acids** to be made from others (section 9.5.2). **Essential amino acids** cannot be made in this way and are vital constituents of the diet (Table 9.1). The average adult man requires about $440\,\mu g$ of protein

**Table 9.1 Essential amino acids**

| | | |
|---|---|---|
| lysine | phenylalanine | leucine |
| threonine | methionine | isoleucine |
| tryptophan | histidine | valine |

Why do men require more protein in their diet than women? Why is more protein required by women during pregnancy? Why do children require a lot of protein in their diet?

$kg^{-1}$ body mass each day. Women need about 10% less as their bodies contain less tissue in which protein synthesis occurs. More protein is required during pregnancy for foetal development. Children have a relatively higher protein requirement than adults because they are constantly forming new tissues throughout the body as they grow and develop.

## 3 Lipids

**Fats** and **oils** are the main lipids we eat. They can be used as respiratory substrates and account for about 45% of the energy in the average diet in Britain. Some lipids are sources of **essential fatty acids** such as arachidonic acid. Many fatty acids are **non-essential** because they can be synthesised in our bodies. The synthesis of cell membranes requires lipids as well as proteins. The fatty membranes of the myelin sheath around certain nerve cells are important in impulse transmission (Chapters 7 and 14). Surplus fats are stored beneath the skin and around internal body organs such as the kidneys. As well as being a foodstore, fat acts as a heat insulator which helps to keep a constant body temperature (Chapter 17).

### 9.1.2 Vitamins

Vitamins are a range of substances necessary for good health. They affect a variety of body functions and are required only in trace quantities. Some vitamins are soluble in fats and are absorbed from the gut dissolved in lipids. **Fat-soluble vitamins** can be stored in the liver and therefore need not be consumed regularly. Other vitamins are water-soluble and are absorbed into the blood dissolved in water. **Water-soluble vitamins** are readily excreted in urine and must, therefore, be consumed regularly.

### 1 Fat-soluble vitamins

**i. Vitamin A (retinol)** is made from the plant pigment $\beta$-carotene and is needed for synthesis of the visual pigments. Deficiency of vitamin A can lead to night blindness (Chapter 15). Vitamin A is also essential for maintaining epithelial tissues in good condition. Infection of the epithelium of the respiratory tract often results from serious lack of vitamin A in the body.

**ii. Vitamin D** is a group of several sterols including **ergocalciferol** (vitamin $D_2$). Ergocalciferol is produced by the action of ultraviolet radiation from the sun on ergosterol in the skin. Vitamin $D_2$ stimulates the absorption of calcium from the gut. Deficiency of vitamin $D_2$ can lead to the release of calcium from the bones under the influence of parathyrin (Chapter 16). Malformation of the skeleton due to vitamin $D_2$ deficiency is called rickets.

**iii. Vitamin E** The function of vitamin E in the human body is not fully understood. However, a deficiency of it may lead to degeneration of muscle tissue, destruction of liver cells and anaemia.

**iv. Vitamin K** is necessary for the production of prothrombin in the liver. Deficiency of vitamin K thus leads to impaired coagulation of the blood (Chapter 10).

### 2 Water-soluble vitamins

**i. Vitamin B** is a complex of several vitamins, most of which are components of co-enzymes (Chapter 3). Vitamin $B_1$ (**thiamin**) is required to make the co-enzyme of carboxylase enzymes. Deficiency of vitamin $B_1$ leads to beri-beri which is characterised by paralysis, oedema and heart failure.

Vitamin $B_2$ (**riboflavin**) is part of the co-enzyme FAD used as a hydrogen acceptor in Krebs' tricarboxylic acid cycle (Chapter 5). Vitamin $B_6$ is required for protein metabolism including transamination (section 9.5.2). Vitamin $B_{12}$ (**cyanocobalamin**) is necessary for the normal maturation of red blood cells and the health of the nervous system and mucosae. Deficiency of vitamin $B_{12}$ causes anaemia and degeneration of the nerve cord (Chapter 10). **Nicotinamide** and **nicotinic acid** are other B-group vitamins. They are required to make the hydrogen acceptor co-enzymes NAD and NADP. Folic and pantothenic acids also belong to this group. **Folic acid** is essential for normal red blood cell maturation. **Pantothenic acid** is a component of acetyl co-enzyme A (Chapter 5).

**ii. Vitamin C (ascorbic acid)**  is thought to act as an electron carrier in respiration. It also stimulates synthesis of collagen fibres. Deficiency of vitamin C causes scurvy, symptoms of which include breakdown of connective tissues and blood vessels. Scurvy was once common among sailors who were deprived of fresh fruit containing vitamin C while on long voyages. Bleeding of the gums and loosening of the teeth are symptoms of scurvy. Internal bleeding also occurs in many tissues. Where it occurs near the surface, bruised patches appear in the skin, a symptom often seen in elderly people whose consumption of fresh fruit and vegetables may be inadequate. Recent evidence suggests that a high daily intake of vitamin C in the diet may have many beneficial effects, including reduction of blood cholesterol.

### 9.1.3 Inorganic nutrients

**Minerals** are inorganic nutrients which participate in a wide variety of body functions. Some minerals called **macronutrients** are needed in relatively large quantities. Others called **trace elements** are needed in small amounts.

### 1 Macronutrients

**Sodium** ions are more abundant than any other cation in the body fluids. Sodium ions play an important role in the electrochemical activities of nerves and muscles (Chapter 14). Sodium and chloride ions are the main determinants of the osmotic potential of tissue fluid and blood plasma. **Potassium** ions are the most abundant of cations in cells. Potassium ions are involved, like sodium, in nerve and muscle impulse transmission. **Calcium** and **phosphate** ions are the main mineral constituents of bone and teeth. Calcium ions are also required for coagulation of blood and for muscle contraction.

### 2 Trace elements

Trace elements must be provided in the diet to maintain health. Large quantities of trace elements can be poisonous. **Cobalt** is a constituent of vitamin $B_{12}$ (cyanocobalamin) and is vital for the production of haemoglobin and red blood cells. **Iodine** is a constituent of the hormone thyroxin and is absorbed from the blood by the thyroid gland. Lack of iodine in the diet leads to decreased thyroxin production and the enlargement of the thyroid gland called a goitre (Chapter 16). **Copper** is vital for the activation of a variety of enzymes and for the production of haemoglobin. **Iron** is a constituent of haemoglobin and myoglobin (Chapter 12) and cytochromes (Chapter 5). Reserves of iron are stored in the liver (section 9.5.2). **Zinc** is an activator of several enzymes, including carbonic anhydrase, and is required for synthesis of the hormone insulin (Chapter 16).

### 9.1.4 Diet

The food we eat constitutes our **diet**. Different foods contain different nutrients. To obtain a balanced diet, that is, all the nutrients in the amounts required to maintain health, we must eat a variety of different foods.

Among the most important foods are **cereals** such as wheat, maize and rice. They contain many vitamins, much starch and protein, but are deficient in lysine, an essential amino acid. Cereals are also poor sources of minerals (Table 9.2).

**Table 9.2 Nutritional make-up (per 100 g) of some of the main constituents of the human diet (data from Taylor, after *Manual of Nutrition*, HMSO 1976)**

| Item of diet | Energy/ kJ | Protein/ g | Fat/g | Carbohy-drate/g | Minerals (Ca + Fe)/mg | Vitamins A/$\mu$g | D/$\mu$g | B$_1$/mg | B$_2$/mg | Nicotinic acid/mg | C/mg |
|---|---|---|---|---|---|---|---|---|---|---|---|
| apples | 197 | 0.3 | 0 | 12.0 | 4.3 | 5 | 0 | 0.04 | 0.02 | 0.1 | 5 |
| bananas | 326 | 1.1 | 0 | 19.2 | 7.4 | 33 | 0 | 0.04 | 0.07 | 0.8 | 10 |
| beef | 940 | 18.1 | 17.1 | 0 | 8.9 | 0 | 0 | 0.06 | 0.19 | 8.1 | 0 |
| bread, white | 1068 | 8.0 | 1.7 | 54.3 | 101.7 | 0 | 0 | 0.18 | 0.03 | 2.6 | 0 |
| bread, wholemeal | 1025 | 9.6 | 3.1 | 46.7 | 31.0 | 0 | 0 | 0.24 | 0.09 | 1.9 | 0 |
| butter | 3006 | 0.5 | 81.0 | 0 | 15.2 | 995 | 1.25 | 0 | 0 | 0.1 | 0 |
| cabbage | 66 | 1.7 | 0 | 2.3 | 38.4 | 50 | 0 | 0.03 | 0.03 | 0.5 | 23 |
| carrots | 98 | 0.7 | 0 | 5.4 | 48.6 | 2000 | 0 | 0.06 | 0.05 | 0.7 | 6 |
| cheese, cheddar | 1708 | 25.4 | 34.5 | 0 | 810.6 | 420 | 0.35 | 0.04 | 0.50 | 5.2 | 0 |
| cod | 321 | 17.4 | 0.7 | 0 | 16.3 | 0 | 0 | 0.08 | 0.07 | 4.8 | 0 |
| cream, double | 1848 | 1.8 | 48.0 | 2.6 | 65.0 | 420 | 0.28 | 0.02 | 0.08 | 0.4 | 0 |
| eggs | 612 | 12.3 | 10.9 | 0 | 56.1 | 140 | 1.50 | 0.09 | 0.47 | 3.7 | 0 |
| liver | 1020 | 24.9 | 13.7 | 5.6 | 22.8 | 6000 | 0.75 | 0.27 | 4.30 | 20.7 | 20 |
| margarine | 3019 | 0.2 | 81.5 | 0 | 4.3 | 900 | 8.00 | 0 | 0 | 0.1 | 0 |
| milk | 274 | 3.3 | 3.8 | 4.8 | 120.1 | 40 | 0.03 | 0.04 | 0.15 | 0.9 | 1 |
| oranges | 150 | 0.8 | 0 | 8.5 | 41.3 | 8 | 0 | 0.10 | 0.03 | 0.3 | 50 |
| parsnips | 210 | 1.7 | 0 | 11.3 | 55.6 | 0 | 0 | 0.10 | 0.09 | 1.3 | 15 |
| peas | 208 | 5.0 | 0 | 7.7 | 14.2 | 50 | 0 | 0.25 | 0.11 | 2.3 | 15 |
| potatoes, boiled | 339 | 1.4 | 0 | 19.7 | 4.5 | 0 | 0 | 0.08 | 0.03 | 1.2 | 10 |
| rice | 1531 | 6.2 | 1.0 | 86.8 | 4.4 | 0 | 0 | 0.08 | 0.03 | 1.5 | 0 |
| sugar, white | 1680 | 0 | 0 | 100.0 | 1.0 | 0 | 0 | 0 | 0 | 0 | 0 |
| tomatoes | 52 | 0.8 | 0 | 2.4 | 13.4 | 117 | 0 | 0.06 | 0.04 | 0.7 | 21 |

**Nuts** and **pulses** (leguminous seeds) such as beans, lentils and peas are valuable sources of protein. Soya beans and groundnuts also contain much oil.

**Vegetables** include cabbage, carrots, cassavas, cauliflowers, potatoes, swedes, turnips and yams. Most root vegetables contain much carbohydrate and fibre but are poor in protein, fat and vitamins. Leafy vegetables contain many vitamins and minerals and are a good source of fibre too. Over-cooking, however, often destroys and removes water-soluble vitamins (section 9.1.2).

Much of the carbohydrate in the diet of the Western world is provided by refined **sugar** which is little more than pure carbohydrate. Honey, brown sugar, treacles and molasses are less refined and contain many minerals.

Most **fruits** contain much carbohydrate and water; many also provide vitamin C.

**Fats** and **oils** are concentrated sources of energy. They may also contain fat-soluble vitamins (section 9.1.2). Generally fats of animal origin such as butter, lard and suet are more saturated than those from plants (Chapter 2). The high incidence of coronary disease in many parts of the Western world has been linked to excessive dietary intake of saturated fats. In much of Africa and southern Italy where plant oils such as olive oil, corn oil and sunflower oil are consumed instead, coronary disease is rare.

**Meat** and **fish** are valuable sources of high quality protein. This means they provide all the essential amino acids we require. Meat and liver contain much iron as well as other minerals and many vitamins. **Eggs** are rich in protein, fat and vitamins. The human diet often contains **milk** from cows, goats and sheep. Milk contains carbohydrate (lactose), fat, protein, vitamins and minerals, especially calcium and phosphate. It provides all the nutritional requirements for a young baby.

## 9.2 Dentition

Ingested food first enters the mouth where it is chewed. **Chewing** makes it easier to swallow food and helps the action of digestive enzymes secreted by the various glands which open into the gut. The cutting and grinding of food which takes place when it is chewed is the function of the **teeth**.

### 9.2.1 Structure and development of teeth

**Tooth buds** in the developing jaw give rise to the teeth. The buds each contain an **enamel organ** in which **ameloblast cells** produce **enamel** made of calcium salts. Enamel is the hardest material in the body and coats the protruding surfaces called the **crowns** of the teeth. Inside the cup-shaped enamel organ, **odontoblast cells** of the dental papilla produce **dentine** (Fig 9.1). During tooth development, sockets of bone grow around the **roots** of the teeth (Fig 9.2).

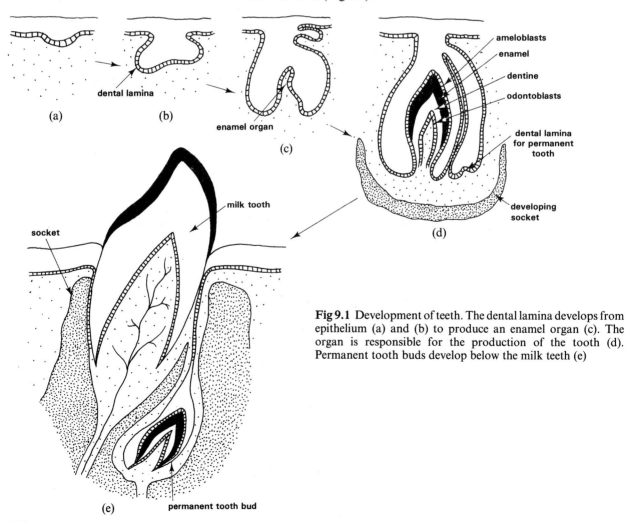

**Fig 9.1** Development of teeth. The dental lamina develops from epithelium (a) and (b) to produce an enamel organ (c). The organ is responsible for the production of the tooth (d). Permanent tooth buds develop below the milk teeth (e)

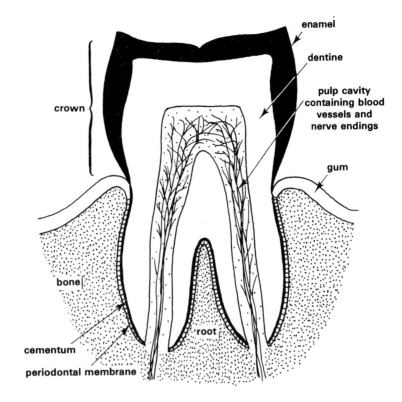

**Fig 9.2** Diagrammatic vertical section through a molar. The root is held in the bony socket by cementum joined to a fibrous connective layer called the periodontal membrane. Elasticity of the fibres in this membrane allows limited movement of the root in the socket, thus acting as a 'shock absorber' during chewing

During our lives we have two sets of teeth. The first are called **milk teeth**. They are later replaced by the **permanent teeth**. Tooth replacement begins usually at about five years of age and takes several years to complete. The final molars are called the **wisdom teeth** and usually emerge in early adult life. Permanent teeth develop from separate rows of tooth buds which arise under the buds from which the milk teeth grow (Figs 9.1 and 9.3).

**Fig 9.3** X-ray photograph showing the dentition of an 8-year-old child. Note the presence of milk (deciduous) teeth as well as permanent teeth (courtesy Mr Gould, Dental Department, General Hospital, Nottingham)

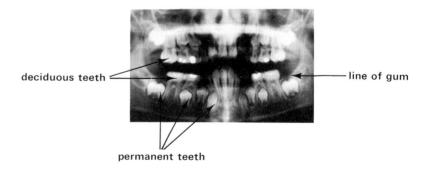

### 9.2.2 Variety and functions of teeth

Like other mammals we are **heterodont**: our dentition includes several kinds of teeth. From the front of the jaw to the rear there are **incisors, canines, premolars** and **molars**. The various kinds of teeth perform different functions.

The fossil record shows that primitive mammals had a dental formula of:

$$\frac{3 \quad 1 \quad 4 \quad 3}{3 \quad 1 \quad 4 \quad 3}$$

That is, there are three incisors, one canine, four premolars and three molars on each side of both upper and lower jaws, a total of 44 teeth.

149

Human evolution has resulted in fewer teeth with a dental formula of:

$$\frac{2 \quad 1 \quad 2 \quad 3}{2 \quad 1 \quad 2 \quad 3}$$

in the adult (Fig 9.4).

**Fig 9.4** Human dentition

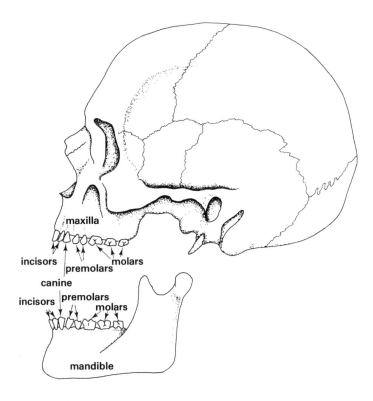

The chisel-like incisors and canines pierce, grip and rip food as it is taken into the mouth. The premolars and molars are used to cut and grind food. They have broad, but irregular surfaces that come together when the lower jaw is raised to the upper jaw during chewing (Fig 9.5). Humans are **omnivores** because we usually eat a mixed diet including meat, fruits and vegetables.

**Fig 9.5** (a) Incisor, (b) canine, (c) premolar, (d) molar

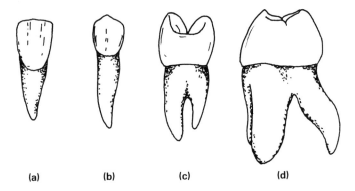

## 9.3 Digestion

Many of the organic nutrients in food, such as proteins and polysaccharides, are substances of high relative molar mass. They are insoluble in water and cannot pass through the membranes of body cells. It is necessary for such substances to be broken down into units small enough to be absorbed into the body fluids in which they can be transported to all parts of the body. The breaking down of food is called **digestion**. It takes place in the gut where food is acted on by hydrolytic enzymes (Chapter 3).

The digestive system is illustrated in Fig 9.6.

**Fig 9.6** Diagram of the digestive system

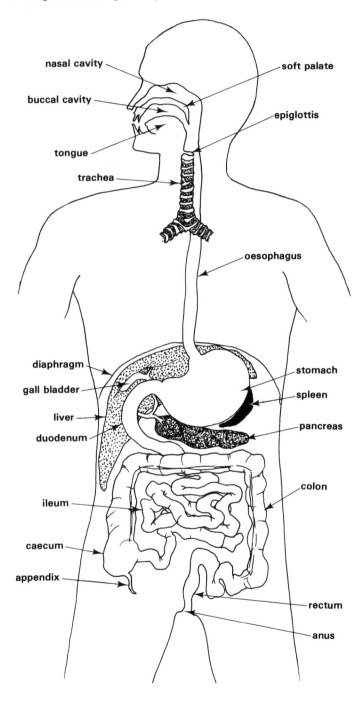

## 9.3.1 Mouth

The smell and sight of food, as well as the mechanical stimulation of food in the mouth, triggers a reflex action which results in the secretion of **saliva**. The reflex ensures a flow of saliva when food is in the mouth. Secretion of saliva slows down when we are not eating. Saliva enters the mouth from three pairs of salivary glands (Fig 9.7).

**Fig 9.7** Position of the salivary glands

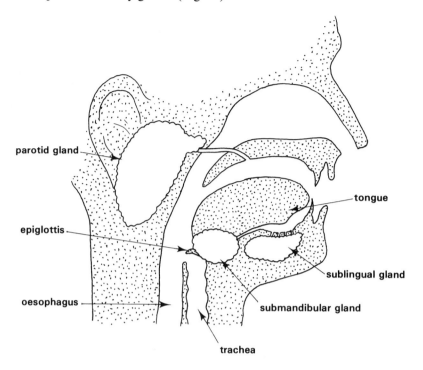

parotid gland

tongue

epiglottis

sublingual gland

oesophagus

submandibular gland

trachea

An enzyme called **salivary amylase** is usually found in saliva. Salivary amylase hydrolyses glycosidic linkages in **starch**, breaking it down to the disaccharide sugar **maltose** (Chapter 2). Salivary amylase probably contributes little to digestion. This is because we hold food in the mouth for only a short time. Salivary amylase works best in neutral conditions and the pH of saliva is about 7. After swallowing, food passes down the oesophagus into the stomach where the pH is much lower. In acid conditions salivary amylase becomes inactive. Saliva **lubricates** the pharynx and oesophagus, making it easier for food to be swallowed.

The food is moulded into a ball (bolus) by the tongue, then pushed into the pharynx. Contraction of the pharyngeal wall pushes the bolus into the oesophagus. Rhythmical contractions of the oesophagus, called **peristalsis**, move the bolus towards the stomach. Behind the bolus the circular muscle layer contracts, while the longitudinal muscle layer relaxes. The effect is to constrict the gut behind the bolus, thus pushing it along. Peristalsis is a property of all parts of the gut and helps to move food along the alimentary canal. During swallowing the larynx is raised and the **glottis** is covered by the **epiglottis**, so preventing food from entering the trachea (Fig 9.8).

**Fig 9.8** Peristalsis

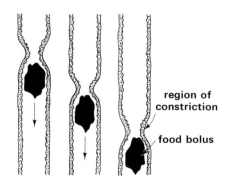

region of constriction

food bolus

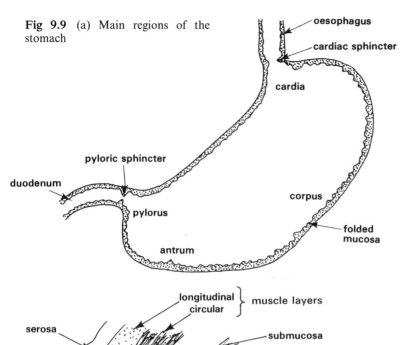

**Fig 9.9** (a) Main regions of the stomach

oesophagus
cardiac sphincter
cardia
pyloric sphincter
duodenum
pylorus
corpus
folded mucosa
antrum

## 9.3.2 Stomach

The first region of the gut where any significant digestion of food takes place is the **stomach.** It is a muscular bag with a volume of about two litres. It has a delicate inner folded membrane called the **gastric mucosa. Gastric juice** is secreted by numerous microscopic **gastric pits** embedded in the mucosa (Figs 9.9 and 9.10). The juice is very acid with a pH of about 2. The acidity is caused by **hydrochloric acid** secreted by **oxyntic cells.** It is thought the main function of gastric acid is to kill microbes which enter the body in our food. **Peptic cells** secrete **pepsinogen** which, on contact with the acid, is converted to the peptidase enzyme **pepsin**. Pepsin hydrolyses peptide linkages in proteins to produce polypeptides.

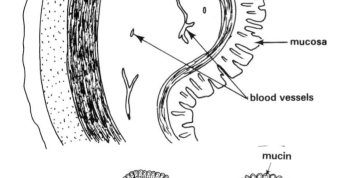

longitudinal
circular } muscle layers
serosa
submucosa
muscularis mucosa
gastric pits
mucosa
blood vessels

**Fig 9.9** (b) Three layers of muscles in the wall bring about stomach movements which mix the food with gastric juice during digestion. See Fig 9.12

**Fig 9.10** (b) Photomicrograph showing the gastric glands in a thin section of cat stomach, × 400

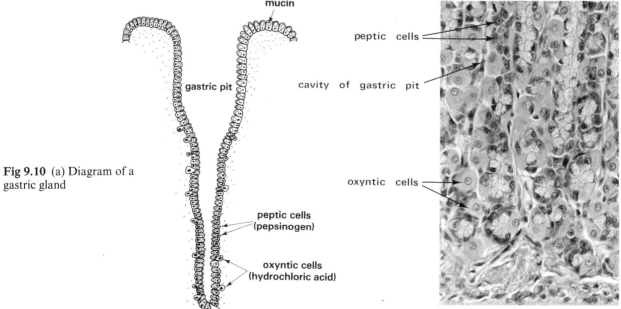

mucin
peptic cells
cavity of gastric pit
gastric pit
oxyntic cells
**Fig 9.10** (a) Diagram of a gastric gland
peptic cells (pepsinogen)
oxyntic cells (hydrochloric acid)

The corrosive and digestive properties of hydrochloric acid and pepsin place the delicate gastric mucosa at risk. Protection is given by a slimy glycoprotein called **mucin**. Mucin is secreted by **goblet cells** in the mucosa and forms a layer of **mucus** over the stomach lining. Mucus protects against mechanical as well as chemical injury. If the protection is not effective the mucosa and stomach wall are attacked by the gastric juice, causing an **ulcer** to form.

Gastric secretion is partly controlled by autonomic reflexes. Secretion of gastric juice begins when food is in the mouth. The presence of food in the stomach triggers the stomach lining to produce a hormone called **gastrin** which enters the blood. Gastrin stimulates continued secretion of gastric juice after the faster nervous mechanism has started it off (Fig 9.11).

**Fig 9.11** Relative importance of nervous (vagus) and hormonal (gastrin) influence on gastric juice secretion

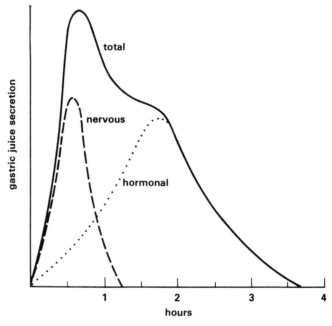

During the three or four hours food is held in the human stomach, rhythmic muscular contractions of the stomach wall churn the food, mixing it with gastric juice. Consequently more protein molecules in the food are brought into contact with gastric enzymes in a given time, so promoting digestion. The food is gradually converted into a creamy, acid suspension called **chyme** (Fig 9.12).

**Fig 9.12** Churning action of the stomach during digestion (from X-ray photographs)

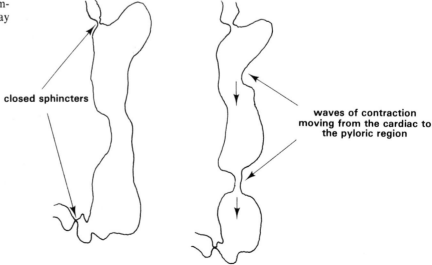

Before removing a sample of gastric juice for analysis, it is necessary to ensure the stomach secretes the juice. How is this achieved?

Sometimes it is necessary to determine the pattern of acid secretion. Several methods have been used to stimulate gastric secretion prior to sampling. Originally a **fractional test meal** consisting of a thin gruel was administered. **Histamine** has also been used for this purpose but has a number of undesirable side effects (Chapter 11). Recently, a short chain polypeptide called **pentagastrin** has come into use. The amount of acid in the sample is determined by titration against sodium hydroxide (Table 9.3). Patients who produce excess gastric acid may be treated by surgery, and by vagotomy, a severing of the branches of the vagus nerve which innervate the stomach. The effectiveness of the treatment may be tested by administering insulin to the patient. The consequent drop in blood glucose concentration (Chapter 16) stimulates the vagus nerve. Despite this, gastric secretion should not be affected.

Table 9.3 **Normal values for gastric acid (mmol per litre)**

|  | Males | Females |
| --- | --- | --- |
| basal (1 hour after stomach emptied following a fast) | 1.3 | 1.1 |
| total secretion during 1 hour after pentagastrin administration | 17.1 | 9.4 |

Fig 9.13 Anatomical relationship between the liver, gall bladder, pancreas, stomach and small intestine (not drawn to scale)

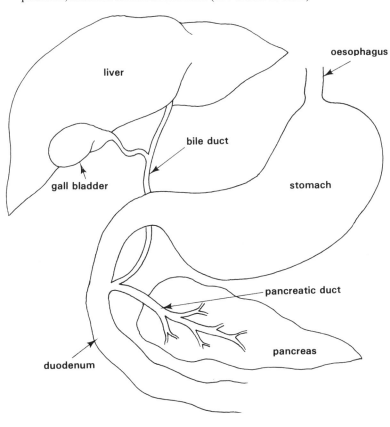

### 9.3.3 Small intestine

During gastric digestion, food is kept in the stomach by the constriction of two **sphincters**. These are circular muscles which seal off the stomach where it is joined to the oesophagus and small intestine (Fig 9.9(a)). Relaxation of the **pyloric** sphincter allows the propulsion of chyme, a little at a time, into the small intestine by peristalsis of the involuntary muscles in the gastric wall.

From the stomach, chyme passes first into the **duodenum** which is about 30 cm long. Beyond the duodenum is the **jejunum**, and then the **ileum** making up the rest of the small intestine (Fig. 9.6). The total length of the small intestine is about 6 m.

Digestion occurs mainly in the small intestine where enzymes complete the chemical breakdown of food into a state suitable for absorption. The enzymes come from two sources, the pancreas and the glands in the wall of the small intestine. Food is also exposed to a nonenzymic fluid called bile made in the liver (Fig. 9.13).

155

## 1 Bile

**Bile** is a greenish fluid containing **bile pigments** which are the excretory products of the breakdown of the haem part of haemoglobin. Though containing no enzymes, bile helps digestion in several important ways. First, it contains sodium hydrogencarbonate which gives it a pH of between 7 and 8. This is the optimum pH for the action of pancreatic and intestinal enzymes. Bile is therefore important in **neutralising** the acid chyme from the stomach. The second digestive function of bile is due to the **bile salts**, sodium and potassium glycocholate and taurocholate. They **emulsify** lipids, causing them to break down into numerous small droplets, about 0.5–1.0 $\mu$m in diameter. Emulsification provides a relatively large surface area of lipid for the action of lipase enzymes and hence speeds up digestion of fats and oils.

*Bile contains no digestive enzymes, yet is important to the process of digestion. What digestive role does bile play?*

## 2 Pancreatic juice

The **pancreas** is situated just beneath the stomach and is connected to the small intestine by a pancreatic duct through which **pancreatic juice** is discharged. The bile duct joins the pancreatic duct. The endocrine component of the pancreas, the islets of Langerhans (Chapter 16), plays no part in digestion.

Pancreatic juice contains four enzymes. **Pancreatic amylase** hydrolyses glycosidic linkages in starch, converting it to maltose in the same way as salivary amylase. Because food remains in the duodenum for some time, there is more opportunity for hydrolysis of starch in the small intestine than in the mouth. **Pancreatic lipase**, probably a group of several lipase enzymes, hydrolyses lipids to fatty acids and glycerol (Fig 9.14).

**Fig 9.14** Summary of the actions of pancreatic amylase and lipase

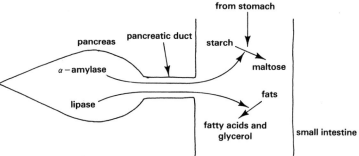

Two inactive enzyme precursors are also found in pancreatic juice. They are **trypsinogen** and **chymotrypsinogen**. Trypsinogen is converted to the active enzyme **trypsin** by an activator called **enterokinase** made in the intestinal glands. Trypsin activates the conversion of chymotrypsinogen to **chymotrypsin**. Trypsin and chymotrypsin are enzymes having an effect on proteins similar to that of gastric pepsin. They bring about a partial breakdown of proteins to polypeptides (Fig 9.15).

**Fig 9.15** Summary of the actions of trypsin and chymotrypsin

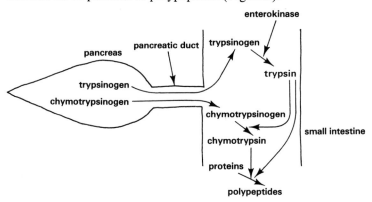

### 3 Intestinal fluid

Embedded in the intestinal wall is a great number of microscopic pits called the **crypts of Lieberkühn**. In the duodenum, coiled **Brunner's glands** in the sub-mucosa secrete alkaline mucus into the crypts. Enzymes from the crypts complete the process of digestion started by the mouth, stomach and pancreas. Intestinal juice, called **succus entericus**, contains a large number of enzymes which act on carbohydrates, polypeptides and lipids. The enzymes and their activities are summarised in Table 9.4. Some of the final stages in the digestion of carbohydrates, fats and polypeptides occur in the cells lining the intestinal mucosa (section 9.4).

**Table 9.4 Summary of the action of enzymes secreted by the small intestine**

| Enzymes | Substrates | Products |
|---|---|---|
| amylase | starch | maltose |
| maltase | maltose | glucose |
| sucrase (invertase) | sucrose | glucose and fructose |
| lactase | lactose | glucose and galactose |
| peptidases | polypeptides | amino acids |
| lipases | fats | fatty acids and glycerol |

The mechanisms controlling the release of bile, pancreatic juice and succus entericus are complex. They are triggered by the presence of chyme in the small intestine. Chyme stimulates the release of at least three hormones from the intestinal mucosa into the blood. **Pancreozymin** (cholecystokinin-pancreozymin, CCK-PZ) triggers the release of bile from its temporary store, the gall bladder, and the release of pancreatic enzymes. **Secretin** stimulates the flow of an alkaline fluid from the pancreas. **Gastric secretin** from the stomach wall has an effect similar to secretin. Finally **enterocrinin** stimulates the intestinal glands to secrete their digestive juice (Fig 9.16).

**Fig 9.16** Summary of the actions of gastrointestinal hormones involved in the release of bile and the digestive juices of the stomach, pancreas and small intestine

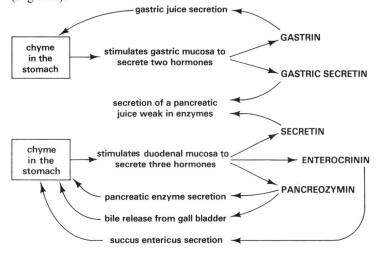

### 9.3.4 Intestinal microbes

There are countless microbes on our bodies and in our digestive system. They perform a variety of functions that are poorly understood. Nonetheless they are known to be vital to our normal development.

There are normally relatively few microbes in the stomach, duodenum, jejunum and upper ileum. Many of those present have probably been swallowed. Gastric juice is an effective barrier to the entry of unwanted microbes into the gut. Microbes indigenous to the digestive system are frequently present in the lower ileum and are always found in the caecum and colon. They usually closely adhere to the mucosal surface where they metabolise bile pigments and acids and produce gases such as carbon dioxide and hydrogen sulphide. They also degrade enzymes secreted in the various digestive juices. Some intestinal bacteria can synthesise vitamins such as K, $B_{12}$ thiamin and riboflavin (section 9.1.2).

Occasionally the intestinal microbes may cause disease. They can remove excessive quantities of nutrients from the gut and cause mild toxaemias. There is some evidence of a possible link between the activities of some kinds of intestinal microbes and cancer of the colon.

**Fig 9.17** (a) Diagrammatic structure of the small intestine wall

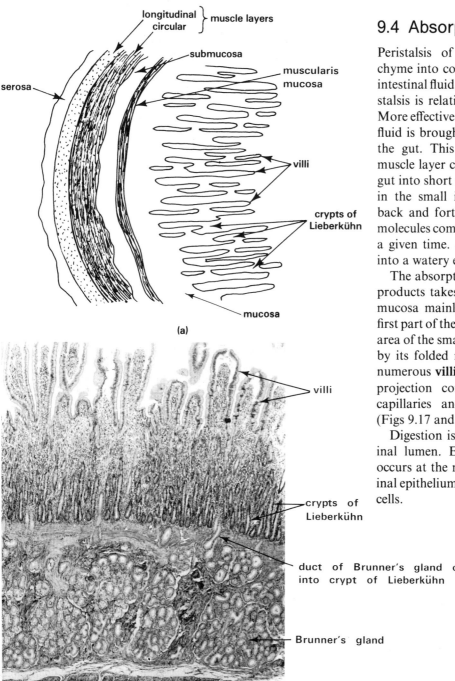

(a)

villi

crypts of Lieberkühn

duct of Brunner's gland opening into crypt of Lieberkühn

Brunner's gland

**Fig 9.17** (b) Photomicrograph of a transverse section of cat duodenum, × 45

## 9.4 Absorption

Peristalsis of the small intestine brings chyme into contact with fresh secretions of intestinal fluid. However, movement by peristalsis is relatively slow, about $1-2\,\mathrm{cm\,s^{-1}}$. More effective mixing of food with intestinal fluid is brought about by **segmentation** of the gut. This happens when the circular muscle layer contracts locally, pinching the gut into short segments. Consequently food in the small intestine is squeezed rapidly back and forth. In this way more enzyme molecules come into contact with the food in a given time. As a result, chyme is turned into a watery emulsion called **chyle**.

The absorption of nearly all the digestive products takes place through the extensive mucosa mainly in the duodenum and the first part of the jejunum. The internal surface area of the small intestine is greatly enlarged by its folded nature and even more so by numerous **villi**. Each villus is a microscopic projection containing blood and lymph capillaries and covered with epithelium (Figs 9.17 and 9.18).

Digestion is not completed in the intestinal lumen. Enzymic hydrolysis of foods occurs at the mucosal surface of the intestinal epithelium and even inside the epithelial cells.

158

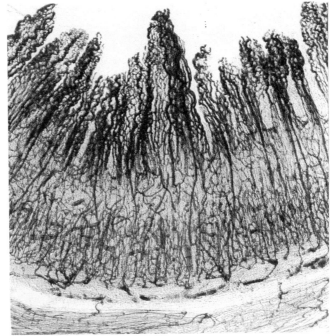

**Fig 9.17** (c) Photomicrograph of a transverse section of cat ileum. The blood vessels have been impregnated with a dye, × 20

**Fig 9.18** Diagram of a villus

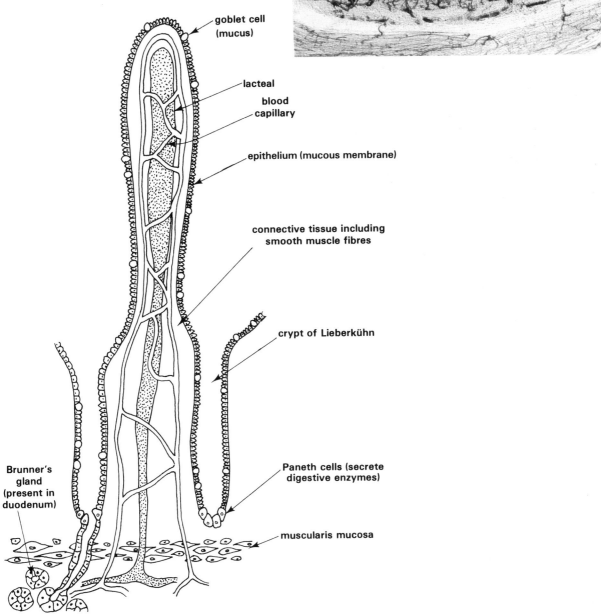

goblet cell (mucus)

lacteal

blood capillary

epithelium (mucous membrane)

connective tissue including smooth muscle fibres

crypt of Lieberkühn

Brunner's gland (present in duodenum)

Paneth cells (secrete digestive enzymes)

muscularis mucosa

### 9.4.1 Fate of lipids

**Emulsification** increases the surface area of fats and oils exposed to the action of pancreatic and intestinal lipases. Bile salts, in combination with lecithin or polar lipids such as monoglycerides, act as powerful emulsifiers. Fatty acids derived from dietary lipids are insoluble in water. They are **solubilised** through the formation of small, stable **micelles** of about 4–5 nm diameter. Each micelle consists of a shell containing monoglycerides, free fatty acids, cholesterol and fat-soluble vitamins. In this form the products of fat digestion are transported to the surfaces of the intestinal epithelial cells into which they are absorbed. Absorption is thought to be by diffusion. Di- and triglycerides are formed inside the epithelial cells and incorporated into envelopes of protein, cholesterol and phospholipid called **chylomicrons** which can be more than 100 nm across. Chylomicrons enter the intestinal **lacteals** which are lymphatic capillaries (Figs 9.18 and 10.48).

### 9.4.2 Fate of protein

Free amino acids and short-chain polypeptides remain in the small intestine after the action of gastric and pancreatic proteolytic enzymes. Complete hydrolysis of polypeptides involves intestinal enzymes which are active in the plasma membrane of the epithelium and in the epithelial cytoplasm. Polypeptides are hydrolysed to free amino acids which are transferred into the blood vessels of the villi. Amino acid transfer is enhanced by the presence of sodium ions ($Na^+$). Some long-chain polypeptides and even traces of intact dietary protein may be absorbed into the blood from the intestinal lumen.

### 9.4.3 Fate of carbohydrate

Carbohydrate can only be absorbed into the blood in appreciable quantities in the form of monosaccharides. Most carbohydrate digestion occurs in the jejunum, mainly at the epithelial surface. Pancreatic amylase has been found on the jejunal surface. The transfer of the products of carbohydrate digestion into the blood involves a special mechanism that is not fully understood. It requires metabolic energy and may be enhanced by sodium ions.

Some of the factors affecting the absorption of dietary constituents are summarised in Table 9.5.

**Table 9.5 Factors affecting the absorption of nutrients**

| | |
|---|---|
| Glucose | Active absorption may be linked with the transport of sodium ions across the membranes of mucosal cells. |
| Amino acids, some peptides | Active absorption may be affected by absorption of sugars and may also be linked with sodium ion transport through the mucosa. |
| Vitamin $B_{12}$ | Absorption dependent upon presence of intrinsic factor (IF) in gastric juice. |
| Vitamin K | Much produced by intestinal microbes. Absorption promoted by bile. |
| Calcium ions | Active absorption promoted by vitamin $D_2$ and probably also by parathyrin. |
| Iron ions | Absorbed in the iron(II) state. Rate of absorption depends on degree of saturation of transferrin with iron in the blood. |
| Water | Passive absorption by osmosis depends upon solute absorption. |

### 9.4.4. Colon

It is in the large intestine, especially the colon, that the absorption of water takes place. The digestive secretions of the gut add several litres of fluid to the contents of the intestine. If most of the water were not reabsorbed, the loss would seriously reduce the body's water content. Faecal water accounts, on average for about 4–8% of the total water loss in man. Sodium, chloride and hydrogencarbonate ions are also absorbed from the contents of the colon.

In recent years a lot of interest has been expressed in the **fibre** content of our diet. Fibre is a complex of many substances of vegetable origin, mainly polysaccharide carbohydrates such as cellulose, hemicelluloses, pectins, gums and mucilages. Non-carbohydrate components of fibre include lignin. Most Western diets are relatively low in fibre compared with those of rural populations in Africa. In the West there is a higher risk of intestinal diseases such as constipation, diverticular disease, haemorrhoids (piles) and cancer of the colon. Why this should be so is not understood, but it has been shown that fibre decreases intestinal transit time and increases the faecal mass. Fibre absorbs water and swells, filling and stretching the gut. A stretched gut stimulates peristalsis, hence waste food is held for a shorter time in the large intestine. Also, fibre combines with fatty acids and lowers their rate of absorption. A full gut triggers the satiety ('full-up') reflex too, thus suppressing the desire to eat.

Unabsorbed matter including the excretory products of bile is passed to the exterior through the **anus** as **faeces**.

## 9.5 The liver

Several chapters in this book refer to **homeostasis**, the maintenance of a steady internal environment. Regulating the composition of blood and tissue fluid is an important aspect of homeostasis. The composition of blood leaving the gut in the hepatic portal vein is largely determined by the diet and thus the substances absorbed from the small intestine. However, the blood which enters the general circulation is often very different in composition. The hepatic portal vein carries blood to the liver. One of the functions of the liver therefore is to regulate the composition of blood. The liver is the largest organ in the body and performs many functions, not all of which are concerned with homeostasis.

**Fig 9.19** (a) Diagram of a liver lobule

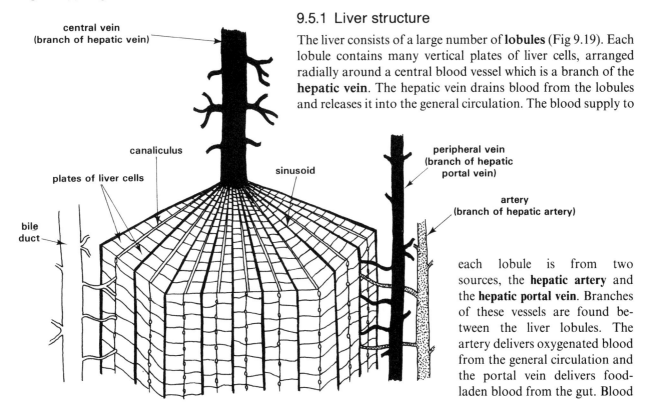

central vein
(branch of hepatic vein)

canaliculus

plates of liver cells

sinusoid

peripheral vein
(branch of hepatic
portal vein)

artery
(branch of hepatic artery)

bile
duct

### 9.5.1 Liver structure

The liver consists of a large number of **lobules** (Fig 9.19). Each lobule contains many vertical plates of liver cells, arranged radially around a central blood vessel which is a branch of the **hepatic vein**. The hepatic vein drains blood from the lobules and releases it into the general circulation. The blood supply to each lobule is from two sources, the **hepatic artery** and the **hepatic portal vein**. Branches of these vessels are found between the liver lobules. The artery delivers oxygenated blood from the general circulation and the portal vein delivers food-laden blood from the gut. Blood from the hepatic artery and hepatic portal vein flows between the plates of liver cells inwards to the central vein in channels called **sinusoids** (Fig 9.19).

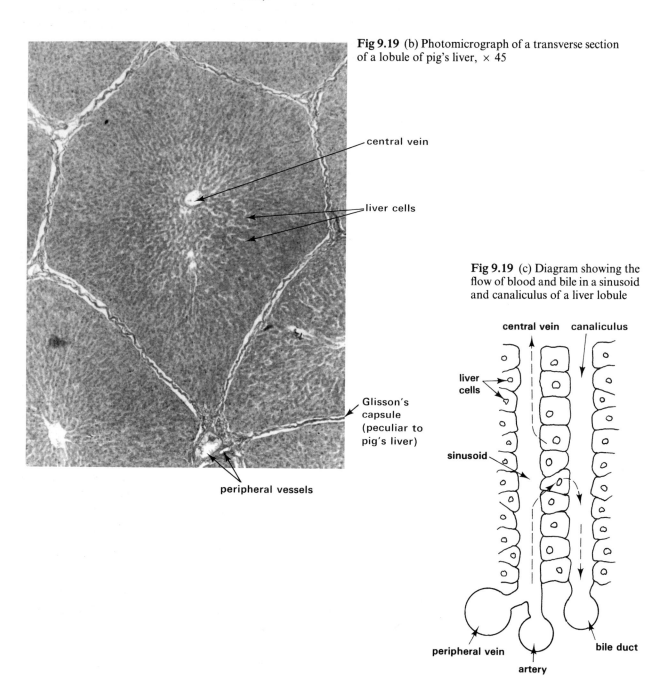

**Fig 9.19** (b) Photomicrograph of a transverse section of a lobule of pig's liver, × 45

central vein

liver cells

**Fig 9.19** (c) Diagram showing the flow of blood and bile in a sinusoid and canaliculus of a liver lobule

Glisson's capsule (peculiar to pig's liver)

peripheral vessels

central vein    canaliculus

liver cells

sinusoid

peripheral vein        bile duct

artery

Between the plates of liver cells are other channels called **canaliculi** which receive **bile**. The bile moves outwards to the periphery of the lobules where it collects into **bile ducts**. Bile is stored temporarily in a sac-like **gall bladder** before its periodic release into the small intestine.

### 9.5.2 Liver functions

The functions of the liver are numerous and vital to life. For convenience they are described under three main headings.

## 1 Metabolism of absorbed food

**i. Carbohydrate metabolism**   Soluble sugars, mainly glucose, are carried from the gut to the liver in the hepatic portal vein. Following a carbohydrate-rich meal the concentration of glucose in blood going to the liver is relatively high. When absorption is completed the glucose level drops. However, the concentration of glucose in the general blood circulation remains fairly stable. This is because excess glucose absorbed after a meal is converted in the liver cells to a storage polysaccharide called **glycogen**. The conversion is controlled by the hormone **insulin** from the islets of Langerhans in the pancreas (Chapter 16).

Glucose carried in the general blood circulation is used for tissue respiration throughout the body. It is replaced mainly from the liver's store of glycogen. The liver thus regulates the concentration of glucose in the general circulation even though the rate of glucose absorption from the gut varies.

**ii. Protein metabolism**   Unlike carbohydrate, protein is not stored in the body. Some amino acids from the gut pass through the liver and enter the **amino acid pool** in the general circulation. Amino acids are absorbed by the body cells and used for protein synthesis. The liver cells are important sites of **protein synthesis**, producing the plasma proteins. A stable **protein pool** is established in the blood. Synthesis is balanced by an equal breakdown of plasma proteins which have a limited lifetime in the circulation.

Plasma proteins at the end of their lifetime, together with excess amino acids absorbed from the gut, are deaminated by liver cells. **Deamination** is the removal of amino groups and their conversion to ammonia. The remains of the acid molecules enter the Krebs tricarboxylic acid cycle and are used to produce respiratory energy. Ammonia is very toxic and enters a sequence of reactions called the **ornithine cycle** which converts ammonia to the less toxic **urea** (Fig 9.20). Urea is released into the general circulation and is excreted by the kidneys (Chapter 13). About half of the twenty amino acids used for protein synthesis are non-essential: they can be produced from other amino acids in the liver by **transamination**. Liver cells contain a number of transaminase enzymes which transfer amino groups from amino acids to carboxylic acids, thus producing new amino acids. For example, glutamic-oxaloacetic transaminase (GOT) catalyses the following reaction:

$$\text{glutamic} + \text{oxaloacetic} \quad \overset{\text{GOT}}{\rightleftharpoons} \quad \text{aspartic} + \alpha\text{-ketoglutaric}$$
$$\text{acid} \qquad\qquad \text{acid} \qquad\qquad\qquad \text{acid} \qquad\qquad \text{acid}$$
$$\text{(amino acid)} \quad \text{(carboxylic acid)} \qquad \text{(amino acid)} \quad \text{(carboxylic acid)}$$

Such enzymes enable the liver cells to make a wide range of amino acids, even though the variety of amino acids in the diet is limited.

**Fig 9.20** Ornithine cycle

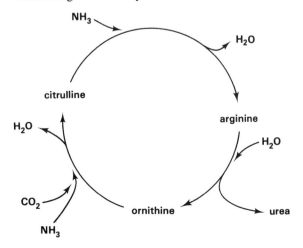

The liver has been described as a filter. How does the liver help to regulate the concentrations of carbohydrates, fats and proteins in the systemic blood?

**iii. Lipid metabolism** Fats and oils absorbed from the gut can be converted to glycogen which is stored in the liver. Fat-soluble vitamins such as vitamins A and D are also stored in liver cells. **Cholesterol**, a sterol carried in the blood, is used in a variety of syntheses, particularly cell membrane production. The liver makes cholesterol which is important when the dietary intake is inadequate. Excess cholesterol absorbed from the gut is eliminated in bile. Gross excesses of cholesterol, however, may be precipitated as gallstones in the bile duct and gall bladder.

## 2 Iron metabolism

Some of the liver cells lining the sinusoids are part of the body's **reticulo-endothelial system**. The system removes particles of debris from the blood. The **Küpffer cells** are phagocytic liver cells which remove old red cells from the blood. After old red cells have been engulfed, the haemoglobin is split into two, an iron–globin complex and an iron-less haem group. The haem is then converted to **bilirubin**. Bilirubin combines with glucuronic acid to form the bile pigment **bilirubin diglucuronide** which is excreted in bile. About 7–8 g haemoglobin are removed from the circulation in this way each day.

Any physical obstruction, such as **gallstones** in the bile ducts, or an increased breakdown of red cells, results in an increase in the concentration of bilirubin in blood. The skin takes on a yellow colour and the condition is called **jaundice**. The iron–globin complex is metabolised further in the liver. The globin is broken down into amino acids which are used for synthesis of proteins such as those of the blood plasma. The iron is retained for the production of fresh haemoglobin. Haemoglobin synthesis and red cell production (erythropoiesis) occur in the liver of the foetus. In the adult these processes are restricted to the red bone marrow (Chapter 10). Iron released from broken-down haemoglobin becomes attached to a plasma protein called **transferrin** and is transported in the blood to the bone marrow. Excess iron is stored in the liver cells as **ferritin** and **haemosiderin**.

The total iron content of the adult body is normally between 3 and 5 g. Of this, about 1.5–5 g is in haemoglobin and about 1–1.5 g is stored in the liver. Pregnancy and lactation can lower a woman's total iron by as much as 20%. This is why iron-containing tablets are given to pregnant women. Fig 9.21 summarises some of the factors involved in iron metabolism.

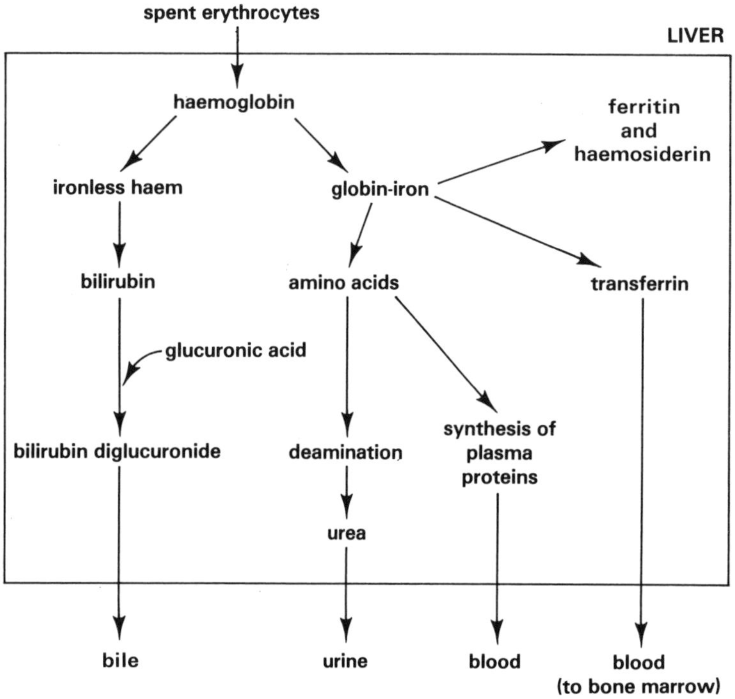

Fig 9.21 Summary of haemoglobin breakdown in the liver and the fate of the products

### 3 Detoxification

Many chemical substances which pass through the liver in the blood are modified by the liver cells. The substances include a variety of hormones. For example, much of the insulin from the pancreas is broken down by enzymes in the liver. Many sex hormones are also inactivated in the liver and excreted in the bile or released into the blood and excreted by the kidneys. Some chemicals are destroyed by liver cells, other are combined with various substances to render them less toxic. For example, benzenecarboxylic acid, a commonly used food preservative, is joined to the amino acid glycine to form *N*-benzoylglycine which is excreted in the urine. Certain chemicals such as tetrachloromethane, trichloromethane and ethanol damage the liver cells.

Within limits, then, the liver acts as a filter, removing toxic substances from the blood, making them less harmful and preparing them for excretion. The liver performs a variety of other functions which are described elsewhere in the book. Notable among them is the contribution the liver makes to heat production in the body (Chapter 17).

# 10 Circulatory systems

Metabolites move through tissues and cytoplasm mainly by diffusion (Chapter 1). Diffusion is relatively slow however, and alone could not supply the body's cells with all their requirements rapidly enough to sustain life. Similarly the waste products of metabolism could not be eliminated from the cells fast enough by diffusion alone. Toxic wastes would accumulate and metabolic processes would eventually stop. The presence of a mass transport system greatly speeds the movements of metabolites and wastes. This function is performed especially by **blood**.

## 10.1 Blood functions

An adult may have about 5 dm³ of blood which moves in a system of vessels called the **blood vascular system**. Blood consists of cells suspended in a liquid called **plasma**. All the blood constituents have a limited life-time in the circulation and there is constant renewal of spent cells and plasma. Blood cells normally occupy between about 42 % and 48 % of the total blood volume. The cells are on average about 7 µm in diameter. They can be divided into three main categories: **red cells, white cells** and **platelets** (Fig 10.1).

**Fig 10.1** Photomicrograph of a normal blood smear, × 1250

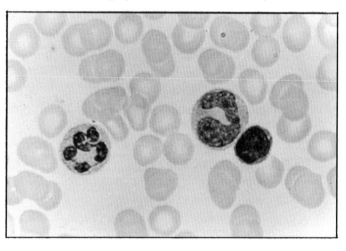

**Fig 10.2** Red blood cells: (a) plan, (b) section

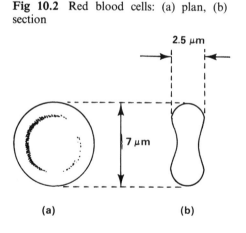

(a)                             (b)

(c) scanning electronmicrograph, × 1700

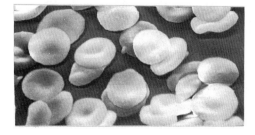

### 10.1.1 Red blood cells

Red blood cells are called **erythrocytes** (eruthros = red). They are by far the most numerous cells in the blood; there are usually more than a thousand times as many red cells as white cells. Each red cell is a biconcave disc which lacks a nucleus. Electron microscopy reveals that red cells are usually totally devoid of any sub-cellular structures. They consist of cytoplasm bounded by unit membrane (Fig 10.2). The shape of red cells allows them to move easily through the narrowest blood capillaries without getting trapped (Fig 10.38). The biconcave disc also provides them with a bigger surface area in relation to their volume than they would have if they were spherical. A large surface area is important in cells whose main function involves exchanging oxygen and carbon dioxide with their surroundings.

Normal red blood cells are biconcave discs about 7 $\mu$m in diameter. The surface area is about 120 $\mu$m². Red cells which are spherical and of the same volume have surface area of about 90 $\mu$m². Why do the spherical red cells carry less oxygen than the disc-shaped ones?

Red cells are mainly concerned with the collection of oxygen in the lungs and the delivery of oxygen to the body's tissues. For oxygen carriage, red cells contain the respiratory pigment **haemoglobin** (Chapter 12). Each red cell contains about 30 pg haemoglobin (30 millionths of a millionth of a gram!). There are usually $3.9$–$6.5 \times 10^{12}$ red cells in each dm³ of blood.

Since red cells are without a nucleus and contain no ribosomes, they cannot produce proteins (Chapter 6). Consequently they have no powers of regeneration, growth or repair. Red cells survive for an average of about 4 months in the blood.

The process by which red cells are produced is called **erythropoiesis**. In the foetus erythropoiesis occurs in the yolk sac and then the liver. After birth, erythropoiesis occurs only in the marrow of certain bones (Chapter 8). At an early age almost all the skeleton is involved, but as we get older the number of bones producing red cells decreases. In an adult, erythropoiesis is usually restricted to the ribs, sternum, vertebrae, skull, parts of the pelvic girdle and the ends of long bones such as the femur. The marrow in which erythropoiesis occurs is coloured red by the haemoglobin in the red cells. It is therefore called **red bone marrow**. Marrow not involved in erythropoiesis is yellow in appearance, due mainly to its fat content.

Erythropoiesis is stimulated by a hormone called **erythropoietin**. This is secreted by the kidneys in response to a drop in the **oxygen tension** of the blood. The oxygen tension of the blood is the concentration of oxygen dissolved in the plasma. The tension may fall normally because of growth, following haemorrhage or at high altitude where atmospheric oxygen is less than at sea level. Increased secretion of erythropoietin leads to an increase in red cell numbers and hence an increase in the ability of the blood to carry oxygen to the tissues (Fig 10.3). The fact that erythropoietin is produced in

**Fig 10.3** Hormone control of erythropoiesis

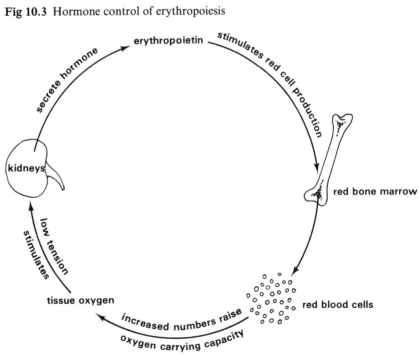

the kidneys explains the drop in red cell numbers and consequent anaemia that is common in people with kidney failure (Chapter 13). After circulating the body for about 4 months, spent red cells are taken out of the blood by phagocytic cells which line blood-filled channels in the bone marrow, spleen and liver (Chapter 9). Once engulfed by phagocytes, the red cells and the haemoglobin they contain are broken down. Much of the material of which they are made is recycled in the body, while some fragments of the haemoglobin are excreted from the liver as pigments in the bile (Chapters 9 and 12).

Normal erythropoiesis requires a number of dietary substances. Important among these are **vitamin $B_{12}$** and **folic acid** (Chapter 9). Deficiency of vitamin $B_{12}$ results in poor erythropoiesis and anaemia. It sometimes happens that an individual may suffer the effects of vitamin $B_{12}$ deficiency, yet ingest a lot of the vitamin in his diet. Absorption of vitamin $B_{12}$ requires the presence of a substance called **intrinsic factor** normally found in gastric juice. People who do not secrete intrinsic factor are thus unable to absorb vitamin $B_{12}$ and suffer from **pernicious anaemia**. People with pernicious anaemia have fewer and larger red cells than normal: they are **macrocytic**. The pattern of red cell maturation is affected and some abnormal marrow cells appear: they are **megaloblastic**. A diet containing large quantities of liver used to be prescribed to treat pernicious anaemia. Nowadays vitamin $B_{12}$ may be injected directly into the blood.

The term **anaemia** refers to a variety of conditions all characterised by a lowering of the blood's ability to carry oxygen. Anaemias can be classified into two main groups, those caused by a decrease in red blood cell and/or haemoglobin production, and those resulting from blood loss (Table 10.1).

**Table 10.1  A classification of anaemias according to their causes**

A  Increased blood loss

    a  Haemorrhage   i  acute, e.g. accident  
                        ii  chronic, e.g. menstruation

    b  Haemolysis      i  intracellular cause, e.g. abnormal haemoglobin variant  
                        ii  extracellular cause, e.g. bacteria

B  Decreased blood production

    a  Nutritional deficiency   e.g. iron, vitamin $B_{12}$, folic acid, protein

    b  Bone marrow failure   i  primary, i.e. congenital  
                        ii  secondary, i.e. acquired, say, by exposure to ionising radiation

In haematology laboratories red blood cells are usually counted accurately and very rapidly by machines such as the Coulter S PLUS III counter (Fig 10.4). The counters can also determine the haemoglobin content of the blood and a range of other parameters simultaneously. Red cells may be counted manually with the aid of a counting chamber called a **haemacytometer**. This consists of a thick glass slide with a central area exactly 0.1 mm below a cover slip placed over the top (Fig 10.5(a)). The central area has a counting grid etched onto it. Blood is diluted 1 in 100 or 1 in 200 in formol citrate. The diluent prevents the blood clotting (section 10.1.3). When viewed microscopically, the red cells can be seen over the counting grid (Fig 10.5(b)). Red cells are counted over an area in the central $mm^2$ of the grid, usually in at least 80 of the smallest squares. The number of red cells in a $dm^3$ of undiluted blood is then calculated as shown on page 169.

**Fig 10.4** Coulter counter, Model S-Plus III. This machine can perform a red cell count and a white cell count and determine ten other blood parameters including haemoglobin concentration in just over 30 seconds

**Fig 10.5** (a) Haemacytometer

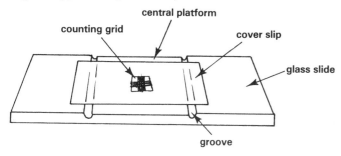

**Fig 10.5** (b) Blood cell counting grid

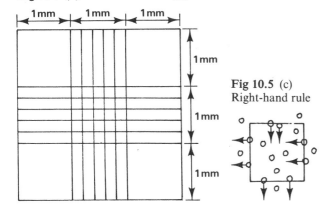

**Fig 10.5** (c) Right-hand rule

Let $n$ = the number of red cells in 80 small squares

The area of each small square = $0.0025 \, \text{mm}^2$

Since the counting chamber is 0.1 mm deep, then the volume over each small square

$$= (0.0025 \times 0.1) \, \text{mm}^3 = 0.00025 \, \text{mm}^3$$

Therefore the volume over 80 small squares

$$= (80 \times 0.00025) \, \text{mm}^3 = 0.02 \, \text{mm}^3$$

Since the sample is diluted, say 1 in 200, then the volume of diluted blood counted over 80 small squares is equivalent to:

$$\frac{0.02}{200} \, \text{mm}^3 = 0.0001 \, \text{mm}^3 \text{ undiluted blood}$$

That is, $n$ cells are counted in $0.0001 \, \text{mm}^3$ undiluted blood

$$= \frac{n}{0.0001} \text{ cells in } 1 \, \text{mm}^3 \text{ blood}$$

$$= \frac{n}{100} \times 10^{12} \text{ cells per dm}^3$$

169

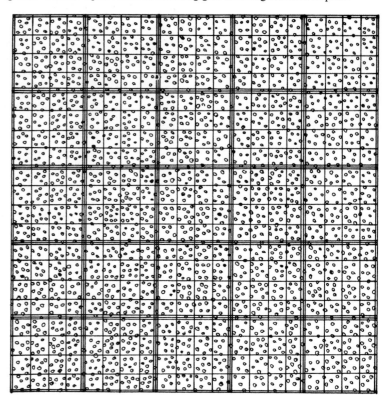

Look carefully at Fig 10.6 and perform a red cell count for yourself. How many cells are there in 80 small squares of the counting grid? To make sure you do not miss any cells or count any cells twice, operate the **right-hand rule**. That is, include in your count for a particular square any cells straddling the top or right-hand margins, but exclude any cells straddling the bottom or left-hand margins (Fig 10.5(c)). The chances are that your result will be different if you repeat your count for 80 different small squares. Why should this be so? The main reason is that the cells settle randomly in the counting chamber when it is filled. Each time you count the red cells in a given volume of diluted blood your result should be accurate for that particular volume of blood. This may not be a truly representative sample of the whole blood. Therefore your calculated number of red cells per $dm^3$ of undiluted blood is only approximate. To take account of this **field error**, instead of saying that there are, say, $n$ cells in 80 small squares, we can say that the real value probably lies within the range:

$$n \pm \sqrt{n}$$

where $\sqrt{n}$ is the **standard error of distribution**. There is a 95% chance that the real value lies within the range $n \pm \sqrt{2n}$. The bigger the sample we count in, the more representative it is of the whole blood and hence the more accurate our estimate will be. Cell counting machines count the cells in larger samples than are used in a haemacytometer. They also perform the counts much more quickly. It is for these reasons that they are widely used.

Another aspect of red blood cells which has been useful in investigating disease is the rate at which they settle under the influence of gravity. This is called the **erythrocyte sedimentation rate, ESR**. A sample of blood is diluted 4 to 1 in sodium citrate solution to prevent clotting. It is then allowed to stand undisturbed in a vertical tube. The tube is 30 cm long, with

What is the red cell count for the blood shown in the counting chamber in Fig 10.6? Take account of field error and express the result as a range of values in units of 'numbers $* 10^{12}$ per litre'.

an internal diameter of 2.5 mm and is graduated in mm divisions. After one hour at 25°C, a reading is taken of how many mm the red cells have settled. Normally this is between 0 and 7 mm. In some kinds of infection, kidney disease and degenerative disorders, the value for the ESR may be greater.

The reliability of the ESR is widely questioned because of the non-specific nature of the measurement. Many factors can affect the results, especially the concentration in the plasma of globulin proteins called **agglomerins**. They increase the tendency of the red cells to aggregate into **rouleaux** (Fig 10.7) which increase the rate of settling. In some inflammatory conditions the concentration of agglomerins in the blood rises. One such protein is called **C-reactive protein, CRP**. The quantity of CRP can be determined by other techniques which are now replacing ESR measurements.

**Fig 10.7** (a) Arrangement for measuring the erythrocyte sedimentation rate, ESR

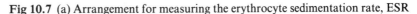

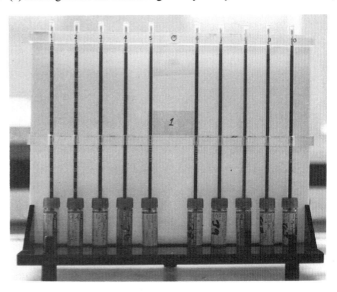

**Fig 10.7** (b) Photomicrograph of blood film showing rouleaux, × 800

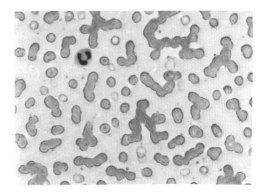

Yet another important feature of red blood cells in the diagnosis of disease is their ability to resist haemolysis (Chapter 1). The **osmotic fragility test** is designed to determine the resistance of red cells to haemolysis when subjected to a range of saline solutions of different osmotic potential. 0.05 cm³ heparinised blood are added to 10 cm³ saline solution in tubes over a range of concentration between 0.0 and 0.9 % NaCl. After 30 minutes the blood–saline mixtures are centrifuged

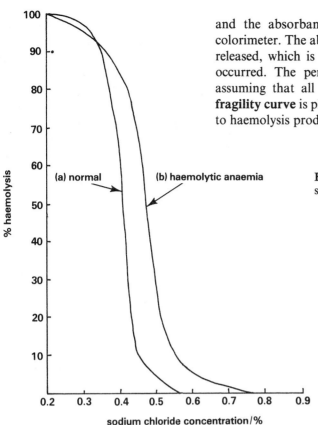

and the absorbance of the supernatant fluids is determined with a colorimeter. The absorbance is dependent on the amount of haemoglobin released, which is proportional to the degree of haemolysis which has occurred. The percentage haemolysis is calculated for each mixture assuming that all the cells haemolyse in distilled water. An **osmotic fragility curve** is plotted from the results. Red cells which are more prone to haemolysis produce a curve to the right of normal (Fig 10.8).

**Fig 10.8** Osmotic fragility curves: (a) normal, (b) for a patient suffering from haemolytic anaemia

### 10.1.2 White blood cells

White blood cells are called **leucocytes** (leukos = white). There are usually $4-11 \times 10^9$ white cells in each $dm^{-3}$ of blood. Like red cells, they originate in the red bone marrow. White cells are produced by a process called **leucopoiesis** and are of five different kinds (Fig 10.9). They are all concerned in one way or another with protecting the body against infection and disease.

**Fig 10.9** Photomicrographs of blood smears showing the main kinds of white cells: (a) neutrophil, (b) lymphocyte, (c) monocyte, (d) eosinophil; × 1500. Basophils appear similar to eosinophils

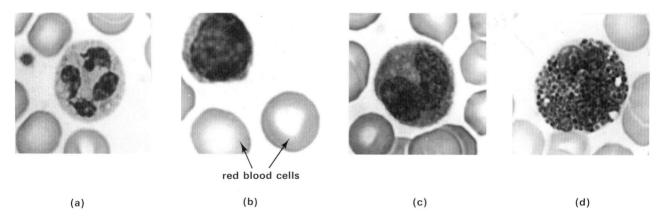

red blood cells

(a)          (b)          (c)          (d)

### 1 Neutrophils and monocytes

These cells can form cytoplasmic extensions called **pseudopodia** around particles such as bacteria. The particles are then engulfed whole, a process called **phagocytosis** (Fig 10.10). Neutrophils and monocytes can locate bacteria at a distance and move towards them, a form of behaviour called **chemotaxis**. They respond to chemical substances which are emitted from

**Fig 10.10** Phagocytosis: (a) a neutrophil engulfing a chain of bacteria

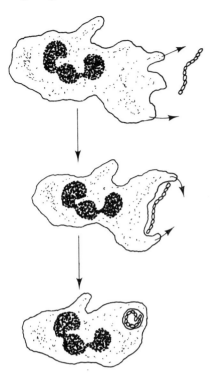

the bacteria into the surroundings. Once inside the phagocytes, the engulfed bacteria are prevented from further growth or are killed by the action of enzymes secreted onto the bacteria from the phagocytes' lysosomes. If the bacteria are resistant or virulent they may destroy the phagocytes. Other defences, such as antibodies (Chapter 11), may be necessary to assist the phagocytes in destroying the bacteria. The numbers of circulating neutrophils and monocytes may increase greatly during an infection. The phagocytes can move out of the blood and into the tissues. They do this by squeezing in between the cells making up the walls of the blood capillaries (section 10.2.3).

**Fig 10.10** (b) Photomicrograph showing neutrophils containing phagocytosed bacteria, × 1400

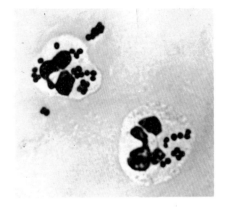

## 2 Lymphocytes

These cells originate in the bone marrow but are produced in large numbers in the nodes of the lymphatic system (Fig 10.45). Lymphocytes produce and may carry **antibodies** and are part of the body's **immune response** to infection (Chapter 11).

## 3 Basophils and eosinophils

Basophils are thought to be **mast cells** which have entered the blood. Mast cells are found in connective tissues (Chapter 7). The numbers of eosinophils may rise greatly during an **allergic response** (Chapter 11).

Normally the numbers of circulating white blood cells are regulated by the mechanisms which control leucopoiesis. However, an uncontrolled massive increase in the rate of leucopoiesis may occur and result in **leukaemia**. It is possible to assess the relative numbers of circulating white cells by a procedure called **differential leucocyte counting**. A thin and even smear is made of the blood to be examined on a microscope slide. After suitable staining, the different kinds of white cells can be identified. The smear is scanned with a microscope (Fig 10.11). About 100 white cells are counted

**Fig 10.11** Photomicrograph showing a blood smear from a patient suffering from lymphatic leukaemia, × 800. Compare with Fig 10.1

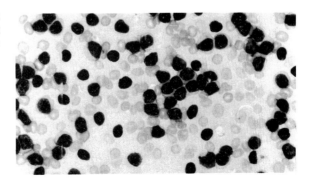

173

In what ways do the white blood cells help to protect the body against infection by microorganisms?

from all over the smear and the numbers of the different kinds are recorded as percentages (Table 10.2).

**Table 10.2 Typical results for differential leucocyte counts on blood from three people**

| White blood cells | Percentages of the different white cells | | |
| --- | --- | --- | --- |
| | Normal person | Patient with chronic lymphatic leukaemia | Patient suffering an allergic response |
| neutrophils | 65 | 3 | 51 |
| monocytes | 8 | 1 | 4 |
| lymphocytes | 24 | 96 | 27 |
| eosinophils | 2 | — | 18 |
| basophils | 1 | — | — |

### 10.1.3 Platelets

**Platelets** are called **thrombocytes**. There are normally $150–400 \times 10^9$ platelets in each $dm^3$ of blood. They are made in the red bone marrow by a process called **thrombocytopoiesis**. Large cells called **megakaryocytes** break into fragments which become platelets (Fig 10.12). Platelets are involved in the **coagulation** of blood.

**Fig 10.12** Photomicrographs showing: (a) a megakaryocyte in a bone marrow smear from a guinea pig, × 750; (b) a blood smear showing platelets, × 1500

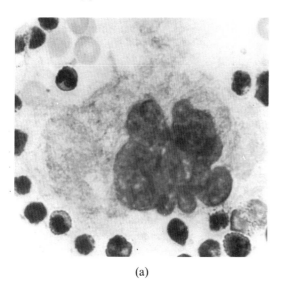

(a)

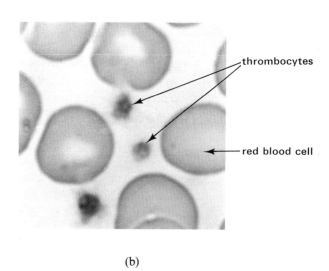

(b)

If blood is drawn into a syringe, or if the vessel through which the blood is flowing is damaged, platelets **agglutinate** by sticking together. Platelets also stick to the walls of the syringe or to the damaged parts of the blood vessels. Agglutinated platelets form a **platelet plug**. In small vessels such as arterioles and capillaries, this plug may be sufficient to stop blood loss from the vessel while repairs are made. Agglutinated platelets also break open. Platelet **lysis** releases into the surrounding blood plasma a number of substances called **platelet factors** (Table 10.3). One of the factors is called **serotonin** (5-hydroxytryptamine) which causes **vasoconstriction** of the damaged vessel. It may help to reduce the blood flow through the damaged region. Other platelet factors react with substances in the plasma called **blood factors** (Table 10.3(a)) and initiate a sequence of chemical reactions which results in the formation of a blood **clot** (Fig 10.13(a)). The first stage in this process is the production, at the site of injury, of a lipoprotein called **thromboplastin**. It may originate from undamaged tissues by a process

**Table 10.3 Coagulation factors**

| | |
|---|---|
| I | Fibrinogen |
| II | Prothrombin |
| III | Thromboplastin |
| IV | Calcium ions |
| V | Labile factor |
| VII | Serum prothrombin conversion accelerator (SPCA) |
| VIII | Anti-haemophilic globulin (AHG) |
| IX | Christmas factor |
| X | Prower–Stuart factor |
| XI | Plasma thromboplastin antecedent (PTA) |
| XII | Hageman factor |
| XIII | Fibrin stabilising factor (FSF) |

**Fig 10.13** (a) Blood coagulation: main stages in the process

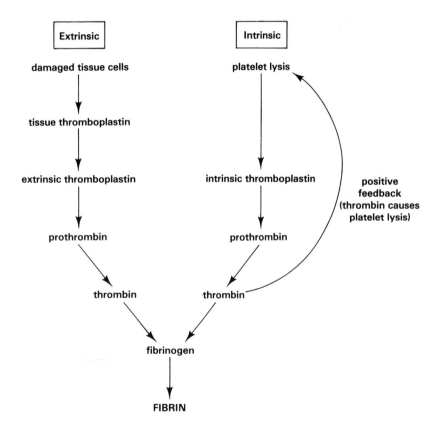

called the **extrinsic mechanism** that is outside the blood vessels. Thromboplastin is also derived from the agglutination and lysis of platelets in the blood by the **intrinsic mechanism** that is inside the blood vessels. A number of factors are required to produce thromboplastin. They include the anti-haemophilic factor (factor VIII).

In the presence of other factors, notably **calcium ions** (factor IV), thromboplastin acts on **prothrombin**, an inactive plasma protein produced in the liver. When activated, prothrombin is converted to **thrombin**. Thrombin in turn acts on another plasma protein called **fibrinogen**, also from the liver. Fibrinogen is converted to **fibrin** which is insoluble and deposits as strands attached to solid particles such as platelets and damaged vessel walls. Blood cells become attached to the fibrin strands and a clot is formed (Fig 10.13(b)). The clot usually retracts and forms a solid plug in the vessel. Here it prevents further blood loss while the injury is repaired. The clot is eventually dissolved by the action of the enzyme **fibrinolysin** which is released by basophils.

**Fig 10.13** (b) Blood coagulation: scanning electronmicrograph showing some of the components of a blood clot, × 1000

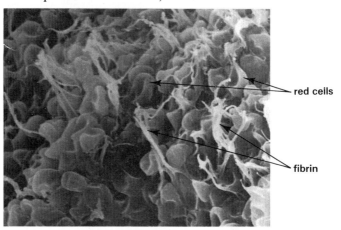

Just as it is important for blood to coagulate at the site of the injury, it is equally important that blood does not coagulate in a healthy vessel. Platelet agglutination is prevented in healthy vessels by substances secreted from the vessel wall. Most notable among these substances is **prostacyclin**. If a vessel is injured, the damaged part does not produce prostacyclin and the platelet agglutinate. Prostacyclin may be used to prevent platelet agglutination in blood which is circulated outside the body (extracorporeal circulation) such as in a kidney machine (Chapter 13).

Sometimes the damage to a vessel wall is less obvious than a cut or laceration. Cholesterol may deposit in the walls and platelets may agglutinate inside the vessel. This leads to an intravascular clot called a **thrombus**. **Thrombosis** is the condition caused by such coagulation. If a thrombus forms in arteries leading to a vital organ, then part or all of the organ may be suddenly deprived of oxygen. Parts of the organ may die, **infarction**, or the whole organ may stop working and death might follow very quickly. In coronary thrombosis, for example, the coronary arteries become blocked (Fig 10.17(a)).

Blood may continue to flow around the clot and circulation continues. However, if the clot becomes dislodged it may move along with the blood flow as an **embolus**. Eventually an embolus will stick in a small artery leading to a body organ. Thus **embolism** within an artery causes the same effect as a thrombosis. The effect differs according to the site of origin of the clot. For example, a clot arising in a leg vein may break loose and be carried along the inferior vena cava (Fig 10.39) and into the right side of the heart. From there the embolus may go to one of the lungs, causing a pulmonary embolism (Fig 10.14). Similarly an embolus may arise in the heart. If this occurs in the left side the embolus may be pumped along a carotid artery (Fig 10.35) and into the brain causing a cerebral embolism.

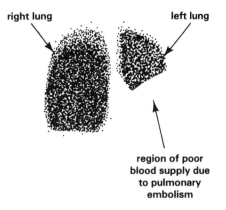

**Fig 10.14** Drawing of a perfusion scan (anterior view) of the lungs of a patient suffering from pulmonary embolism. Compare with Fig. 12.5. Explanation of perfusion scanning technique is in Section 12.1.2

right lung    left lung

region of poor blood supply due to pulmonary embolism

Intravascular blood clots can be prevented with **anticoagulants**. **Heparin** is often used and prevents coagulation by neutralising thrombin. Heparin is a natural product and is found in mast cells and basophils (section 10.1.2). The effects of heparin are only temporary since pro-thrombin is produced by the liver continuously. A longer-lasting effect may be attained by the use of **warfarin** which blocks the production of prothrombin in the liver. Patients are often treated with a combination of herapin and warfarin. Many other substances may be used to prevent coagulation *in vivo* (Table 10.4).

**Table 10.4 The modes of action of some anticoagulants commonly used in medicine**

| Anticoagulant | Mode of action |
| --- | --- |
| heparin | neutralises prothrombin |
| warfarin | blocks prothromin production in the liver |
| fluoride | inhibits the actions of enzymes |
| ethylenediamine tetraacetic acid EDTA (sequestrine) citrate | remove free calcium ions from blood plasma – they become incorporated into salt molecule |

How can blood clot inside intact blood vessels? In what ways can anticoagulants be used to prevent intravascular coagulation? Which anticoagulants cannot be used inside the body? Why is blood stored for transfusion anticoagulated with acid citrate dextrose?

Blood taken for analytical purposes often requires treatment to prevent coagulation. If a sample is allowed to coagulate and the resulting clot is removed, the remaining fluid is called **serum**. Samples may be anti-coagulated with a variety of substances that combine with calcium ions in the plasma (Table 10.4). Blood taken for transfusion must be treated with an anticoagulant and also dextrose, a source of energy which preserves the cells in a viable state for as long as possible. Blood treated with acid-citrate dextrose and stored at 4 °C should remain viable for up to 28 days.

### 10.1.4 Plasma proteins

Blood plasma contains a large quantity of proteins normally amounting to about 70 to 90 g in each $dm^3$. Nearly all the plasma proteins are made in the liver. They circulate in the blood only for a limited period after which they are broken down, also in the liver (Chapter 9). The most abundant plasma protein is **albumin**, accounting for about 36 to 52 g per $dm^3$. The rest consists of **globulins**, accounting for about 24 to 37 g per $dm^3$. When subjected to electrophoresis (Chapter 2) the plasma proteins separate into

177

several fractions (Fig 10.15). Each of the plasma proteins has a specific function. Prothrombin and fibrinogen are involved in coagulation (section 10.1.3). The gamma-globulin fraction contains antibodies called **immunoglobulins** (Chapter 11). Most plasma proteins also act as carriers of metabolites. For example, transferrin carries iron (Table 10.5).

**Fig 10.15** Main fractions of plasma proteins separated by electrophoresis, normal pattern (courtesy Mr R. Dainty, Department of Biochemistry, University of Nottingham Medical School)

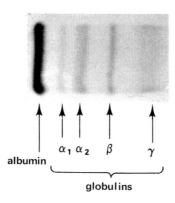

**Table 10.5 Some of the main metabolites transported in the blood by plasma proteins**

| Protein | Metabolites |
| --- | --- |
| albumin | hormones, e.g. thyroxin and steroid hormones<br>drugs, e.g. aspirin, penicillin<br>vitamins A and C<br>acetylcholine, bilirubin, calcium, copper, zinc |
| $\alpha_1$-lipoprotein | phospholipids, cholesterol, hormones,<br>vitamins A and E |
| transcortin | cortisol |
| $B_{12}$-binding protein | vitamin $B_{12}$ |
| thyroxin-binding protein | thyroxin |
| $\alpha_2$-macroglobulin | insulin |
| $\beta$-lipoprotein | phospholipids, cholesterol, hormones,<br>vitamins A and E, free fatty acids |
| transferrin | iron |
| haemopexin | haem |

Collectively all the plasma proteins exert a **colloid osmotic pressure**. This is important in controlling the interchange of water and dissolved solutes between the blood and the tissue fluid (section 10.3.1) and the ultrafiltration of blood in the kidneys (Chapter 13). Another collective property of the plasma proteins is their contribution to the viscosity and density of the blood which are important in determining the pattern of blood flow in the vessels (section 10.2.5).

## 10.2 Blood circulation

The pumping action of the heart and the structure and properties of the different blood vessels allows rapid and uninterrupted delivery of blood to all the body's organs.

### 10.2.1 The heart

The heart is a four-chambered muscular pump. It is situated in the thorax between the lungs and is surrounded by a membranous sac called the **pericardium** (Fig 10.16). It is useful to think of the heart as two pumps, one on the left dealing with oxygenated blood and one on the right dealing with deoxygenated blood. The wall of the heart is called the **myocardium** and is largely composed of **cardiac muscle** (Chapters 8 and 14).

**Fig 10.16** Anatomical position of the heart in the chest

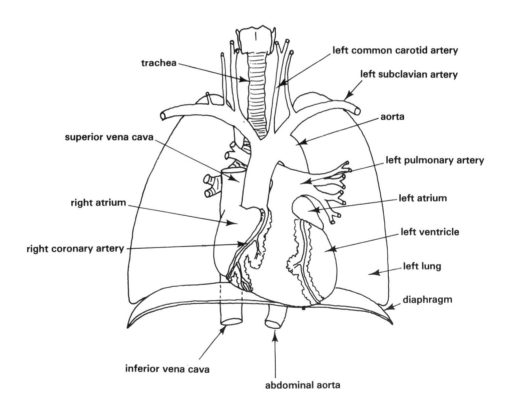

On the left side of the heart an upper chamber called the **left atrium** receives blood from the lungs. Beneath the left atrium is a larger chamber called the **left ventricle** which pumps blood into the **aorta** and from there into the general or **systemic circulation**. On the right side the **right atrium** receives blood from the systemic circulation. The **right ventricle** pumps the blood into the pulmonary trunk which divides into the right and left **pulmonary arteries**. They deliver the blood to the lungs for reoxygenation. On each complete circuit around the body blood travels through the heart twice (Fig 10.17). The myocardium of the ventricles is much thicker than that of the atria; the left ventricle is thickest of all. The thickness of the myocardium reflects the force of pumping done by each chamber. The blood supply to the myocardium is delivered by two **coronary arteries** which branch from the aorta where it leaves the left ventricle (Fig 10.17). Blood flowing through the myocardium returns to the right atrium through **cardiac veins**.

**Fig 10.17** (a) Ventral view of the heart showing coronary blood supply

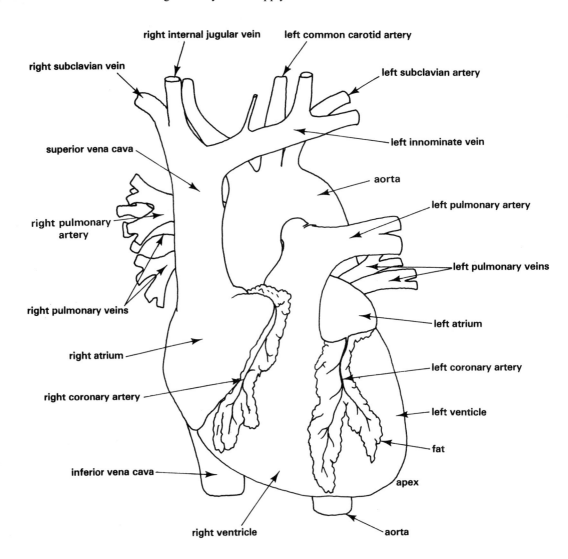

right internal jugular vein

left common carotid artery

right subclavian vein

left subclavian artery

superior vena cava

left innominate vein

aorta

right pulmonary artery

left pulmonary artery

left pulmonary veins

right pulmonary veins

left atrium

right atrium

left coronary artery

right coronary artery

left venticle

fat

inferior vena cava

apex

right ventricle

aorta

Between the atria and ventricles are two atrioventricular valves, the **mitral** (bicuspid) on the left and the **tricuspid** on the right. They are made of non-conductive, fibrous material. They do not contract at all but open and close passively in response to changes in blood pressure occurring in the atria and ventricles. The valves prevent a back-flow of blood when the ventricles relax (Fig 10.18).

**Fig 10.17** (b) Vertical section of the heart

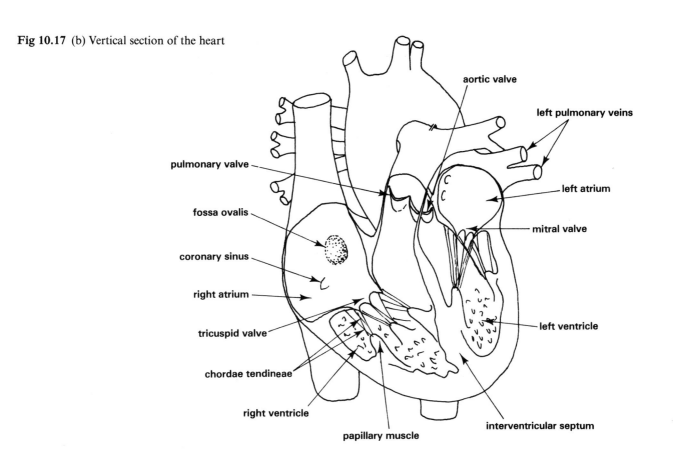

aortic valve

left pulmonary veins

pulmonary valve

fossa ovalis

coronary sinus

right atrium

tricuspid valve

chordae tendineae

right ventricle

papillary muscle

left atrium

mitral valve

left ventricle

interventricular septum

**Fig 10.18** Mitral valve: (a) open, (b) closed

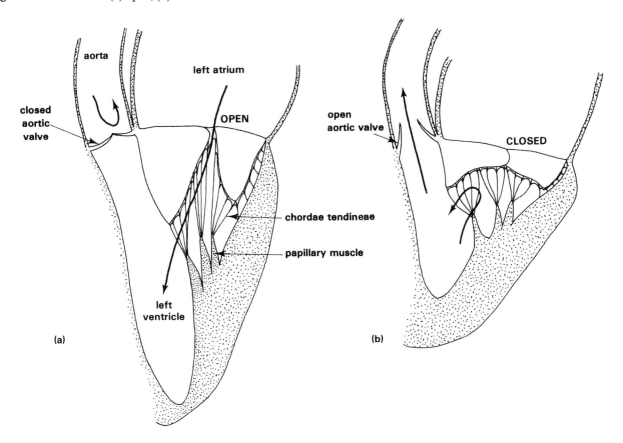

aorta

left atrium

closed
aortic
valve

OPEN

chordae tendineae

papillary muscle

left
ventricle

(a)

open
aortic valve

CLOSED

(b)

The **cardiac cycle** is the sequence of events in the heart which result in effective forward blood flow. It is conventional to begin with atrial contraction, called **atrial systole**. The walls of both atria squeeze on the blood they contain, forcing it through the atrioventricular valves and into the ventricles. At about the same time, the ventricles enlarge to fill with blood. This is **ventricular diastole**. When full, the ventricles begin to squeeze on the blood they contain. **Ventricular systole** forces blood into the aorta and the pulmonary arteries. Contraction of the ventricles also pushes blood against the atrioventricular valve leaflets, causing them to close. The valves are restrained by fibres called **chordae tendineae** which anchor the rims of the valve leaflets to **papillary muscles** projecting into the ventricles (Fig 10.18). Contraction of the papillary muscles during ventricular systole pulls on the chordae tendineae and prevents the valves from turning inside out.

The entrances to the aorta and pulmonary trunk contain **semi-lunar valves**, the **aortic** and **pulmonary valves** respectively. Each consists of three cup-shaped flaps of fibrous tissue thickened along their inner rims. During ventricular systole increased pressure in the ventricles forces the semi-lunar valves open. Blood fills the flaps of the valves, which then close. They are prevented from inverting by the thickening on their rims (Fig 10.19).

**Fig 10.19** Semilunar valve: (a) open, (b) closed

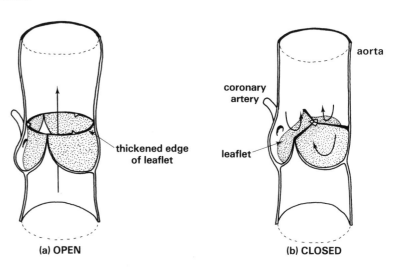

(a) OPEN      (b) CLOSED

**Fig 10.20** Phonocardiogram

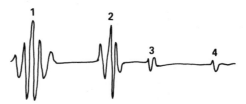

The **heart sounds** can be heard with the aid of a stethoscope. They are the result of the arrest of moving blood when the valves close (Fig 10.20). Extra heart sounds called **murmurs** may be heard and may be caused by the rush of blood into the heart chambers. Turbulence may result from the valve opening becoming narrow or leaky when closed. Movements of the valves during the cardiac cycle can be observed using a technique called **echocardiography**. Ultrasound of very high frequency is generated by a probe and transmitted through the chest wall towards the heart. Ultra-

**Fig 10.21** (a) Transmission of ultrasound through heart including the mitral valve

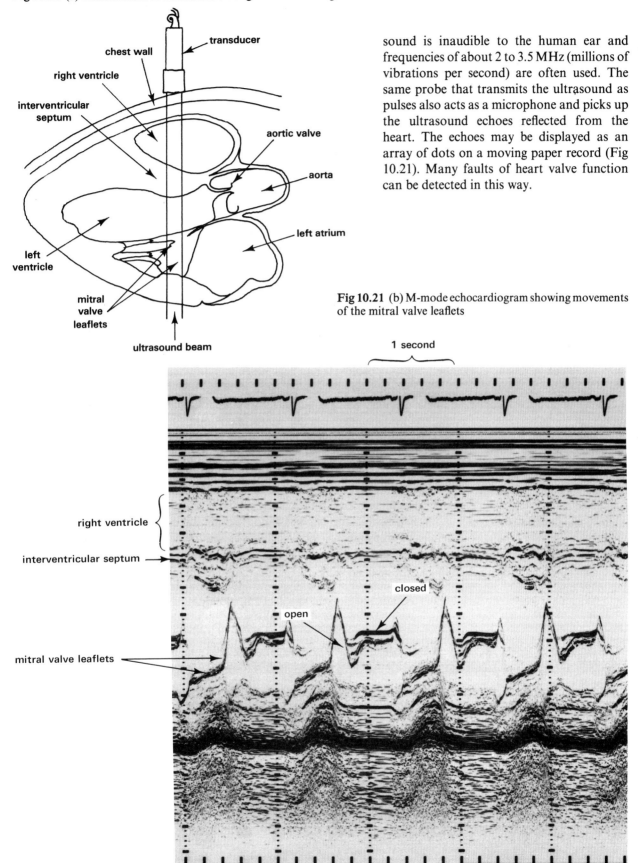

sound is inaudible to the human ear and frequencies of about 2 to 3.5 MHz (millions of vibrations per second) are often used. The same probe that transmits the ultrasound as pulses also acts as a microphone and picks up the ultrasound echoes reflected from the heart. The echoes may be displayed as an array of dots on a moving paper record (Fig 10.21). Many faults of heart valve function can be detected in this way.

**Fig 10.21** (b) M-mode echocardiogram showing movements of the mitral valve leaflets

When the ventricles contract they each expel about 70 cm³ blood. This is called the **stroke volume**. The **cardiac output** is defined as:

**the volume of blood pumped by each ventricle in one minute.**

This is given by:

$$\text{cardiac output} = \text{stroke volume} \times \text{heart rate}$$

where the **heart rate** is expressed as the number of cardiac cycles per minute.

Typical values in resting adults are:

$$\text{cardiac output} = 70 \text{ cm}^3 \times 70 \text{ beats per minute}$$
$$= 4900 \text{ cm}^3 \text{ per minute}$$
$$= 4.9 \text{ dm}^3 \text{ per minute.}$$

Since the total blood volume may only be about 5 dm³, it can be seen that the whole blood is circulated about once every minute, even in resting conditions. During exercise or emotional excitement the cardiac output rises greatly (section 10.2.3).

### 10.2.2 The cardiac impulse

Contractions of the heart muscle during the cardiac cycle are triggered off by an electrical stimulus called the **cardiac impulse**. This originates in a region of the right atrium called the **sino-atrial (SA) node** or **pacemaker**. The impulse spreads out into the walls of the atria causing their contraction, atrial systole. The cardiac impulse is prevented from spreading directly to the ventricles by non-conductive material called the **fibrous ring**, parts of which form the atrioventricular and semi-lunar valves. The impulse is then picked up by a second node called the **atrioventricular (AV) node**, also in the wall of the right atrium (Fig 10.22). From here it is conducted through the fibrous ring along a short **bundle of His**. The bundle divides in the interventricular septum into **left** and **right bundle branches**. The impulse is conducted to the apex of the heart where the bundle branches turn upwards along the endocardium of the ventricles. They then pass through the ventricular myocardium along a finely branched network of **Purkinje fibres** (Fig 10.22), causing the ventricles to contract. A wave of repolarisation, electrical recovery from the impulse (Chapter 14), follows along the same pathway before the next impulse is generated.

**Fig 10.22** Conductive pathway of the cardiac impulse (modified from Julian)

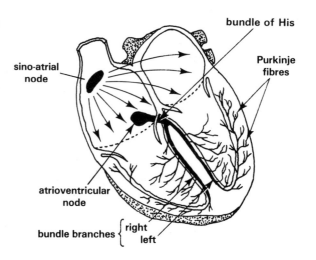

sino-atrial node

bundle of His

Purkinje fibres

atrioventricular node

bundle branches { right left

The conduction of the cardiac impulse through the heart can be recorded by **electrocardiography, ECG**. Electrodes are placed at conventional sites on the skin and voltage differences between selected pairs of the electrodes are measured and recorded. Four electrodes are placed on the limbs, one on each wrist and one on each ankle. One of the ankle electrodes is connected to earth. Six electrodes are usually also placed around the heart on the chest (Fig 10.23). The ECG can be taken between right and left wrists (Lead I), right wrist and left ankle (Lead II) or left wrist and left ankle (Lead III).

**Fig 10.23** (a) Positions for the electrodes used in electrocardiography

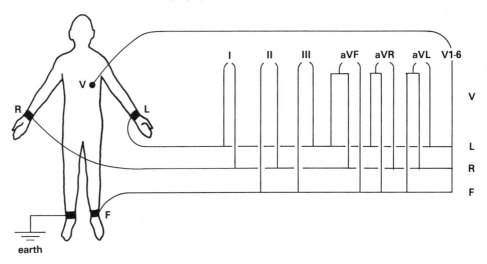

**Fig 10.23** (b) Patient 'wired up' for an ECG

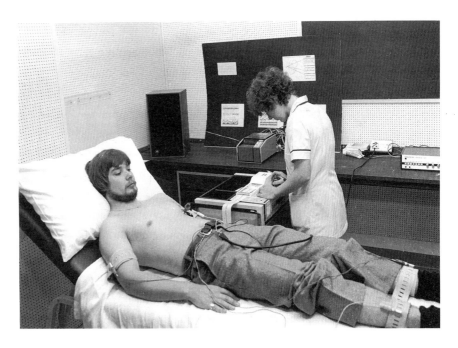

These are the standard limb leads. The ECG can also be taken between any of the three limb electrodes and an electrode made up of the remaining two limb electrodes. These are called the augmented limb leads and provide measurements at angles intermediate between the standard limb leads. This can be appreciated by thinking of the limb electrodes as forming the points of an imaginary triangle with the heart at the centre. This is called **Eindhoven's triangle** (Fig 10.24). The chest leads record the ECG in the transverse plane, that is the transmission of the cardiac impulse outwards from the heart towards the chest wall (Fig 10.25).

**Fig 10.24** Eindhoven's triangle

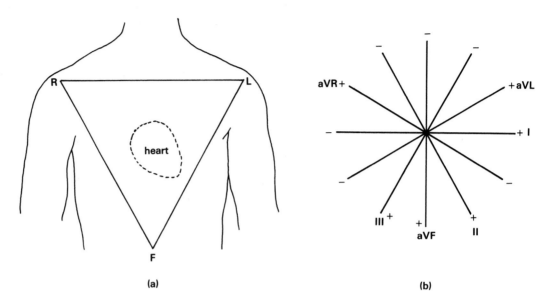

(a)

(b)

**Fig 10.25** Positions of ECG chest leads relative to the heart

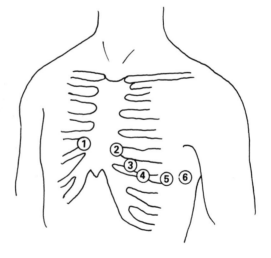

The **electrocardiogram** consists of a number of **waves** and **complexes** that correspond to definite electrical events during the cardiac cycle (Fig 10.26). Atrial depolarisation produces the **P wave**, ventricular depolarisation produces the **QRS complex** and ventricular repolarisation produces

**Fig 10.26** Normal electrocardiogram (Lead II)

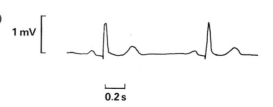

1 mV

0.2 s

**Fig 10.27** Some of the important time intervals between ECG events

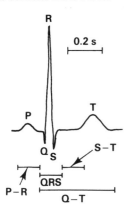

the **T wave**. Atrial repolarisation is masked by ventricular depolarisation which occurs at the same time. The time intervals between the waves are also important and can be measured easily from the ECG (Fig 10.27). Abnormal activity of the heart can often be detected from the ECG pattern (Fig 10.28).

**Fig 10.28** Some abnormal ECGs (a) atrial flutter (b) 2° atrio-ventricular block (c) complete (3°) atrio-ventricular block

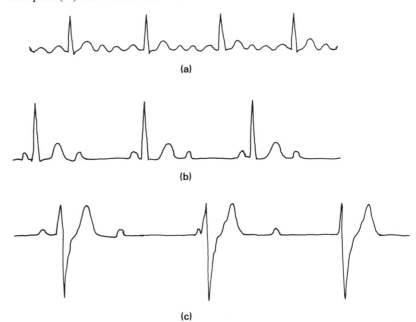

**Fig 10.29** Neuronal pathways controlling the heart (after Green)

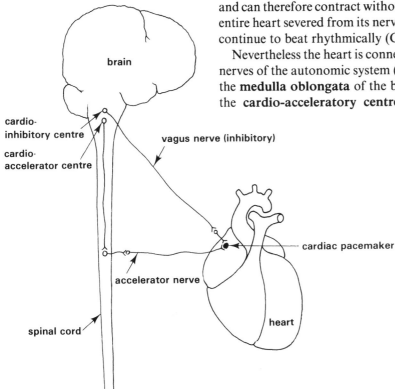

## 10.2.3 Control of cardiac activity

Cardiac muscle is **myogenic** because it generates impulses spontaneously and can therefore contract without any external stimulus being applied. An entire heart severed from its nerve supply and removed from the body may continue to beat rhythmically (Chapter 14).

Nevertheless the heart is connected to sympathetic and parasympathetic nerves of the autonomic system (Figs 10.29 and 14.34). The nerves arise in the **medulla oblongata** of the brain. The sympathetic pathway begins in the **cardio-acceleratory centre**, the parasympathetic pathway in the **cardio-inhibitory centre**. Activation of the cardiac pacemaker by the **cardiac nerve** (sympathetic) increases the rate at which cardiac impulses are generated and hence the rate of heart beat. Branches of the cardiac nerve also stimulate the ventricular walls and increase the power of contraction, hence the stroke volume and cardiac output also increase (section 10.2.1). Activation of the pacemaker by the **vagus nerve** (parasympathetic) decreases the heart rate.

Cardiac output increases when we engage in physical exercise. How is the heart controlled and what causes cardiac output to fall after the exercise stops?

Although cardiac muscle is myogenic, the nerves supplying the heart allow cardiac activity to be modified in different circumstances. For example, during exercise cardiac output increases, thus supplying extra oxygen to the skeletal muscles and removing the extra carbon dioxide produced by them as a result of the exercise. Conversely cardiac output drops during periods of rest. The maximum amount by which the cardiac output can increase is called the **cardiac reserve**. It is usually about 400 % but may be about 600 % in an athlete.

There are several mechanisms which control the heart's activity.

### 1 Starling's law of the heart

Starling's law states that:

> **the power of cardiac contraction is directly related to the length of the cardiac muscle fibres.**

If more blood enters the heart from the veins, the cardiac muscle fibres in the myocardium are stretched more and this induces the fibres to contract with greater force.

### 2 Baroreceptors

Small receptors sensitive to stretching are found in the walls of the aortic arch, the carotid sinuses, the venae cavae and the right atrium (Fig 10.30). They are called **baroreceptors**. If blood pressure increases in any of these vessels, the baroreceptors are activated and they transmit sensory impulses to the cardiac centres. Cardiac activity is decreased appropriately.

**Fig 10.30** Position of the baroreceptors (stippled) in the cat

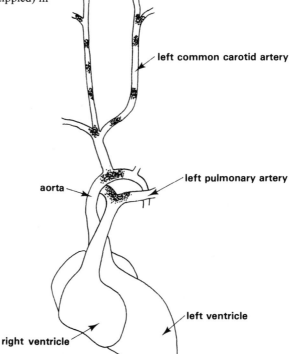

left common carotid artery

left pulmonary artery

aorta

left ventricle

right ventricle

### 3 Adrenalin

In times of stress adrenalin is secreted into the blood by the medullae of the adrenal glands (Chapter 16). Adrenalin increases the excitability of the pacemaker and speeds up the heart rate. A similar substance called **noradrenalin** is released onto the pacemaker by the sympathetic (cardiac) nerve endings on the arrival of impulses. The inhibitory parasympathetic (vagus) nerve releases acetylcholine onto the pacemaker, resulting in a decrease in heart rate (Chapter 14).

### 4 Other factors

A drop in oxygen tension or a rise in carbon dioxide tension of the blood, directly affect the cardio-acceleratory centre and result in an increased heart rate. The same effect results from a drop in blood pH.

A high concentration of potassium ions in the blood interferes with the nerve impulses that control the pacemaker, causing the heart to slow down. Raised levels of blood sodium ions also cause the heart to slow down because excess sodium ions interfere with the action of calcium ions in muscle contraction (Chapter 14).

Emotional excitement and a rise in body temperature cause the heart rate to speed up. Conversely, feelings of grief and depression may slow the heart rate down.

### 5 Artificial cardiac pacing

Should the pacemaker become faulty, the heart may require additional stimulation. This may be provided by an **artificial pacemaker** (Fig 10.31). Artificial pacemakers are devices designed to deliver small electrical pulses to electrodes implanted in the myocardium. They can be set to deliver pulses at different rates and are powered by small replaceable batteries. Research is being carried out into ways of making artificial pacemakers respond to different levels of body activity. If this could be achieved, rapid changes in pacing rate which would normally happen in different circumstances of exercise or emotional excitement could be brought about.

**Fig 10.31** (a) An artificial cardiac pacemaker. (b) Chest X-ray showing pacemaker in position

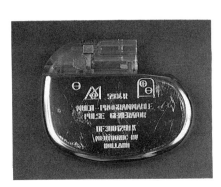

(a)

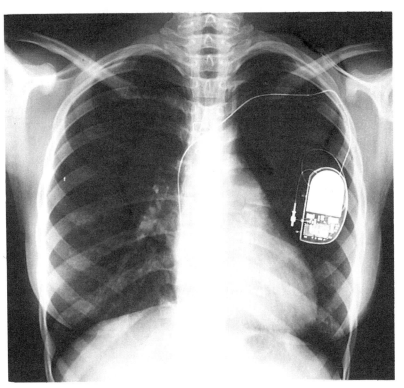

(b)

### 10.2.4 Blood pressure

The pressure at which blood flows in the circulatory system is generated mainly by the pumping action of the ventricles. Pressure-sensitive devices called **pressure transducers** may be connected by long flexible tubes called **catheters** to the heart chambers. Using such devices it has been discovered that the pressure inside the left ventricle during systole usually rises to a maximum of about 15.79 kPa (120 mmHg) and then drops to zero during diastole. The pressures in the right ventricle are somewhat lower, between about 3.29 kPa (25 mmHg) systolic and zero diastolic pressures. The fluctuations in blood pressure in the main arteries supplied by the ventricles, however, are not quite the same. During ventricular systole, when the semi-lunar values are open, the aortic pressure reaches 15.79 kPa (120 mmHg) and the pulmonary artery pressure reaches 3.29 kPa (25 mmHg). However, during ventricular diastole, when the semi-lunar valves close, the aortic pressure drops only to about 10.53 kPa (80 mmHg) and the pulmonary artery pressure drops to about 1.1 kPa (8 mmHg). How can this be so, since during diastole the ventricles are not forcing blood into the arteries?

The explanation lies in the fact that the walls of the aorta and pulmonary arteries are elastic (section 10.2.5). During ventricular systole they stretch, whereas during ventricular diastole they recoil, thus continuing to propel blood through the vessels. As the blood flows from arteries into arterioles and then into capillaries, the pressure drops. A major cause of this **peripheral resistance** is friction. The narrower the vessels are through which the blood flows, the greater is the resistance. The vessels offering the greatest resistance are the arterioles. They contain smooth muscle (Chapter 8) in their walls. When stimulated by sympathetic nerves the smooth muscle contracts, causing a narrowing of the arterioles, **vasoconstriction**. When sympathetic stimulation stops, the arteriolar muscle relaxes and **vasodilation** occurs (Fig 10.37). This activity affects peripheral resistance and hence blood pressure. It is controlled by a sympathetic nerve pathway originating in the **vasomotor centre** in the medulla oblongata of the brain. Sensory impulses arising in the baroreceptors (Fig 10.30) activate the vasomotor centre and blood pressure is increased or decreased according to the body's requirements (Fig 10.32).

**Fig 10.32** Main factors controlling blood pressure

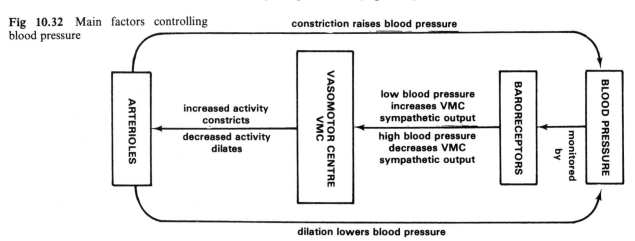

Blood pressure, of course, is also affected by changes in the cardiac output. Fig 10.33(a) correlates some of the main events of the cardiac cycle and the pressure changes which occur in the heart chambers and elastic arteries. Blood spurting through arteries near the surface of the skin may be detected as the **pulse**. Pulse rate corresponds to heart rate.

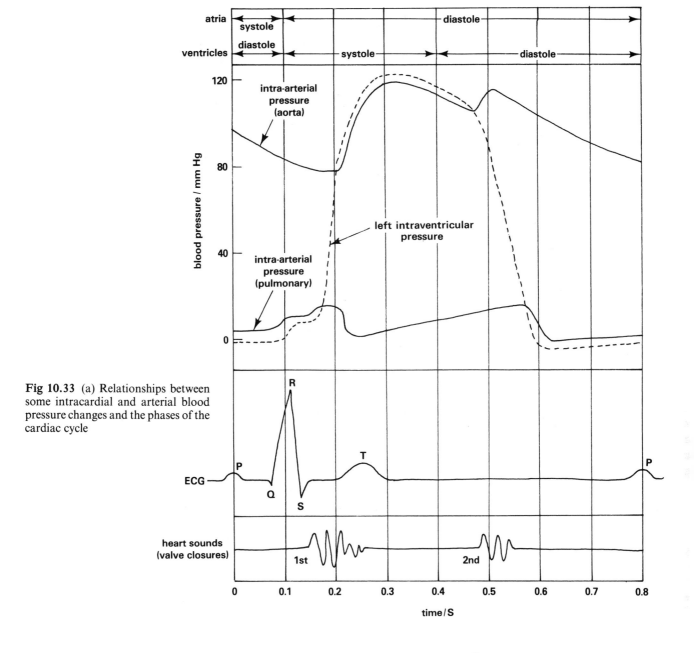

**Fig 10.33** (a) Relationships between some intracardial and arterial blood pressure changes and the phases of the cardiac cycle

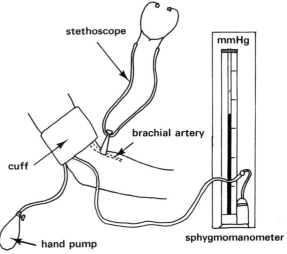

**Fig 10.33** (b) Sphygmomanometer (from Green, J.H. *An Introduction to Human Physiology*, Oxford Medical Publications)

Blood pressure is usually measured with the aid of a **sphygmomanometer** (Fig 10.33(b)). A cuff is wrapped around the arm, just above the elbow. A rubber bag inside the cuff is inflated by squeezing a hand pump. The pressure of the air inside the bag is raised to about 26.32 kPa (200 mmHg) which is enough to flatten the brachial artery in the arm. Blood flow in the artery is thus stopped. The microphone of a stethoscope is placed over the brachial artery, just distal to the cuff. The pressure inside the cuff is slowly lowered by means of a valve. A pressure is reached at which the blood can just squeeze through the artery each time the left ventricle contracts. The artery makes tapping sounds, named after **Korotkov** who first described them in 1905. The pressure under the cuff corresponds to the systolic blood pressure. The cuff pressure is now lowered still further until the blood flows through the artery even between ventricular contractions. The Korotkov sounds suddenly become muffled and disappear. The cuff pressure at this point corresponds to the diastolic blood pressure. It is usual to express the blood pressure as 'systolic over diastolic'. Typical values in a resting adult are 15.79/10.53 kPa (120/80 mmHg).

### 10.2.5 Blood vessels

Blood flows from the ventricles into **arteries**, then **arterioles, blood capillaries** and **venules**, and is finally returned to the heart in **veins**. Blood vessels have properties that help the circulation and allow the blood to perform many of its functions (Fig 10.42). The walls of the larger blood vessels conform to a basic structural pattern (Fig 10.34). There are three main layers. On the inside the **tunica intima** (t. interna) consists of a single layer of flat cells, the **endothelium**, supported by a basement membrane and **connective tissue** containing **collagen** (Chapter 7). The middle layer is the **tunica media** which contains smooth muscle and fibres of collagen and **elastin**. The tunica media varies in thickness in different vessels. It is completely absent in blood capillaries but is the thickest layer in the elastic arteries such as the aorta (section 10.2.4). Collagen provides strength and prevents excessive stretching of the blood vessels. In contrast, elastin can be stretched, thus allowing vasodilation. The outermost layer is called the **tunica adventitia** (t. externa). It also contains collagen fibres and is

**Fig 10.34** Structure of the wall of (a) elastic and (b) muscular arteries

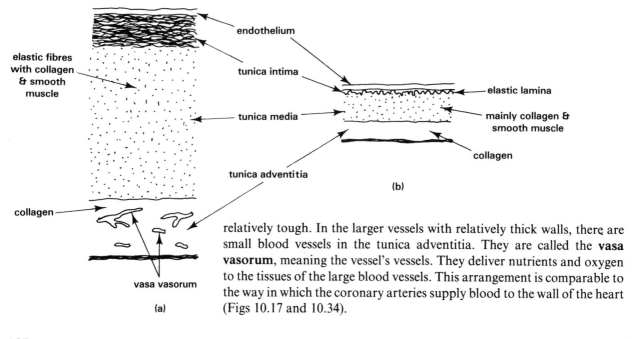

relatively tough. In the larger vessels with relatively thick walls, there are small blood vessels in the tunica adventitia. They are called the **vasa vasorum**, meaning the vessel's vessels. They deliver nutrients and oxygen to the tissues of the large blood vessels. This arrangement is comparable to the way in which the coronary arteries supply blood to the wall of the heart (Figs 10.17 and 10.34).

## 1 Arteries

The arrangement of the main arteries in the systemic circulation is illustrated in Fig 10.35. The function of the arteries is to distribute blood from the heart to the organs throughout the body. **Elastic arteries**, especially the aorta and pulmonary arteries, also assist circulation of blood during ventricular diastole (section 10.2.4). Beyond the elastic arteries blood enters **muscular arteries**. The tunica media of muscular arteries contains less elastin and more collagen. For this reason they are less dilatable than the elastic arteries. Sometimes arteries become diseased. Deposits called **atheromatous plaques** may form in the tunica intima. They cause a narrowing, **stenosis**, of the affected vessel. Blood flow is impaired (Fig 10.36). The deposits may also give rise to intravascular thrombi (section 10.1.3). High concentrations of cholesterol and triglycerides in the blood are thought to cause such deposits.

**Fig 10.35** Diagram showing the main arteries of the blood vascular system

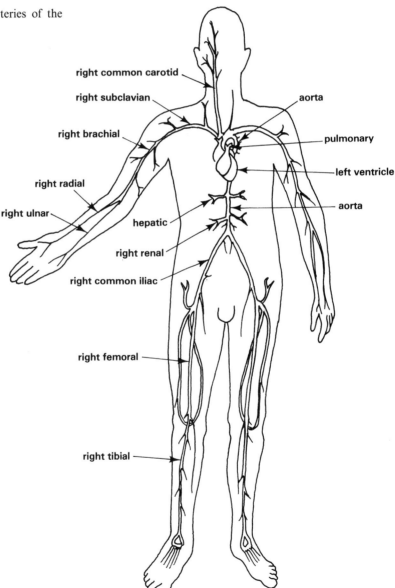

**Fig 10.36** Stenosis of an artery due to atheromatous plaques

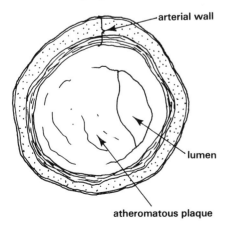

## 2 Arterioles

Arterioles are found inside the tissues. They are the terminal branches of arteries. Smooth muscle fibres are relatively abundant in arteriole walls. By contracting or relaxing, the fibres can alter the diameter of the **lumen** through which the blood flows, and thus the blood pressure (section 10.2.4 and Fig 10.37).

**Fig 10.37** Photomicrographs of (a) dilated and (b) constricted arterioles seen in transverse section, × 400 and × 270 respectively

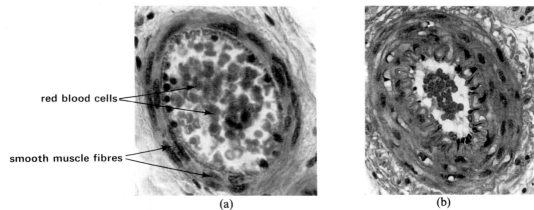

red blood cells

smooth muscle fibres

(a)                (b)

## 3 Blood capillaries

Capillaries are the smallest blood vessels, having a diameter as small as 3 to 4 $\mu$m. This is about half the diameter of a red blood cell (Fig 10.38). Some blood capillaries are up to ten times larger than this and are called **sinusoids**, such as those in the liver (Fig 9.19). The capillary wall consists of a single layer of endothelium and presents little barrier to the movement of many metabolites across the wall. The main function of blood capillaries is to allow an orderly interchange of metabolites between blood and tissue fluid (section 10.3.1). The capillary system in any organ is much branched and is often called the **capillary bed**. No cell in the organ is more than a few cells distant from the nearest blood capillary. Once in the tissue fluid, metabolites move to and from the cells largely by diffusion (Chapter 1).

**Fig 10.38** Electronmicrograph showing a blood capillary in transverse section, × 6000

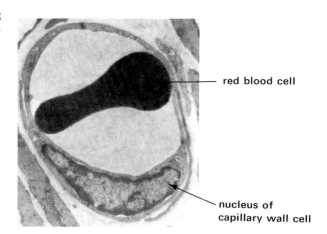

red blood cell

nucleus of capillary wall cell

## 4 Venules and veins

The positions of the main veins in the systemic circulation are illustrated by Fig 10.39. Venules and veins collect blood of low pressure and return it to the heart. Blood in the veins flows in steady streams rather than spurts as in arteries. The blood in the centre of a vein flows fastest, whereas that just

**Fig 10.39** Diagram showing the main veins of the blood vascular system

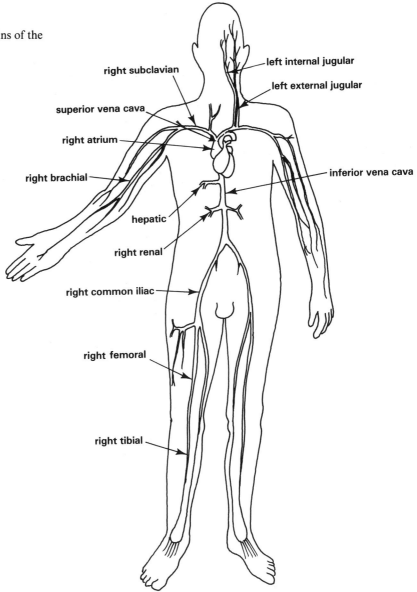

inside the vein wall flows very slowly. This is called **laminar flow** (Fig 10.40(a)). It is affected by blood velocity, density and viscosity and the diameter of the containing vessel. These factors determine **Reynold's number, Re** as follows:

$$\text{Re} = \frac{VD\rho}{\mu}$$

where $V$ = velocity/cm s$^{-1}$
$D$ = diameter of vessel/cm
$\rho$ = density, and
$\mu$ = viscosity

When the value for Reynold's number is below about 2000 then blood flow is laminar. If for any reason the value exceeds 2000, turbulence occurs (Fig 10.40(b)). In the inferior vena cava Reynold's number is about 1750.

**Fig 10.40** (a) Laminar and (b) turbulent blood flow

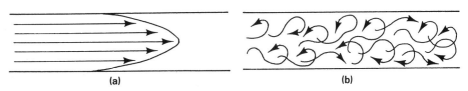

(a)          (b)

Several factors cause venous blood to flow towards the heart. One is the negative pressure created in the thorax every time air is drawn into the lungs (Chapter 12). When exhaling air from the lungs the thoracic pressure is raised. Blood in the veins is also prevented from flowing away from the heart by semi-lunar valves (Fig 10.41). Many large veins, especially in the limbs, lie between skeletal muscles. Contraction of the muscles squeezes the veins and forces blood through them towards the heart.

**Fig 10.41** Roles of semi-lunar valves and the contraction of skeletal muscle in regulating blood flow through veins

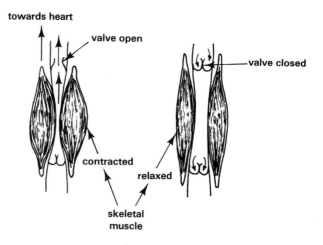

The important properties of the different blood vessels are summarised in Fig 10.42.

**Fig 10.42** Main properties of the different kinds of vessels in the blood vascular system

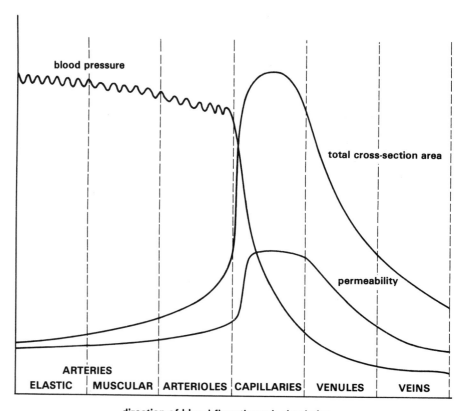

## 10.3 Tissue fluid and lymph

Blood distributes oxygen and metabolites to the body's tissues and removes their products including wastes. However, tissue cells are not in direct contact with the blood. They are bathed in **tissue fluid** which is an intermediary between the blood and the cells.

### 10.3.1 Interchange between blood and tissue fluid

Metabolites pass between the blood and the tissue fluid through the walls of the blood capillaries. Capillary walls are freely permeable to substances of relatively small molecular size. The walls of other blood vessels are relatively impermeable to the blood constituents (Fig 10.42). Substances of relative molar mass greater than about 65 000 and most kinds of blood cell normally remain inside the capillaries. The molecules of most blood proteins are too large to pass into the tissue fluid.

Interchanges between blood and tissue fluid are brought about by the interaction of the **hydrostatic pressure (HP)** and the **colloid osmotic pressure (COP)** of the blood and tissue fluid. Hydrostatic pressure is the pressure exerted by a liquid against its surroundings. Colloid osmotic pressure is created by molecules of large relative molar mass, especially proteins which attract water (Chapter 1). The actions of these forces are summarised in Fig 10.43. Because of the pumping action of the heart and recoil of the elastic arteries, blood leaving arterioles has a relatively high hydrostatic pressure at about 3.29 kPa (25 mmHg). Plasma proteins, especially albumin, cause the colloid osmotic pressure of the blood to be about 3.68 kPa (28 mmHg). In the tissue fluid the hydrostatic pressure is estimated at about −0.83 kPa (−6.3 mmHg), slightly less than atmospheric pressure. The colloid osmotic pressure in the tissues is also low at about 0.66 kPa (5 mmHg). This is because only few proteins can pass through capillary walls from the blood into the tissue fluid.

**Fig 10.43** Summary of the forces affecting the interchange of materials between blood and tissue fluid (figures from Guyton)

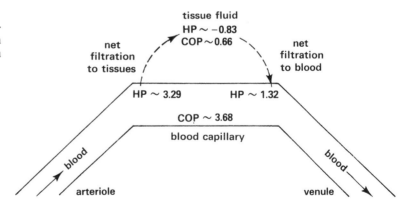

The resultant of the pressures is called **filtration pressure (FP)**:

$$FP = \left[ \begin{array}{c} HP\ of \\ blood \end{array} - \begin{array}{c} HP\ of \\ tissue\ fluid \end{array} \right] - \left[ \begin{array}{c} COP\ of \\ blood \end{array} - \begin{array}{c} COP\ of \\ tissue\ fluid \end{array} \right]$$

$$= [25 - (-6.3)] - (28 - 5)$$

$$= 31.3 - 23 = 8.3\ mmHg$$

The positive filtration pressure at the arteriolar end of the capillary network causes water, oxygen and other metabolites to be forced from the blood into the tissue fluid.

As the blood flows through to the venous end of the capillary network the hydrostatic pressure drops to about 1.32 kPa (10 mmHg). All other pressures remain fairly constant and so the filtration pressure is now given by:

$$FP = [10 - (-6.3)] - (28 - 5)$$
$$= 16.3 - 23$$
$$= -6.7 \, \text{mmHg}$$

Because the filtration pressure at the venous end of the capillary network is negative, water, excess metabolites, tissue products and wastes are forced from the tissue fluid into the blood. However, there is a greater loss from the blood to the tissue fluid at the arteriolar end of the capillaries than returns to the blood in the venous capillaries. For this reason the tissues gradually accumulate fluid at the blood's expense.

### 10.3.2 Lymph

The excess fluid absorbed from the blood in the tissues enters **lymphatic capillaries** (Fig 10.44). They drain the excess fluid, now called **lymph**, into a system of vein-like vessels known as the **lymphatic system** (Fig 10.45). Lymph is produced at a rate of about 1.5 cm³ each minute. The main lymphatic vessels contain semi-lunar valves which ensure lymph flows in one direction. Lymph is pushed through the lymphatic system when the surrounding tissues and muscles squeeze on the lymphatic vessels.

**Fig 10.44** Diagrammatic relationship between the blood vascular and lymphatic systems

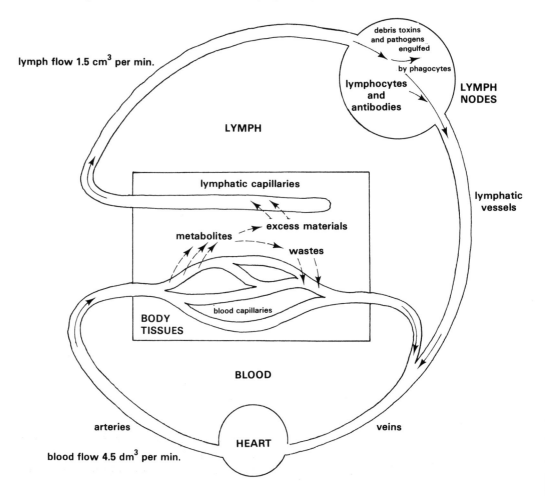

**Fig 10.45** Main vessels of the lymphatic system. Those of the left limbs are omitted for clarity

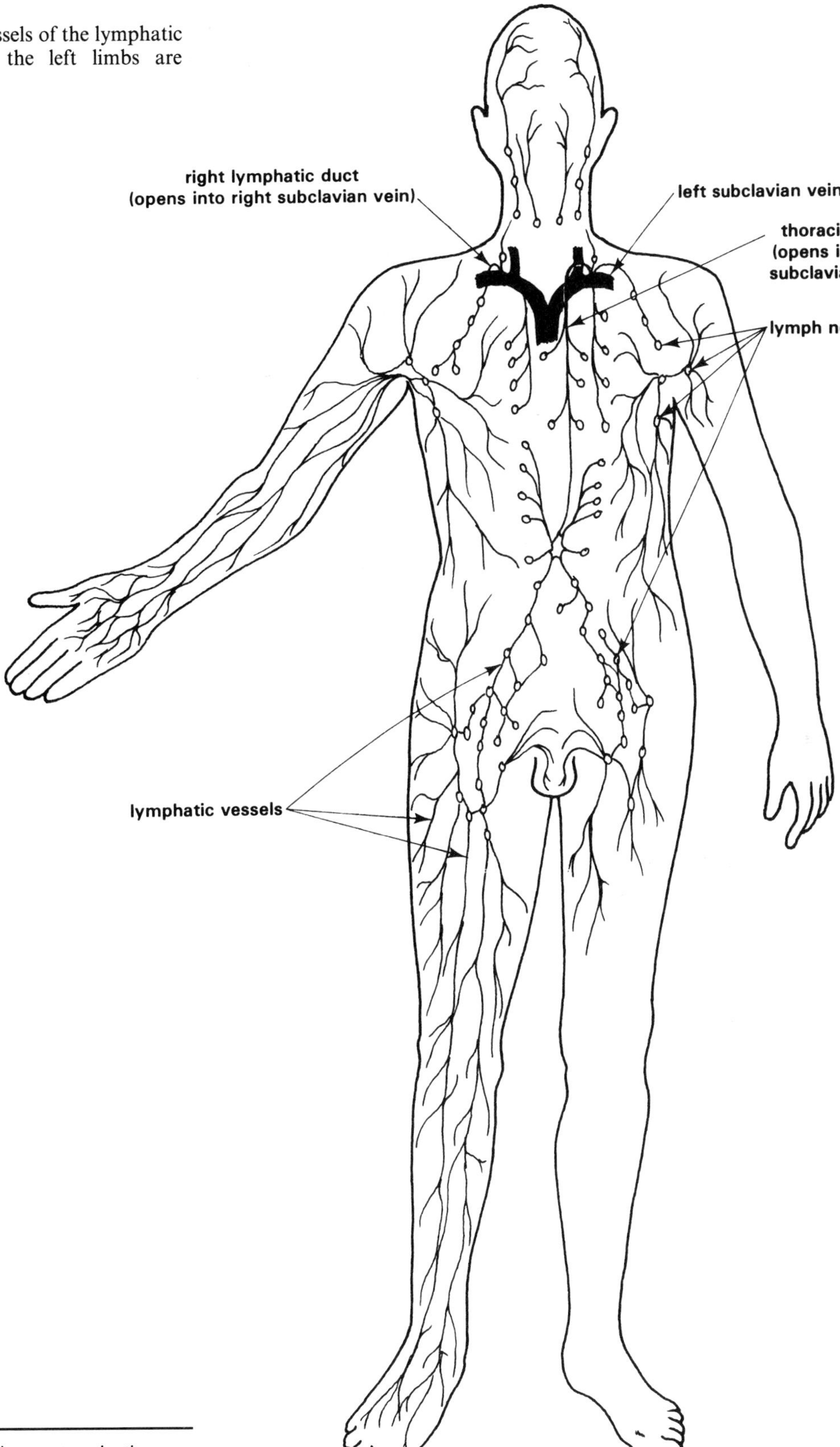

right lymphatic duct
(opens into right subclavian vein)

left subclavian vein

thoracic duct
(opens into left subclavian vein)

lymph nodes

lymphatic vessels

Water containing nutrients enters the tissues from the blood capillaries. A smaller volume of water containing various substances, including metabolic wastes, is reabsorbed by the blood capillaries. What happens to the rest of the fluid? How does it return to the blood?

Lymph eventually returns to the blood. This is because the thoracic lymph ducts open into the subclavian veins. Thus a constant blood volume is maintained and the tissues are not saturated with excess fluid. If lymph return is blocked, such as by parasitic worms in **elephantiasis**, then tissue fluid accumulates. This is called **oedema** (Fig 10. 46). Oedema can also result from imbalances in the hydrostatic and colloid osmotic pressures of the blood and tissue fluids, or from changes in the permeability of blood capillary walls. Severe protein deficiency is sometimes seen in the diets of children in some underdeveloped countries. The deficiency leads to **kwashiorkor**. Among the many harmful effects is severe oedema caused by a low colloid osmotic pressure in the blood and excess retention of fluid in the tissues.

**Fig 10.46** Elephantiasis

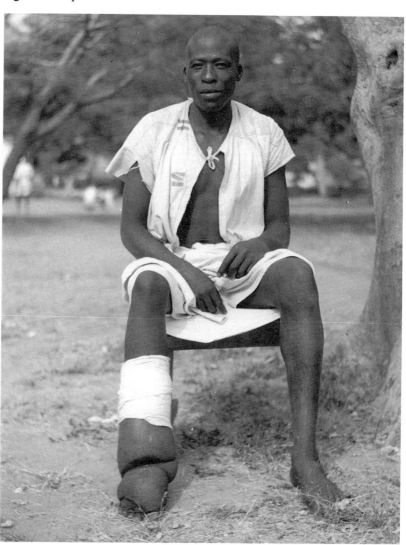

On its return journey to the blood, lymph flows through one or more **lymph nodes**. They remove particulate debris, microbes and toxins from the lymph and release **antibodies** and **lymphocytes** into the lymph (Chapter 11).

# 11 Immune system

**Immunity** is the body's ability to resist disease. Much of the body's resistance to pathogenic (disease-causing) microbes is brought about by several general body functions. Such resistance is called **innate immunity**. Mechanisms of innate immunity include phagocytosis (Chapters 7 and 10), the acidity of gastric juice, the hydrolytic action of the enzyme muramidase in tears and resistance of the epidermis of skin to penetration (Chapter 17).

**Acquired immunity** results from the actions of **antibodies**. Antibodies are proteins produced in the lymph nodes in response to the presence of **antigens** to provide **active immunity**. A foetus or the new-born infant may also obtain antibodies from its mother, across the placenta or in breast milk respectively. Antibodies obtained in this way provide **passive immunity**. They are not produced by the individual's own immune system. Passive immunity is short-lived because the transferred antibodies are soon destroyed in the liver. For this reason new-born babies are usually resistant to a variety of diseases for only about three months. At this age a programme of vaccination is normally begun to give active immunity against serious microbial diseases (section 11.1.6). Active immunity is usually long-lasting. The antibody content of the blood rises after initial exposure to an antigen, but usually falls quite quickly afterwards. Following subsequent exposure to the same antigen the antibody level rises further and stays high for a long time (Fig 11.1).

**Fig 11.1** Changes in the level of antibody in the blood following two injections of antigen

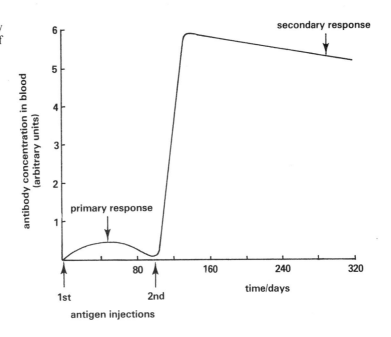

## 11.1 Acquired immunity

Newborn babies are resistant to a variety of diseases. How is this when they have not yet been exposed to the organisms that cause the diseases? How do the babies lose their resistance after about three months? How can the resistance be restored by the use of vaccines?

Any substance which triggers off an active immune response is called an antigen. Antigens are mainly proteins or even protein fragments. **Non-self** antigens form a part of the outer surface of bacteria, fungi and viruses, or may be secreted as toxins by pathogenic microbes. **Self** antigens are found in the membranes of our body cells. We do not normally produce antibodies in response to self antigens. Blood transfusion and organ transplantation are ways in which our antigens could be introduced into someone else. To the recipient, our antigens are non-self. An immune response normally follows exposure to non-self antigens. Any particular antibody reacts with only one kind of antigen. Exposure to a different antigen induces the production of a different antibody. Like enzymes, antibodies are specific.

Clearly the ability to recognise and distinguish between self and non-self antigens is vital to the proper functioning of our immune system. **Lymphocytes** are stimulated by antigens to produce an immune response. They originate in the bone marrow from **stem cells** similar to those which give rise to the other blood cells (Chapter 10). However, unlike the other blood cells, lymphocytes migrate to the lymph nodes where they mature and are produced in large numbers throughout life. The advantage of the lymph nodes being the sites of sensitivity to non-self antigens is that the lymphatic system drains lymph from all the body's tissues. Antigens anywhere in the body should therefore be detected.

There are two kinds of lymphocyte, **T-cells** and **B-cells**. T-cells are so called because they are processed in the **thymus** gland before entering the lymph nodes. They are thymus-dependent lymphocytes. Removal of the thymus gland before birth prevents T-cell formation. The thymus gland is prominent early in life, reaching its maximum size during puberty (Fig 11.2). By the time physical maturity is established the thymus is usually atrophied. T-cells are produced throughout life in the lymph nodes from the lymphocytes that were processed in the thymus.

**Fig 11.2** Position of the thymus gland in a child

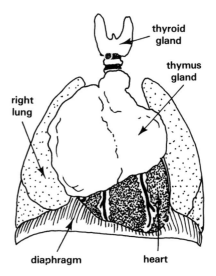

B-cells are thought to be processed in the **bone marrow**. T-cells account for about 80 % of lymphocytes in the blood, the rest are B-cells. T- and B-cells are involved in two distinct immune mechanisms called the cellular and humoral responses respectively.

### 11.1.1 Cellular immune response

The **cellular** immune or **cell-mediated response** is usually brought about by the presence in our bodies of cells with non-self antigens on their surfaces. Such cells may be part of a transplanted organ. They may be cancer cells or self cells whose antigens have been changed to non-self after a viral infection.

Non-self antigens entering the lymph nodes **sensitise** T-cells. Large numbers of these cells are then quickly produced by mitosis, many of them entering the bloodstream. Such a population of identical cells is called a **clone**.

Several types of T-cells can be recognised in a clone. **Killer** T-cells attach to the invading cells onto which they secrete a number of **cellulotoxic** substances. The substances may be enzymes from the killer cells' lysosomes (Chapter 6). Other substances released by killer T-cells attract **macrophages** and activate phagocytosis. Macrophages are thought to be monocytes which have invaded the tissues (Chapters 7 and 10). Other substances secreted by T-cells are called **lymphokines** such as **interferon**. It prevents viral replication.

**Helper** T-cells assist the plasma cells produced from B-lymphocytes to secrete antibodies (section 11.1.2).

**Suppressor** T-cells suppress the activity of killer T-cells and B-cells. Interaction between suppressor T-cells and helper T-cells regulates the immune response.

**Memory** T-cells retain the ability to recognise the non-self antigen in the future. Hence, subsequent exposure to the same antigen normally brings about a rapid cellular response. Immunity of this kind is thus conferred for a long time, often for life (Fig 11.3).

**Fig 11.3** Summary of the cellular response

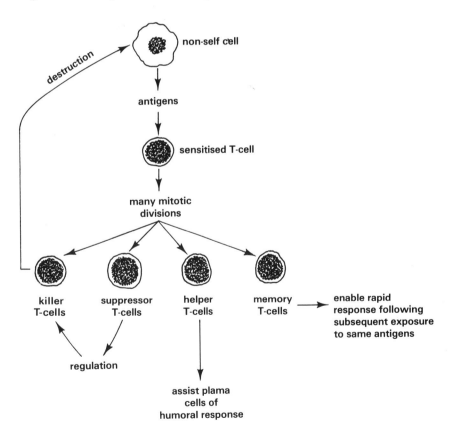

## 11.1.2 Humoral immune response

The **humoral response** usually occurs when pathogenic microbes get into our bodies. On entering the lymph nodes they activate B-cells to divide

**Fig 11.4** Electronmicrograph of a plasma cell, × 15 000

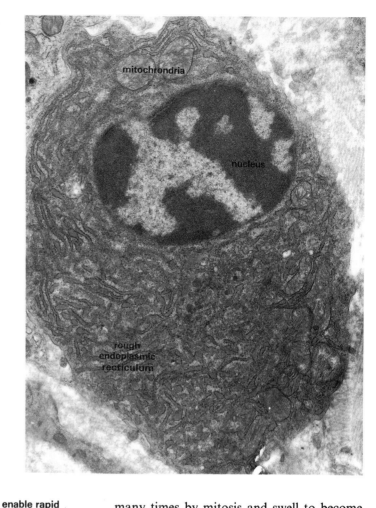

**Fig 11.5** Summary of the humoral immune response

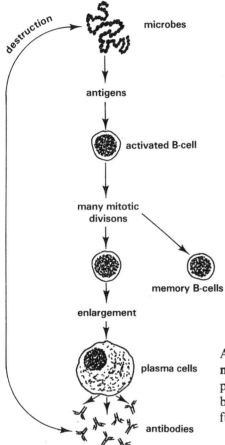

many times by mitosis and swell to become large **plasma cells** (Fig 11.4). Plasma cells produce large quantities of antibody specific to the antigen which triggered the response. Antibodies produced like this enter the blood and circulate as **immunoglobulins** (Fig 11.5). They form the **gamma-globulin** fraction of the plasma proteins (Chapter 10). There are several categories of immunoglobulins called IgG, IgM, IgA, IgE and IgD. Their characteristics and main functions are summarised in Table 11.1 and Fig 11.6.

**Table 11.1 Main categories of immunoglobulins**

| | |
|---|---|
| IgG | Represents about 85 % of the total immunoglobulin fraction of plasma proteins. Relative molar mass is about 160 000. IgG includes many different antibodies, e.g. anti-D (section 11.2.2). |
| IgM | Consist of a cluster of five IgG-like antibodies joined at the centre. IgM includes anti-A and anti-B antibodies (section 11.2.1). |
| IgA | Much less abundant than IgG in blood but is found in tears, secretions of the naso-pharynx and in breast milk. May provide passive immunity for the new-born as well as protecting the eyes and respiratory tract from infection. |
| IgE | Includes the reagin-type antibodies involved in allergic responses (section 11.1.5). IgE antibodies may cross the placenta to provide passive immunity for the foetus. |
| IgD | A category of immunoglobulin of uncertain function. |

**Fig 11.6** Immunoglobin structures

(a) Structure of IgG. The antigen-binding sites are variable thus allowing different antibodies to attach to different antigens. The antigen-binding site of any one antibody is fixed thus explaining the specificity of antigen-antibody reaction

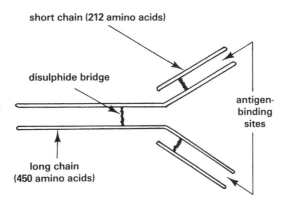

short chain (212 amino acids)

disulphide bridge

antigen-binding sites

long chain (450 amino acids)

(b) Structure of IgM

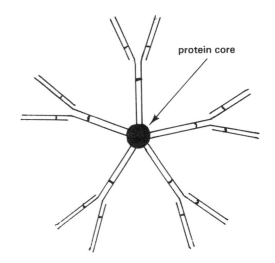

protein core

(c) Electronmicrograph of IgM antibody in mouse serum, × 1 000 000 (courtesy Dr R Dourmashkin)

Blood plasma contains at least 15 different proteins collectively called **complement**. They may be activated by the binding of immunoglobulin molecules to surface antigens such as may be found on bacteria. When activated, the complement proteins interact to help the humoral immune response. One reaction causes destruction of the bacterial cell wall, leading to its lysis and death (Fig 11.7). Complement may also attach to

**Fig 11.7** Electronmicrograph of the surface of a red blood cell membrane showing the holes produced by the combined action of antibody and complement, × 400 000 (courtesy Dr R Dourmashkin)

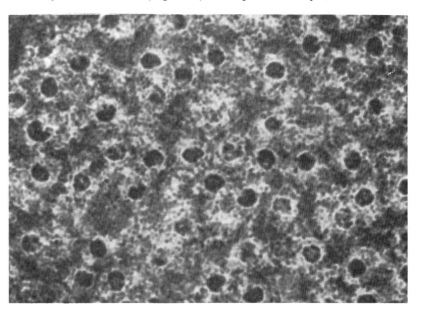

macrophages and neutrophils, promoting phagocytosis. Some of the activated B-cells do not develop into plasma cells. They remain in the lymph nodes as **memory B-cells**. Like memory T-cells they can recognise the original non-self antigens in the future. They bring about a rapid humoral immune response following subsequent infection involving the same antigen. In recent years a technique has been developed to fuse lymphocytes with tumour cells. The resulting **hybridomas** can be grown indefinitely. They secrete antibody specific to the antigen that activated the lymphocytes used in the hybridomas. Such **monoclonal antibody** can be purified and used to provide immunity to certain diseases.

### 11.1.3 Antigen-antibody reactions

Antibodies react with antigens in a variety of ways. The reactions, which usually destroy the antigens, include the following.

**i. Neutralisation** results in the **inactivation** of toxins produced by pathogenic bacteria.

**ii. Precipitation** Each IgG antibody molecule can react with two molecules of antigen. Hence, many cross-linkages can be made to form a large **lattice** precipitate (Fig 11.8). Antibodies which cause precipitation to occur are called **precipitins**. Lattices are readily phagocytosed by macrophages.

**Fig 11.8** Antigen–antibody lattice formation

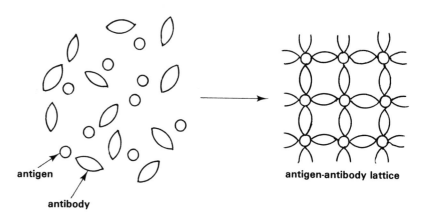

antigen

antibody

antigen-antibody lattice

Precipitation reactions form the basis of a number of assay techniques. The simplest technique is illustrated in Fig 11.9. Agarose is first prepared containing **antiserum** against the antigen to be assayed. Antiserum for the test is prepared by injecting the antigen into an experimental animal. The animal recognises the antigen as non-self and produces antibodies against it. Agarose impregnated with the antiserum is then poured onto a plastic plate to form a thin jelly-like layer. Wells are cut into the agarose. Into one well is placed a known volume of serum from a patient under test. Into each of the other wells is placed a similar volume of reference serum containing known concentrations of the antigen. The test and standard sera diffuse out of the wells and into the agarose. Circular zones of precipitation are formed around each well as the diffusing antigen reacts with the antiserum in the agarose (Fig 11.9(c)). The diameter of each precipitation zone is proportional to the concentration of the antigen in the well. The concentration of the antigen test sample is determined by comparing the diameter of its precipitation zone with those produced by the standard solutions (Fig 11.10).

**Fig 11.9** Immunodiffusion assay

(a) Antigen placed in wells cut into antibody-impregnated agarose

(b) Antigen diffuses into agarose

(c) Rings of precipitated antigen--antibody complex

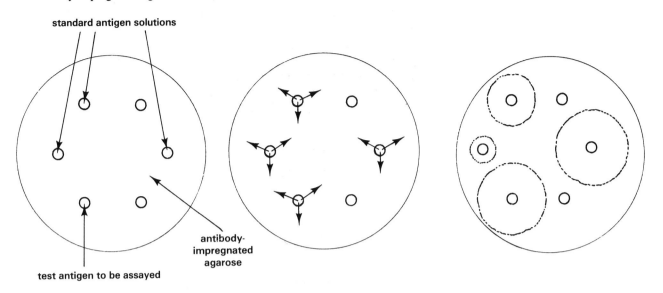

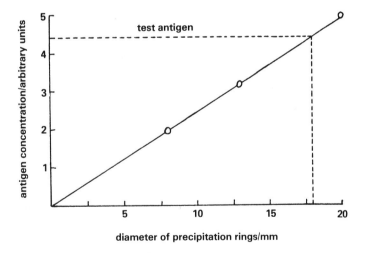

**Fig 11.10** Calibration curve constructed from standard antigen concentrations (2, 3 and 5 units) and the diameters of their precipitation rings (Fig 11.9(c)). The concentration of a test antigen solution may be obtained from the curve using the diameter of its precipitation ring

**iii. Agglutination** **Agglutinin** antibodies cause cells coated with non-self antigens to clump together. The clumping is called **agglutination** and the antigens are called **agglutinogens**. Agglutinated bacteria are susceptible to phagocytosis. Agglutination reactions also form the basis of blood grouping procedures (section 11.2.3). Modern pregnancy tests are based on **agglutination inhibition**. Human chorionic gonadotropin (HCG) appears in the urine of pregnant women. In the test, anti-HCG antibody is added to the urine. Small latex particles coated with HCG are added next.

Any HCG in the urine binds with the antibody. Consequently the HCG attached to the latex has no antibody with which to produce agglutination of the particles. Agglutination occurs however if the urine lacks HCG (Fig 11.11).

**iv. Lysis** Antibodies attached to antigens on cell surfaces may cause the cells to rupture. This is called **lysis** and leads to cell death. Complement proteins help bring about lysis (Fig 11.7).

**Fig 11.11** Sequence of events in an agglutination inhibition test: (a) antibody added to test sample; (b) latex particles coated with antigen are added; (c) agglutination of latex particles – negative, agglutination inhibited – positive; (d) results of a pregnancy test (courtesy Wellcome Diagnostics).

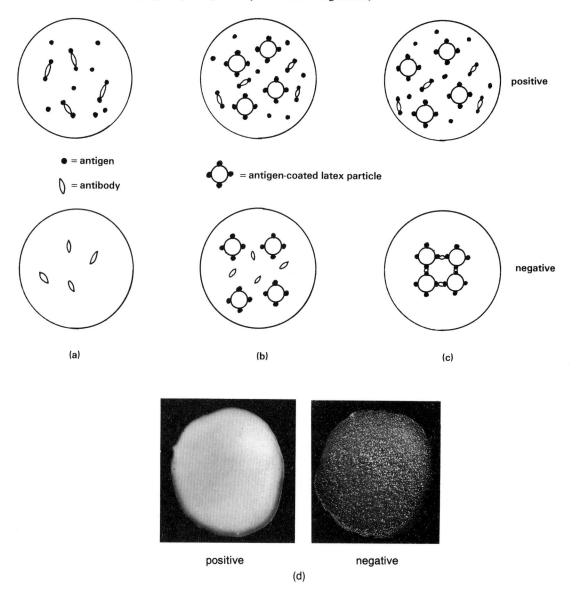

### 11.1.4 Autoimmune responses

Self antigens do not normally initiate an immune response. Sometimes, however, immune responses result in the destruction of self cells and tissues. They are called **autoimmune responses**. They may be caused by changes in self antigens brought about by the actions of viruses. There are many diseases which have recently been explained on this basis. For example, diabetes mellitus results from the body's inability to produce enough insulin (Chapter 16). This may follow destruction by antibodies of $\beta$-cells in the islets of Langerhans. Another autoimmune disease of the endocrine system is myxodema, caused by antibody activity in the thyroid gland.

The most frequent cause of kidney failure in Britain is glomerulonephritis, an autoimmune disease. Another autoimmune response is antibody neutralisation of receptor sites on muscle membranes (Chapters 7 and 14). This leads to poor neuromuscular transmission and a muscle wasting disease called myasthenia gravis.

### 11.1.5 Allergy

About 30 % of the population of Britain has an **allergy**. Allergies include hay fever, asthma, childhood eczema and food allergies. An allergy is an immune response to an antigen called an **allergen**, to which most people show no reaction. Allergens are found on pollen grains, fungal spores, house dust, feathers, fur and in a variety of foods. Allergens on pollen for example, become attached to the mucus membranes in the breathing passages. The presence of an allergen in a person who suffers from an allergy stimulates the production of antibodies called **reagins**. Reagins belong to the IgE category of immunoglobulins (Table 11.1, p. 204). They circulate in the blood and become attached to **mast cells** throughout the body, particularly in the skin and mucus surfaces of the mouth, nose and breathing passages (Chapter 7). The reagins can remain for years in these tissues which are said to be **hypersensitive**. Later, whenever there is exposure to the allergen, an allergen–reagin reaction takes place. The reaction triggers a vigorous response involving the rupture of the mast cells and the release of **histamine**. Histamine causes inflammation of the affected tissues, constriction of the bronchi leading to breathing difficulties, and excessive secretion of mucus. Eosinophils increase in number during allergic responses. They are thought to have an anti-inflammatory effect by absorbing histamine from the tissues (Chapter 10).

Some people gradually become accustomed to an allergen if it is presented to them in gradually increased doses. They become **desensitised**. Children often grow out of an allergy for this reason. Sensitivity to different allergens may be determined by performing a **skin prick test**. Small amounts of allergens are scratched into the skin. A weal appears around the scratch in response to a substance to which the person is allergic. The patient may then be dosed with the allergen to produce desensitisation.

There is considerable interest in the possible link between certain food allergens and hyperactivity in children. Some highly-strung children with undesirable behaviour have calmed down after removal of the allergens from their diet.

### 11.1.6 Vaccines

Among the main weapons in our armoury against disease are **vaccines**. Vaccines are made from microbes having the antigens which stimulate the body's immune system. The organisms are treated so that they do not give rise to disease when administered to a patient. The antibodies produced as a result of vaccination give immunity against disease following subsequent exposure to disease-causing microbes. Unfortunately it is not yet possible to make vaccines against all diseases caused by microbes.

In 1797 Edward Jenner, a Gloucestershire country doctor, was the first to use a vaccine successfully in preventing a human disease. He noticed that dairymaids who had milked cows suffering from cowpox were far less susceptible to the much more virulent and often fatal smallpox, although they often showed mild symptoms of the disease, such as hand sores. Jenner removed some of the liquid from a sore on the hand of a dairymaid and scratched it into the skin of a young boy, James Phipps. Later Jenner inoculated Phipps with material from the sore of someone suffering from smallpox. The boy was found to be immune to smallpox. Jenner had used an antigen to produce immunity against a dangerous pathogen. It is only relatively recently that mass vaccination programmes have eliminated smallpox all over the world. The reaction of the body to cowpox vaccine is an example of **active immunisation**. The body actively makes its own antibodies against the antigen.

Passive immunity is important in a foetus which absorbs antibodies from its mother across the placenta. The mother's milk also contains antibodies which provide passive immunity in the gut of new-born infants.

In Britain, children are normally offered a vaccination programme which gives them protection against childhood diseases that were once widespread killers (Table 11.2).

**Table 11.2 Vaccination schedule recommended in Britain by the Department of Health and Social Security**

| Age | Vaccine | Notes |
|---|---|---|
| During the 1st year of life. | diphtheria, tetanus, pertussis and poliomyelitis | 1st dose at 3 months of age; 2nd dose 6–8 weeks later; 3rd dose after a further interval of 4–6 months. |
| During the 2nd year of life. | measles | |
| At school entry or entry to nursery school. | diphtheria, tetanus, poliomyelitis | Allow an interval of at least 3 years after completing the basic course. |
| Between 11 and 13 years of age. | tuberculosis (BCG) | Leave an interval of at least 3 weeks between BCG and rubella vaccination. |
| Between 10 and 13 years of age (girls only). | rubella | |
| Between 15 and 19 years of age or on leaving school. | poliomyelitis, tetanus | |

Vaccines are produced in several ways including:

**1** The microbes may be killed in such a way that their antigens are not affected and can bring about the required immune response. Vaccines of this type are used to provide immunity against influenza, typhoid fever and cholera.

**2** Some diseases are caused by **toxins** produced by the infecting microbes. Toxins may be treated to make harmless **toxoids** which can initiate the immune response. Vaccines of this kind include diphtheria and tetanus toxoids.

**3** Pathogens may be **attenuated** (weakened) by injecting them into some other animal such as cattle or horses. Living attenuated organisms contained in vaccines multiply within the body but do not cause disease. They include oral polio vaccine, smallpox and measles vaccines and BCG vaccine against tuberculosis. *BCG* is derived from *Bacillus Calmette Guérin*, the strain of bacterium used in the vaccine's manufacture.

The first vaccines were used by Edward Jenner in the late eighteenth century. He deliberately infected a young boy with smallpox, one of the main killer diseases of his day. Jenner's pioneering work has now led to the eradication of smallpox from the human population. Was Jenner right to put a young life at risk?

### 11.1.7 Aids

In recent years a disease of the immune system has spread throughout the world and given rise to much concern. The disease is called AIDS which stands for Acquired Immune Deficiency Syndrome. Aids is caused by an RNA retrovirus, that is, a virus which injects its own RNA into the nucleus of the host cell (section 25.1.1). The virus is called HIV which stands for Human Immunodeficiency Virus. The host cells are Helper T-lymphocytes. These cells normally help in both the humoral immune response (antibody production) and the cellular immune response. People who have been infected with HIV usually produce HIV antibody. However, the virus can exist in the presence of the antibody without losing its ability to produce Aids or to infect others.

HIV has been found in many different body fluids of infected people. Transmission of the virus, though, occurs via blood or semen. Routes of transmission include anal intercourse between homosexual men, vaginal intercourse between an infected bisexual man and a woman, antenatal infection of a foetus by an infected mother, transfusion of unscreened blood products, and the sharing of infected needles by drug addicts.

The majority of people who are infected show no symptoms and so there are no reliable figures for people carrying the virus. Official figures in the UK suggest that up to July 1986 over 4000 people had been infected with HIV. However, it is estimated that by November 1986 over 30 000 people may have been infected. By the end of November 1986, there had been 599 confirmed cases of Aids in the UK since 1983; 296 of these had died.

Infection with HIV does not automatically mean that Aids will develop. The incubation period is between 15 months and 5 years. People with the disease suffer from opportunistic diseases and may develop a rare vascular tumour called Kaposi's sarcoma.

## 11.2 Blood groups

People can be classified into one of several **blood groups** depending on the presence or absence of certain antigens (agglutinogens) on the red blood cells. Antibodies (agglutinins) may also be present in the plasma. A large number of blood group systems has been described (Table 11.3). The **ABO** and **Rhesus** systems are particularly important. Transfusion from a person of one blood group into someone of another group may be fatal.

**Table 11.3 Red cell antigens of the main blood group systems in man**

| System | Antigens |
|---|---|
| ABO | $A_1$ $A_2$ $A_3$ $A_x$ and others |
| Rhesus | D C c $C^W$ $C^X$ E $D^U$ and others |
| MNSs | M N S s |
| P | $P_1$ $P^k$ |
| Lutheran | $Lu^a$ $Lu^b$ |
| Kell | K K $Kp^a$ $Kp^b$ $Js^a$ $Js^b$ |
| Lewis | $Le^a$ $Le^b$ |
| Duffy | $Fy^a$ $Fy^b$ |
| Diego | $Di^a$ $Di^b$ |
| Yt | $Yt^a$ $Yt^b$ |
| I | I i |
| Xg | $Xg^a$ |
| Kidd | Kj |

### 11.2.1 ABO system

Our red cells have one, both or neither of two agglutinogens called **A** and **B**. Correspondingly, our plasma contains one, both or neither of two agglutinins called **anti-A** or $\alpha$ and **anti-B** or $\beta$. The agglutinins are immunoglobulins of the IgM category (Table 11.1 and Fig 11.6, pp. 204–5). They appear soon after birth and are present throughout life. They are called **isoantibodies** as distinct from **immune antibodies** which appear after an immune response. Production of isoantibodies may decline and disappear in old people.

Cells with A agglutinogen belong to **group A**, those with B belong to **group B** and those with both agglutinogens belong to **group AB**. Cells with neither agglutinogen belong to **group O**. Anti-A and anti-B agglutinins are distributed in such a way that they do not normally come into

**Table 11.4 The main antigens and antibodies of the human ABO blood group system**

| Blood group: | O | A | B | AB |
|---|---|---|---|---|
| Red cell antigen | — | A | B | A + B |
| Plasma antibody | anti-A + anti-B | anti-B | anti-A | — |
| British population/% | 46.7 | 41.7 | 8.6 | 3 |

described groups A, B and O in 1900 and the AB group in 1901. His discoveries led to a dramatic rise in the success rate of blood transfusions. Techniques were devised to establish the ABO groups of donors and recipients for transfusion (section 11.2.3). Only blood of the same group within the ABO system as the recipient is normally transfused. **Incompatible** transfusions may bring A (or B) agglutinogen into contact with anti-A (or anti-B) agglutinin. If this happens, the transfused cells are made to agglutinate by the antibody and they haemolyse. The **transfusion reaction** is normally fatal.

People of group O have red cells which possess neither A nor B and have been described as **universal donors**. Similarly, people of group AB have plasma containing neither anti-A nor anti-B. They have been called **universal recipients**. These terms have fallen into disuse nowadays, since the term *universal* means *in all possible circumstances* and takes no account of blood group systems other than ABO.

After Landsteiner's discoveries it was noticed that the cells from some group A people react more strongly than those of others when mixed with anti-A. This led to the discovery of sub-groups of A and AB called $A_2$ and $A_2B$ (Table 11.5). They were first described in 1911 by Von Dungern and Hirszfeld.

**Table 11.5 Sub-groups of the ABO blood group system**

| Group | Sub-group | Approximate percentage of UK population |
|---|---|---|
| A | $A_1$ | 34 |
| | $A_2$ | 8 |
| AB | $A_1B$ | 2.6 |
| | $A_2B$ | 0.4 |

### 11.2.2 Rhesus system

The rhesus blood group system was first described by Landsteiner and Levine in 1940. They injected some red cells from rhesus monkeys into rabbits. They then exposed human red cells to blood serum extracted from the rabbits. The cells from about 85 % of the people whose blood was tested agglutinated. Their cells must therefore have had the same agglutinogen as that on the monkey cells. This was called the **rhesus factor** or agglutinogen D. It is now known that there are many variants of D (Table 11.3). Cells that possess D are **rhesus positive, Rh +**, those without it are **rhesus negative, Rh −**. There is no corresponding isoantibody. However, transfusion of Rh + cells into a Rh − individual may stimulate a humoral immune response in the recipient. The **anti-D** agglutinin produced causes agglutination and haemolysis of the transfused cells, a transfusion reaction. By accounting for the rhesus as well as the ABO groups of patients requiring blood, adverse reactions following transfusions have become very rare (section 11.2.3).

Complications arise sometimes when a Rh − woman bears a Rh + foetus. During normal pregnancy there is no mixing of the foetal and maternal blood cells, although the two circulations run close together in the placenta (Chapter 19). However, when the child is born, the severe contractions of the uterine wall may squeeze significant numbers of foetal red cells into the mother's blood. Some days later the mother's immune system may produce anti-D antibody. Anti-D belongs to the IgG category of immunoglobulins (Table 11.1 and Fig 11.6). It is capable of crossing the placental membranes and entering the blood of any foetus the woman may bear in the future. If the foetus is Rh + then the anti-D causes a transfusion reaction in the foetal circulation. The reaction is called **haemolytic disease of the newborn, HDNB**. It can be treated successfully by giving the foetus several transfusions while it is developing inside the uterus.

An alternative is to avoid the problem of HDNB by injecting a Rh− mother with anti-D immunoglobulin immediately she has given birth to her Rh + child. Any of the foetal cells which may have entered her blood are thus destroyed before they trigger the immune response.

HDNB arises in only about 10 % of cases like that described above. It is mainly because D-carrying foetal cells may be destroyed by the mother's own ABO isoantibodies. If the child belongs to group A, for example, and the mother is group O, then her anti-A haemolyses the foetal cells, regardless of rhesus group (Table 11.4).

HDNB caused by ABO incompatibility between mother and foetus is rare. The ABO isoantibody molecules (IgM) are far too large to cross the placental membranes.

Haemolytic disease of the newborn, HDNB, can occur in the second or subsequent Rhesus positive foetus of a Rhesus negative woman. Why does HDNB only occur in about 10% of such cases?

### 11.2.3 Blood grouping procedures

Techniques for establishing blood groups involve exposing batches of red cells to a variety of antisera containing known specific blood group agglutinins. Agglutination of the cells indicates the agglutinogens they possess and hence the groups to which they belong.

**i. ABO grouping** A simple technique can be performed on a glass slide. It is rapid and useful in emergencies. Four clean slides are arranged as in Fig 11.12. The first slide is used to test the patient's red cells, the other slides are used as controls. Red cells are suspended in isosmotic saline (0.9 % aqueous) to wash any antibodies off them and to provide an approximately 10 % (by volume) suspension. The patient's red cells and the control cells are mixed in appropriate known antisera and left undisturbed for a few minutes. The slides are then examined for agglutination either with the naked eye or with the aid of a microscope. The controls are checked first. In

Fig 11.12 ABO grouping, slide method

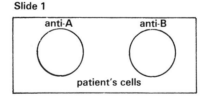

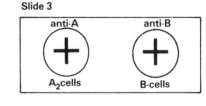

Fig 11.13 ABO grouping, slide method results for a group A patient

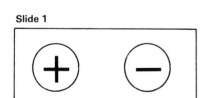

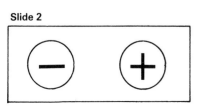

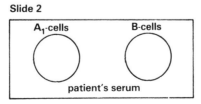

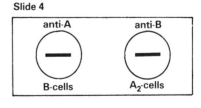

the **positive controls**, $A_2$ cells and B cells are mixed with anti-A and anti-B agglutinins respectively and should agglutinate. In the **negative controls**, $A_2$ cells and B cells are mixed with anti-B and anti-A respectively and should not agglutinate. Fig 11.13 illustrates the results which might be expected for a group A patient. Agglutination reactions may be strong or

213

weak and have to be classified. Fig 11.14 illustrates the microscopic appearance of the main categories of agglutination.

**Fig 11.14** Some of the main categories of agglutination used in blood grouping procedures: (a) complete, (b) plus – clumps of agglutinates each of more than 20 cells, (c) weak – small clumps of 4 to 6 cells

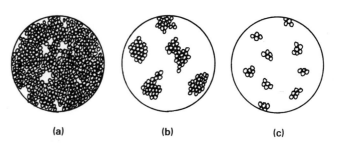

(a)        (b)        (c)

Difficulties may arise in blood grouping procedures. For example, infected blood cells may agglutinate spontaneously. Sometimes the cells string together into **rouleaux**, not by agglutination but because of an excess of globulin proteins in the antiserum (Fig 10.7(b)).

A more reliable technique for ABO grouping is performed in small glass tubes. The method takes up to 2 hours. It includes an **auto-control** in which the patient's cells are added to his or her own serum; agglutination should not occur in this tube.

---

What pattern of agglutination would you expect in an ABO slide test for a group AB patient? What about patients of the other ABO groups?

---

**ii. Rhesus grouping**  Unlike the isoantibodies of the ABO system, anti-D is an immune antibody. Anti-D is an IgG immunoglobulin and is usually incapable of causing the agglutination of D-cells (Rh+) on its own. For this reason it is called an **incomplete** antibody. The cells become coated (sensitised) by anti-D. Agglutination may occur when another protein is added, usually albumin.

In a simple test, a drop of 20% bovine albumin is added to a drop of antiserum containing anti-D. To this is added a drop of the patient's cells in saline (30–40 % suspension). After 3 to 5 minutes incubation at room temperature, agglutination indicates rhesus positive cells. As with ABO methods, more complex and reliable tube methods exist for rhesus grouping. They also involve positive, negative and auto-controls.

## 11.2.4 Transfusion

In 1658 Denis, a physician at the court of the French king Louis XIV, transfused blood from a sheep into a youth. Many of his transfusion attempts were unsuccessful but in 1818 in London, Blundell successfully transfused patients. However, before Landsteiner's discoveries of the ABO blood groups at the beginning of this century blood transfusion was a risky procedure. Many recipients died from the transfusion reaction (section 11.2.1).

During the First World War, anticoagulation of blood with sodium citrate was discovered (section 10.1.3). It was then possible to store donated blood for transfusion at a later date. Before this discovery, direct person-to-person transfusion was the standard procedure.

Nowadays, collection, anticoagulation, storage and transfusion of blood are performed with a single container called a **blood pack** (Fig 11.15). It is designed to keep the blood away from the air and any other sources of contamination. It has two outlet ports at one end which is dome-shaped to

**Fig 11.15** Blood pack

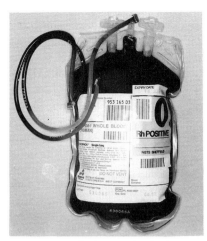

allow all the blood to run out of the pack during transfusion. An **administration set** is attached to an outlet port. Blood passes through a filter and drip chamber, then along a tube and into the recipient's blood vascular system, often via a vein in the hand or forearm. The filter collects debris which accumulates in stored blood, such as fibrin strands, platelets and white cell aggregates. The nylon mesh filter has an average pore size of about 200 $\mu$m and normal blood cells pass through easily. Red cells are about 7 $\mu$m in diameter.

Packed red cells are used in transfusion. They are separated from plasma by centrifuging the pack and squeezing the plasma into another pack. Whole blood is transfused when treating for haemorrhage (bleeding) when the replacement of blood volume is important. Other blood components may be separated for transfusion in special circumstances. They include platelets and white cells such as lymphocytes.

Blood for transfusion has to be checked for any unwanted contaminants before it is used. Viral hepatitis, syphilis and acquired immune deficiency syndrome (AIDS) are among the diseases which have been contracted following transfusion.

## 11.3 Tissue typing

Just as the antigens on red blood cells can be used to categorise people within the major blood groups, antigens on other body cells may be used to determine their **tissue type**. They are called **histocompatibility antigens**. Transplantation of incompatible antigens leads to an immune response and **rejection** of the transplant. The histocompatibility antigens are determined by several hundred genes at loci called **HLA** on chromosome 6. HLA stands for **human lymphocyte antigens**.

HLA-A and HLA-B are the loci of antigens which produce the strongest transplantation reactions and are thus most often used in tissue typing. Typing can be performed by exposing some of the patient's lymphocytes to a range of specific HLA antisera and complement. The presence on the lymphocytes of an HLA antigen specific to the antibody in the antiserum leads to lysis of the lymphocytes. Lysis is detected by use of a blue dye which is absorbed only by lysed cells. The different HLA-A and HLA-B antigens important in transplantation have been numbered. They include HLA-A1, 2, 3, 9, 10, 11, 28 and 29 and HLA-B5, 7, 8, 12, 13, 14, 18 and 27. There are two antigen specifications for each locus. Consequently, an individual who is heterozygous for each locus has, for HLA specificities, two HLA-A and two HLA-B from each parent. There are other HLA loci, such as HLA-C.

They determine relatively weak antigens and are less important in transplantation.

There is very great variation in HLA types between different people. Even closely related persons are rarely identical in this respect. The closer the HLA match between donor and recipient in transplantation, the greater is the likelihood of success. Despite efforts including international cooperation to match donors to recipients across the world, immune rejection is still the main hazard in transplantation. In addition to tissue typing, **immunosuppression** is used to help prevent rejection. After receiving the transplant the patient may be treated with a variety of drugs that suppress the immune response. Unfortunately, the drugs are usually not specific in their actions. The patient's natural defences to many pathogenic microbes may be lowered. Contraction of otherwise trivial diseases may thus threaten life.

Nonetheless, many people who were near to death have been able to live many years of active life with a transplant. In 1985, a man from South Wales who was given the heart of a 16-year-old donor a year earlier, entered and completed the Boston marathon in America.

The classification of people into different tissue types depends on the same principles as classifying them into different blood groups. Why is tissue-typing more difficult than blood grouping? What problems are involved in organ transplantation that do not occur in blood transfusion?

# 12 Gas exchange system

Energy required by the body is released from organic molecules in tissue respiration (Chapter 5). Aerobic respiration requires oxygen and releases carbon dioxide as a waste product. These respiratory gases are exchanged between the atmosphere and the blood in the lungs by external respiration and between the blood and the tissues by internal respiration. The gas exchange system includes external respiration, the transport of respiratory gases in the blood, and internal respiration.

## 12.1 The lungs

There are two lungs. Each is approximately conical and separated from the other in the thorax by the heart (Figs 10.16 and 12.1). The lungs are surrounded by the two **pleural membranes**. The outer **parietal pleura** is attached to the inner wall of the thorax, the inner **visceral pleura** covers the lungs. Between the pleurae is a lubricating **pleural fluid** which enables the pleurae to slide easily over each other during breathing. On the surface of the left lung there is a depression called the **cardiac notch** which accommodates the leftward-pointing heart.

**Fig 12.1** Anatomy of the respiratory system

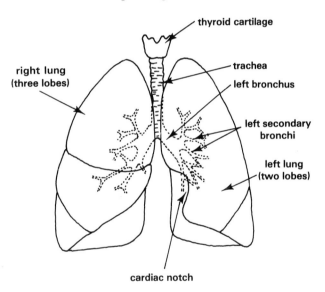

- thyroid cartilage
- right lung (three lobes)
- trachea
- left bronchus
- left secondary bronchi
- left lung (two lobes)
- cardiac notch

The lungs are divided by fissures into **lobes**, two on the left and three on the right. Beneath the lungs and separating the thorax from the abdomen is a tough sheet consisting of fibrous tissue and muscle and called the **diaphragm**. Beneath the diaphragm in the abdominal cavity is the liver, most of which is on the right side (Fig 9.6). Because of this and the cardiac notch in the left lung, the right lung is usually shorter and wider than the left lung.

### 12.1.1 Air supply to the lungs

The route along which air passes to the lungs begins in the mouth and nasal cavities. At the back of the throat is the **pharynx** through which food passes from the mouth into the oesophagus (Chapter 9) and air into the **larynx**. The pharynx also acts as a resonating chamber and is important in the production of speech sounds. The **auditory (eustachian) tubes** connect the middle ear cavities to the pharynx (Chapter 15). The larynx connects the pharynx and **trachea**. The larynx is the voice box which contains several cartilages. The largest of these is the thyroid cartilage. This produces the bulge on the front of the neck called Adam's apple. Another cartilage called the **epiglottis** lies on top of the larynx and normally prevents food entering the trachea during swallowing.

Inside the larynx there are several folds of tissue called the **vocal cords**. Contractions of laryngeal muscles stretch the vocal cords and push them into the air passageway. Air directed at the cords causes them to vibrate. In this way sounds are generated and transmitted into the pharynx, mouth and nasal cavities. The pitch of the sound may be altered by changing the tension on the vocal cords. If they are stretched tight, high pitched sounds are produced. Lower pitched sounds are produced when the cords are slackened.

The trachea leads into the thorax where it branches into left and right **bronchi**, one to each lung. The walls of the trachea and bronchi contain rings of cartilage which keep the air passages open (Figs 12.1–3). The bronchi divide into secondary bronchi which extend into the lobes of the lungs. In each lobe the secondary bronchi divide into a large number of tertiary bronchi. These extend to the segments of the lobes. Further branching results in a fine network of terminal **bronchioles**. The bronchioles terminate in many **alveolar air sacs**, which lead into small air-filled sacs called **alveoli** (Fig 12.2). It is estimated that the lungs contain about 300 million alveoli. The terminal and respiratory bronchioles, alveoli and surrounding blood and lymphatic vessels are surrounded by sheets of elastic connective tissue (Chapter 7). They divide the lobes of the lungs into a large number of **lobules** (Fig 12.2(a)).

The epithelium lining the trachea, bronchi and bronchioles is ciliated (Fig 12.3). Mucus is secreted onto the ciliated surface and traps any particles which are breathed in. The cilia waft the particles and mucus up into the pharynx from where it may be swallowed or coughed out.

**Fig 12.2** Microscopic structure of lungs

(a) Diagram of part of a lobule

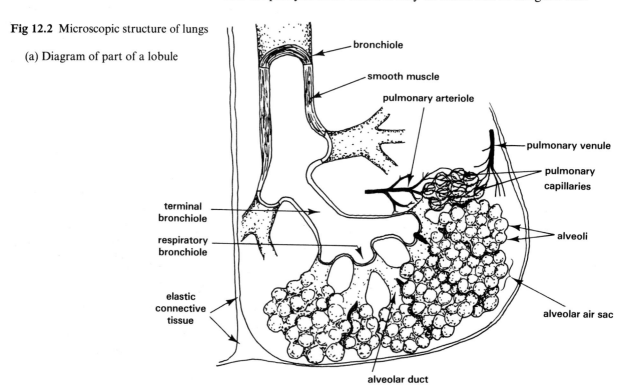

**Fig 12.2** (b) Photomicrograph of a thin section of cat lung, × 200

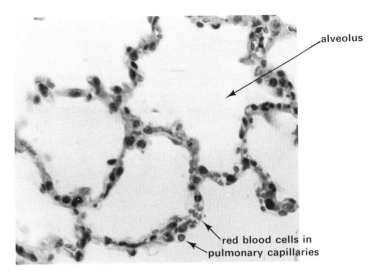

alveolus

red blood cells in pulmonary capillaries

**Fig 12.3** (a) Photomicrograph of a transverse section of trachea, × 6

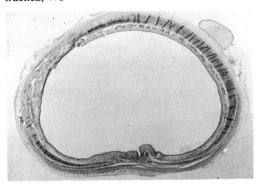

**Fig 12.3** (b) Section part of the trachea and cartilage, × 300

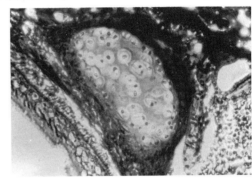

## 12.1.2 Blood supply to the lungs

Deoxygenated blood is delivered from the right ventricle of the heart through two **pulmonary arteries**, one to the left lung and one to the right lung. Inside the lungs the arteries branch many times. In each lobule arterioles supply blood to an extensive system of pulmonary blood capillaries surrounding each alveolus (Figs 12.2 and 12.4). Following oxygenation in the capillaries the blood enters a system of venules which

**Fig 12.4** Blood supply to the lungs: (a) arteries, (b) veins. Outlines of heart chambers are not drawn to scale

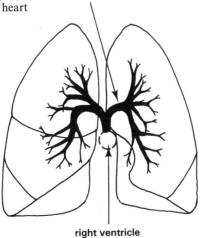

left pulmonary artery

right ventricle

(a)

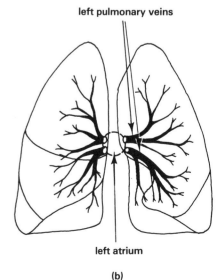

left pulmonary veins

left atrium

(b)

Blood flow through the lungs, pulmonary perfusion, is vital to lung function. How can doctors investigate pulmonary perfusion?

drains the blood from the lobules into **pulmonary veins**. Two of these transfer oxygenated blood from each lung to the left atrium of the heart (Figs 10.17 and 12.4). At any one time there may be about $1\,dm^3$ ($=1$ litre) blood in the pulmonary circulation of an adult. Blood flow through the lung tissue is called **pulmonary perfusion** and may be examined with the aid of radionuclides. One technique involves an intravenous injection of microparticles labelled with technetium–99 m, $^{99}Tc^m$ which emits gamma rays. The microparticles may be aggregates of human albumin with a density similar to that of blood cells but an average diameter of between 15 and $70\,\mu m$. The particles flow in the blood to the lungs and become trapped in the pulmonary capillaries. The average diameter of the capillaries is between 5 and $7\,\mu m$. Because relatively few particles are administered, only about 0.1 % of the pulmonary capillaries become obstructed. The gamma rays emitted from the trapped particles are detected by a gamma camera (Fig 13.20(a)). The images obtained in this way are called **perfusion scans** and indicate the distribution of blood in the lungs. Lowered gamma emission in an affected region indicates poor perfusion. This may be caused by an intravascular blood clot, **pulmonary embolism** (section 10.1.3 and Fig 12.5). The trapped particles disperse in 1–2 hours.

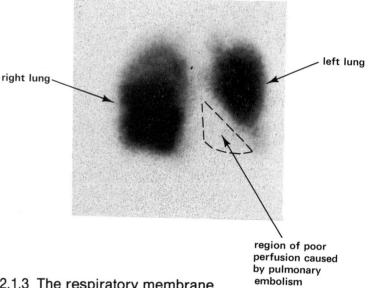

**Fig 12.5** Lung perfusion scan (anterior view) showing the effects of pulmonary embolism (courtesy Mr T. Johnson, Pilgrim Hospital, Boston, Lincolnshire)

right lung

left lung

region of poor perfusion caused by pulmonary embolism

### 12.1.3 The respiratory membrane

The gases breathed into the respiratory bronchioles and alveoli are separated from the blood in the pulmonary capillaries by the **respiratory membrane** which consists of the wall of the alveolus and the wall of the blood vessel. Nevertheless, the respiratory membrane is only about $0.2$–$0.5\,\mu m$ thick (Fig. 12.6). Gases pass between the alveolar air and the blood by diffusion (Chapter 1). Diffusion is a relatively slow process so the thinness of the respiratory membrane greatly assists gas exchange.

The alveolar nature of lung tissue provides a huge surface area of respiratory membrane – about $70\,m^2$ in an adult male. The large area is necessary for the exchange of gases in the volumes required by the body.

The alveolar lining is a single layer of squamous epithelium (Chapter 7). It is seated on an **elastic** basement membrane. The alveolar epithelium includes **septal cells** which secrete a phospholipid called **surfactant** ($=$surface active agent). This lowers the surface tension of the water layer lining the alveoli and thus prevents the alveoli from collapsing. The alveoli must be moist so that oxygen in the air can dissolve before it diffuses through the respiratory membrane. Oxygen passes much less easily through a dry membrane.

What features of the respiratory membrane enable gas exchange between inhaled air and the pulmonary blood?

220

**Fig 12.6** Main constituents of the respiratory membrane

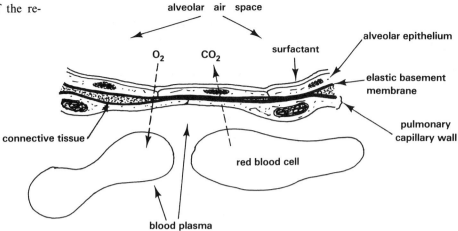

The alveolar wall also contains **macrophages** (Chapters 7 and 11) which are phagocytic cells similar to blood monocytes (Fig 10.9). They engulf dust and microbes which enter the alveoli in the inspired air.

### 12.1.4 Breathing

Breathing is the cyclical alternation of **inspirations** (inhalations) and **expirations** (exhalations) by which the lungs are ventilated. It is brought about by changing the pressure of air in the lungs relative to that of the atmosphere. The muscles which bring this about are found between the ribs, the external and internal **intercostal muscles**, and in the **diaphragm**.

When the external intercostal muscles contract and the internal intercostal muscles relax, the ribs are drawn obliquely upwards and outwards. At the same time the diaphragm muscle contracts, pushing the diaphragm down towards the abdomen. The pleural surface of the lungs is kept in contact with the pleura on the chest wall by surface tension so that the lungs are made to expand. This increases the volume of the air spaces inside the lungs. Consequently the air pressure inside the alveoli drops from atmospheric, say 100 kPa (760 mmHg) to about 99.74 kPa (758 mmHg). Provided there is no obstruction in the ventilation pathway, air is sucked into the lungs from outside. This is inspiration (Fig 12.7(a)).

**Fig 12.7** Breathing: (a) inspiration, (b) expiration

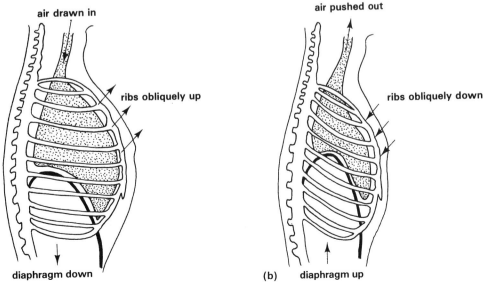

When the external intercostal muscles and the diaphragm relax there is a recoil of the elastic fibres in the stretched lungs. The fibres are present in the elastic basement membrane of the alveolar epithelium (Fig 12.6) and the connective tissue surrounding each lobule (Fig 12.2(a)). Consequently the lung tissue springs back to its original non-stretched state and the ribs and diaphragm move to their original positions. The pressure of the alveolar air is thus raised to about 100.39 kPa (763 mmHg), just above atmospheric pressure. Air is forced out of the lungs into the atmosphere. This is expiration (Fig 12.7(b)).

The values given above for the pressure changes are approximate and apply to normal quiet breathing in an adult. Forced expiration can be brought about by conscious contraction of the internal intercostal muscles. This pushes the ribs inwards and forces more air out of the lungs than is expelled by quiet expiration. The pressure of the fluid between the pleural membranes is always slightly less than that of the alveolar air. This pressure difference helps to keep the alveoli inflated.

Since air moves into and then out of the lungs, ventilation is said to be **tidal**. The volume of air breathed in and then out during one ventilation cycle is called the **tidal volume**. This can be about 400 cm$^3$ in a healthy adult (the value varies widely between different individuals). The extra air we can draw into our lungs by breathing in deeper is called the **inspiratory reserve volume, IRV**. The extra air we can force out after a normal expiration is called the **expiratory reserve volume, ERV**. All the volumes together add up to the maximum possible tidal volume and is called the **vital capacity, VC**. This varies between different individuals but is often between about 3 and 5 dm$^3$ in most healthy adults.

The volumes of air breathed in and out in different circumstances may be measured using a **spirometer** (Fig 12.8). The subject breathes into sealed bellows through a mouthpiece and a system of pipes. As the bellows inflate and deflate, a recording pen traces out the breathing movements on a moving paper chart. The machine is calibrated so that the volumes of gases breathed can be determined accurately from the recording. Fig 12.8(c) illustrates a spirometer trace showing a variety of lung volumes and

**Fig 12.8** (a) Spirometer (courtesy P.K. Morgan Ltd, Gillingham, UK)

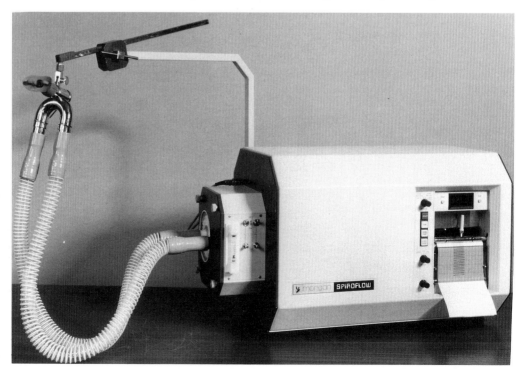

**Fig 12.8** (b) Spirometer mechanism

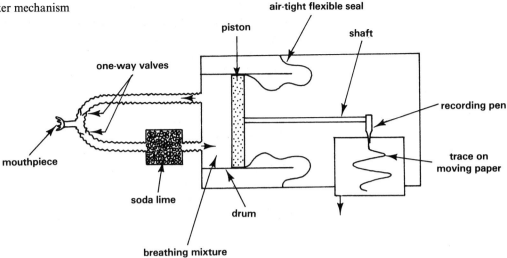

**Fig 12.8** (c) Spirometer trace showing the main lung volumes of a normal adult male

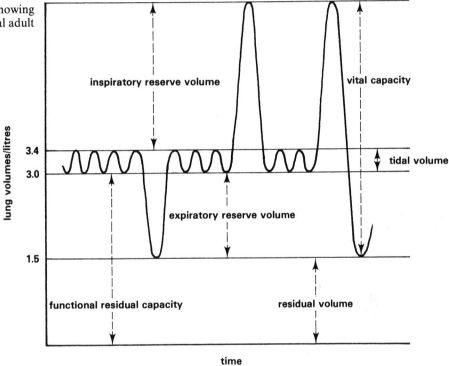

capacities. However, the trace shown has a base line drawn on it so that the **total lung capacity** can be seen. This cannot be measured using a spirometer in the way described above. When we breathe out as hard as we can, the lungs do not collapse. There remains a **residual volume** of gases in the alveoli. This may be between about 1.0 and 2.5 dm³ in a healthy adult. The total lung capacity can be between about 3.5 and 8.0 dm³ in a healthy adult. Like most other aspects of the body the values for lung volumes and capacities vary with the age, sex and height of the individual.

A proportion of the air breathed in does not reach the parts of the lungs where gas exchange occurs. The air filling the terminal bronchioles, bronchi, trachea, larynx, pharynx and nasal cavities does not contribute to gas exchange. The air is said to occupy **dead space** and can be about 150 cm³ in a healthy adult.

As air is breathed into the lungs from the atmosphere its composition changes. Water from the alveolar lining evaporates into the alveolar air.

Also, fresh air drawn in is diluted with air in the residual volume which has already exchanged gases with the blood. Consequently the oxygen content of the alveolar air is about 67 % of that in the atmosphere (Table 12.1). Nevertheless there is usually more than enough oxygen in the alveoli to saturate the blood.

**Table 12.1 The composition of inspired air (atmosphere) and alveolar air**

| | Inspired air | | Alveolar air | |
|---|---|---|---|---|
| Constituent | % gases | partial pressure /kPa (mmHg) | % gases | partial pressure /kPa (mmHg) |
| oxygen | 21 | 21 (160) | 14 | 13.16 (100) |
| carbon dioxide | 0 | 0 | 6 | 5.26 (40) |
| nitrogen | 79 | 79 (600) | 80 | 75.39 (573) |
| water vapour | — | — | — | 6.19 (47) |
| total barometric pressure | | 100 (760) | | 100 (760) |

The volume of oxygen used by an individual can also be measured with a spirometer. The instrument can be arranged so that the subject breathes pure oxygen and rebreathes from the spirometer. Carbon dioxide in the expired air is absorbed by soda-lime (Fig 12.8(b)). When the breathing movements are recorded for a few minutes, a trace may be obtained similar to that shown in Fig 12.9. Notice that the average level of the trace changes as the recording proceeds. This is because the volume of gas expired into the spirometer diminishes as oxygen is absorbed by the blood. The slope of the trace is a measure of the rate at which oxygen is consumed.

**Fig 12.9** Spirometer trace obtained continuously over a period of two minutes. The slope of the trace is a measure of oxygen consumption by the subject. In the example shown the level of the trace drops by the equivalent of 720 cm³ oxygen in 1 minute.

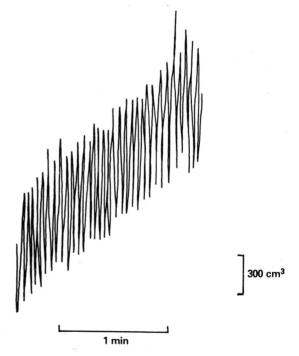

300 cm³

1 min

How much oxygen is consumed each minute by the person who breathed into the spirometer to produce the trace shown in Fig 12.9? In what ways could breathing be modified in order to increase oxygen consumption during physical exercise?

Ventilation of the lungs may be examined with the aid of radionuclides. The subject may be asked to inspire an inert gas such as krypton–81m or an aerosol labelled with technetium–99m. The radiation emitted by the radionuclide in the air passages is used to produce an image called a **ventilation scan** (Fig 12.10). A comparison of ventilation and perfusion scans can often be used to determine the nature and extent of certain lung diseases.

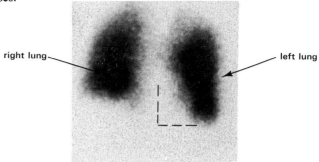

**Fig 12.10** Normal lung ventilation scan (anterior view). Compare with Fig 12.5 (courtesy Mr T. Johnson, Pilgrim Hospital, Boston, Lincolnshire)

right lung — left lung

### 12.1.5 Control of breathing

The muscles which bring about breathing movements are activated by nerve pathways which originate in a region of the hindbrain called the **respiratory centre**. The centre comprises three distinct areas called the **medullary rhythmicity area**, the **apneustic area** and the **pneumotaxic area** (Fig 12.11). The medullary rhythmicity area controls the basic rhythm of breathing. Nerve impulses from the rhythmicity area bring about the rhythmic cycle of inspiration and expiration. It is thought that the rhythmicity area contains two nerve circuits. One circuit is responsible for causing inspiration and inhibiting expiration, the other acts in the opposite way (Fig 14.27).

**Fig 12.11** Main areas of the respiratory centre

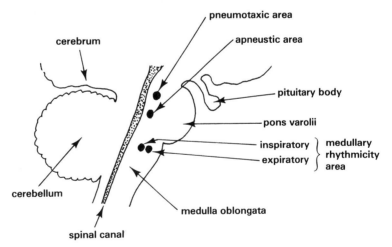

pneumotaxic area

apneustic area

cerebrum

pituitary body

pons varolii

inspiratory ⎱ medullary
expiratory ⎰ rhythmicity area

cerebellum

medulla oblongata

spinal canal

During quiet breathing, inspiration lasts about 2 seconds and expiration about 3 seconds. This rhythm produces about 12 ventilations a minute. In certain circumstances the quiet breathing rhythm can be modified. We may breathe faster and deeper, for example during physical exercise or emotional excitement. We may just decide to breathe at a different rate or even to hold our breath. Stimulation of the rhythmicity area by nerve impulses from the apneustic area can alter the depth of breathing. The tidal volume is increased. Stimulation of the rhythmicity area by impulses from the pneumotaxic area can alter the breathing rate. Suppose we breathe twice as deeply and three times as rapidly as during quiet breathing, the total volume of gases exchanged between the atmosphere and the lungs would increase by six times. The **minute volume** or **pulmonary ventilation, $\dot{V}$**

(the dot above the V indicates a measurement made over a period of time) is the total volume of gases breathed in one minute.

$$\text{Pulmonary ventilation} = \text{tidal volume} \times \text{breathing rate}$$

Typical adult values during quiet breathing might be:

$$\begin{aligned}
\dot{V} &= 400\,\text{cm}^3 \times 15 \text{ ventilations per minute} \\
&= 6000\,\text{cm}^3 \text{ per minute} \\
&= 6\,\text{dm}^3 \text{ per minute}
\end{aligned}$$

During exercise these values may rise to:

$$\begin{aligned}
\dot{V} &= 800\,\text{cm}^3 \times 25 \text{ ventilations per minute} \\
&= 20\,000\,\text{cm}^3 \text{ per minute} \\
&= 20\,\text{dm}^3 \text{ per minute}
\end{aligned}$$

If we subtract the dead space from these volumes we obtain the **alveolar ventilation, a$\dot{V}$**. This is the volume of gases that enters the alveoli each minute.

The respiratory centre responds to a variety of sensory signals. These signals indicate changes in the body that demand an alteration of breathing. The most important of these is the concentration of hydrogen ions in the blood which may rise because of an increase in tissue respiration in the muscles during exercise. A rise in blood carbon dioxide, called **hypercapnia**, causes an increase in the concentration of hydrogen ions in the blood (section 12.2.1). The respiratory centre is sensitive to a rise in hydrogen ion concentration in the blood flowing through it. It responds by causing an increase in the rate and depth of breathing. The response is appropriate as the extra carbon dioxide produced during exercise must be excreted by the lungs. Also more oxygen is required to sustain the increase in tissue respiration. When the deeper, faster breathing has eliminated the extra carbon dioxide from the blood, the respiratory centre reverts to its resting rhythm and quiet breathing resumes.

Another factor that can affect breathing is the oxygen tension in the blood. In the walls of the aortic arch and carotid bodies are **chemoreceptors** which are sensitive to changes in blood oxygen tension. However, they respond only to a relatively large drop in oxygen tension. This is less important than the respiratory centre's direct response to a rise in blood hydrogen ion concentration.

**Baroreceptors**, also in the aortic arch and carotid bodies, are sensitive to changes in blood pressure (section 10.2.3 and Fig 10.30). Sensory nerve impulses are transmitted to the respiratory centre from the baroreceptors. A sudden rise in blood pressure may thus result in a decrease in breathing rate, and a drop in blood pressure may result in an increased breathing rate (Fig 12.12).

What would be the values for alveolar ventilation during quiet breathing and exercise, using the figures for breathing rate and tidal volume given on the right? Assume the dead space is 150 cm³.

**Fig 12.12** Factors affecting the respiratory centre

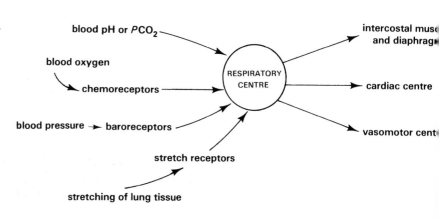

226

A rise in body temperature, as occurs during severe exercise or fever, causes an increase in the breathing rate. Conversely, a fall in body temperature slows the rate of breathing. A sudden drop in temperature caused by plunging into cold water, for example, may cause breathing to stop altogether, **apnoea**.

**Stretch receptors** are present in the walls of the bronchi and bronchioles. When stimulated by overstretching during excessive inspiration, sensory impulses are transmitted from the receptors along afferent branches of the vagus nerve to the respiratory centre. There the inspiratory and apneustic areas are inhibited and expiration occurs. This mechanism is called the **Hering–Breuer reflex**. It may be a protective device that prevents excessive inflation of the lungs. It is not thought to be important in the normal regulation of breathing.

The control of breathing is a complex interaction of very many factors. The complete picture is not yet fully understood. A person who has stopped breathing, say following electric shock, may sometimes be revived by mouth-to-mouth resuscitation (Fig 12.13). The patient's lungs are inflated with air expired from the lungs of the person applying the first aid.

**Fig 12.13** Mouth-to-mouth resuscitation. (a) The patient's mouth is held open and the nose cleared while the head is held back. (b) The resuscitator blows air into the patient's mouth, thus inflating the lungs. The procedure is repeated about 12 times each minute.

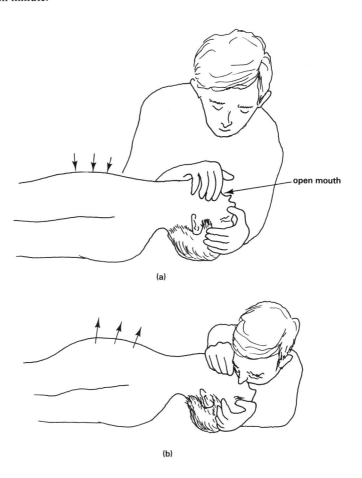

open mouth

(a)

(b)

Although expired air contains only about 16 per cent oxygen, some of it is absorbed by the patient's blood and may be sufficient to supply vital organs, especially the brain, until normal breathing resumes.

## 12.2 Transport of respiratory gases

Oxygen diffuses through the respiratory membrane from the alveolar air and into the blood. The solubility of oxygen in blood plasma, essentially water, is very low. At body temperature only about $0.3 \, cm^3$ oxygen can dissolve in $100 \, cm^3$ plasma. However, an adult man at rest uses about $250 \, cm^3$ oxygen each minute. During exercise the oxygen consumption may exceed $1000 \, cm^3$ a minute. This is much more than can be transported simply dissolved in blood plasma. The **oxygen carrying capacity** of blood is about $20 \, cm^3$ gaseous oxygen per $100 \, cm^3$ whole blood. The presence in the red blood cells of the respiratory pigment **haemoglobin** accounts for about 98 % of the oxygen transported to the tissues. It is carried as oxyhaemoglobin.

### 12.2.1 Haemoglobin

Haemoglobin has a complex molecular structure consisting of four **sub-units**. Each sub-unit contains a group called **haem** attached to a protein called **globin** (Fig 12.14). Haem consists of four **porphyrin** rings arranged around an atom of **iron**. The iron must be in the reduced iron(II) state for the pigment to carry oxygen. Oxidation of the iron to iron(III) prevents oxygen carriage. Haem is attached to the amino acid histidine in the globin.

**Fig 12.14** Haemoglobin structure

(a) Sub-unit

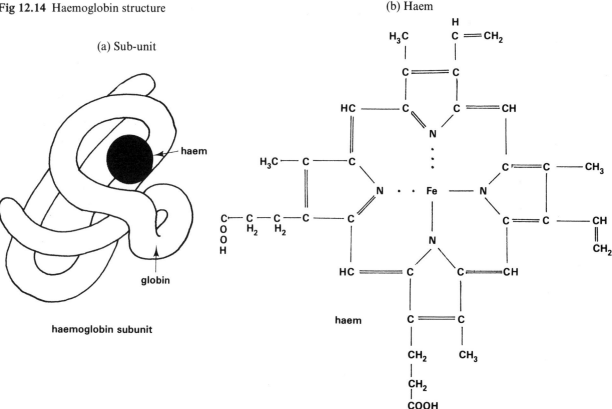

(b) Haem

haemoglobin subunit

Each sub-unit may combine with one molecule of oxygen by a process called **oxygenation**. The derivative formed is called **oxyhaemoglobin**, but the iron in the haem is not oxidised. Oxygen is released from the haemoglobin by **deoxygenation**. Since there are four sub-units, up to four molecules of oxygen can load onto each molecule of haemoglobin (Fig 12.15).

**Fig 12.15** Oxygenation of haemoglobin

$$\text{haemoglobin} + O_2 \underset{\text{deoxygenation}}{\overset{\text{oxygenation}}{\rightleftharpoons}} \text{oxyhaemoglobin}$$

$$Hb_4 + O_2 \rightleftharpoons Hb_4O_2$$
$$Hb_4O_2 + O_2 \rightleftharpoons Hb_4O_4$$
$$Hb_4O_4 + O_2 \rightleftharpoons Hb_4O_6$$
$$Hb_4O_6 + O_2 \rightleftharpoons Hb_4O_8$$

In standard conditions of temperature ($37\,°C$) and pressure ($100\,kPa$; $760\,mmHg$) each gram of haemoglobin may combine with $1.36\,cm^3$ oxygen. In an adult male there are usually between $13\,g$ and $18\,g$ haemoglobin in every $100\,cm^3$ blood. In women the figure is normally between $11.5$ and $16.5\,g$. Haemoglobin loads with oxygen in the lungs and then, usually less than a minute later, it unloads the oxygen to the body tissues. *Why does the affinity of haemoglobin for oxygen change so rapidly?*

There are three main factors: oxygen tension, carbon dioxide tension and temperature. Sometimes the presence of poisons such as carbon monoxide can disrupt oxygen transport by haemoglobin.

### 1 Oxygen tension

The **tension** of a gas in solution is determined partly by the **partial pressure** of the gas in contact with the liquid in which the gas is dissolved. For example, the partial pressure of oxygen in the alveolar air is about $13.16\,kPa$ ($100\,mmHg$). Oxygen diffuses into the blood plasma until it reaches equilibrium with that in the alveoli. That is, the **oxygen tension, $PO_2$** of the blood rises to about $13.16\,kPa$ ($100\,mmHg$). There is a direct relationship between the $PO_2$ in the blood and the saturation of haemoglobin with oxygen. This relationship can be expressed as a graph called the **oxygen dissociation curve** for haemoglobin (Fig 12.16).

The curve shows almost total saturation of haemoglobin with oxygen at oxygen tensions from about $9.21\,kPa$ ($70\,mmHg$) up to $13.16\,kPa$ ($100\,mmHg$). In other words, the partial pressure of oxygen in the lungs is more than adequate to fully charge the blood with oxygen. The partial

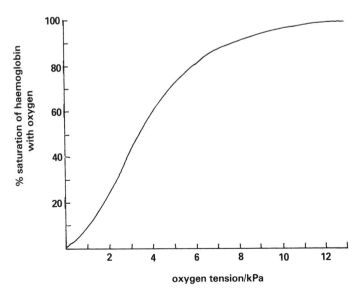

**Fig 12.16** Haemoglobin–oxygen dissociation curve (standard conditions of temperature $37°C$, $CO_2$ $5.26\,kPa$, pH $7.4$)

229

pressure of oxygen in the atmosphere falls with a rise in altitude (Fig 12.17). A man breathing air at about 3 km altitude can still almost fully oxygenate his haemoglobin. (Remember the oxygen content of alveolar air is less than that in the atmosphere.)

Fig 12.17 Relationship between the partial pressure of oxygen in the atmosphere and altitude

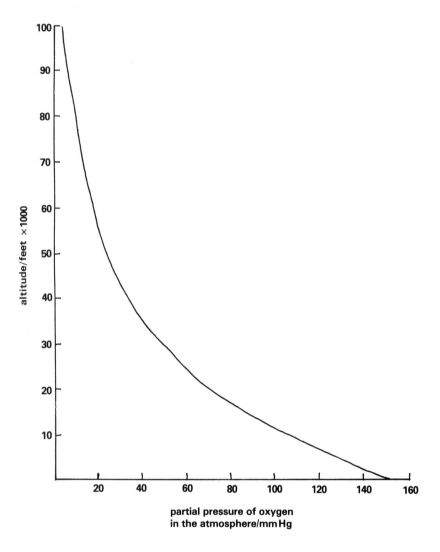

partial pressure of oxygen
in the atmosphere/mm Hg

In the body's tissues the $PO_2$ falls as oxygen is consumed by the process of tissue respiration. The $PO_2$ in blood returning to the lungs may be about 5.26 kPa (40 mmHg) in which case haemoglobin is 75 % saturated (Fig 12.16). That is because of a slight drop in $PO_2$, 25 % of the haemoglobin in the blood has been induced to unload its oxygen to the tissues. The more oxygen consumed by the tissues, the greater is the fall in $PO_2$ and hence more oxygen is released to the tissues from the blood. Some tissues, such as those in the liver or muscles, use more oxygen than others. It is appropriate that the oxygen requirement of any given tissue is one of the factors that triggers the amount of oxygen released from haemoglobin.

For $PO_2$ values between about 1.32 kPa (10 mmHg) and 5.26 kPa (40 mmHg) the dissociation curve is very steep. Between these limits, a slight rise (or fall) in $PO_2$ causes a relatively great oxygenation in the lungs (or deoxygenation in the tissues).

In the lungs, haemoglobin associates with oxygen very rapidly to form oxyhaemoglobin. In the tissues, less than a minute later, oxyhaemoglobin dissociates just as rapidly, thus making oxygen available to tissue respiration. How does the affinity of haemoglobin for oxygen change so quickly?

## 2 Carbon dioxide tension

The **carbon dioxide tension, $PCO_2$** of arterial blood delivered to the tissues is about 5.26 kPa (40 mmHg). As a result of tissue respiration, not only does the $PO_2$ fall in the tissues but the $PCO_2$ rises. The $PCO_2$ of venous blood returning to the lungs may be about 6.05 kPa (46 mmHg). The rise in $PCO_2$ lowers the affinity of haemoglobin for oxygen in the tissues, causing a proportion of the oxyhaemoglobin in the blood to unload its oxygen. The oxygen dissociation curve shifts to the right. This is called the **Böhr shift** (Fig 12.18).

**Fig 12.18** Effect of increased carbon dioxide tension on the position of the haemoglobin–oxygen dissociation curve

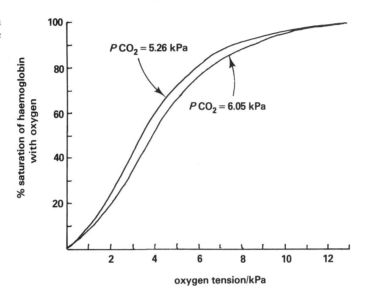

Suppose that in a tissue somewhere in the body at a given instant the $PO_2$ is 6.58 kPa (50 mmHg) and the $PCO_2$ is 5.26 kPa (40 mmHg). Reading off the oxygen dissociation curve in Fig 12.18 we see that the haemoglobin is about 83 % saturated with oxygen. If the $PCO_2$ rises to 6.05 kPa (46 mmHg) but the $PO_2$ remains unchanged then, reading off the Böhr-shifted curve, we see the haemoglobin is about 79 % saturated.

Now suppose that, at the same time as the $PCO_2$ rises, the $PO_2$ falls to 5.26 kPa (40 mmHg). The curves show us that the saturation of haemoglobin with oxygen falls even further to 70 %. That is, if the $PCO_2$ rises from 5.26 kPa (40 mmHg) to 6.05 kPa (46 mmHg) *and* the $PO_2$ falls from 6.58 kPa (50 mmHg) to 5.26 kPa (40 mmHg), then 13 % of the haemoglobin is induced to unload its oxygen to the tissues. Of course, the greater the respiratory activity in the tissues, the more oxygen is made available from the blood because of the $PO_2$ and $PCO_2$ effects. We can explain the $PCO_2$ effect by following the fate of carbon dioxide in the blood. Carbon dioxide diffuses into the blood and into the red cells. Inside the red cells there is an enzyme called **carbonic anhydrase** which catalyses the reaction between carbon dioxide and water to produce carbonic acid.

$$CO_2 + H_2O \overset{\text{carbonic anhydrase}}{\rightleftharpoons} \underset{\text{carbonic acid}}{H_2CO_3}$$

This is a weak acid and a small proportion of the $H_2CO_3$ molecules dissociate into hydrogen ions and hydrogencarbonate (bicarbonate) ions.

$$H_2CO_3 \rightleftharpoons H^+ + HCO_3^-$$

Suppose that respiration causes the oxygen tension to fall from 6.58 kPa (50 mmHG) to 5.26 kPa (40 mmHg) and the carbon dioxide tension to rise from 5.26 kPa (40 mmHg) to 6.05 kPa (46 mmHg). What effect would these changes have on the release of oxygen to the tissues from the blood?

231

The hydrogencarbonate ions diffuse out of the red cells into the plasma, while the hydrogen ions combine with oxyhaemoglobin forming haemoglobinic acid. The hydrogen ions displace oxygen molecules attached to the pigment.

$$H^+ + \underset{\text{oxyhaemoglobin}}{HbO_2} \rightleftharpoons \underset{\text{haemoglobinic acid}}{HHb} + O_2$$

The oxygen thus unloaded diffuses out of the red blood cells into the tissues (Fig 12.19). Strictly speaking, the effect of $PCO_2$ in causing unloading of oxygen from haemoglobin is a hydrogen ion effect. The hydrogen ion concentration or pH of the blood is therefore important for oxygen carriage by haemoglobin (section 12.3).

**Fig 12.19** Summary of red cell chemistry related to the carriage of respiratory gases

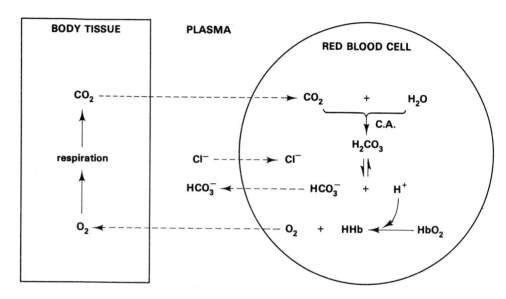

### 3 Temperature

A rise in blood temperature lowers the affinity of haemoglobin for oxygen, thus causing unloading from the pigment (Fig 12.20). Increased tissue respiration which occurs in skeletal muscles during exercise generates heat. The consequent temperature rise causes the release of extra oxygen from the blood.

**Fig 12.20** Effect of increased temperature on the position of the haemoglobin–oxygen dissociation curve

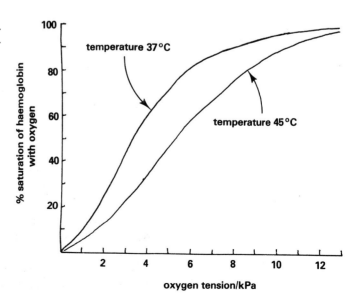

## 4 Carbon monoxide

Haemoglobin reacts with carbon monoxide more than 200 times more readily than with oxygen. **Carbon monoxide haemoglobin** (carboxyhaemoglobin) is formed in the red cells. If there are significant quantities of carbon monoxide in the air we breathe, then oxygen carriage by haemoglobin diminishes. Carbon monoxide is a constituent of the exhaust from internal combustion engines. It is also found in tobacco smoke. A cigarette smoker may have between 5 % and 15 % carbon monoxide haemoglobin in the blood. This leaves only 85 % to 95 % of the haemoglobin capable of carrying oxygen. Carbon monoxide poisoning results from excessive exposure to the gas. The consequent oxygen starvation in the tissues is called **anoxia**.

### 12.2.2 Myoglobin

Muscles contain a respiratory pigment called **myoglobin**. Each molecule of myoglobin consists of a haem group and a globin chain, similar to a haemoglobin sub-unit. The red colour of muscle is due to myoglobin. Some muscle does not contain myoglobin and is consequently white in appearance, such as fish meat or breast meat in poultry. Myoglobin combines with oxygen from the blood to form **oxymyoglobin**. The oxygen dissociation curve for myoglobin is far to the left of that for haemoglobin (Fig 12.21). It indicates that myoglobin loads with oxygen very readily in the same conditions of $PO_2$, $PCO_2$ and temperature in which oxyhaemoglobin unloads oxygen. Similarly the curve shows that oxymyoglobin only dissociates when the $PO_2$ is very low indeed. Oxymyoglobin acts as an emergency reserve of oxygen to be made available to the muscles only when the blood supply of oxygen is very low (section 14.1.6).

**Fig 12.21** Myoglobin–oxygen dissociation curve. Compare with the position of the haemoglobin–oxygen dissociation curve (standard conditions)

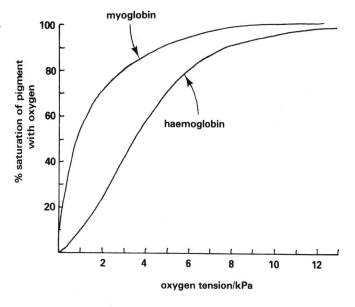

### 12.2.3 Foetal haemoglobin

The blood of a foetus contains a variant of haemoglobin slightly different in molecular structure from that in an adult's blood. The amino acid constitution of the globin chains in two of the sub-units of **adult haemoglobin** is slightly different in **foetal haemoglobin**. They are called $\gamma$-sub-units. The $\alpha$-sub-units are the same in both pigments (Fig 12.14). The effect of these structural changes is that foetal haemoglobin has a greater affinity for oxygen than has adult haemoglobin. This property is reflected in the oxygen dissociation curves for the two haemoglobin variants (Fig 12.22). The curve for foetal haemoglobin is to the left of that for adult haemoglobin. This means that the foetal pigment can load with oxygen in the same conditions in which the adult pigment unloads. This, of course, occurs in the **placenta** (section 19.9). If the foetal blood contained the same pigment as in the mother's blood, then the foetus would be unable to obtain sufficient oxygen. If the foetus is a female, then later on in life she may also be a mother. Her foetal haemoglobin must consequently be replaced by the adult variant. This occurs in all normal individuals, both male and female, at birth. Since red cells survive in the circulation for only about four months, then after this time all the foetal haemoglobin should have disappeared.

**Fig 12.22** Haemoglobin–oxygen dissociation curves for adult and foetal haemoglobins

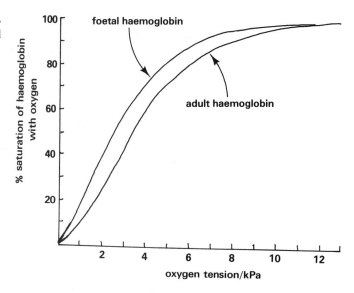

### 12.2.4 Abnormal haemoglobin variants

Sometimes the molecular structure of haemoglobin may change producing an abnormal variant. Most of these carry oxygen less well than the normal pigment and result in **anaemia** (Table 10.1). Abnormal variants may result from a suppression of part of the process by which haemoglobin is synthesised in the bone marrow. This suppression may lead to **thalassaemia** and is usually an inherited defect. Thalassaemia major occurs in homozygous individuals who have inherited the gene which causes the defect from both parents. Thalassaemia major is often fatal early in life. On the other hand, thalassaemia minor is less severe and appears in heterozygous individuals who have inherited the gene from one parent only (Chapter 20).

Abnormal haemoglobin variants may result from a structural alteration in the pigment molecule. A classic example is **haemoglobin S** (Chapter 20). In normal adult haemoglobin, the sixth amino acid in the globin chains of the $\beta$-sub-units is glutamic acid. In haemoglobin S this amino acid is substituted by valine. The relatively small change in structure causes considerable functional changes in the pigment. When haemoglobin S unloads oxygen to the tissues it becomes insoluble. It crystallises in the red cells, causing them to buckle into a typical sickle shape (Fig 12.23). This is why the pigment variant is called haemoglobin S – 'S' for sickle cells. Sickle cells haemolyse more readily than normal ones, thus leading to anaemia. Worse still, sickle cells that have not haemolysed may become lodged in narrow blood capillaries. Blocked capillaries can lead quickly to anoxia from which sufferers may die. People who are homozygous for the gene

Normal adult haemoglobin contains 574 amino acid residues in four globin chains. In haemoglobin S one of the acids, glutamic acid, in two of the chains is substituted by valine. What effect does this small structural change have on the ability of haemoglobins to carry oxygen?

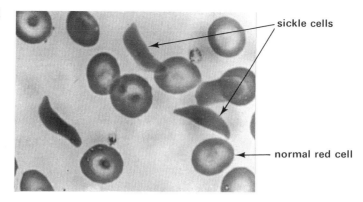

**Fig 12.23** Photomicrograph of a blood smear showing sickle cells, × 750

sickle cells

normal red cell

which causes haemoglobin S have **sickle cell disease**. Heterozygotes have **sickle cell trait**. The trait is less severe and relatively common among people living in areas of the world where malaria is prevalent. It seems that sickle cell trait gives a measure of resistance to one species of malarial blood parasite (Chapter 25).

### 12.2.5 Carbon dioxide carriage

Carbon dioxide produced in the tissues by respiration is transported in the blood to the lungs for excretion. Most of the carbon dioxide taken into the blood is converted to **hydrogencarbonate (bicarbonate) ions, $HCO_3^-$** in the red cells (Fig 12.19). The ions diffuse into the plasma and account for about 86 % of all the carbon dioxide carried in the blood. About 7 % of the carbon dioxide is carried in the blood in simple solution and 7 % of the gas combines with blood proteins and haemoglobin to form **carbamino compounds**. Carbon dioxide haemoglobin (carbaminohaemoglobin) may be formed in the red cells in the tissues. The transport of carbon dioxide in blood is related to the $PCO_2$ and may be expressed as the **carbon dioxide dissociation curve** for whole blood (Fig 12.24).

**Fig 12.24** Carbon dioxide dissociation curves for whole blood. The more oxyhaemoglobin there is, the less haemoglobin is available to combine with carbon dioxide

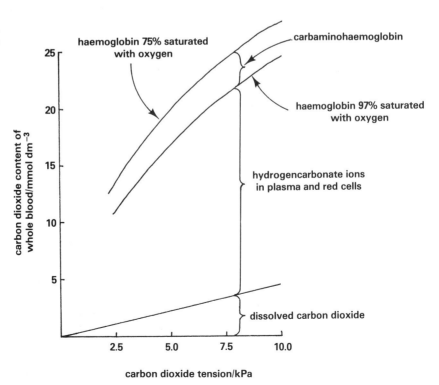

haemoglobin 75% saturated with oxygen

carbaminohaemoglobin

haemoglobin 97% saturated with oxygen

hydrogencarbonate ions in plasma and red cells

dissolved carbon dioxide

carbon dioxide content of whole blood/mmol dm⁻³

carbon dioxide tension/kPa

## 12.3 Acid–base balance

The maintenance of a normal blood pH is vital to many body functions including the transport of respiratory gases. pH is a measure of hydrogen ion concentration (Chapter 1). The effect of hydrogen ion concentration on oxyhaemoglobin is illustrated by the oxygen dissociation curves (Fig 12.18). There is a definite relationship between hydrogen ion concentration, the carbon dioxide tension and concentration of hydrogencarbonate ions in the blood. The relationship is given by **Henderson's equation**:

$$[H^+] = \frac{24 \times PCO_2}{[HCO_3^-]}$$

In normal blood the pH ranges between about 7.35 and 7.45 (arterial) and 7.30 and 7.41 (venous). Any significant change in pH outside these relatively narrow ranges causes disturbances in blood gas carriage. **Acidaemia** results when the blood pH falls below normal, **alkalaemia** results when the blood pH rises above normal. The limits of blood pH normally compatible with survival are about 6.8 to 7.7.

The main source of hydrogen ions in the body is carbonic acid made from carbon dioxide released in tissue respiration (Fig 12.19). Some of the hydrogen ions are removed from free solution and no longer contribute to pH (Fig 12.25). In this way the blood can **buffer** itself against any substantial change in pH. Such buffering in the blood is only a temporary compensation for the production of acid. Excretion of carbon dioxide by the lungs is the way in which the body finally compensates for acid production in the tissues. Hence inadequate excretion of carbon dioxide may lead to disturbances in blood pH.

(a)     HA ⇌ H⁺ + A⁻
acid    hydrogen ion    conjugated base

$H_2CO_3$ ⇌ $H^+$ + $HCO_3^-$
carbonic acid      hydrogencarbonate

(b)

HHb ⇌ $H^+$ + $Hb^-$
haemoglobinic acid     haemoglobin

**Fig 12.25** Buffer solutions may consist of a weak acid and the conjugated base of that acid: (a) general equation, (b) carbonic acid and haemoglobin buffer pH changes in tissues and blood

Other acids may enter the body in foods such as vinegar and carbonated drinks. Diets containing a lot of meat result in the formation of acids such as sulphuric acid and phosphoric acid. Vegetarian diets contain many alkalis. The kidneys also play a vital role in the regulation of blood pH (Chapter 13).

# 13 The renal system

The complex and numerous activities of the body require a stable environment in which to take place. Even slight chemical or physical changes can upset the smooth functioning of the body. Many of the physiological activities which regulate the constitution of the body's internal environment do so to within relatively narrow limits. Maintaining a stable internal environment is called **homeostasis**.

The very functions and activities which need a stable environment change it continuously. Metabolism involves the production of wastes. If wastes accumulate in the body they can become toxic. Consequently the body has to eliminate or **excrete** its wastes. **Excretion** is

> the elimination of any substances which are present in the body's tissues in concentrations exceeding normal levels, whether metabolic wastes or not.

Among the organs concerned with excretion and homeostasis are the kidneys which form part of the **renal system**.

## 13.1 Anatomy of the renal system

The **kidneys** are paired organs found in the abdominal cavity. They are usually embedded in fat and held firmly in position by the peritoneum, a thin layer of tissue lining the abdominal cavity (Fig 13.1).

**Fig 13.1** The renal system

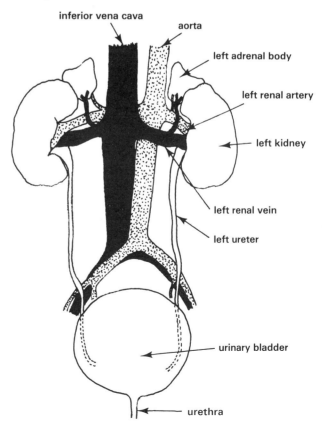

inferior vena cava

aorta

left adrenal body

left renal artery

left kidney

left renal vein

left ureter

urinary bladder

urethra

Each kidney receives blood through a **renal artery** and is drained of blood by a **renal vein.** In kidneys the blood circulates in a network of arterioles, capillaries and venules which surrounds numerous microscopic urinary tubules called **nephrons**. The nephrons remove excess and unwanted materials from the blood. The excretory products collect as **urine** which passes from the kidneys through the **ureters**. Urine is stored temporarily in the **urinary bladder** before it is eliminated from the body. Emptying of the bladder is called **micturition** and is controlled by the autonomic nervous system (Chapter 14). The exit from the bladder into the **urethra** is closed by contraction of rings of muscle called the **bladder sphincters**. As the bladder fills, cells in its wall sensitive to stretching trigger off a reflex action which results in relaxation of the bladder sphincter. Simultaneous contraction of the smooth muscle in the bladder wall forces the urine out through the urethra. Micturition can be controlled by voluntary nervous activity which is learned by humans in early life (Fig 13.2).

**Fig 13.2** Autonomic supply to the urinary bladder and urethra

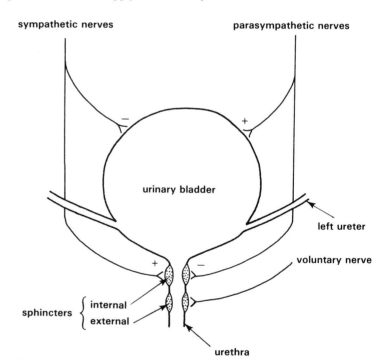

Each human kidney is a compact organ measuring about 7–10 cm long and 2.5–4 cm wide in an adult. Inside, the tissues are in distinct regions. There is an outer dark **cortex** and an inner lighter **medulla** (Fig 13.3). The internal appearance is due to the arrangement of the blood vessels and nephrons which make up most of the organ.

There are about a million nephrons in each kidney. Nephrons are very small thin-walled tubules between 2 and 4 cm long (Fig 13.4). At one end is a cup-shaped **Bowman's capsule** which encloses a small group of capillaries called a **glomerulus**. A Bowman's capsule and the glomerulus together are called a **Malpighian body**. The capsule leads into a coiled structure called the **proximal convolution** which opens into the **loop of Henle**. The loop of Henle consists of a descending limb and an ascending limb and leads into a second coiled tube, the **distal convolution**. The distal convolutions of several nephrons join a common **collecting duct** and many collecting ducts lead through the medulla to the **renal pelvis**.

There are two types of nephrons depending on the length and nature of their loops of Henle. **Cortical nephrons** have 'short-reach' loops which project to the boundary between the outer and inner zones of the medulla.

**Fig 13.3** Diagrammatic vertical section of a kidney. Note the arrangement of the medulla into a series of pyramids

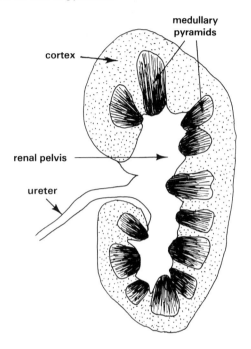

**Fig 13.4** (a) Photomicrograph of a thin section of kidney in the cortical region, × 300

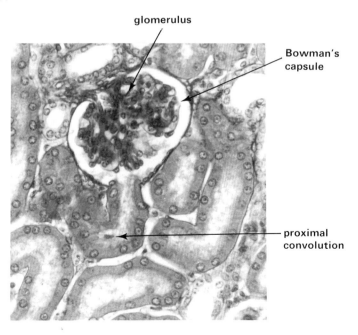

**Juxtamedullary nephrons** have 'long-reach' loops which extend deeper into the medulla, usually to the tips (papillae) of the pyramids (Fig 13.4(b)). In human kidneys about 85 % of the nephrons are cortical and 15 % are juxtamedullary. The proportions are different in the kidneys of other mammals. In some desert-living rodents, for example, nearly all the nephrons have very long loops of Henle (Fig 13.14).

**Fig 13.4** (b) Cortical and juxtamedullary nephrons and associated blood vessels

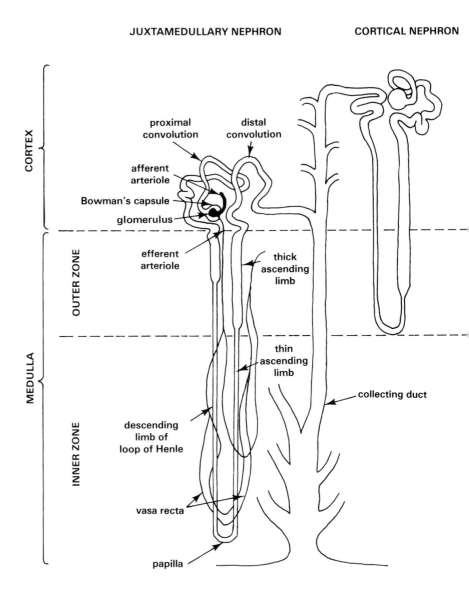

JUXTAMEDULLARY NEPHRON    CORTICAL NEPHRON

## 13.2 The functions of nephrons

Much of what is known of the functions of kidneys come from analysis of the fluid inside nephrons. Delicate micropipettes are inserted at various points into nephrons of experimental animals. The fluid is drawn off carefully and analysed to determine what changes have occurred during its passage through a nephron (Fig 13.5).

**Fig 13.5** Removal of a sample of filtrate for analysis

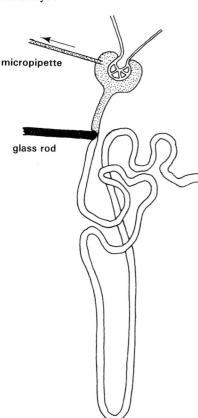

micropipette

glass rod

## 13.2.1 Ultrafiltration

The first activity of a nephron is the **ultrafiltration** of blood brought to the Bowman's capsule by arterioles. The arterioles branch into the capillaries of the glomeruli, tightly nestled in the capsules. About 20 % of the blood plasma which enters the kidneys is filtered. Electron microscope studies of the capsules show that the only effective barrier between the blood in the glomeruli and the cavity of the capsules is a thin porous **basement membrane** (Fig 13.6). The basement membrane of the capsule is permeable to some blood constituents but not to others. The hydrostatic pressure of blood in the glomerular capillaries is relatively high, at about 5.92 kPa (45 mmHg). This is partly because the diameter of the afferent arterioles is

**Fig 13.6** (a) Diagram of the barrier between the glomerular blood and the filtrate in Bowman's capsule

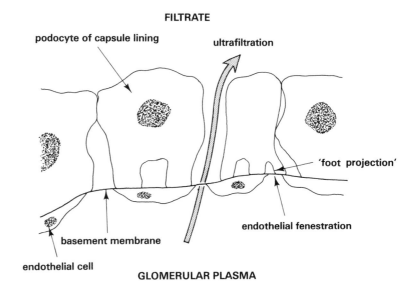

FILTRATE

podocyte of capsule lining

ultrafiltration

'foot projection'

endothelial fenestration

basement membrane

endothelial cell

GLOMERULAR PLASMA

**Fig 13.6** (b) Electronmicrograph showing a section through a podocyte and a glomerular capillary, × 12 000

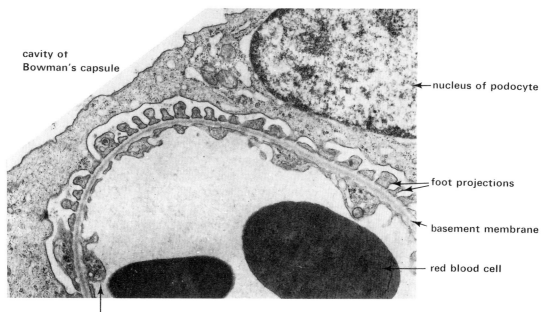

cavity of Bowman's capsule

nucleus of podocyte

foot projections

basement membrane

red blood cell

endothelium of glomerular capillary

greater than that of the efferent arterioles. Blood pressure is normally maintained by the pumping action of the heart. If the blood pressure falls too low, for example when much blood is lost in an accident, temporary use of a kidney machine may be required until the patient's blood volume and pressure are restored (section 13.5.1).

Because the blood in the glomerulus has a high hydrostatic pressure it is filtered through the basement membrane. Blood cells and substances of high relative molar mass, such as most of the plasma proteins, are too large to pass through the pores of the basement membrane. The chemical composition of the glomerular **filtrate** is thus virtually the same as the plasma minus its proteins. The composition of urine, however, is very different. Hence the filtrate is modified considerably while passing along the nephrons. The volume of urine excreted is also much less than the volume of filtrate produced in a given time (Table 13.1).

**Table 13.1 Some blood constituents and the quantities filtered and reabsorbed by the kidneys in a day**

| Constituents | Amount in filtrate/g | Amount in urine/g | Percentage reabsorption |
|---|---|---|---|
| sodium ions | 600 | 6 | 99 |
| potassium ions | 35 | 2 | 94 |
| calcium ions | 5 | 0.2 | 96 |
| glucose | 200 | 0 | 100 |
| urea | 60 | 35 | 42 |
| water | $180 \, dm^3$ | $1.5 \, dm^3$ | 99 |

### 13.2.2 Direct secretion

About 80 % of the blood plasma which enters the kidneys is not filtered from the glomeruli into the Bowman's capsules. However, some substances in the blood may be discharged into the nephrons by direct secretion, mostly into the proximal convolutions, as well as by ultrafiltration (Fig 13.7). Substances excreted in this way include uric acid. **Direct secretion** enables greater quantities of such wastes to be eliminated than by ultrafiltration alone (section 13.4).

### 13.2.3 Selective reabsorption

Changes in the composition of the filtrate begin in the proximal convolutions. Here the epithelial cells of the nephron wall reabsorb a large proportion of the filtrate, passing it back into the blood flowing in the surrounding vessels. **Selective reabsorption** of individual substances is at a rate just sufficient to maintain normal concentrations in the blood. Any excesses stay in the nephron.

Since the transfer of materials from the filtrate into the blood is against a concentration gradient, selective reabsorption is active (Fig 13.8). The energy for **active uptake** is provided by respiration in the nephron's cells which contain many mitochondria. The efficiency of reabsorption is helped by the presence of numerous **microvilli** which greatly enlarge the surface area through which the materials pass (Fig 13.9).

In humans about 120 cm³ of water pass into the nephrons every minute. Of this, about 100 cm³ per minute are reabsorbed passively from the proximal convolutions. The concentration gradient of solutes between the remainder of the filtrate in the proximal convolution and the blood in surrounding capillaries promotes **osmosis**. The osmotic reabsorption of

**Fig 13.7** Means by which substances enter the nephrons from blood

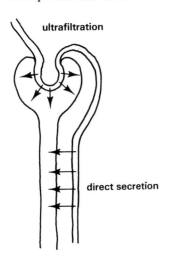

ultrafiltration

direct secretion

Our kidneys filter about 120 cm³ of fluid each minute. However, only about 1 cm³ of urine is produced each minute. What happens to the other 119 cm³ of fluid?

**Fig 13.8** Active solute reabsorption from the proximal convolutions against a concentration gradient as the levels of reabsorbed materials exceed those remaining in the filtrate

**Fig 13.9** Electronmicrograph of a cell lining the proximal convolution of a nephron, × 5800

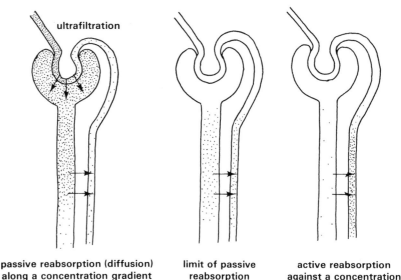

passive reabsorption (diffusion) along a concentration gradient

limit of passive reabsorption

active reabsorption against a concentration gradient

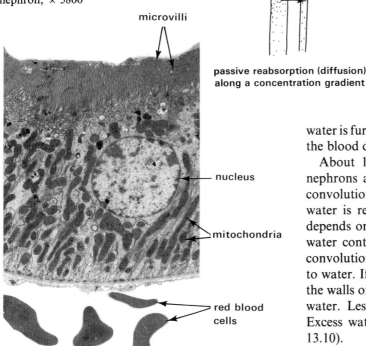

microvilli

nucleus

mitochondria

red blood cells

water is further assisted by the colloid osmotic pressure of the blood due to the plasma proteins (Chapter 1).

About 19 cm³ of every 20 cm³ of water left in the nephrons are reabsorbed every minute from the distal convolutions and collecting ducts. The extent to which water is reabsorbed from these parts of the nephrons depends on the body's state of hydration. If the body's water content is below normal, the walls of the distal convolutions and collecting ducts become very permeable to water. If the body contains sufficient water, however, the walls of the distal convolutions are less permeable to water. Less water is then reabsorbed into the blood. Excess water remains in the urine to be excreted (Fig 13.10).

**Fig 13.10** Some stages in the establishment of the medullary gradient by the loop of Henle of a cortical nephron. (b) and (d) show the effects of water reabsorption from the descending limb and solute reabsorption from the ascending limb. (a) and (c) show the effect of fluid flow within the loop. Figures are in units of mosmol per kg H₂O. (Modified from *Best and Taylor's Physiological Basis of Medical Practice*, eleventh edition, 1985, Williams and Wilkins)

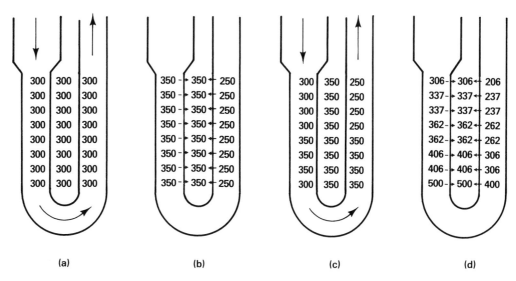

(a)          (b)          (c)          (d)

### 13.2.4 Role of the medulla in controlling water retention

The loops of Henle are important in creating a concentration gradient of osmotically active solutes in the medullary tissue fluid, that is, the **peritubular fluid**. The creation of the gradient depends on the flow of fluid in opposite directions in adjacent limbs of the loops of Henle. For this reason it is called a **counter-current multiplier**. It is the **medullary gradient** which enables water to be reabsorbed from the nephrons beyond the proximal convolutions.

In all nephrons, the descending limb of the loop is permeable to water but relatively impermeable to solutes. In cortical nephrons, establishing a concentration gradient requires an active transport mechanism along the whole length of the ascending limb. Sodium chloride is pumped out of the limb into the surrounding peritubular fluid. The increased solute concentration in the peritubular fluid causes water to leave the descending limb by osmosis. The fluid in the descending limb is thus concentrated. The fluid moves along the loop and more sodium chloride is actively removed from it in the ascending limb. More water is then removed passively from the descending limb, and so on (Fig 13.10). In this way a gradient is established in the peritubular fluid of between about 300 mosmol per kg $H_2O$ (isotonic with blood plasma) at the top, to about 600 mosmol per kg $H_2O$ around the hairpin of the loop.

In juxtamedullary nephrons, the descending limb of the loop of Henle is longer. Also there is a thin segment of the ascending limb which is not found in cortical nephrons (Fig 13.4(b)).The thin segment cannot force sodium chloride actively into the peritubular fluid, although solutes may be reabsorbed passively by diffusion. As in cortical nephrons, sodium chloride is pumped out of the thick segment of the ascending limb. Water then enters the peritubular fluid by osmosis from the descending limb. This dilution of the peritubular fluid creates a concentration gradient for urea which diffuses into the peritubular fluid from the medullary collecting ducts. In this way the medullary gradient in juxtamedullary nephrons is established by active sodium chloride reabsorption in the outer zone, and also by passive urea reabsorption in the inner zone (Fig 13.11).

**Fig 13.11** The role of urea in establishing the medullary gradient (inner zone) around juxtamedullary nephrons. Although some urea enters the thin ascending limb of the loop of Henle there is a net loss of solute (NaCl) by diffusion. Figures are in units of mosmol per kg $H_2O$. (Modified from *Best and Taylor's Physiological Basis of Medical Practice*, eleventh edition, 1985, Williams and Wilkins)

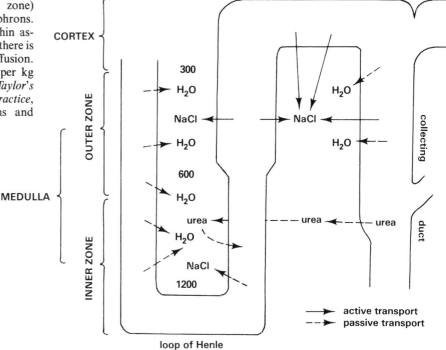

The gradient surrounding juxtamedullary nephrons is from about 300 mosmol per kg $H_2O$ at the top to about 1200 mosmol per kg $H_2O$ around the hairpin of the loop. Clearly the greater the concentration gradient between the fluid in the descending limb and the peritubular fluid (horizontally) the greater the reabsorption of water. Long loops of Henle establish a greater concentration gradient in the medulla (vertically) than short ones.

The movement of water from the descending limb into the peritubular fluid could dilute or even 'wash out' the medullary gradient. However, the gradient is maintained by another counter-current mechanism in the medullary blood vessels. The **vasa recta** vessels bring blood into the medulla from the efferent arterioles of the juxtamedullary nephrons. Blood flows relatively slowly in the vasa recta which carry about 10 % of the total kidney blood flow. Like the loops of Henle, the vasa recta turn and then pass back into the cortex (Fig 13.4(b)). Blood entering the vasa recta from the cortex is isotonic with the peritubular fluid in the cortex, that is about 300 mosmol per kg $H_2O$. The walls of the vasa recta are very permeable to water and solutes of small relative molar mass. As the blood flows deeper into the medulla it becomes surrounded by an increasing concentration of solute. Water leaves the blood by osmosis and solutes enter the blood from the peritubular fluid. The blood becomes more concentrated. This exchange continues as the blood approaches the hairpin of the vasa recta. The osmolality of the blood is always slightly different from that of the peritubular fluid because of the time taken to reach equilibrium and the fact that the blood is continuously flowing through the gradient.

The blood turns the hairpin and begins to flow back towards the cortex. As it ascends the medullary gradient, water is absorbed by osmosis from the peritubular fluid; solutes leave the vasa recta by diffusion (Fig 13.12). This passive **counter-current exchange** preserves the medullary gradient.

There is a slightly greater absorption of water by the ascending vasa recta than is lost to the peritubular fluid from the descending vasa recta. This is because the plasma proteins in the vasa recta attract water (colloid osmotic pressure). The proteins cannot leave the blood. They counteract the osmotic loss from the descending vasa recta but add to the osmotic gain in the ascending vasa recta (Fig 13.13). Consequently there is a net loss of water from the kidney's medulla

**Fig 13.12** Water and solutes exchange between the blood in the vasa recta and the peritubular fluid, thus maintaining the medullary gradient (from *Best and Taylor's Physiological Basis of Medical Practice*, eleventh edition, 1985, Williams and Wilkins)

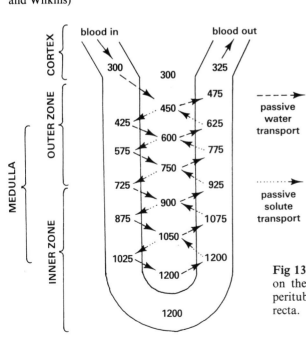

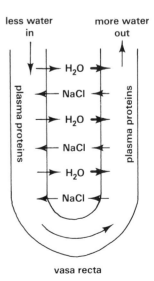

**Fig 13.13** The effect of plasma proteins on the exchange of water between the peritubular fluid and the blood in the vasa recta.

while most of the solutes responsible for the medullary gradient remain. Although some solutes are removed in the blood the gradient is not washed out. The mechanism allows water which is absorbed from the descending limb of the loop of Henle and the collecting duct to be retained in the body.

Sodium chloride is actively reabsorbed from the fluid in the distal convolution and collecting duct. In the presence of **vasopressin** (also called **antidiuretic hormone, ADH**), the collecting duct becomes permeable to water and urea. Water is reabsorbed by osmosis, urea by diffusion. The urine is thus further concentrated. In the absence of vasopressin less water and urea are reabsorbed and more are excreted.

The urea reabsorbed into the peritubular fluid stimulates water reabsorption by osmosis from the descending limbs of the loops of Henle in juxtamedullary nephrons. The fluid in the loops is thus concentrated. When it enters the solute-permeable thin segments of the ascending limbs, sodium chloride is passively reabsorbed. Some of the urea in the peritubular fluid may also enter the ascending thin segments by diffusion to be recycled to the collecting ducts (Fig 13.11).

The ability to concentrate urine is important to mammals living in arid conditions. It is therefore of interest that the nephrons of many desert mammals have unusually long loops of Henle (Fig 13.14).

**Fig 13.14** Relative proportions of the loops of Henle in the kidneys of three mammals living in different conditions

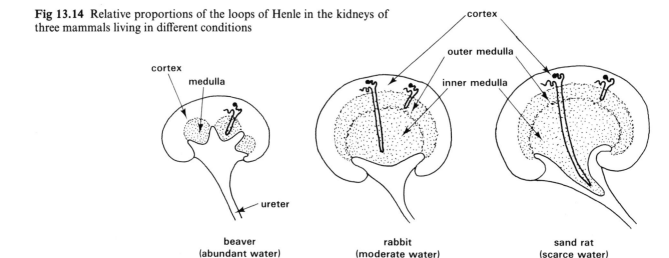

beaver
(abundant water)

rabbit
(moderate water)

sand rat
(scarce water)

### 13.2.5 Regulation of nephron function

Active reabsorption of some of the solutes in the renal filtrate is regulated by hormones (Chapter 16). Reabsorption of water from the proximal convolutions is by osmosis and normally accounts for the reabsorption of over 80 % of the water in the filtrate. Water reabsorption from the distal parts of the nephrons is also by osmosis. The extent of water reabsorption depends on the body's state of hydration and the permeability of the walls of the distal convolutions and collecting ducts.

Beneath the **hypothalamus** in the brain and projecting downward from it is an endocrine gland called the **pituitary body** (Chapter 16). The hypothalamus contains the **osmoregulatory centre** which is sensitive to the concentration of sodium chloride and hence the water content of blood flowing through it. If the water content is low, the sodium chloride concentration is high. The pituitary body, stimulated by the hypothalamus, releases **vasopressin (VP)**. Vasopressin increases the permeability to water of the walls of the distal convolutions and collecting ducts, encouraging reabsorption of water into the blood. Consequently a small volume of concentrated urine is eliminated, a condition called **antidiuresis**. Conversely, if the body's water content is high, VP output diminishes. The rate of water reabsorption from the distal ends of the nephrons then slows down. Urine flow increases and the urine becomes diluted, a condition called **diuresis** (Fig 13.15).

**Fig 13.15** Mechanism controlling water reabsorption from the distal parts of the nephrons

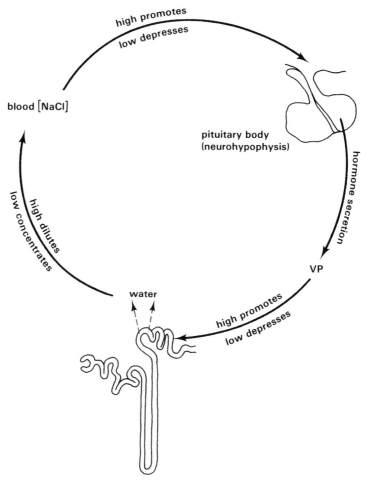

How does the body detect and monitor its water content? How is the information translated into permeability changes in the walls of the distal convolutions and collecting ducts?

Reduced VP output can result in a daily urine production of up to ten times the average 1.5 dm³. The condition is called **diabetes insipidus** and is quite distinct from diabetes mellitus (Chapter 16).

The role of the hormone renin in the regulation of the body's water content is described in Chapter 16.

## 13.3 The role of the kidneys in acid–base balance

Buffers such as the carbonic acid–hydrogencarbonate acid–base pair (Chapter 1) are limited in the extent to which they can cope with excess hydrogen ions in the body. The final regulation of the pH of body fluids is carried out by the lungs and kidneys. Gas exchange in the lungs disposes of carbon dioxide (Chapter 12). The build-up of carbonic acid in the tissues with accompanying high concentration of hydrogen ions is thus prevented.

Relatively small but significant quantities of acids enter the body daily in food. These and other acids are disposed of by the kidneys in such a way as to keep basic ions like sodium ($Na^+$) in the body. Acid excretion in the kidneys takes place in the distal convolutions of nephrons. Here, inside the tubule cells the enzyme **carbonic anhydrase** acts on carbonic acid.

Hydrogen ($H^+$) and hydrogencarbonate ($HCO_3^-$) ions are formed. The carbonic acid comes from the carbon dioxide produced by the tubule cells as they respire (Fig 13.16). The hydrogen ions are secreted into the tubule

**Fig 13.16** Production of hydrogencarbonate and hydrogen ions from respiration

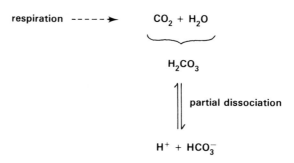

respiration $----\rightarrow$ $CO_2 + H_2O$

$H_2CO_3$

$\updownarrow$ partial dissociation

$H^+ + HCO_3^-$

where they meet and react with disodium hydrogenphosphate in the glomerular filtrate. The hydrogen ions replace some of the sodium ions, producing sodium dihydrogenphosphate. Sodium ions are then actively reabsorbed into the blood where they combine with the hydrogencarbonate ions from the tubule cells (Fig 13.17).

**Fig 13.17**

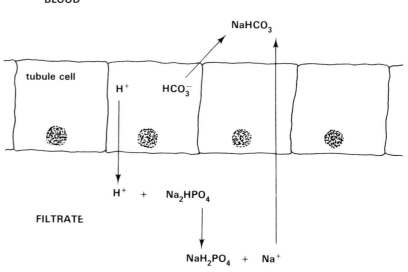

**Fig 13.18**

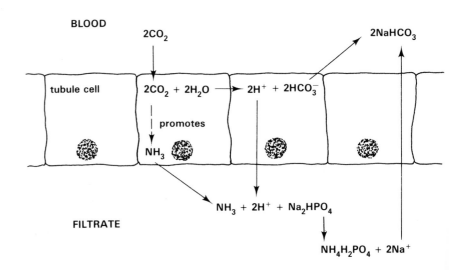

If the acid content of the blood is very high, the resulting low pH of the glomerular filtrate stimulates production of ammonia by the tubule cells. The ammonia completely replaces sodium ions from disodium hydrogenphosphate in the filtrate, resulting in increased acid excretion (Fig 13.18).

The pH of human blood lies normally between 7.35 and 7.45. A drop in pH below 7.30, a condition called **acidosis**, or a rise above 7.50 called **alkalosis**, results in metabolic malfunction. So sensitive are the body's physiological processes that failure to regulate the pH of the blood between about 7.0 and 8.0 is usually fatal.

## 13.4 Measurement of kidney function

An important aspect of kidney function is renal **clearance**. Clearance is defined as

> the volume of blood which is theoretically completely cleared of a given substance in one minute.

Most substances are filtered into the Bowman's capsules and then partially reabsorbed into the blood mainly from the proximal convolutions. Consequently, no blood is actually completely cleared of its constituents in this way.

The rate at which ultrafiltration occurs is called the **glomerular filtration rate (GFR)**. The GFR is normally about $120\,cm^3$ per minute in an adult. It can fall to a value as low as $30\,cm^3$ filtered each minute before any symptoms appear. Thus the kidneys can work at only 25 % maximum capacity and still clear the blood effectively. However, if the GFR falls further, then usually the problem can be managed by the selection of a diet designed to prevent accumulation of toxins in the blood. For example, a low protein diet avoids excessive amounts of blood urea. Should the GFR fall below about $3\,cm^3$ per minute, then dietary controls are no longer effective. This is called **kidney failure** and replacement of kidney function is necessary (section 13.5).

There are two main approaches to the measurement of kidney function.

### 13.4.1 Chemical methods

Renal clearance may be measured by comparing the concentrations of selected substances in blood and urine.

If $U$ = the concentration of the selected substance in urine,
and $V$ = the volume of urine produced in one minute,
then $U \times V$ = the quantity of the selected substance excreted in one minute.

Since clearance is the volume of blood from which this quantity is removed (cleared) then clearance ($C$) is given by

$$C = \frac{U \times V}{P}$$

where:

$P$ is the blood (plasma) concentration of the selected substance expressed in the same units as for $U$.

$V$ may be determined by collecting all the urine voided over a 24-hour period and by dividing the volume by 1440 (the number of minutes in 24 hours).

What is the clearance of urea in a person whose blood contains 30 mg urea per $100\,cm^3$, the urine contains 1800 mg per $100\,cm^3$ and urine is produced at a rate of $0.81\,cm^3$ per minute?

249

The clearance of urea, for example, is normally about 75 cm³ blood cleared each minute. That for glucose and water is zero and about 1 respectively (Fig 13.19).

**Fig 13.19** Clearance of (a) urea, (b) glucose, (c) water, (d) creatinine

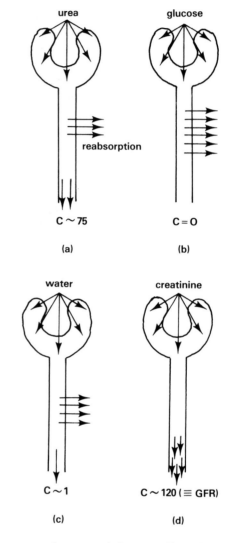

Some substances, such as creatinine, are filtered and totally excreted. They are not reabsorbed at all, nor do they enter the nephrons to any great extent by direct secretion (Fig 13.19(d)). Hence, the volume of blood cleared of creatinine is the volume of blood filtered. Clearance of creatinine may thus be taken as a measure of GFR.

Table 13.2 illustrates some typical clearance calculations. Notice the very high clearance value for para-amino hippuric acid (PAH). This is because

**Table 13.2 Clearance calculations for some constituents of blood and urine**

| Constituents | U | V | P | C |
|---|---|---|---|---|
| urea | 750 | 3 | 30 | 75 |
| creatinine | 12 | 1 | 0.1 | 120 |
| glucose | 0 | 1 | 4.5 | 0 |
| para-aminohippuric acid | 1625 | 2 | 5 | 650 |

**Fig 13.20** Clearance of para-amino-hippuric acid, PAH

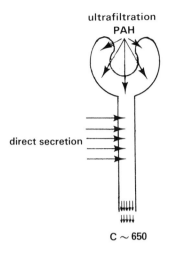

C ~ 650

PAH enters the nephrons by direct secretion as well as ultrafiltration (Fig 13.20). Substances which are excreted in this way are used in the radionuclide method for measuring kidney function.

### 13.4.2 Renograms

Substances with very high clearance values such as PAH (Fig 13.20) may be labelled with a radionuclide and their progress through the renal system followed using a suitable radiation detector. A common method involves the use of diethylenetriaminepenta-acetic acid (DTPA) labelled with gamma-ray-emitting technetium, $^{99}Tc^m$. The radionuclide is administered by injection into the blood and the renal system is imaged with a gamma camera (Fig 13.21). The radiation data collected by the gamma camera can

**Fig 13.21** (a) Gamma camera (courtesy ELSCINT).

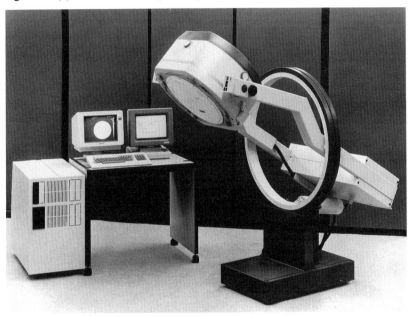

**Fig 13.21** (b) Anatomical features included in the gamma camera image (from O'Reilly, Shields and Testa, *Nuclear Medicine in Urology and Nephrology*, Butterworths)

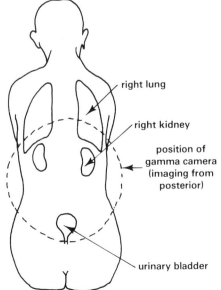

251

be displayed as a series of visual images (Fig 13.22). More usefully the data can be plotted as a **renogram** for each kidney (Fig 13.23). On the renogram, the arrival of the radionuclide in each kidney's blood supply is marked by a sudden increase in radiation. This is called the **vascular phase** of the renogram. Filtration and direct secretion of the radionuclide into the

**Fig 13.22** Sequence of gamma camera images showing the accumulation of radiation from a radionuclide of Technetium as it passes through the renal system: (a) 1 minute, (b) 8 minutes, (c) 16 minutes after start of test

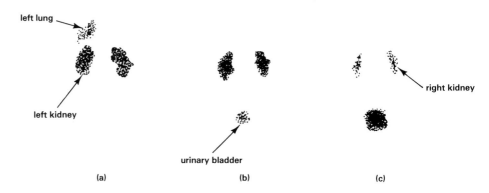

**Fig 13.23** Renograms

(a) Note how the blood radiation drops as the radionuclide is taken into the nephrons. The radiation in the urinary bladder rises as the radionuclide leaves the kidneys

(b) Diagram showing the main phases

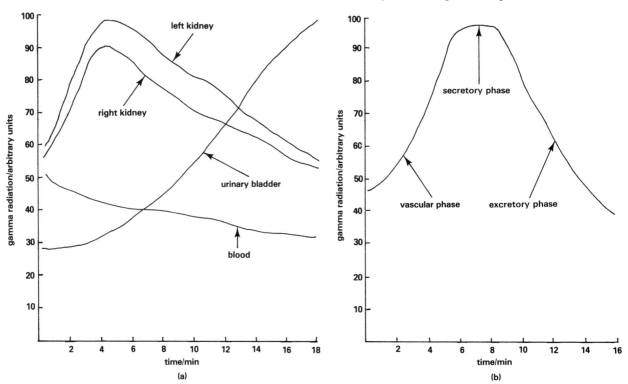

nephrons takes place, during which the radiation emitted from the kidneys remains fairly constant for a short period called the **secretory phase** of the renogram. Rapid clearance of the radionuclide is marked by a drop in radiation called the **excretory phase** of the renogram. Measurements for left and right kidneys are made separately but simultaneously (Fig 13.23(b)). Any impairment to kidney function such as an obstruction in a ureter or the renal pelvis will show on the renogram (Fig 13.24).

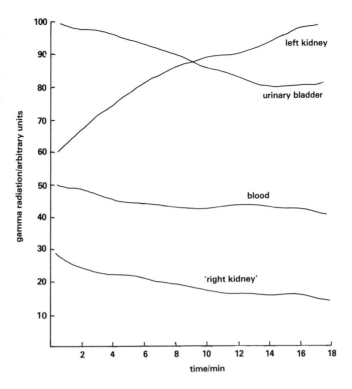

**Fig 13.24** Renogram from a patient whose right kidney has been removed surgically. There is poor clearance from the left kidney. Further investigation showed this to be caused by an obstruction. Radiation from the area where the right kidney used to be is from background tissue

## 13.5 Replacement of kidney function

Kidney failure can be treated in either of two main ways: artificial **dialysis**, that is by **kidney machine**, or kidney **transplantation**.

### 13.5.1 Kidney machines

A kidney machine is a mechanical device through which a patient's blood passes. The blood leaves the body usually from an artery in the forearm and returns to a nearby vein (Fig 13.25). Inside the machine the blood flows

**Fig 13.25** (a) Patient connected to a kidney machine

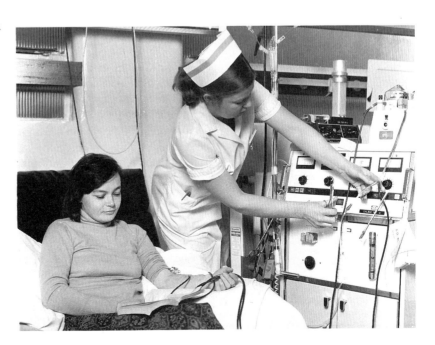

**Fig 13.25** (b) Diagrammatic arrangement of the apparatus causing dialysis of a patient's blood in a kidney machine

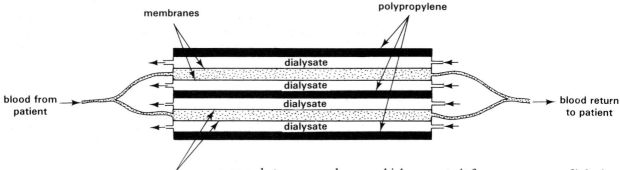

over or between membranes which separate it from an aqueous **dialysing fluid** containing dissolved sugars and salts in concentrations normally found in blood. Soluble constituents in the blood in excess of normal concentrations diffuse across the membrane into the dialysing fluid. In this way wastes like urea which accumulate in the body are extracted. Blood cells and proteins remain in the blood. The process is called renal dialysis (Fig 13.26).

**Fig 13.26** Mechanism of dialysis. Normal (for the blood) concentrations of dialysable substances in the dialysing fluid promote the diffusion of excesses from the blood across the membrane. Dialysis continues until the concentrations of dialysable substances on either side of the membrane are equal

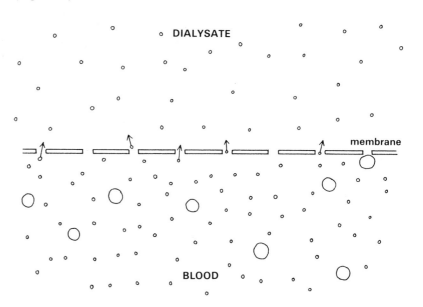

By using a clip to partially obstruct the tube carrying blood back to the patient's vein the blood pressure in the machine is increased. This makes it possible to remove water from the blood by ultrafiltration. Water is removed from the blood in a comparable way in a real kidney.

The pH of the blood falls from 7.4 to about 7.3 between dialysis treatments. This is because the body's store of hydrogencarbonate ions is used to buffer the blood and in so doing is lowered to about 75 % of its normal value. The kidney machine replaces hydrogencarbonate used up in this way. Earlier machines had sodium hydrogencarbonate in the dialysing fluid but sodium ethanoate is used now. The ethanoate diffuses in the blood and is metabolised in the body to hydrogencarbonate.

A patient usually spends several hours twice a week connected to a kidney machine, during which time the dialysing fluid drains the blood of excess and toxic consistuents. However, man-made machines are not perfect. As we saw earlier in the chapter, the activities of kidneys are regulated in response to changes in the body's internal environment. As yet kidney machines are not sufficiently sophisticated to be self-regulating.

Kidney machines do not replace all the kidney functions. Why are patients with failed kidneys often anaemic?

Furthermore, the machines are not designed to perform functions other than excretion. Kidneys functioning abnormally usually stop secreting the hormone **erythropoietin** which is necessary for normal red blood cell production (Chapter 10). Because of this, patients requiring dialysis are often anaemic. Nevertheless, despite their shortcomings kidney machines have prolonged the lives of a large number of people.

### 13.5.2 Kidney transplantation

Transplantation involves the transfer of a healthy kidney from one person called the **donor** into the body of the patient whose kidneys have failed, the **recipient**. The problems associated with kidney machines are eliminated, since the transplanted kidney takes over all the functions of the failed kidneys. The surgical procedures of kidney transplantation are relatively straightforward (Fig 13.27). However, as with the transplantation of any body organs, a major problem which has to be overcome is **immune rejection** (Chapter 11). The recipient's immune system recognises the transplant as non-self and reacts by the cellular immune response. T-lymphocytes invade the tissues of the transplanted organ and destroy it. The rejection problem can be largely avoided by matching as closely as possible the tissue type of the donor to that of the recipient. If relatively few of the transplant's antigens are non-self to the recipient, the immune response is less likely. In addition to tissue type matching, the recipient may be treated with immunosuppressant drugs. These inhibit the immune response. However, immunosuppression lowers the body's general defences, and antibodies which would normally protect the recipient against infections are not produced. The patient is thus prone to infections. Treatment usually includes a combination of tissue type matching and limited immunosuppression (Chapter 11).

**Fig 13.27** A transplanted kidney connected to the recipient's blood and urinary systems

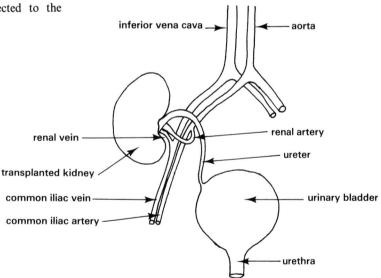

The main problem for kidney transplantation today is the shortage of suitable donor kidneys. There is an international organisation which keeps on file the tissue type of patients who require new kidneys. When kidneys become available, say from road traffic accident victims, their tissue type is determined and the international agency is informed immediately. Kidneys may then be flown from one country to another in order to provide a transplant for a suitable patient who may be thousands of miles away.

# 14 Nervous system

Any body system concerned with coordination must have a number of components to perform certain basic functions (Fig 14.1). There must be a **sensor** which can detect the stimulus. Sensors in the nervous system are called **receptors** and are modified nerve cells. They include rods and cones in the eyes, hair cells in the ears, stretch receptors in the lungs, skeletal muscles and the walls of major blood vessels and, in the skin, receptors sensitive to heat, touch, pressure and pain. Details of the actions of some of these receptors are described in Chapters 15 and 17.

**Fig 14.1** Basic components of a feed-back control mechanism

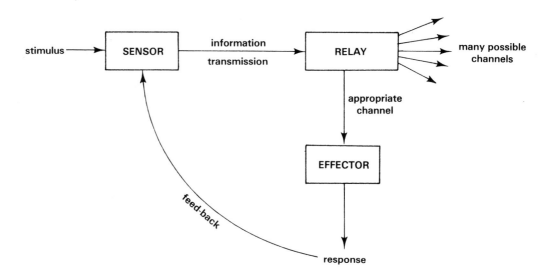

Receptors are **transducers** which convert the stimulus energy into a means whereby information about the stimulus can be transmitted within the body. In the nervous system this is done by nerve impulses transmitted along distinct pathways composed of interconnecting nerve cells called **neurons**. Another basic component is a **relay** which directs the information (impulses) along an appropriate pathway. The relay acts rather like a post office sorter who directs letters to their correct destination indicated by the addresses on the envelopes. In the nervous system the **central nervous system, CNS**, acts as the relay (section 14.2.2).

Eventually the information reaches an **effector**. This is an organ in the body that responds to the stimulus. Effectors are either muscles or glands. Responses are either contractile or secretory.

For activity to be coordinated the responses to stimuli must be appropriate. A response can be appropriate in two ways. The **kind** of response is important. For example, when we hear a warning signal such as a fire bell, we ought to move away from the hazard. The **degree** of response is important too. We should move away from a hazard *quickly*. On another occasion it may be appropriate to move slowly. A system of coordination must therefore incorporate a means of detecting its own response as well as the original stimulus. Further response can then be modified if necessary.

## 14.1 Excitable cells

The ability to transmit impulses is found only in neurons and muscle cells (Chapter 7). These cells are **excitable**.

### 14.1.1 The resting state

All living cells maintain an uneven distribution of different inorganic ions across their membranes. The extracellular concentration of these ions is different from that in the cytoplasm (Table 14.1). This state of affairs is

**Table 14.1 Concentrations of various ions in cells and tissue fluid**

| Ion | Cytoplasm/ m mol dm$^{-3}$ | Tissue fluid/ m mol dm$^{-3}$ |
|---|---|---|
| sodium | 16 | 140 |
| potassium | 100 | 4.4 |
| chloride | 4 | 103 |
| hydrogencarbonate | 8 | 27 |

created by the action of the cell membrane which pumps some ions out of the cell and pumps others into the cytoplasm. Metabolic energy provided by ATP is required for the pumping process. Important among the membrane pumps is the **sodium–potassium exchange pump**. It discharges sodium ions ($Na^+$) to the outside and potassium ions ($K^+$) to the inside of the cell. The ions are exchanged in equal numbers on a one-for-one basis. The pump creates concentration gradients down which diffusion occurs (Chapter 1). Sodium ions slowly diffuse back into the cytoplasm, and potassium ions slowly leak out of the cell. However, the cell membrane is about fifty times more permeable to potassium ions than it is to sodium ions. Consequently the potassium ions diffuse out of the cell more rapidly than the sodium ions diffuse in. As both ions have a positive charge the outer surface of the cell membrane is positive relative to the inside. The membrane is **polarised** because an electrical potential difference exists across it (Fig 14.2).

**Fig 14.2** Establishment of transmembrane potential by the combined actions of the sodium–potassium exchange pump and diffusion

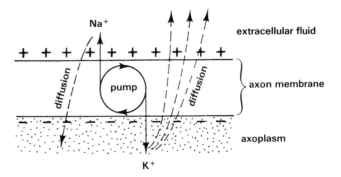

When the cell is not transmitting an impulse, the transmembrane potential is called the **resting potential**. The size of the resting potential varies from cell to cell, but for most excitable cells it is about $-70\,mV$. The minus sign indicates that the inside of the cell is negative relative to the outside. It is possible to measure the transmembrane potential directly by connecting an intracellular electrode and an extracellular electrode to an oscilloscope which acts as a sensitive voltmeter. The intracellular electrode is usually made from a very narrow glass tube tapered to a microscopic tip.

It is filled with an electrolyte such as potassium chloride solution. The tip of the electrode is then inserted through the membrane, with the aid of a microscope and a device called a micromanipulator (Fig 14.3).

**Fig 14.3** Measuring the transmembrane potential of an axon

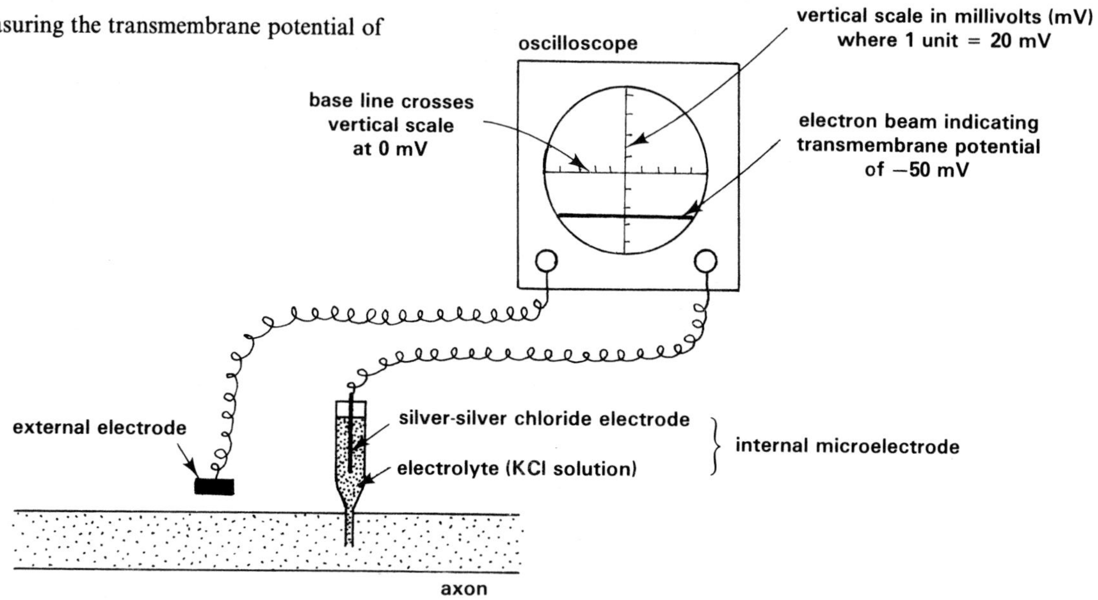

oscilloscope

vertical scale in millivolts (mV) where 1 unit = 20 mV

base line crosses vertical scale at 0 mV

electron beam indicating transmembrane potential of −50 mV

external electrode

silver-silver chloride electrode

electrolyte (KCl solution)

internal microelectrode

axon

## 14.1.2 Depolarisation

Cell membranes can be stimulated in various ways, including the application of an electrical discharge through the membrane or the addition of chemical substances to the membrane surface. The primary effect of such stimulation is to cause, at the site of stimulus, a sudden and very brief change in the permeability of the membrane to sodium ions. The membrane becomes more permeable to sodium than to potassium ions. Some of the extracellular sodium ions enter the cytoplasm by diffusion more easily than before. The ions move down a concentration gradient and are also attracted into the cell by the negative charge on the inside of the membrane.

The entry of sodium ions into the cell disturbs the resting potential. A point may eventually be reached when enough sodium ions have entered the cytoplasm to balance the positive charge remaining outside. The entry of more sodium ions causes the inside of the cell to become positive relative to the outside. The membrane is now **depolarised** (Fig 14.4). When enough sodium ions enter the cell to change the transmembrane potential to a certain **threshold** level, an **action potential** arises which generates an **impulse**. At threshold, pores called **sodium gates** open in the membrane and allow a sudden flood of sodium ions into the cell. When the action potential reaches its peak the sodium gates close and no more sodium ions

**Fig 14.4** Depolarisation of a neuron membrane

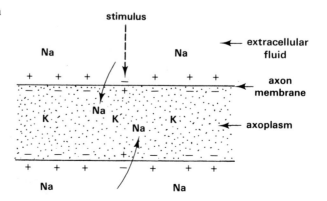

stimulus

Na                    Na

extracellular fluid

axon membrane

K        K        K

axoplasm

Na

Na                    Na

258

enter. If the depolarisation is not great enough to reach threshold, then an action potential and hence an impulse are not produced. This is called the **all or none** phenomenon. If a stimulus stronger than that necessary to produce an impulse is applied, the frequency of impulse production generally increases (Fig 14.5). In this way the nervous system discriminates between strong and weak stimuli. For most excitable cells the threshold is about $-60\,\text{mV}$. This is $10\,\text{mV}$ less than the resting potential of $-70\,\text{mV}$.

**Fig 14.5** All-or-none effect. Raising the stimulus strength from zero does not produce an impulse until a stimulus strength results in the same action potential

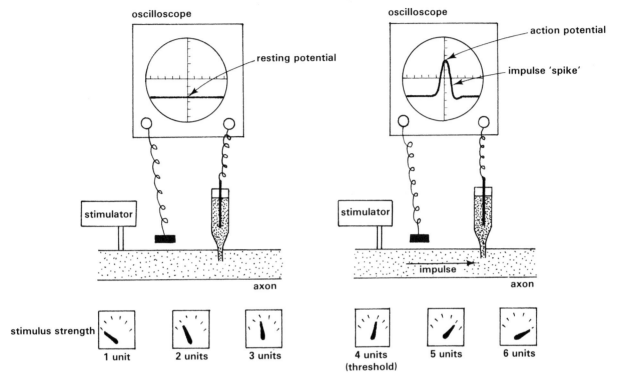

Fig 14.6 illustrates the voltage changes that occur across a neuron's membrane as an impulse passes by.

**Fig 14.6** Changes in transmembrane potential during impulse transmission

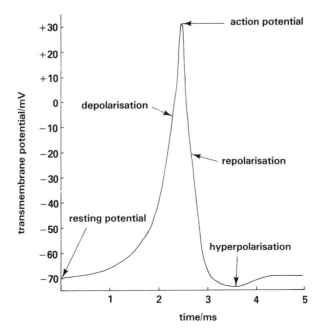

### 14.1.3 Impulse conduction

Entry of sodium ions into an excitable cell at the point of stimulation sets up a **current** of ions in the extracellular fluid and the cytoplasm (Fig 14.7). The electric current crosses the membrane a short distance in both directions from the point at which the stimulus was applied. Electric current is also a stimulus, so the cell is further stimulated where the current crosses the membrane. The membrane's permeability to sodium ions changes, depolarisation occurs and another action potential is set up. This is followed in turn by more current flow and stimulation of the membrane even further away from the original point of stimulation. This sequence of electrochemical events constitutes the **impulse**.

**Fig 14.7** Impulse transmission along an axon

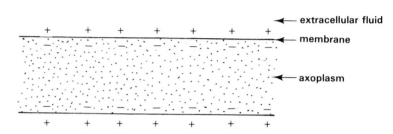

(a)  The resting state

(b)  Depolarisation of the membrane causing the flow of electric current

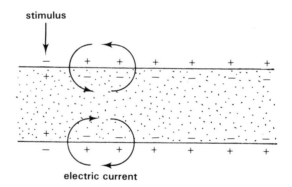

(c)  Electric current stimulates the membrane nearby causing further depolarisation

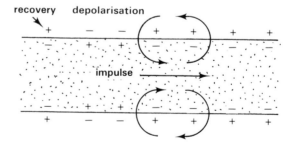

(d)  The process repeats itself along the axon

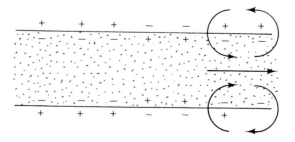

As neurons are usually stimulated at one end (either the cell body or dendrites), impulses travel in one direction along them. In contrast, muscle fibres are normally stimulated in the middle and so impulses travel in both directions.

Impulses travel very rapidly along neurons. It was once believed that the rate of nerve impulse conduction was comparable with the speed of light! In our bodies impulses can be conducted at speeds of up to about $100 \, \text{m s}^{-1}$. However, many nerve fibres conduct more slowly than this. Two main factors affect impulse velocity. They are myelination and fibre diameter. The presence of a **myelin** sheath (Fig 7.18) greatly increases the velocity at which impulses are conducted along the axon of a neuron. The electro-chemical events of impulse propagation can occur only where the axon membrane is exposed to the extracellular fluid at the **nodes of Ranvier**. The current generated at one node stimulates the next node a relatively long distance away, perhaps as much as 1 mm. This is called **saltatory conduction** (Fig 14.8). In unmyelinated fibres the entire axon membrane is exposed and impulse conduction is slower.

**Fig 14.8** Saltatory conduction

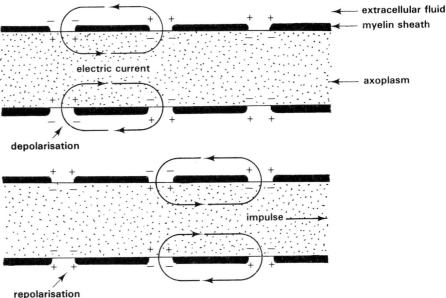

**Fig 14.9** Photomicrograph of a transverse section of human median nerve. Note the variety of axon diameters, × 800

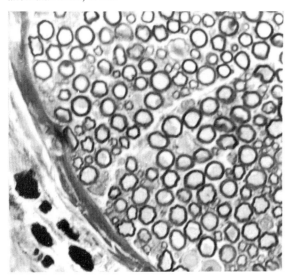

In a limb nerve there are very many neurons. Some conduct impulses down the limb away from the central nervous system (motor neurons) whilst others conduct impulses up the limb towards the central nervous system (sensory neurons). In transverse section it can be seen that the axons vary in diameter (Fig 14.9). Fibres of broad diameter conduct impulses more rapidly than narrow ones. The rate at which impulse conduction occurs can be determined by applying a small electric voltage to the surface of one leg at the knee. The voltage generates impulses in the

261

common peroneal nerve which conducts to the extensor digitorum brevis muscle in the foot (Fig 14.10(a)). A second similar stimulus is then applied to the nerve a measured distance nearer to the muscle at the instep. Arrival of the impulses is detected by a recording electrode placed over or inserted into the muscle and connected to an oscilloscope (Fig 14.10(b)). From the measurements the time can be calculated for the impulse to travel from the first to the second stimulus point. For the common peroneal nerve in the leg the impulse velocity is normally between 40 and 60 m s$^{-1}$. Similar measurements for the median and ulnar nerves in the arms produce a velocity range of about 50 to 70 m s$^{-1}$.

**Fig 14.10** Measurement of impulse conduction velocity

(a) Electrode positions on leg

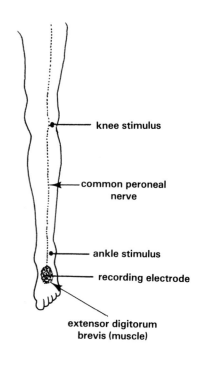

(b) Oscilloscope recordings. The time taken for the impulse to travel from the knee to the ankle is calculated from the two traces (courtesy Dr Smith, Dept. of Neurophysiology, Queen's Medical Centre, Nottingham)

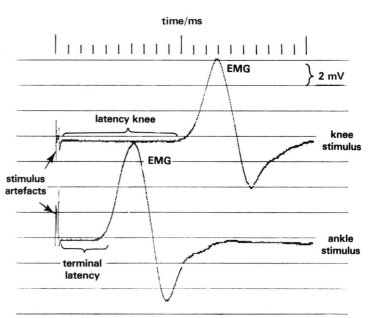

Use the oscilloscope recordings in Fig 14.10(b) to determine the velocity of impulse conduction in the common peroneal nerve.

latency knee = 9.2 ms
terminal latency = 2.7 ms
latency difference = 6.5 ms
ankle-knee distance = 33 cm

262

### 14.1.4 Repolarisation

An impulse is essentially a disturbance of the resting potential of an excitable cell's membrane. Before a second impulse can be generated in a cell the membrane must first recover its resting potential. The recovery period in cells is called the **refractory period**. During this time the membrane repolarises. Repolarisation occurs in two stages. First the membrane's permeability changes back to its original state, allowing potassium ions to diffuse out of the cell more easily. The negative charge outside the depolarised cell membrane also attracts the potassium ions. The movement outwards of potassium ions restores the resting electric potential (Fig 14.6).

The second stage of recovery is marked by restoration of the ionic balance across the membrane. It is brought about by the sodium–potassium exchange pump (Fig 14.11). During the first stage of recovery it is impossible for a second impulse to be generated because the

Fig 14.11 Membrane recovery

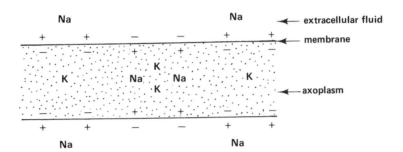

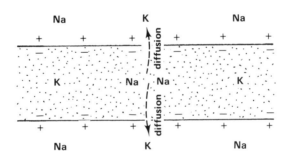

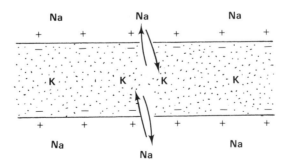

membrane is not yet repolarised. The time required for repolarisation is called the **absolute refractory period**. However, during the second statge it is possible to generate an impulse before membrane recovery is completed. This can occur if a stimulus is applied which is strong enough to produce the threshold transmembrane potential (Fig 14.12). The second stage of recovery is called the **relative refractory period**.

Refraction between successive impulses limits the frequency with which impulses can be conducted. Most neurons can conduct up to about 100 impulses per second.

**Fig 14.12** Absolute and relative refraction. Impulse generation is possible during the relative refractory period providing the stimulus intensity is great enough (after Vander, Sherman and Luciano)

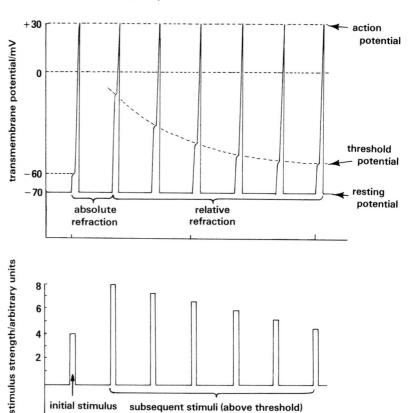

### 14.1.5 Synapses and neuroeffector junctions

Nerve pathways consist of at least two neurons joined end to end. The junctions of neurons are called **synapses** (Figs 7.15 and 14.13). They are similar in structure and function to the junctions between the last neuron in a pathway and an effector to which impulses are transmitted. Each axon branches and terminates in many bulb-like structures called **boutons**. Electron microscopy reveals many mitochondria and small membranous vesicles inside each bouton. Between the bouton and the cell body or dendrites of the postsynaptic neuron, there is a narrow gap called the **synaptic cleft**, usually about 10 to 20 nm across (Fig 14.13).

The vesicles contain substances called **transmitters** which are involved in the transmission of impulses from the presynaptic neurons, across the synaptic clefts and to the postsynaptic neurons. The arrival of an impulse at an axon terminal causes some of the vesicles to fuse with the presynaptic

**Fig 14.13** Electronmicrograph of a synapse, × 24 000. See Fig 7.15

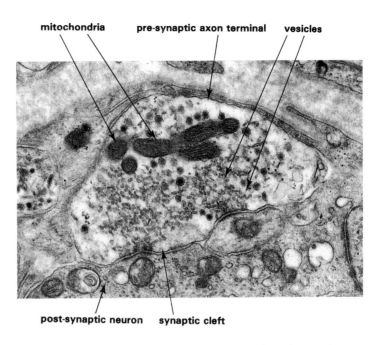

membrane. The contents of the vesicles are released onto the post-synaptic membrane. Here there are specific **receptor sites** to which the transmitter substance attaches, causing depolarisation of the postsynaptic membrane. An **excitatory postsynaptic potential (EPSP)** is generated and an impulse travels along the postsynaptic neuron (Fig 14.14).

**Fig 14.14** Synaptic transmission

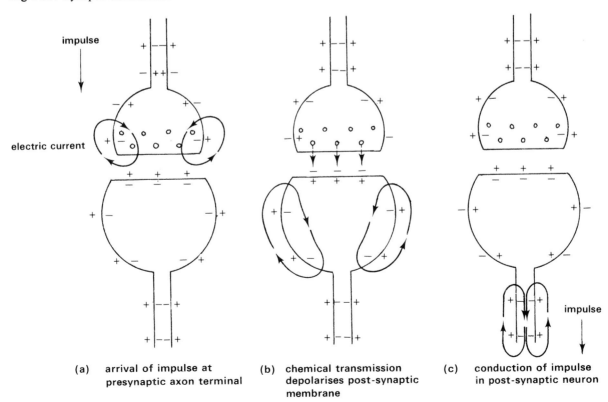

(a) arrival of impulse at presynaptic axon terminal

(b) chemical transmission depolarises post-synaptic membrane

(c) conduction of impulse in post-synaptic neuron

Many potent transmitter substances have been isolated from nerve endings. Most widespread in the nervous system is **acetylcholine**. Another called **noradrenalin** is found in sympathetic nerve endings (section 14.2.3). Following their release into the synaptic clefts, transmitters are quickly degraded by enzymes. Acetylcholine for example is broken down by **acetylcholinesterase** into choline and ethanoic acid. The products cannot cause depolarisation. It is for this reason that the effects of the transmitters are only temporary. It means that the arrival of an impulse at a synapse does not result in the generation of many impulses in the postsynaptic neuron. This is important since the frequency of impulses conducting along a pathway reflects the intensity of the original stimulus at the beginning of the pathway. An uncontrolled increase in impulse frequency at synapses could result in the perception of a stronger stimulus than was present in the first place. We would not wish to be deafened by a quiet noise or blinded by dim light.

Active transmitter is resynthesised from the inactive products reabsorbed by the neurons and packaged in new vesicles. The energy necessary for resynthesis is provided by the relatively large numbers of mitochondria in the boutons (Fig 14.15). Some synapses in the central nervous system are inhibitory. On the arrival of an impulse, **inhibitory synapses** release transmitters that prevent the generation of impulses in the postsynaptic neuron. They do this by **hyperpolarising** the postsynaptic membrane rather than depolarising it. The permeability of the postsynaptic membrane to potassium ions increases. Consequently more of these ions diffuse out of the cytoplasm and an **inhibitory postsynaptic potential (IPSP)** arises (Fig 14.16). Inhibitory synapses are important in nerve pathways and control antagonistic muscles (section 14.2.2).

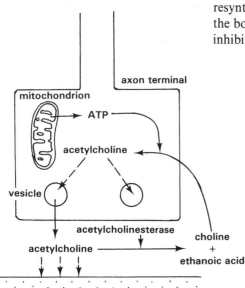

**Fig 14.15** Cycle of release, breakdown, reabsorption and resynthesis of synaptic transmitter

**Fig 14.16** Action of an inhibitory synapse. Inhibitory transmitters include $\gamma$-aminobutyric acid and glycine

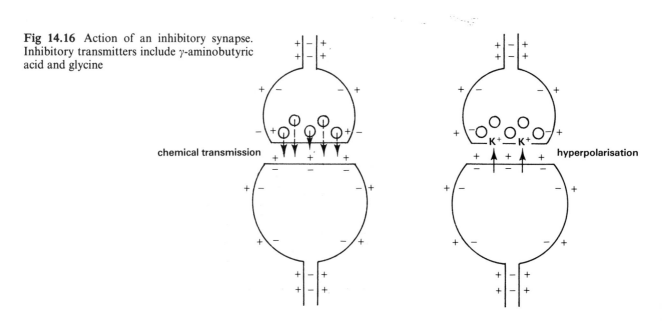

Motor neurons terminate as **motor end plates** on muscle fibres (Fig 14.17). Electron microscopy reveals synapse-like structures including mitochondria, vesicles and a narrow cleft between the axon terminal and the muscle fibre. Stimulation of the muscle by a nerve is achieved by chemical transmitters released from vesicles. Acetylcholine is the main neuromuscular transmitter. Noradrenalin is the neuromuscular transmitter at sympathetic nerve endings.

### 14.1.6 Muscle contraction

In striped and cardiac muscle the fibres contain many **myofibrils**. These are divided internally by membranous partitions called Z-lines into a large number of **sarcomeres** (Figs 7.23 and 14.18). Projecting into the **sarcoplasm** of each sarcomere is an array of **thin filaments** consisting of the protein **actin**. Thin filaments are between 5 and 8 nm in diameter. Suspended between them and in the centre of the sarcomeres is an array of **thick filaments** consisting of the protein **myosin**. Thick filaments are between 12 and 18 nm in diameter. Both kinds of filaments run parallel to the longitudinal axis of the myofibril (Fig 14.18(b)).

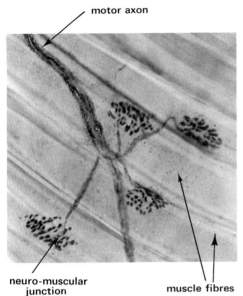

**Fig 14.17** Photomicrograph showing the junctions between motor neuron endings and striped muscle fibres, × 450

motor axon

neuro-muscular junction

muscle fibres

**Fig 14.18** Sarcomeres
(a) Electronmicrograph, × 65 000

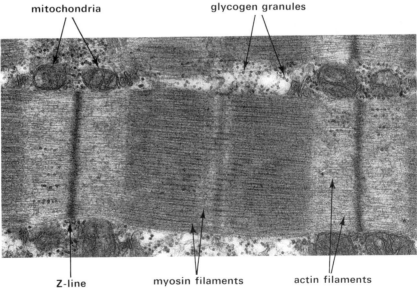

mitochondria

glycogen granules

Z-line

myosin filaments

actin filaments

(b) Protein filaments

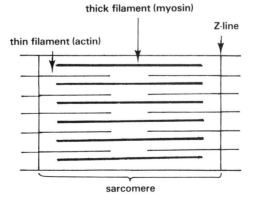

thick filament (myosin)

Z-line

thin filament (actin)

sarcomere

Each thin filament consists of very many molecules of actin arranged in long threads that spiral around each other in pairs (Fig 14.19(a)). Along each thread are many sites where the myosin molecules can become attached to the actin. Another protein called **tropomyosin** is arranged along the actin threads and may cover the myosin-binding sites.

**Fig 14.19** Muscle proteins: (a) thin filaments of actin, (b) thick filaments of myosin

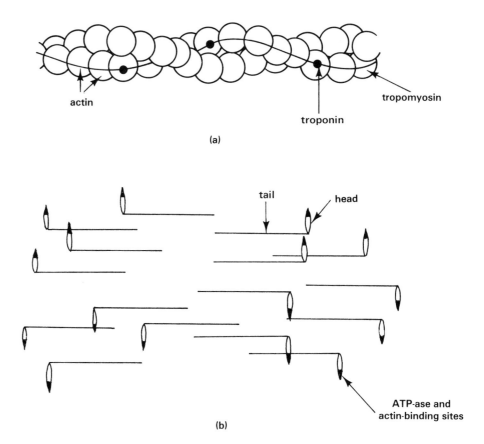

A thick filament contains about 200 molecules of myosin. Each molecule has a bulbous head and a filamentous tail. They are arranged so that the heads project from the sides and at the ends of the thick filaments (Fig 14.19(b)). On the tips of the myosin heads there are sites which bind them to the thin filaments. About half the myosin heads are attached to thin filaments at any one time. Also present on the myosin heads are molecules of the enzyme **ATP-ase**. They catalyse the removal of terminal phosphate from ATP, producing ADP (Chapter 5). The energy released is used to bring about muscle contraction. When muscle contracts the myosin heads detach from the thin filaments and rotate around in a spiral fashion. Other myosin heads nearer to the Z-lines attach to the thin filaments, drawing them towards the centre of the sarcomeres (Fig 14.20).

This **sliding filament** model explains why the Z-lines of each sarcomere are drawn towards each other during muscle contraction. The sarcomeres and hence the myofibrils and fibres shorten. At the level of structure revealed by the light microscope the I-bands and the H-zones become narrower or even disappear when contraction occurs. The A-bands, representing the thick filaments remain unchanged (Fig 7.23).

In the relaxed state, the myosin-binding sites on the thin filaments are covered up by the tropomyosin threads (Fig 14.19(a)). When the muscle is stimulated by a nerve impulse, muscle impulses are generated in the myofibril membranes. The muscle impulses depolarise the myofibril membranes. Sodium and **calcium ions** enter the sarcoplasm. Calcium ions are normally excluded from the sarcoplasm by a membrane pump. The calcium ions displace the tropomyosin threads, thus exposing the myosin-binding sites on the thin filaments. Contraction can now occur. This is equivalent to uncovering a door lock so that a key can be inserted, turned and the door opened.

**Fig 14.20** Sliding filaments during muscle contraction

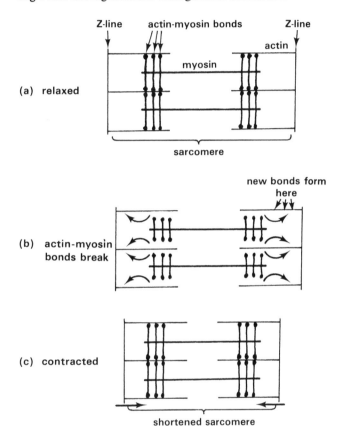

Excessive fluctuations in the calcium ion content of body fluids lead to neuromuscular disturbances (Chapter 16). Muscle is adapted in a number of ways to provide the energy required for contraction. Its red colour is due to the presence of **myoglobin** (Chapter 12). The pigment may become oxygenated to oxymyoglobin by collecting oxygen from oxyhaemoglobin in the muscle's blood supply. Oxymyoglobin is induced to dissociate, releasing its oxygen to the muscle when the oxygen tension falls so low that the blood cannot supply enough oxygen to the muscle (Fig 12.24). Oxymyoglobin is therefore an emergency store of oxygen. Some muscles contain no myoglobin and appear white in colour. Fish contain a lot of white muscle. Another example is breast meat in poultry in contrast to the leg meat which contains myoglobin and is red.

The main respiratory substrate is glucose. Muscle contains the polysaccharide **glycogen** (Fig 14.18(a)) which can be hydrolysed, thus providing glucose for respiration in addition to that supplied in the blood (Chapter 16).

Since all the components for the contraction mechanism are present inside the myofibrils, why are muscles not contracted all the time? How does the arrival of impulses at the neuromuscular junctions trigger off contraction? How does contraction occur simultaneously thoughout the entire muscle?

269

Muscle also contains **creatine** which can be phosphorylated by the transfer of phosphate from ATP. When ATP is used to provide energy for muscle contraction the ADP so formed is immediately rephosphorylated by phosphocreatine. In this way a supply of ATP is maintained (Fig 14.21). Phosphocreatine thus acts as an energy store in muscle fibres in addition to ATP (Chapter 5).

**Fig 14.21** The role of creatine in the provision of energy for muscle contraction

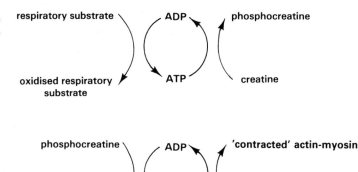

**Fig 14.22** Diagram of the main divisions of the nervous system

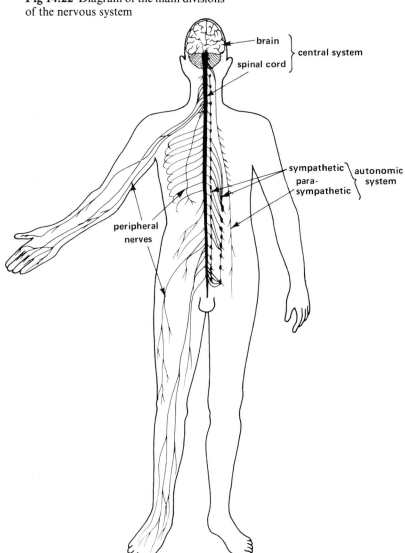

## 14.2 Nervous integration

The nervous system contains countless millions of neurons which are joined by synapses into a large number of circuits or **pathways**. The precise ways in which neuronal pathways are constructed can affect the nature and destination of the information transmitted. Any one neuron may receive impulses from many other neurons that synapse with it. Synapses are formed on either the dendrites or the soma of the postsynaptic neuron (Fig 7.14). Similarly the axon of any one neuron may branch many times and terminate at synapses on many other neurons. The pathways along which impulses are transmitted may thus be complex, and impulses may be sent to any part of the body. In addition, some pathways may be closed at times by the action of inhibitory synapses.

The nervous system consists of several distinct divisions. Each division performs its own particular function and contributes to overall nervous integration. The **central nervous system (CNS)** consists of the brain and spinal cord. The rest of the nervous system is called the **peripheral system**. It includes **somatic nerves** and the **autonomic nervous system**. The autonomic system may be subdivided further into the **sympathetic** and **parasympathetic** divisions (Figs 14.22 and 14.34).

### 14.2.1 Neuronal pathways

Several kinds of neuronal pathways can be described.

### 1. Diverging pathways

In a **diverging pathway** the route along which impulses are transmitted divides, enabling information from a single source to be transmitted to a number of destinations (Fig 14.23). There are many instances of nervous coordination in which a variety of responses are appropriate reactions to a single stimulus. An example is the body's responses to changes in blood and skin temperature (Chapter 17).

**Fig 14.23** Diverging neuronal pathway

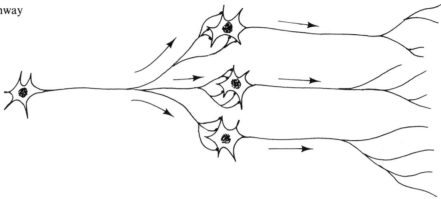

At the branches in diverging pathways at least two different zones of activity can be seen (Fig 14.24). Look at the pathways from neuron 1 to neuron B and from neuron 2 to neuron E. There are relatively many synapses joining these neurons *in line* at regions called **discharge zones**. Impulses are readily transmitted through discharge zones. Now look at the pathways from neuron 1 to neurons A and C and from neuron 2 to neurons D and F. There are relatively fewer synapses joining these neurons in regions called **facilitated zones**. Synaptic transmission here may be too

**Fig 14.24** Discharge zones (dz) and facilitated zones (fz) in diverging neuronal pathways

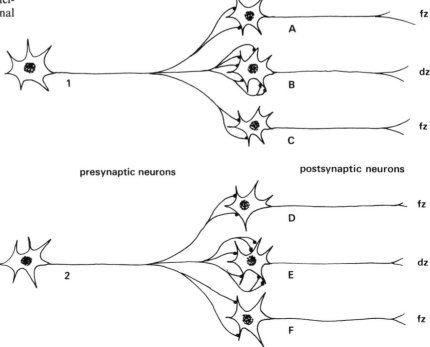

271

weak to generate action potentials in the postsynaptic neurons. However, the postsynaptic membranes may be rendered more susceptible to depolarisation when activated by subsequent stimuli. Hence, impulses are transmitted through facilitated zones only if the presynaptic stimulation is sufficiently frequent.

## 2 Converging pathways

In a **converging pathway**, impulses from two or more sources are channelled to the same destination in the body (Fig 14.25). This enables an organ to respond to more than one stimulus. An example is the effect on the heart's activity of changes in the pressure and the gas content of the blood (Chapter 10).

**Fig 14.25** Converging neuronal pathway

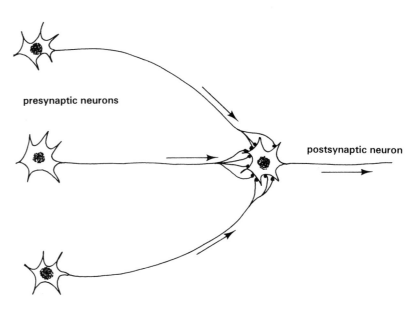

presynaptic neurons

postsynaptic neuron

The more synapses there are in a neuronal pathway, the longer it takes for impulses to travel the entire length of the pathway. This is because the secretion of chemical transmitters at the synapse is slow relative to impulse conduction along an axon. Examine Fig 14.26. The neuron on the left diverges to several pathways which then converge on the neuron on the right. Because of the different numbers of synapses in the pathways, many impulses arrive at the neuron on the right at different times. The frequency of output is thus greater than the frequency of input.

**Fig 14.26** If the input to this circuit is a single impulse, what do you think would be the output?

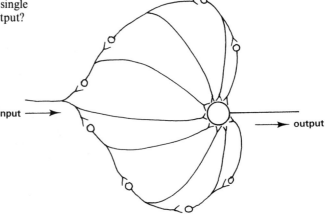

input →

output →

## 3 Reverberating pathways

In a **reverberating pathway** impulses are routed back into the same circuit time and time again (Fig 14.27). Re-routing continues until the neurons or their synapses fatigue. Reverberating pathways are thought to be responsible for the actions of the medullary rhythmicity area in controlling the basic rhythm of breathing (Fig 12.14).

**Fig 14.27** Reverberating neuronal pathway. Once activated by an input, the pathway continues to transmit impulses until it fatigues

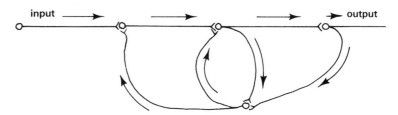

## 14.2.2 Nerve cord and reflexes

The great majority of body functions controlled by the nervous system involve **reflexes**. Reflex actions are very rapid, automatic responses to stimuli. Reflex pathways involve the central nervous system but do not necessarily involve conscious awareness. Thus we do not have to think about our limb movements while we are walking, or our rib and diaphragm movements while breathing or our heart's actions in pumping blood. These activities are controlled rapidly and effectively for us by reflex actions. We can thus carry out a conversation with someone, even while we are walking, breathing and maintaining our circulation.

Many reflex pathways involve the spinal cord, a posterior extension of the brain which runs the length of the back. It is enclosed and protected by the vertebrae. The cord is about 130 mm across in the cervical (neck) region and tapers to about 70 mm across in the sacral (pelvic girdle) region (Fig 14.28). Between the vertebrae pairs of spinal nerves arise, one on each side of the spinal cord. Spinal nerves contain both **sensory (afferent) neurons** and **motor (efferent) neurons**. The sensory neurons transmit impulses to the spinal cord from receptors. The motor neurons transmit impulses from the spinal cord to effectors.

**Fig 14.28** Photomicrograph of the spinal cord of a cat, lumbar region, × 25

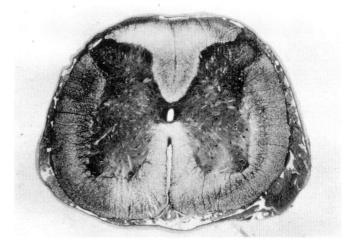

**Fig 14.29** Structure of the spinal cord. Thick arrows indicate the direction of impulses

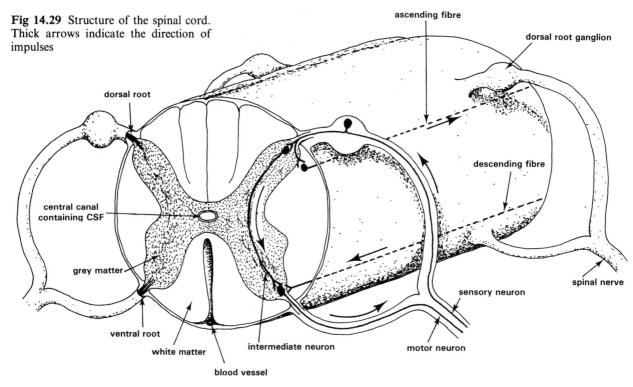

Each spinal nerve joins the spinal cord at two points. Sensory fibres are found in the **dorsal root**, motor fibres in the **ventral root** (Fig 14.29). Within the spinal cord the nervous tissue is distributed in two regions. The central **grey matter** contains transverse fibres, the outer **white matter** contains tracts of longitudinal fibres that transmit impulses up and down the cord. The simplest reflex pathways involve only two neurons, joined by a single synapse. An example is the **stretch reflex** (Fig 14.30). Striped muscle contains small receptors called **spindles** that generate sensory impulses when stretched. Impulses are transmitted along sensory neurons and enter the spinal cord through the dorsal root. The sensory neurons terminate in the grey matter where they synapse with motor neurons. The latter transmit impulses out of the nerve cord, through the ventral root and back into the same spinal nerve that carried the sensory impulses. The motor neurons terminate at neuromuscular junctions on the stretched muscle. Contraction of the muscle counteracts the stretching. A **knee-jerk** reflex occurs when the patellar ligament connecting the knee-cap (patella) to the tibia in the lower leg is struck with a special hammer. The patella is pulled and stretches the quadriceps femoris muscle in the upper leg. The reflex results in contraction of the quadriceps muscle and the lower leg jerks outwards.

**Fig 14.30** Stretch reflex

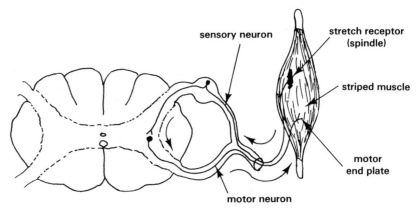

274

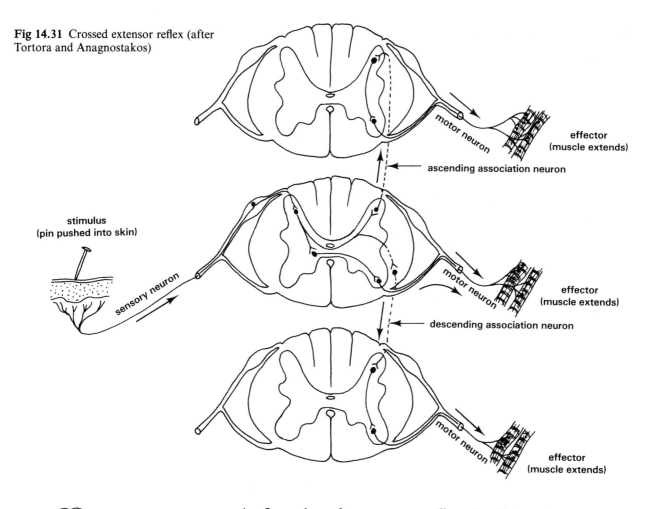

**Fig 14.31** Crossed extensor reflex (after Tortora and Anagnostakos)

effector (muscle extends)

ascending association neuron

stimulus (pin pushed into skin)

sensory neuron

motor neuron

effector (muscle extends)

descending association neuron

motor neuron

effector (muscle extends)

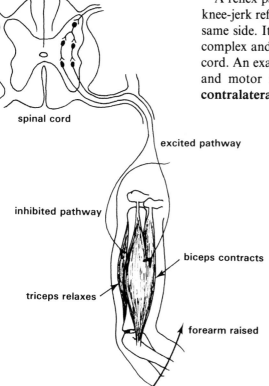

spinal cord

excited pathway

inhibited pathway

biceps contracts

triceps relaxes

forearm raised

A reflex pathway from receptor to effector is called a **reflex arc**. In the knee-jerk reflex the sensory and motor neurons join the spinal cord on the same side. It is called an **ipsilateral** reflex arc. Some pathways are more complex and include longitudinal tracts in the white matter of the spinal cord. An example is the **crossed extensor reflex** (Fig 14.31). The sensory and motor neurons are on opposite sides of the spinal cord in this **contralateral** reflex arc.

Antagonistic activity of muscles (Chapter 8) regulates movement and posture. Contraction of one muscle is usually accompanied by relaxation of another. During such activity the simultaneous contraction of both sets of muscles is prevented by the actions of inhibitory synapses (Fig 14.32 and section 14.1.5).

**Fig 14.32** Reciprocal inhibition. During body movements involving the actions of antagonistic muscles, one muscle is excited while the other is inhibited. Here the forearm is flexed by contraction of the biceps while the triceps relaxes. Neuronal pathways involving excitatory and inhibitory synapses are thought to control this type of activity

275

We are born with certain neuronal pathways ready to function immediately. A new-born baby automatically sucks at objects placed in its mouth. This **innate reflex** is vital for the baby to feed. If a baby's body is supported and its feet allowed to touch a solid surface, it will usually move its legs in a characteristic walking action. A baby's hands grip tightly onto objects placed in them. Very young infants acquire additional reflexes that override the innate reflexes at certain stages of development. Observation of innate and acquired reflexes can help in determining the neurological development of babies (Fig 14.33).

**Fig 14.33** Some infantile reflexes (after Williams and Wendell-Smith)

grasp reflex

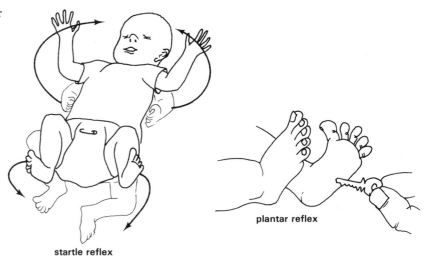

startle reflex

plantar reflex

It is an interesting exercise to look for Pavlov's conditioning in advertisements. How many of them describe the true nature of the product and how many associate it with something attractive but quite unrelated to the product?

In 1910, Pavlov demonstrated **conditioned reflexes**. He noticed that hungry dogs salivated when presented with food. He also observed that hungry dogs salivated when exposed to a secondary stimulus such as ringing bells at the same time as the primary stimulus, food. Pavlov discovered that, in time, the dogs learned to associate the secondary stimulus with food. They salivated when bells were rung even if food was absent. Knowledge of such **conditioning** is effectively used by the advertising industry.

### 14.2.3 Autonomic system

The autonomic nervous system consists of two sets of nerves, **sympathetic** and **parasympathetic** (Fig 14.34). The two sets generally have the opposite effects (Table 14.2). Many organs have double innervation, being supplied by both sympathetic and parasympathetic nerves. The balance between the activity of the two sets of nerves results in coordinated regulation of the organs, usually by reflex action. As with peripheral reflexes, the central nervous system is involved in autonomic activity. Many of the centres coordinating autonomic functions are situated in the hind-brain. They include the cardiac and vasomotor centres (Chapter 10), the respiratory centre (Chapter 12) and the thermoregulatory centres (Chapter 17). These receptors detect body changes that require adjustment of the cardiovascular and gas exchange systems and the skin. Appropriate impulses are conveyed to these systems along autonomic nerves. Unlike somatic reflexes, autonomic reflexes are less easily controlled by conscious activity of the brain. Although it is possible to train an individual to reduce the heart rate, for instance, autonomic function relies almost entirely on reflexes. The ability to control breathing consciously is due to the fact that the muscles involved are supplied by somatic as well as autonomic nerves.

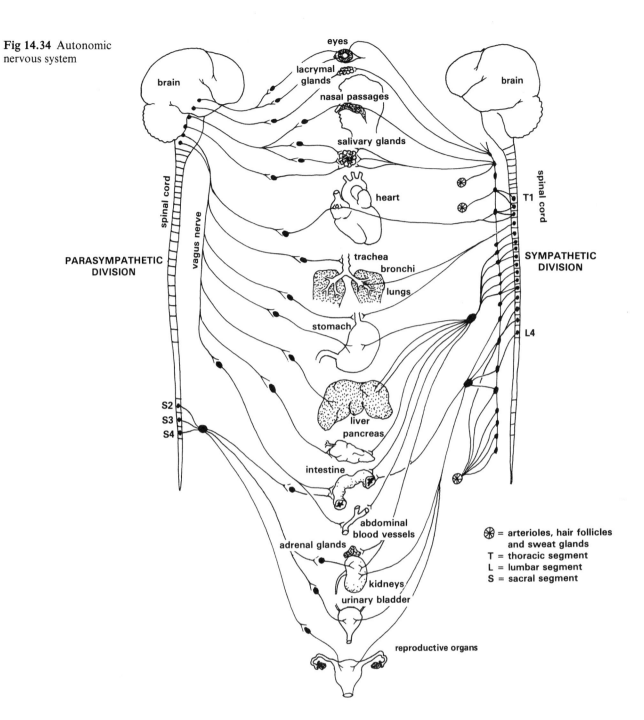

**Fig 14.34** Autonomic nervous system

**PARASYMPATHETIC DIVISION**

**SYMPATHETIC DIVISION**

eyes
lacrymal glands
nasal passages
salivary glands
heart
trachea
bronchi
lungs
stomach
liver
pancreas
intestine
abdominal blood vessels
adrenal glands
kidneys
urinary bladder
reproductive organs

brain
spinal cord
vagus nerve
S2
S3
S4

brain
spinal cord
T1
L4

⊛ = arterioles, hair follicles and sweat glands
T = thoracic segment
L = lumbar segment
S = sacral segment

**Table 14.2 Some of the main effects of the autonomic nervous system**

| Target | Sympathetic effect | Parasympathetic effect |
|---|---|---|
| iris of eye | pupil dilation | pupil constriction |
| bronchi | — | constriction |
| heart | increased cardiac output | decreased cardiac output |
| gut sphincters | increased tone | relaxation |
| urinary bladder | relaxation of urinary bladder wall | contraction of urinary bladder wall |
| sweat glands | stimulates secretion | — |
| salivary glands | decreases secretion | increases secretion |
| stomach | decreases secretion | increases secretion |
| pancreas | decreases secretion | increases secretion |
| genitalia | — | vasodilation in erectile tissue |

The opposite effects of sympathetic and parasympathetic nerves result from the different chemical transmitters secreted at their neuroeffector junctions. Acetylcholine is the transmitter secreted by parasympathetic nerve endings. Parasympathetic nerves are said to be **cholinergic**. Sympathetic nerves secrete noradrenalin and are said to be **noradrenergic** (Fig 14.35).

Adrenalin as well as noradrenalin is secreted into the blood from the medullae of the adrenal glands which are modified portions of the sympathetic system. In conditions of stress, adrenalin secretion increases. The significance of this response is discussed in Chapter 16.

**Fig 14.35** Cholinergic and adrenergic nerves

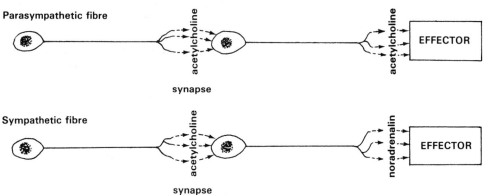

### 14.2.4 Brain

The adult human brain contains more than a thousand million neurons. It has been estimated that in the cerebral cortex alone there are about $10^{2\,783\,000}$ synapses! Considering these numbers of neurons and synapses in an organ weighing only about 1.3 kg, it is evident how complex the

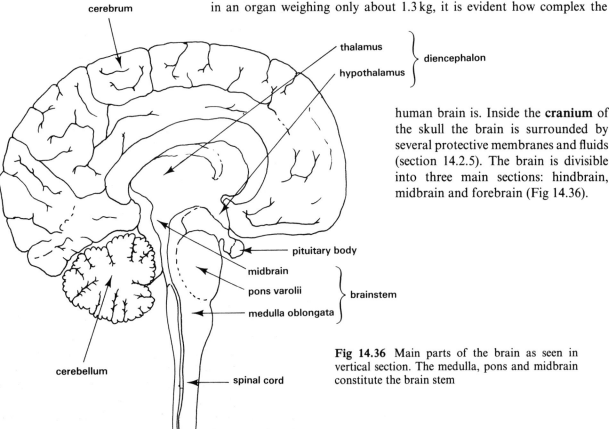

human brain is. Inside the **cranium** of the skull the brain is surrounded by several protective membranes and fluids (section 14.2.5). The brain is divisible into three main sections: hindbrain, midbrain and forebrain (Fig 14.36).

**Fig 14.36** Main parts of the brain as seen in vertical section. The medulla, pons and midbrain constitute the brain stem

## 1 Hindbrain

There are three main parts to the hindbrain, the **medulla oblongata**, the **pons varolii** and the **cerebellum**. The medulla contains many tracts of longitudinal nerve fibres connecting the spinal cord with the rest of the brain. Many of the fibres cross over from left to right and right to left. The medulla also contains a number of **reflex centres**. The **vital centres** include the cardiac, vasomotor and respiratory centres. (Chapters 10 and 12). The **non-vital centres** are concerned with the reflex control of swallowing, vomiting, hiccupping, coughing and sneezing. **Cranial nerves** VIII, IX, X, XI and XII originate in the medulla (Fig 14.37 and Table 14.3). The pons is on the ventral surface of the medulla. It acts as a bridge between various parts of the central nervous system. There are longitudinal fibres connecting the spinal cord and medulla with the rest of the brain. There are also transverse fibres between the pons and the cerebellum. Cranial nerves V, VI, VII and branches of VIII also originate in the pons (Fig 14.37 and Table 14.3, p.280).

**Fig 14.37** Cranial nerves (brain viewed from underneath)

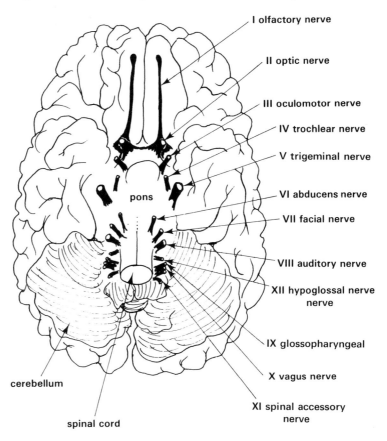

The cerebellum consists of two hemispheres attached to the rest of the hindbrain by three bundles of nerve fibres. The surface of the cerebellum has many ridges called **folia**. It is a motor region concerned with subconscious skeletal movements involved in posture and balance. Sensory information is supplied to the cerebellum from the vestibular apparatus of the inner ear (Chapter 15). The cerebellum is essential for smooth precise movement and delicate manipulations such as playing a piano, typing speech and writing. Without cerebellar control, such movements lack precision and are erratic in range and direction.

**Table 14.3 Summary of the actions of the cranial nerves**

| Nerve | Function | Origin |
|---|---|---|
| I Olfactory | Sensory–smell | Olfactory mucosa |
| II Optic | Sensory–vision | Retina |
| III Oculomotor | Motor–eyelids & eyeball muscles, ciliary & iris muscles | Midbrain |
| | Sensory | Eyeball muscles |
| IV Trochlear | Motor–eyeball muscles | Midbrain |
| | Sensory | Eyeball muscles |
| V Trigeminal | Motor–chewing | Pons varolii |
| | Sensory | Eyelids, eyeballs, lachrymal glands, nasal cavity, forehead, scalp, palate, pharynx, teeth, lips, tongue, cheek |
| VI Abducens | Motor–eyeball muscles | Pons varolii |
| | Sensory | Eyeball muscles |
| VII Facial | Motor–facial, scalp & neck muscles, lachrymal & salivary glands | Pons varolii |
| | Sensory–taste | Taste buds on tongue |
| VIII Vestibulocochlear | Sensory–balance | Vestibular apparatus |
| | Sensory–hearing | Cochlea |
| IX Glossopharyngeal | Motor–swallowing | Medulla oblongata |
| | Sensory–taste | Taste buds on tongue |
| X Vagus | Motor–pharynx, larynx, respiratory tract, lungs, heart, gastrointestinal tract & gall bladder | Medulla oblongata |
| | Sensory | Organs supplied by motor fibres |
| XI Accessory | Motor–pharynx, larynx & palate | Medulla oblongata & cervical portion of spinal cord |
| | Sensory | Muscles supplied by motor fibres |
| XII Hypoglossal | Motor–tongue | Medulla oblongata |
| | Sensory | Tongue muscles |

## 2 Midbrain

As you sit reading this book you are bombarded by a variety of environmental stimuli. The more successfully you concentrate on your reading, the less aware you are of outside distractions. When studying, unwanted stimuli may include the noise of traffic, people talking, a radio or television. You also constantly receive signals from your skin in response to the touch of your clothes or the chair you are sitting on. In other words, of the great number of environmental stimuli you receive at any one time, you may be consciously aware of only a few, whether you are studying or not. This **filtering** prevents overloading your conscious awareness. Most people experience difficulty in concentrating on more than one thing at a time. Confused behaviour results from overloading the brain with too many stimuli. Filtering sensory signals is a function of the **reticular formation** that lies mainly in the midbrain. The reticular formation activates the forebrain with appropriate signals.

Activation of the cerebral cortex by the reticular formation is essential for **wakefulness**. When activation stops, a state of **sleep** follows. The significance of sleep is not fully understood. Nevertheless, sleep is vital for normal activity of the brain. Lack of sleep and excessive nervous fatigue can lead to serious mental disturbances. Sleep deprivation has been used to lower the mental resistance of prisoners prior to interrogation.

On the dorsal surface of the midbrain are four rounded bodies called the **corpora quadrigemina**. They control movements of the eyeballs, head and trunk. Cranial nerves III and IV originate in the midbrain (Fig 14.38 and Table 14.3).

**Fig 14.38** Brain viewed from above. Note the extensive folds and grooves

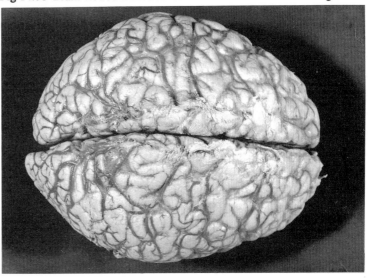

### 3 Forebrain

The forebrain consists of two main parts, the **diencephalon** and the **cerebrum**. The diencephalon contains the **thalamus** and **hypothalamus** (Fig 14.36).

The thalamus directs sensory impulses from the lower parts of the brain and the spinal cord to appropriate parts of the cerebrum. It is organised into several masses called **nuclei** which act as the relays for particular sensory pathways (Table 14.4). The thalamus also relays motor impulses to the spinal cord from the cerebrum. Limited sensory awareness of pain, temperature, touch and pressure is provided by the thalamus.

Just beneath the thalamus is the hypothalamus. It contains reflex centres linked to the autonomic system. They include the **thermoregulation centres** (Chapter 17). The **feeding centre** is stimulated when the stomach is empty and gives us the sensation of hunger. The **satiety centre** is stimulated when the stomach is full and inhibits further feeding. The **thirst centre** makes us feel thirsty when stimulated by angiotensin which appears in the blood when the body is short of water (Chapter 16).

Many of the **pituitary hormones** are thought to originate in the hypothalamus. The pituitary body projects beneath the hypothalamus (Chapter 16). The hypothalamus is also concerned with feelings of rage and aggression. One of the hypothalamic centres acts with the reticular formation to regulate wakefulness and sleep.

By far the largest and most highly developed part of the brain is the cerebrum. Most cerebral activity occurs in the outer **cortex** of grey matter about 2 to 4 mm thick. The great mass of the cerebrum is white matter. Because of relatively greater growth of the cortex, the cerebral surface is highly folded. Between the folds, called **gyri**, are very many shallow

**Table 14.4 Main sensory functions of the thalamic nuclei**

| Nucleus | Sensory pathways relayed |
| --- | --- |
| Median geniculate | hearing |
| Lateral geniculate | vision |
| Ventral posterior | general sensations and taste |

281

grooves called **sulci** and deeper **fissures**. Consequently much of the grey matter is hidden beneath the surface (Fig 14.38). The cerebrum is divided into two **hemispheres** by a prominent longitudinal fissure. The two hemispheres are connected by a bundle of transverse fibres called the **corpus callosum**. Each cerebral hemisphere is divided into four **lobes** (Fig 14.39). They are the **frontal** at the front, the **parietal** towards the top of the head, the **temporal** on the side and the **occipital** at the rear. The **basal ganglia** are paired masses of neurons in the cerebral hemispheres. They contain sensory and motor fibres connecting the cerebral cortex with the **brainstem** and spinal cord. The brainstem is the midbrain, pons and medulla. The basal ganglia control subconscious skeletal movements such as arm swinging while walking.

**Fig 14.39** Lobes of the cerebral hemispheres

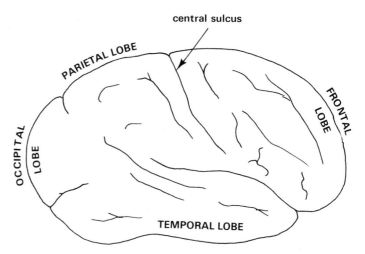

The **limbic system** consists of parts of the cerebral hemispheres and diencephalon. It controls the emotional aspects of behaviour such as pleasure, anxiety and pain. The limbic system is also a major centre for **memory** and overall control of **behaviour**.

**Fig 14.40** Some of the main functional areas of the cerebral cortex

The cerebral cortex can be divided into three main kinds of functional areas. They are **sensory**, **motor** and **association** areas (Fig 14.40). The

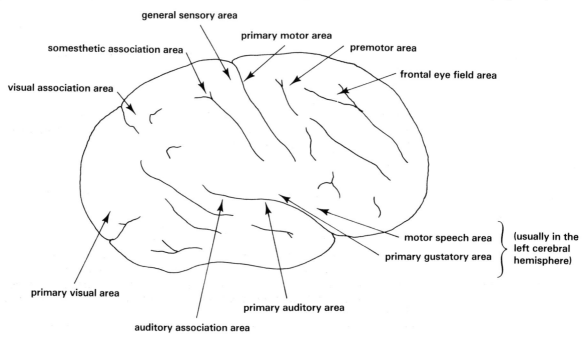

sensory areas are the sites of reception, correlation and interpretation of sensory information, that is, **perception**. The motor areas are the regions from which motor pathways originate. Voluntary motor activity is controlled by a strip of cerebral cortex just anterior to the central sulcus. Different parts of the body are controlled by specific parts of the voluntary motor area. Similarly, sensory information from particular parts of the body is interpreted in specific parts of the **general (somesthetic) sensory area** just posterior to the central sulcus (Fig 14.41). The functions of the main regions of the cerebral cortex are summarised in Table 14.5.

**Fig 14.41** Localised regions of the general sensory and primary motor areas

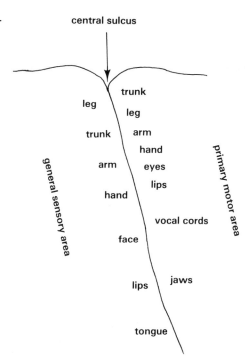

**Table 14.5 Summary of the functions of the main areas of the cerebral cortex**

| Area | Function |
| --- | --- |
| General sensory (somesthetic) | Receives sensations from skin, muscle and visceral receptors from all over the body. |
| Somesthetic association | Receives signals from general sensory area and thalamus – interpretation and integration. Memory of past sensory experiences. |
| Primary visual | Vision – shape and colour. |
| Visual association | Receives signals from primary visual area. Recognition of objects by reference to memory of past visual experiences. |
| Primary auditory | Interpretation of fundamental characteristics of sound such as pitch and intensity. |
| Auditory association | Translation of speech into meaningful ideas. |
| Primary gustatory | Taste. |
| Primary olfactory | Smell. |
| Gnostic | Receives signals from all other sensory areas, enabling an overall impression of sensory experience. Transmits impulses to other areas to enable the appropriate responses. |
| Primary motor | Voluntary motor pathways to specific skeletal muscles. |
| Premotor | Control of complex sequences of motor actions such as writing or playing a piano. |
| Frontal eye field | Controls voluntary eye scanning movements such as are used during reading. |
| Motor speech (Broca's) | Translation of thoughts into speech. Generation of speech by controlling the actions of the larynx, throat, mouth and muscles responsible for breathing. This area is usually located in the left cerebral hemisphere. |

### 14.2.5 Cerebrospinal fluid

The brain and spinal cord are surrounded by several layers called **meninges** which separate the central nervous system from the bone of the cranium and vertebrae. On the outside the **dura mater** consists of dense fibrous connective tissue. Beneath the dura is a delicate layer called the **arachnoid**. On the inside the **pia mater** adheres to the surface of the brain and spinal cord (Fig 14.42).

**Fig 14.42** Cranial meninges

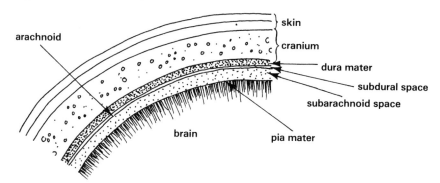

Cerebrospinal fluid (CSF) is an ultrafiltrate of blood. It is formed in the **choroid plexus** of the brain (Fig 14.43). CSF circulates in the ventricles of the brain, the **subarachnoid spaces** between the arachnoid and pia mater and in the central canal of the spinal cord. Cilia project into the ventricles and central canal and their beating may help keep CSF moving. CSF returns to the blood through the **arachnoid villi** and into a vein called the **superior sagittal sinus** which runs along the top of the brain in the mid line (Fig 14.43). Valves in the arachnoid villi prevent blood entering the CSF.

CSF performs two main functions.

**i. Mechanical support**  The brain and spinal cord are suspended in CSF. In this way the CNS is buffered against mechanical injury.

**ii. Pressure compensation**  CSF is produced from and returns to the blood passively in response to pressure differences between the blood and CSF. The effects of increases in blood pressure distending intracranial vessels can thus be reduced as more CSF is produced.

### 14.2.6 Electroencephalography

**Electroencephalography** is the recording of the electrical activity of the brain. Hans Berger recorded the first human electroencephalogram (EEG) in 1929. The activity recorded as an EEG is thought to originate from excitatory and inhibitory post-synaptic potentials (section 14.1.5) in the cerebral cortex. The number of synapses in the cerebral cortex is astronomical. The axon of each neuron may branch and connect with up to 5000 other fibres. Any one neuron may have many hundreds of synapses on it (Fig 7.14).

An EEG is recorded by placing electrodes at measured sites over the scalp. The number of electrodes is usually between 20 and 23 (Fig 14.44). The potential difference between selected pairs of electrodes is recorded on paper. EEG machines usually have 8 or 16 channels which record the

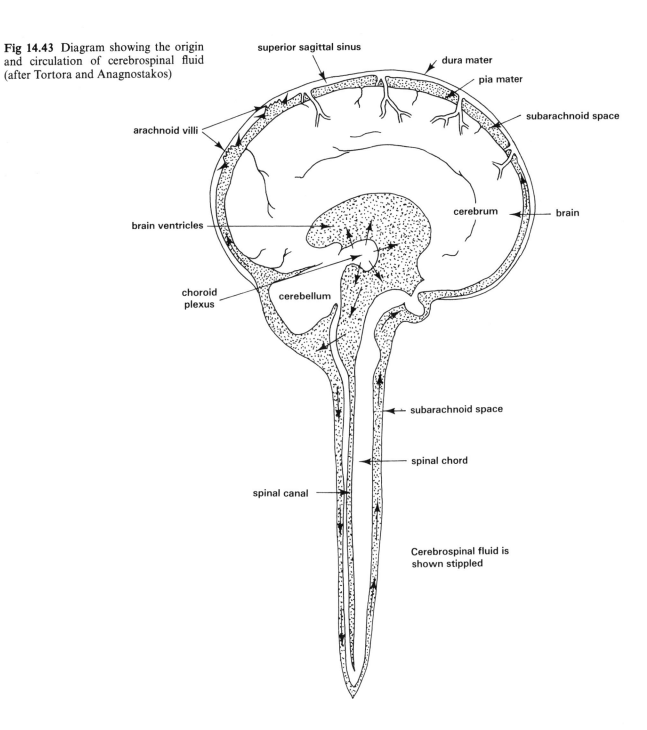

**Fig 14.43** Diagram showing the origin and circulation of cerebrospinal fluid (after Tortora and Anagnostakos)

superior sagittal sinus

dura mater

pia mater

subarachnoid space

arachnoid villi

cerebrum

brain

brain ventricles

choroid plexus

cerebellum

subarachnoid space

spinal chord

spinal canal

Cerebrospinal fluid is shown stippled

**Fig 14.44** Electrodes on the scalp and face of a patient prior to taking an electroencephalogram (Alexander Tsiaras/Science Photo Library)

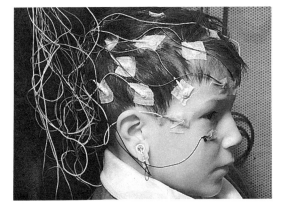

What is electroencephalography? What use is EEG in medicine?

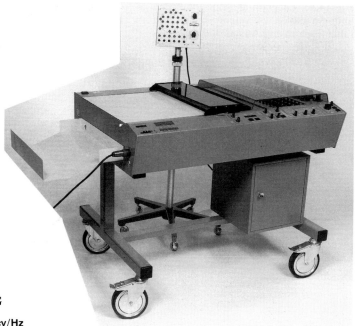

**Fig 14.45** 18-channel EEG machine (courtesy SLE Ltd, Croydon, Surrey)

**Table 14.6 Main categories of frequencies seen in EEG**

| Waveform | | Frequency/Hz |
|---|---|---|
| delta | $\delta$ | < 4 |
| theta | $\theta$ | 4–7.9 |
| alpha | $\alpha$ | 8–13.9 |
| beta | $\beta$ | > 14 |

signals from 8 or 16 different electrode pairs simultaneously (Fig 14.45). Many different electrode pairs may be selected so that data are obtained from all areas of the brain.

The EEG voltage fluctuations can be of different frequencies and are classified into four main groups (Table 14.6). Delta and theta activity predominate in babies and young children. As the individual grows and matures the EEG becomes faster and alpha activity predominates. If delta and theta activity persist in an adult it could indicate abnormality.

In a normal adult the EEG voltage is usually between 30 and 100 $\mu$V. Greater voltages are often recorded in young children. During a routine recording session the patient closes the eyes and is made as comfortable and relaxed as possible. Short recordings are made with the patient's eyes open (Fig 14.46(a)). The patient is usually asked to breathe deeply (hyperventilate) for about 3 minutes. The EEG usually changes greatly in children and is recorded for a couple of minutes until it settles back to its resting state (Fig 14.46(b)). A flashing light (stroboscope) is then placed in front of the patient's eyes. EEG recording is made while the strobe setting is increased from 1 to 30 or 35 flashes per second, with the patient's eyes closed and then open (Fig 14.46(c)). Sometimes EEG recordings are made while the patient is asleep.

**Sleep** has a number of stages. On becoming **drowsy** (stage 1) the alpha activity of the EEG becomes intermittent and there is slow lateral rolling of the eyes. Eventually alpha activity is replaced by low voltage theta activity.

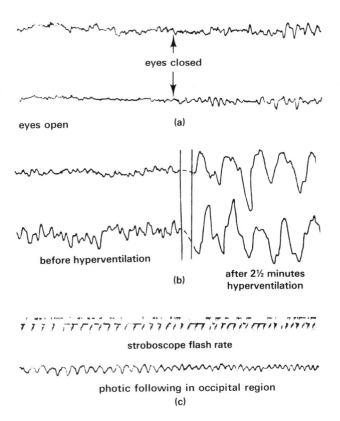

**Fig 14.46** Typical normal EEG recordings (a) with patient's eyes closed and open, (b) following a short period of hyperventilation in an 11-year-old child, (c) during photic (stroboscope) stimulation (courtesy Mrs C. Dunn, Dept. of Neurophysiology, Queen's Medical Centre, Nottingham)

eyes closed

eyes open          (a)

before hyperventilation

(b)      after 2½ minutes hyperventilation

stroboscope flash rate

photic following in occipital region

(c)

Alpha waves may return if the patient is exposed to a sound the ears can hear. When this happens, some people briefly open their eyes. In stage 2 the patient enters into a **light sleep**, the theta waves become well defined and delta waves begin to show. There is little or no eye movement and breathing is regular. In stage 3, sleep deepens and delta activity becomes more prominent. When exposed to loud sounds or calling the subject's name, the patient is fairly easily roused. Stage 4 is **deep sleep**. Breathing and heart rate are regular and relatively slow. The EEG is dominated by delta activity and the patient is difficult to rouse by auditory stimuli. Between 45 and 60 minutes from becoming drowsy there are **rapid eye movements (REM)**. Heart beat and breathing are irregular and sometimes limb movements occur. The voltage diminishes over the whole EEG which becomes lower and faster.

REM sleep is thought to be associated with **dreams**. If normal, healthy subjects are persistently woken when in REM sleep they can become quite paranoid. They may suffer delusions, including unjustified feelings of being persecuted by others. Patients referred for EEG often have a history of blackouts, faints, fits, behavioural changes, personality problems, metabolic disorders, degenerative diseases or infections of the CNS such as meningitis. In most such cases the EEG is abnormal. For example, in primary epilepsy an EEG often shows a characteristic spike-wave pattern, although this may appear only during a fit (Fig 14.47). An EEG showing no waves may be taken as one of the criteria for diagnosing cerebral death.

**Fig 14.47** EEG trace showing 'spike and wave' activity from an individual suffering from petit-mal epilepsy (courtesy Mrs C. Dunn, Dept. of Neurophysiology, Queen's Medical Centre, Nottingham)

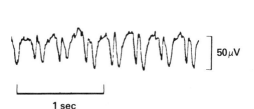

50 μV

1 sec

# 15 The eye and ear

The detection of stimuli depends on the conversion of the stimuli by **receptors** into impulses in the nervous system. The body has a variety of receptors which feed impulses into the nervous system in response to an equal variety of stimuli. Different receptors are sensitive to heat, light, sound, touch, stretch, spatial orientation and to chemicals.

The simplest receptors are single nerve cells which respond directly to a stimulus. Receptors in the skin are of this type (Fig 17.7). Some receptors consist of groups of sensitive neurons such as the cardiac and respiratory centres in the brain (Chapters 10 and 12). Other receptors are grouped in complex **sense organs**. The structure of sense organs causes the stimulus to be channelled into a receptive region of the organ.

The action potentials produced in activated receptors are called **generator potentials**. They lead to the production of impulses in sensory neurons connected to the receptors. The way in which impulses are generated by receptor cells is important in discriminating between strong and weak stimuli. Measurement of action potentials (Chapter 14) in sensory neurons shows no significant difference between one impulse and another. However, a strong stimulus usually causes impulses to be generated rapidly in sensory neurons. The frequency of impulses generated by weak stimuli is lower. The appropriate sensory region of the cerebral cortex of the brain interprets the intensity of a stimulus according to the frequency with which it receives impulses.

Among the most important sense organs are the **eyes** and **ears**.

## 15.1 The eye

We depend greatly on vision to sense our environment. While reading this page, your conscious awareness is almost totally stimulated by visual information.

### 15.1.1 Structure of the eye

The eyes are spherical structures with a wall consisting of three layers, the outer **sclera**, the middle **choroid** layer and the inner **retina** (Fig 15.1). The sclera is a tough, fibrous coating which protects the delicate inner layers. It

**Fig 15.1** (a) Photomicrograph of a longitudinal section of the eye from a monkey, × 4

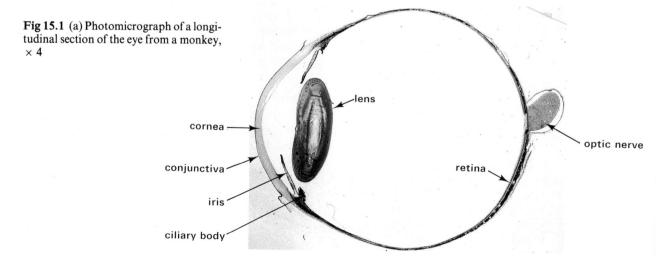

cornea

conjunctiva

iris

ciliary body

lens

optic nerve

retina

**Fig 15.1** (b) Diagram of a vertical section through the eye

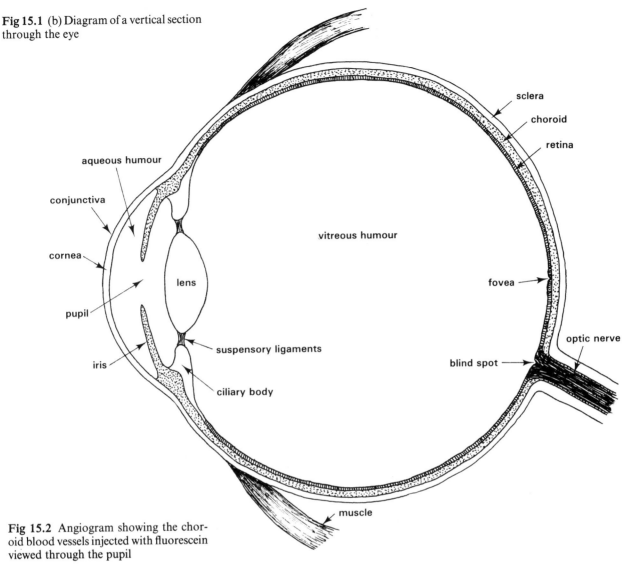

**Fig 15.2** Angiogram showing the choroid blood vessels injected with fluorescein viewed through the pupil

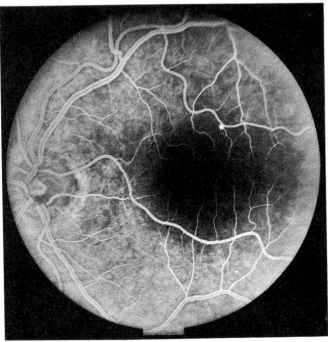

is white in appearance except at the front where there is a transparent area called the **cornea**. The cornea allows light into the eye. Tears secreted by the **lachrymal glands** lubricate the exposed surface of the eye, including the **conjunctiva** which covers the cornea except in the centre. The watery secretion helps to prevent abrasion of the eye's surface by dust particles and helps combat infection of the eye. Periodic closure of the eyelids, blinking, clears away debris.

Inside the sclera is the choroid layer which contains numerous blood vessels (Fig 15.2). At the front of the eye it is modified as the **iris** containing pigments which give the eye its colour. The iris also contains radial bands and a ring of circular smooth muscle. Contraction of the radial muscles and relaxation of the circular muscles cause dilation of an aperture called the **pupil**, in the centre of the iris. Constriction of the pupil occurs when the radial muscles of the iris relax and the circular ones contract. Variation in pupil size is controlled by autonomic reflexes and is usually a response to change in the intensity of light entering the eye. In bright light the pupil constricts and prevents excessive illumination of the interior of the eye. In dim light the pupil dilates, allowing the maximum amount of light to reach the photoreceptor cells.

The photoreceptors are in the retina situated immediately inside the choroid layer. Fibres of sensory neurons lead from the retina at the back of the eye as the **optic nerve** which transmits impulses generated in the retina to the brain. The retina develops in the embryo as an outgrowth of the brain, and can be regarded as a modified part of the central nervous system (Chapter 19).

Suspended in the fluid inside the eye and just behind the pupil is a biconvex, crystalline **lens**. It is held in position by **suspensory ligaments** attached to a ring of smooth muscle called the **ciliary body**.

### 15.1.2 Focusing

Light rays entering the eye are bent or **refracted**. Refraction occurs at three surfaces of the eye before the light reaches the retina. The first of the refracting surfaces is the cornea, then the front surface of the lens and finally the rear surface of the lens (Fig 15.3).

**Fig 15.3** The three main refracting surfaces in the eye

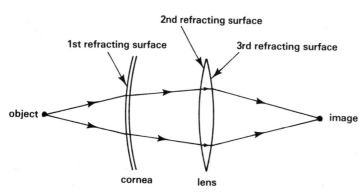

2nd refracting surface

1st refracting surface    3rd refracting surface

object ●

image

cornea    lens

Between the cornea and the lens is a colourless, watery fluid called **aqueous humour**. At the back of the eye between the lens and the retina is the **vitreous humour** made of a gelatinous mucoprotein. The humours are transparent so that transmission of light through the cavities of the eye to the retina is not normally impeded.

In the normal eye light rays are refracted sufficiently to be brought to a point on the retina. In this way the object viewed is brought into focus and a clear image is formed. The **image** is **inverted** by the lens (Fig 15.4). However, objects are seen the right way up because of the way in which the brain interprets images. Experiments have been performed in which human volunteers wore special spectacles, the lenses of which produced upright images on the retina. For a while the subjects in the experiment were confronted with an upside-down world. However, after a few days they became used to the situation and their perception became adjusted so that once more they perceived things the right way up. When the spectacles were removed they again experienced a period of seeing things upside down before normal perception returned. Thus the mechanism of image interpretation, situated in the brain's cerebral cortex, is somewhat flexible.

**Fig 15.4** Light from all points on an object are focused in such a way that the image on the retina is inverted. Only two light rays are shown

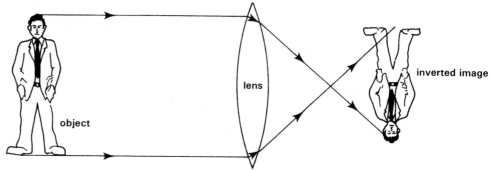

lens

inverted image

object

Light rays from an object near the eye strike the cornea and lens at an acute angle depending on the object's size. If the same object is moved further away from the eye the angle is less acute. Consequently the degree of refraction necessary to **focus** light rays on the retina is greater for close objects than for distant objects. Changes in the degree of refraction of light are achieved by altering the curvature of the lens surfaces (Fig 15.5). The ciliary body contains a ring of involuntary muscles; contraction reduces the tension of the suspensory ligaments which hold the lens in place, and relaxation increases their tension. It is the tension of the ligaments applied to the lens which determines the shape of the lens. When the tension is increased the lens is pulled into a flattened shape suitable for focusing distant objects. When the tension is decreased, the lens becomes a more spherical shape suitable for focusing near objects.

**Fig 15.5** Focusing (accommodation)

(a) Light from a distant object is focused in the retina by a flattened lens

(b) Light from a near object is focused on the retina by a near-spherical lens

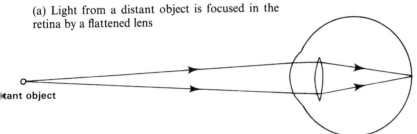

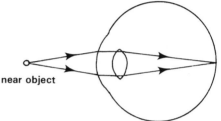

### 15.1.3 Defects of the eye

There are several abnormalities of the eye's focusing mechanism. The commonest are **myopia (short-sightedness), hypermetropia (long-sightedness),** and **astigmatism**.

### 1 Myopia

Myopia results if the lens curvature is too great or the entire eyeball becomes elongated (Fig 15.6(a)). Light rays entering the eye are refracted more than is necessary. Consequently light is focused in front of the retina. By the time the light stimulates the retina it has diverged from the focal point of the lens. The image perceived is thus blurred. The condition is called short-sightedness as objects near the eye are less out of focus than those further away. This is because the light rays from near objects require greater refraction to be focused on the retina than rays from distant objects. Since the lens in a myopic eye refracts light excessively, distant objects appear more blurred than near ones. Myopia can be corrected by placing a **concave lens** in front of the eye. The surface of the concave lens refracts light rays in such a way that the rays diverge slightly from their original path. The lens of the myopic eye now refracts the diverged light rays in to focus on the retina (Fig 15.6(b)).

**Fig 15.6** Myopia
(a) Light is focused at a point in front of the retina

(b) Light may be focused on the retina by placing a concave lens in front of the eye

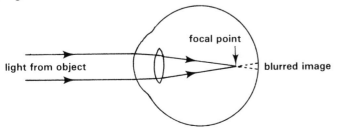

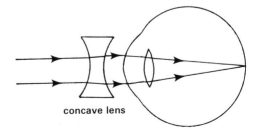

## 2 Hypermetropia

Hypermetropia results when the curvature of the eye's lens is not great enough. Light rays are not refracted enough and would thus be focused behind the retina (Fig 15.7(a)). The condition is called long-sightedness because distant objects are less out of focus than near ones. This happens because light rays from distant objects require less refraction than rays from near objects. Correction of hypermetropia requires placing a **convex lens** in front of the eye. The lens converges light rays before they enter the eye so that the eye's lens focuses the light correctly on the retina (Fig 15.7(b)).

**Fig 15.7** Hypermetropia

(a) Light would be focused at a point behind the retina

(b) Light may be focused on the retina by placing a convex lens in front of the eye

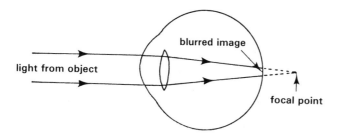

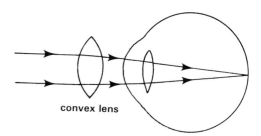

## 3 Astigmatism

Astigmatism occurs if either the cornea or lens is distorted. One part of the focusing mechanism then refracts light rays too much, another not enough. Usually most of the image perceived is out of focus. Light rays from parts of the object are focused in front of the retina, as in myopia. Rays from other parts would be focused behind the retina, as in hypermetropia. Astigmatism can be corrected by placing a lens in front of the eye. The curvature of this lens varies from one part to another to compensate for the eye's deficiencies.

What causes short- and long-sightedness? How can these eye defects be corrected?

## 15.1.4 Photoreception

The transmission of nerve impulses to the brain in response to stimulation of **photoreceptors** in the retina by light is the function of the optic nerves. In human eyes there are more than 100 million of the two types of photoreceptors called **rods** and **cones**.

### 1 Rods and cones

Rods are sensitive to different **intensities** of light. Most mammals have only rods. Cones are sensitive to different **wavelengths** of light and enable us to see things in colour. The assumption that an angry bull will charge a red object is somewhat misplaced, since the retina of a bull's eye has no cones. The bull, if it is annoyed is just as likely to charge an object of some other colour.

### 2 Distribution of rods and cones

The arrangement of photoreceptors in the retina is such that light has to travel through several layers of neurons which are not sensitive to light before reaching the rods and cones: the retina is **inverted** (Fig 15.8).

Outside the photoreceptors is a layer of cells containing the black pigment melanin. Melanin is not sensitive to light but absorbs light rays

**Fig 15.8** Main cellular components of the retina

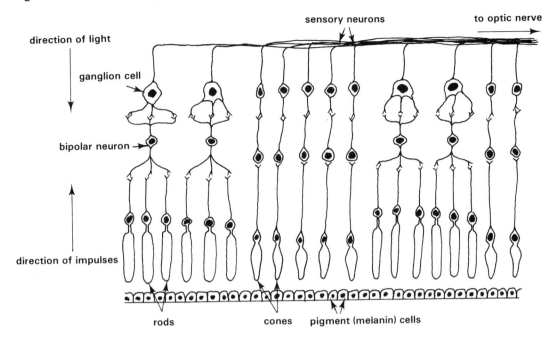

which would otherwise pass through the retina. In this way the formation of hazy images caused by reflection of light to other parts of the retina is prevented. Some mammals have a reflective layer called the **tapetum** in the retina. The reflective 'cat's eyes' along the centre of roads create a similar effect to the reflective tapetum of real cat's eyes. A tapetum is common in nocturnal mammals and enables the maximum use of what little light is available at night.

Many synapses link the photoreceptors with sensory neurons. Impulses generated by the photoreceptors are first transmitted to a small region where the neurons converge to project through the retina into the optic nerve. This region is called the **blind spot**, since no photoreceptors are located there (Fig 15.1). It is possible to demonstrate that there are blind spots in your eyes by referring to the circle and cross illustrated below. Close your right eye and hold the page about 30 cm away from your open left eye. Keeping your left eye focused on the circle, the cross should be visible but slightly less clear. Now move the page slowly towards you and notice that at a distance of about 15 cm from your left eye the cross disappears. At this point light from the cross falls on the blind spot of your left eye. If you now move the page even nearer to your left eye the cross should reappear.

$$+ \qquad\qquad\qquad \bigcirc$$

In this simple demonstration the circle appears clearer than anything around it in the field of vision. The reason for this is that when you direct your eye at the circle the light from it is focused on to a region of the retina called the **fovea**. Only cones are found in the foveas of the eyes of animals which have colour vision. Each cone forms a synapse with very few sensory neurons (sometimes only one). Consequently the signals sent to the brain from each cone come from a small area of the retina on which a small part of the image is focused. The cones can thus discriminate between two points of the image which are close to one another. For this reason cones are said to have high **visual acuity**. On the other hand, many rods synapse with each sensory neuron. The signals transmitted to the brain from the rods

**Fig 15.9** Cones often synapse with single sensory neurons whereas usually many rods synapse with a single sensory neuron. Cones are therefore important in the discrimination of close points in the image

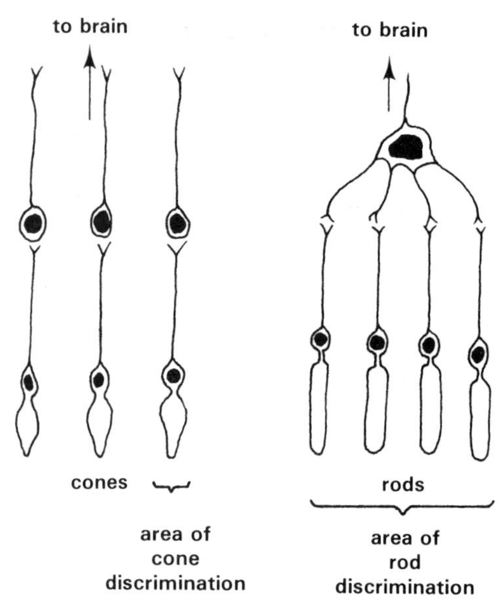

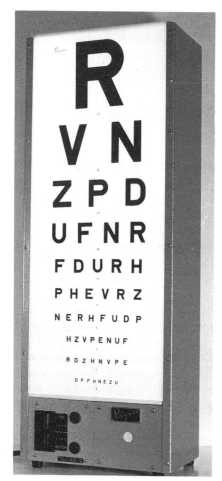

to brain

to brain

cones

rods

area of
cone
discrimination

area of
rod
discrimination

Which small part of your retina gives you the best vision? Which part of your retina cannot 'see' at all?

therefore come from a relatively large area of the image (Fig 15.9). The rods are distributed throughout the retina but are absent from the fovea. The cones in the fovea are much longer and thinner compared with those in other parts of the retina. It enables many cones to be packed together in the fovea. The part of the image focused on the fovea therefore appears much clearer than the rest of the image.

Our distant visual acuity may be examined using the **Snellen** test. The test is based on the ability of each eye to identify correctly typed black letters of different sizes printed on a white card a fixed distance from our eyes in well lit surroundings. Each eye is tested separately. In good illumination, a person with good visual acuity can distinguish between two rays of light that are separated by an angle of $1' (\frac{1}{60}°)$ when they impinge on the lens (Fig 15.10). However, the visual cortex of the occipital lobes of the brain can only perceive the nature of an object from which light

**Fig 15.11** Snellen's test types

**Fig 15.10** Minimum separation of light rays entering the eye that is detectable by the retina

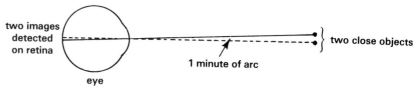

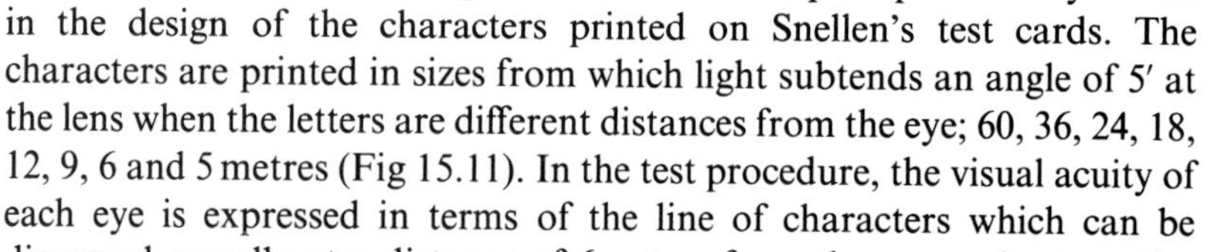

two images
detected
on retina

two close objects

1 minute of arc

eye

impinges on the lens at an angle of 5' or more. This perceptive ability is used in the design of the characters printed on Snellen's test cards. The characters are printed in sizes from which light subtends an angle of 5' at the lens when the letters are different distances from the eye; 60, 36, 24, 18, 12, 9, 6 and 5 metres (Fig 15.11). In the test procedure, the visual acuity of each eye is expressed in terms of the line of characters which can be discerned, usually at a distance of 6 metres from the eye under test. For example, if the top character only can be discerned, the acuity is recorded as 6/60. 6 is the distance from the eye to the character, 60 is the distance at which the light from the character subtends an angle of 5' at the lens. Normal acuity would be 6/6.

## 3 Photosensitive pigments

The functioning of rods and cones depends on **photosensitive pigments**. Electron microscopy has revealed the intricate subcellular structure of rods (Fig 15.12). The outer segments of rods contain a great number of membranous, disc-like **lamellae**. The lamellae contain a photosensitive pigment called **rhodopsin**. Cones contain a similar pigment sometimes called **iodopsin**.

**Fig 15.12** (a) Electronmicrograph of parts of several rod cells from the retina of a cat, × 5800

**Fig 15.12** (b) Diagram of rod cell structure

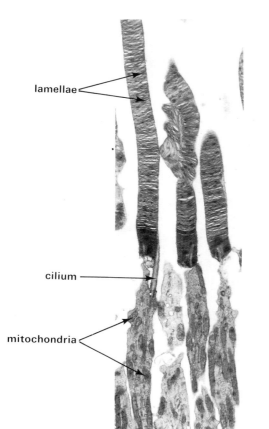

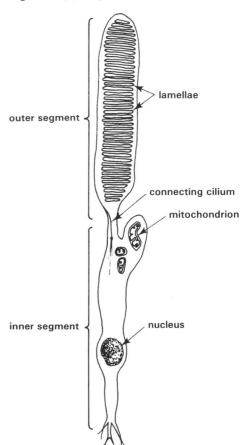

**i. Rhodopsin** consists of a protein called **opsin** attached to **retinal**, a derivative of vitamin A (Chapter 9). When exposed to light rhodopsin is split into opsin and retinal. The retinal molecules change shape and produce a generator potential which is transmitted from the rod as an impulse to a sensory neuron. Strangely, the initial stimulus causes **hyperpolarisation** in the receptor membrane. The usual effect of stimulation is depolarisation (Chapter 14).

Rhodopsin, split in this way, has to be resynthesised to maintain the rod's ability to respond to light. Resynthesis of rhodopsin requires energy and the rods have many mitochondria which make ATP for this purpose. However, rhodopsin resynthesis takes time. It is a common experience to suffer a brief period of poor vision after going from a well-lit room into darkness. When exposed to bright light rhodopsin is broken down rapidly and the reserve of rhodopsin in the rods is low. The eyes are **light-adapted**. If the retina is then exposed to dim light the rods show little response, so vision is poor. The period required to get used to the dark is the time taken for enough rhodopsin to be resynthesised. When the retina is sufficiently sensitive for us to see in dim light the eyes are **dark-adapted**.

295

A brief period of poor vision may also be experienced when you go from a very dark to a brightly lit room. When dark-adapted, the retina can work in dim light. Exposure of a dark-adapted retina to bright light overloads the photoreception mechanism. Light rays, even from the darker areas of an object ih view, stimulate the rods which are now rich in rhodopsin. The eyes are light-adapted when excess rhodopsin is broken down and the retina once more adjusts to working in bright light. Prolonged exposure to very intense light, however, can reduce sensitivity of the retina too much. The rate of rhodopsin resynthesis may then be unable to keep pace with its breakdown. **Snow blindness** is caused by such an effect.

**ii. Cone pigments** are less sensitive to changes in light intensity than rhodopsin, so the cones are of little value in helping us to see in dim light. A popular theory of colour vision suggests that there are three variants of cone pigments, each of which is sensitive to light of the primary colours, red, blue and green (Fig 15.13(a)). For this reason the theory is called the **trichromatic theory** of colour vision. Each type of cone pigment is probably located in different cones. Light with a wavelength between those of the primary colours stimulates combinations of cones. Yellow light for instance, simultaneously stimulates cones which are most sensitive to red

**Fig 15.13** (a) Absorption spectra showing the wavelengths of light (perceived as colours) absorbed most by the iodopsin in the three main types of cones (after Marks, Dobelle, MacNichol, Brown and Wald)

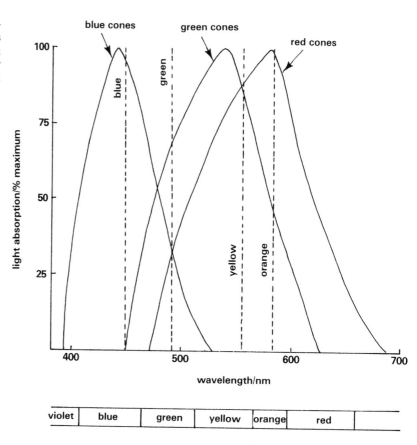

and green light. The impulses generated in the sensory neurons in response to generator potentials in the cones are interpreted by the brain as the appropriate intermediate colour, yellow in this case (Fig 15.13(b)). The interpretation or **perception** of colour pictures seen by our eyes is a complex function of the brain. It is located in the occipital lobes of the cerebral cortex (Chapter 14).

**Fig 15.13** (b) Different colours are perceived in the brain from the sensory information received from the cones. Signals from a combination of different cones produce the sensation of intermediate colours based on red, green and blue detected by the three primary cones

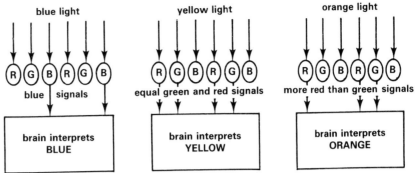

Deficiency of one or more of the three primary colour cones results in **colour blindness**. A glance at Fig 15.13(a) shows that there is some overlap in colour sensitivity between the three types of cones. For example, a green cone is sensitive to red light but its sensitivity to red is far less than that of a red cone. Absence of red cones therefore, means that it is still possible to perceive green, yellow, orange and red. However, the brain cannot distinguish satisfactorily between these colours because there are no impulses from red cones with which to contrast impulses from green cones. A similar effect occurs when green cones are absent. The condition is called **red-green colour blindness**. More rarely, blue cones may be absent causing **blue weakness**.

Colour vision can be investigated with the **Ishihara** test. The test uses a series of pictures each composed of many dots of varying sizes and colours. The dots are arranged in such a way that a person with normal colour vision can see a large numeral in some pictures and a wavy line in others. Persons with red–green deficiences see a different numeral or pattern (Fig 15.14). People with total colour blindness cannot see the numerals or the wavy line.

**Fig 15.14** An Ishihara test plate. (See also the photograph on the back cover)

(a) A person with normal colour vision should see a number '5'

(b) Someone suffering from red–green colour blindness may see a number '2'

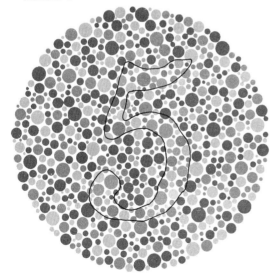

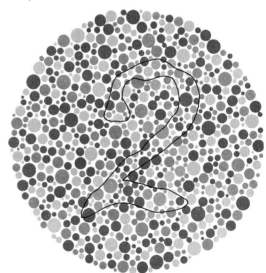

The exact mechanism of colour vision is not known. Many people challenge the simplicity of the trichromatic theory and several alternative theories have been proposed in recent years.

297

## 15.1.5 Perception of distance and size

The region of the environment from which each eye collects light is called the **visual field**. Since both eyes point forwards, there is an overlap between the visual fields of each eye. This is called **binocular vision**. Most of the image perceived by the visual cortex in the brain results from the integration of information from both eyes. Furthermore, signals from both eyes are transmitted to each half of the visual cortex. This is because approximately half the sensory fibres from each eye cross to the other side in the brain (Fig 15.15).

**Fig 15.15** The pathway for sensory impulses from the eyes to the visual cortex in the brain. About half the fibres from each eye cross over to the opposite side in the optic chiasma (modified from Tortora and Anagnostakos)

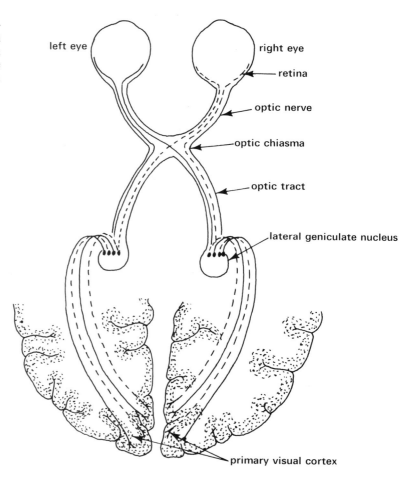

Subtle differences appear between the images from the two eyes because each eye is looking at the environment from a different position. Comparison of the two images in the visual cortex enables us to perceive the shapes, textures, distances and relative movements of objects.

You can appreciate the effect by closing your left eye. Hold one of your fingers vertically about 10 cm in front of your open right eye and line it up with some distant object such as the vertical part of a door frame. Now open your left eye and at the same time close your right eye. Your finger and the distant object will now be misaligned. The effect is called **stereoscopic vision**. It is only significant for objects nearer than about 70 m. Perception at greater distance depends partly on **memory**. We are familiar with most of the objects around us. The smaller they appear, the more distant we assume they are and vice versa. Another effect is called the **moving parallax**. When we scan objects in the visual field, near objects move across the image to a greater extent than distant objects.

Look at the photograph of coloured dots on the back of this book. What number can you see? Now look at Fig 15.14. What can this kind of test tell you about your colour vision?

Visual perception is extremely complex. The brain relies on a variety of sensory information which is integrated in such a way as to provide a complex picture of our surroundings.

## 15.2 The ear

The ability to generate sounds, as well as to receive and interpret them is valuable as a means of communication. Sound waves trigger off the transmission of sensory nerve impulses from the ears to the cerebral cortex of the brain.

The ears also perform another important function. They transmit to the brain information about the head's relative position in space.

### 15.2.1 Sound

Sound is all about us. We make many uses of sound in our everyday lives. We use some sounds, such as sirens, to give warning of danger and others, such as door bells, to attract attention. We sing and make music. Most important of all, we communicate with each other by the spoken word. All these activities require the listener not only to hear the sounds but to discriminate between different sounds in order to obtain meaning from them. So, what is **sound**?

Sound is a series of disturbances of the medium through which it travels. For example, if we pluck a guitar string we cause movement of the air molecules in contact with the string (Fig 15.16). As the string moves into the surrounding air it **compresses** the air molecules. As the string moves back towards its original position, the compressed air molecules have more space to occupy. The air pressure increases and decreases each time the string moves back and forth.

**Fig 15.16** Generation of sound by a vibrating string

(a) regions of compression and rarefaction of air molecules

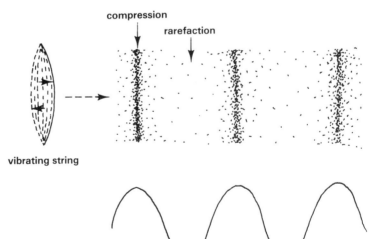

(b) sound waves – fluctuating air pressure changes caused by the vibrating string

Compressed molecules move against neighbouring molecules and compress them, causing compression further away and so on. Compressed molecules then bounce back to their original positions, causing regions of **rarefaction** between the regions of compression. In this way a **sound wave** of compression followed by rarefaction moves away from the original disturbance. It is important to realise that the air molecules oscillate back and forth but the waves of compression they cause move away to neighbouring molecules.

299

**Fig 15.17** Two sound waves of the same amplitude: (a) one of relatively low frequency, (b) one of relatively high frequency

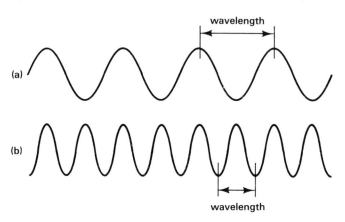

The **frequency** or, as we perceive it, the **pitch** of a sound is determined by the time interval between each wave. This is the **wavelength** (Fig 15.17). The greater the frequency, that is the shorter the wavelength, the higher the pitch of the sound we hear. Our ears are sensitive to sounds of frequencies between approximately 20 and 23 000 Hz. **Hertz** (Hz) is the unit used to express the number of sound waves as **cycles per second**. We are most sensitive to sounds of frequencies between about 1000 and 4000 Hz.

The velocity of sound in air is about 344 metres per second at 20 °C and at a pressure of 1 atmosphere (760 mmHg, 100 kPa). The frequency of a sound produced by a moving object, such as a train, depends on the speed of the object and the direction in which it is moving relative to the listener. You have probably experienced the sudden change in frequency (pitch) of a sound coming from a fast-moving car as it passes you. The frequency drops as the car goes away. It is called the **Doppler effect** (Fig 15.18).

**Fig 15.18** The Doppler effect

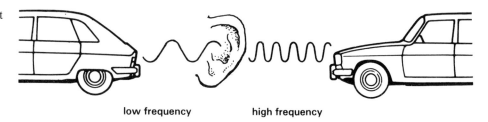

low frequency        high frequency

Wavelength, frequency and velocity of sound are related by the equation:

$$\lambda = c/f$$

where $\lambda$ = wavelength, $c$ = velocity, and $f$ = frequency

From this we can see that sound of high frequency has a short wavelength and vice versa (Fig 15.17).

The **intensity** or, as we perceive it, the **loudness** of sound is determined by the **amplitude** of each sound wave; that is the difference in pressure between maximum compression and rarefaction (Fig 15.19). In the guitar string example of sound production, the distance the string moves back and

**Fig 15.19** Two sound waves of the same frequency: (a) one of relatively low intensity (amplitude), (b) one of relatively high intensity. Compare with Fig 15.16

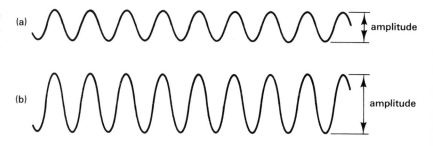

Why does the pitch of the sound made by an approaching train suddenly drop as the train passes and goes away?

forth determines the intensity of the sound produced. Consequently if we pluck a string forcefully we hear a relatively loud sound. The intensity of sound is often expressed in **decibels, dB**. The intensities of two sounds are compared and expressed as a logarithm:

$$dB = 10 \times \log (I/I_0)$$

where $I$ = new intensity, and $I_0$ = original (reference) intensity.

The reference level (original intensity) must be specified. In terms of hearing, a sound of 0 dB is said to be one that can just be heard by a perfect ear. This is the **hearing threshold level, HTL**. Expressing sound intensity in dB gives convenient numbers over the very wide range of intensities to which our ears are sensitive (Table 15.1).

**Table 15.1 Approximate intensities of sounds commonly encountered in everyday life**

$10\,dB = 10 \times$ increase ($10^1$) above 0 dB reference level
$20\,dB = 100 \times$ increase ($10^2$)
$60\,dB = 1\,000\,000 \times$ increase ($10^6$)

| | |
|---|---|
| hearing threshold level | 0 dB |
| whisper | 30 dB |
| normal talking | 60 dB |
| shouting | 90 dB |
| near pain threshold | 120 dB |

The sound in a noisy factory at 120 dB is 1 000 000 000 000 (i.e. $10^{12}$) times louder than a whisper.

### 15.2.2 Structure of the ear

The ear is divided into three main regions, the **outer ear, middle ear** and **inner ear** (Fig 15.20). The outer ear channels sound waves from the surrounding air into the middle ear where the energy of sound waves is converted to mechanical vibrations. In the inner ear nerve impulses are generated in response to vibrations received from the middle ear and to changes in position of the head.

**Fig 15.20** Structure of the ear

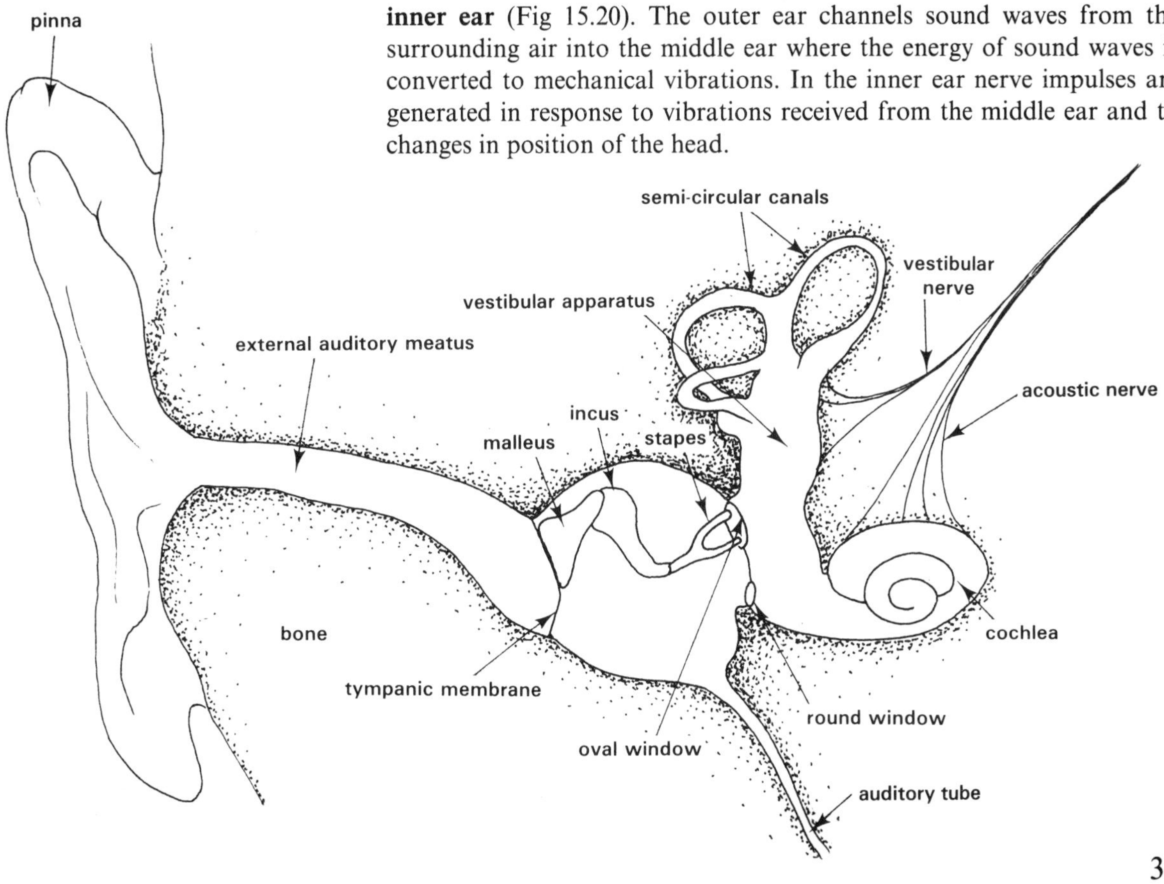

301

### 15.2.3 Hearing

Sound waves enter each ear by a short tube called the **external auditory meatus** (plural meati). The **pinnae** on each side of the head may help in directing sound waves into the meati. At the inner end of each meatus is an elastic membrane called the **tympanic membrane** or **eardrum**.

Bridging the air-filled middle ear are three small bony **ossicles** held in place by muscles and ligaments. The ossicles articulate freely with each other and are the **malleus** (hammer), **incus** (anvil) and **stapes** (stirrup). Sound waves vibrate the tympanic membrane which in turn vibrates the ossicles. The malleus is attached to the tympanic membrane. The stapes is attached to another membrane called the **oval window** which is part of the inner ear. The oval window is less than 5% of the area of the tympanic membrane. Consequently, vibrations of the tympanic membrane are amplified about 20 times in the oval window. Amplification makes it easier for vibrations to pass through the dense fluid in the inner ear.

An air-filled canal called the **auditory tube (Eustachian tube)** connects the middle ear with the pharynx. The air pressure in the atmosphere and in the middle ear is usually the same. Should there be a sudden large increase in external air pressure there is a possibility that the eardrums would burst.

**Fig 15.21** (a) TS diagram of structure of the cochlea

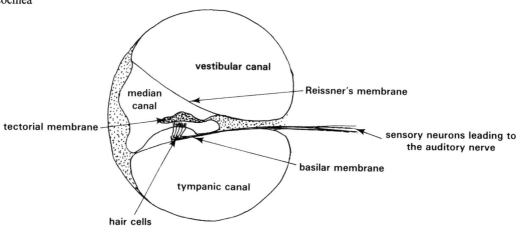

**Fig 15.21** (b) TS photomicrograph of cochlea (guinea pig), × 70

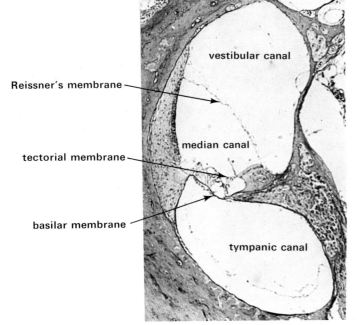

The danger of this happening is usually avoided because air taken in from the outside enters the auditory tubes during swallowing. In this way, air pressure on either side of the eardrums is equalised. The oval window transmits the vibrations of the middle ear ossicles into a coiled, fluid-filled tube called the **cochlea** (Figs 15.20 and 15.21). It contains three longitudinal canals separated from each other by two flexible membranes. The upper **vestibular canal** is connected to the oval window. Between the vestibular canal and the **median canal** is **Reissner's membrane**. The **basilar membrane** separates the median canal from the lower **tympanic canal**.

**Fig 15.21** (c) 'Unrolled' view of cochlea

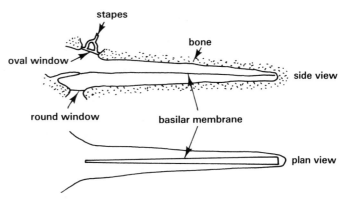

Vibrations of the oval window generate pressure waves in the fluid filling the vestibular canal. The pressure waves pass to the median canal and vibrate the basilar membrane. The tympanic canal is connected to a circular membrane called the **round window** just beneath the oval window. This arrangement allows the pressure waves to transmit through the cochlear fluid. Since liquids are not compressible, when the oval window is pushed inwards the round window pushes outwards and vice versa (Fig 15.22).

**Fig 15.22** Vibrations of the stapes move the oval window back and forth. The oscillations of the oval window push and pull on the cochlear fluids, causing vibrations of the cochlear membranes. Note the effect of the vibrations of the round window.

(a) Effect when the oval window is 'pushed'

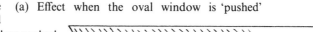

(b) Effect when the oval window is 'pulled'

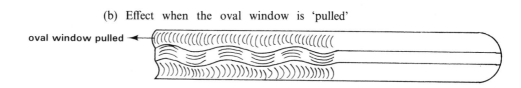

At the tip of the cochlea, called the **apex**, the fluid in the vestibular canal is continuous with that in the tympanic canal through a narrow channel called the **helicotrema**. The helicotrema is thought to allow sudden, large pressure changes to transmit directly to the round window without undue distortion of, and possible damage to the cochlear membranes.

The sensory region of the cochlea is called the **organ of Corti** (Fig 15.21(a)). It contains many **hair cells** rooted in the basilar membrane. The hair cells are arranged in outer and inner ranks. Short hairs called **stereocilia** project into the fluid of the median canal. Some of them are attached to the **tectorial membrane**. Vibrations of the basilar and tectorial membranes cause the stereocilia to distort, resulting in the generation of impulses in the hair cells (Fig 15.23).

**Fig 15.23** Organ of Corti. The median canal contains endolymph. The vestibular and tympanic canals contain perilymph. The fluids differ chemically. Endolymph is similar to cytoplasm. When the stereocilia distort, the hair cells depolarise.

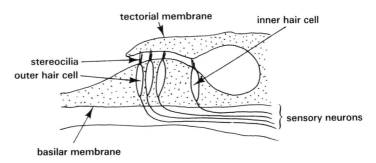

The hair cells are connected by synapses to sensory neurons which transmit the impulses to the brain along the **acoustic nerve**. It is part of cranial nerve VIII and leads to the **primary auditory areas**. They are regions of the cerebral cortex where the sensory impulses from the ears are interpreted as sound (Fig 15.24).

**Fig 15.24** The right primary auditory area. Different regions of the area receive impulses from specific parts of the organs of Corti

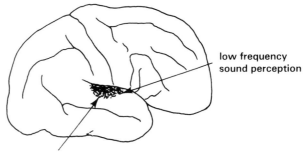

low frequency sound perception

high frequency sound perception

Vibrations of the basilar membrane are crucial to the conversion of sound waves into nerve impulses. The basilar membrane is about 2.5 times wider at the apex than at its base between the oval and round windows. However, the cochlea tapers from base to apex (Fig. 15.21(c)). Sound of short wavelength, that is of high frequency, vibrates a relatively short portion of the basilar membrane. Only the hair cells nearest the oval window are stimulated. The impulses arriving at the brain are interpreted as high-pitched sound. A longer portion of the basilar membrane is vibrated by sound of longer wavelength, that is lower frequency. Hair cells are stimulated further along the basilar membrane. The brain interprets the impulses from the hair cells as a low-pitched sound. Sounds of intermediate wavelength stimulate the basilar membrane, mostly in the middle regions (Fig 15.25).

**Fig 15.25** Vibrations of the basilar membrane in response to two pure tones of 500 Hz and 2000 Hz frequency. Each numbered line represents the position of the membrane at a particular instant of the vibration cycle (from Rosenberg after Eldredge)

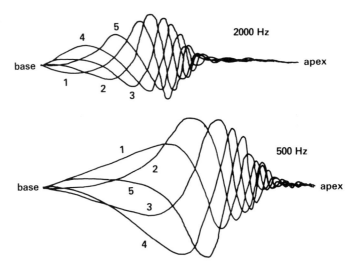

The intensity of sound is translated from the intensity with which the appropriate region of the basilar membrane is vibrated. The greater the amplitude of basilar membrane movement, the more impulses are generated per second in the hair cells and vice versa. Many impulses per second are interpreted in the brain as loud sounds, fewer impulses per second as soft sounds. Interpretation, that is **perception** of sound is thus a function of the brain.

How does the cochlea in the inner ear enable us to distinguish between sounds of different intensities and frequencies?

### 15.2.4 Balance

Just above the cochlea and connected to it by a short tube is the **vestibular apparatus** (Fig 15.26). It consists of two lymph-filled sacs called the **saccule** and the **utricle**. Projecting from the top of the utricle are three **semicircular canals**, also containing lymph.

**Fig 15.26** Vestibular apparatus

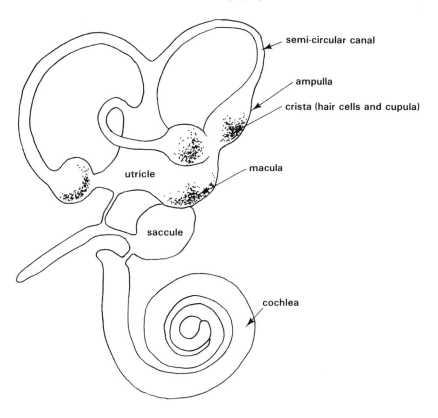

semi-circular canal

ampulla

crista (hair cells and cupula)

utricle

macula

saccule

cochlea

The saccule and utricle contain receptors called **maculae** which are sensitive to gravity. Small hairs project from the receptor cells into the lymph. The hairs are attached to calcium carbonate granules called **otoliths** (Fig 15.27). Gravity causes the otoliths to distort the sensory hairs in a direction determined by the position of the head. In response to the distortion, nerve impulses pass along the **vestibular nerve** to the brain. If the head is moved to a different position, the otoliths distort the sensory hairs in a different direction. Information about the head's new position is interpreted in the brain.

**Fig 15.27** (a) Structure of the macula from the utricle

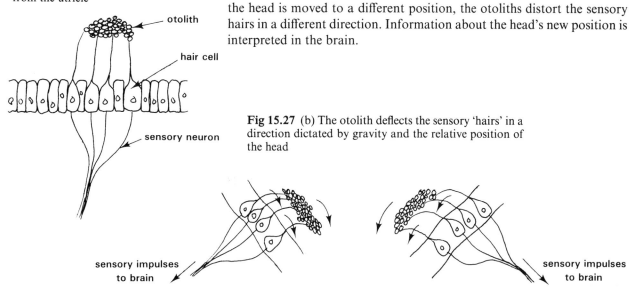

otolith

hair cell

sensory neuron

**Fig 15.27** (b) The otolith deflects the sensory 'hairs' in a direction dictated by gravity and the relative position of the head

sensory impulses to brain

sensory impulses to brain

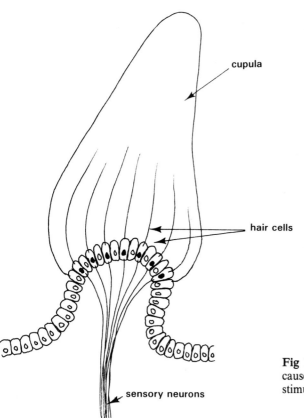

The semicircular canals provide information about head movements, rather than the position of the head when it is stationary. The end of each canal is enlarged to form a **ampulla**. Each contains a receptor, consisting of hair cells similar to those of the maculae. The ampullary hairs, however, project into a gelatinous mass called a **cupula** which is suspended in the lymph inside the ampulla (Fig 15.28(a)).

Any movement of the head of course moves the semicircular canals in the same direction. The lymph inside the canals, however, lags behind and pushes the cupulae in the opposite direction (Fig 15.28(b)). As a result, the hairs projecting into the cupulae are bent and nerve impulses are sent to the brain along the vestibular nerve. There are three semicircular canals and each is arranged at a right angle to the others. Consequently, at least one cupula is stimulated by lymph movements, whatever direction the head is moved in.

Information about the orientation and movement of the head is vital, especially when the whole body is moved. Visual information from the eyes also contributes to the brain's awareness of the head's spatial position. The information is used by the brain to coordinate movement and posture of the body.

Fig 15.28 (b) Movements of the fluid inside the semi-circular canals, caused by movements of the head, displace the gelatinous cupula and stimulate the sensory 'hairs'

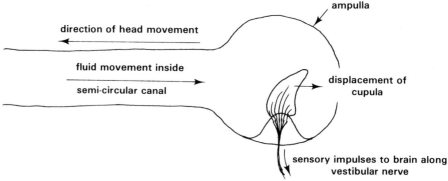

## 15.2.5 Defects of the ear

Several defects of the ear lead to hearing loss or even deafness. **Conductive deafness** is caused by inability of the outer and middle ear to conduct vibrations to the cochlea. One of the commonest causes of conductive deafness is blockage of the external auditory meatus with wax secreted from **ceruminous** glands in the skin lining the meatus. In some people wax accumulates in the meatus and hardens, sometimes pressing against the eardrum. Normal hearing is usually restored after the hardened wax is removed with a special syringe.

Another cause of conductive deafness, is a perforated eardrum. Perforation can be caused by infection in the middle ear or by mechanical injury resulting from a nearby explosion or a sudden blow to the head. Injury to the head can also cause the ossicles of the middle ear to become disconnected from one another, thus breaking the conductive path to the cochlea. Patients with a conductive defect which does not respond fully to treatment may be helped with a **hearing aid**. It is a device which amplifies sound waves before they enter the inner ear (Fig 15.29).

**Fig 15.29** 'Behind-the-ear' hearing aid (from DHSS booklet: General Guidance for Hearing Aid Users)

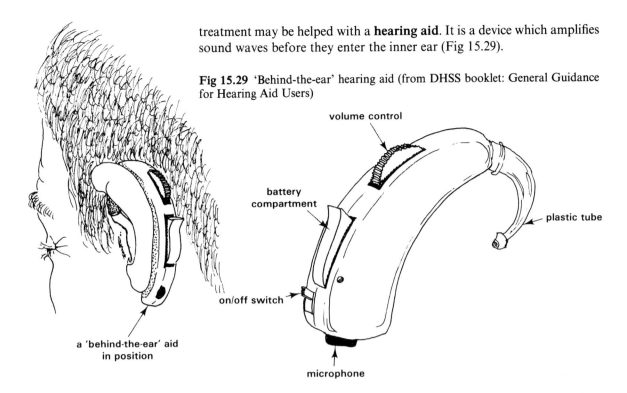

volume control

battery compartment

plastic tube

on/off switch

a 'behind-the-ear' aid in position

microphone

Malfunction of the cochlea and acoustic nerve can be the cause of deafness, even though vibrations are conducted perfectly into the inner ear. Such deafness is called **sensorineural (perceptive) deafness**. Sensorineural deafness can be inherited, though it is often acquired. Acquired forms of the condition can result from infection, head injury, blast from explosions or exposure to excessive noise. Much concern has been expressed in recent years about the possible harmful effects on hearing of the very high noise levels in discotheques, at airports and in places where noisy machinery is used. So serious is the problem that people employed in extremely noisy places are encouraged to wear ear muffs to protect their hearing from permanent damage.

Although the eye and ear are extremely sophisticated sensory organs, it is the interpretation of sensory information which ultimately limits their use. Processing of information from them occurs in the brain by mechanisms which are, as yet, poorly understood.

How can loud music at a disco damage our ears?

# 16 Endocrine system

Many functions of the body are coordinated and controlled by **hormones** which are produced by **endocrine glands**. Endocrine glands are ductless and secrete hormones directly into the body fluids, mainly the blood. Reference has already been made in Chapter 9 to hormones which regulate secretion of digestive fluids. Those which control reproduction are described in Chapter 18. The endocrine glands are distributed throughout the body (Fig 16.1). Because they are transported in the blood, hormones can control body functions taking place some distance from the endocrine glands. This distinguishes endocrine glands from exocrine glands which have ducts to channel their products into nearby regions without using blood for transport (Fig 7.9).

Although only small amounts of hormones are found in blood, their effects on the body tissues are very great. Once in the blood, hormones are bound to plasma proteins which carry them to their sites of activity called **target organs**. Activity of most endocrine glands is controlled by the nervous system (Chapter 14).

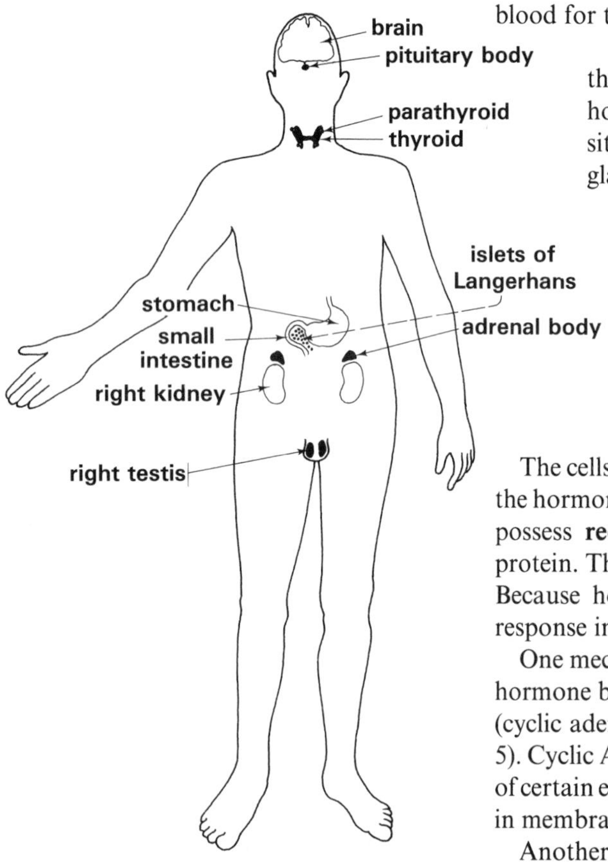

brain
pituitary body
parathyroid
thyroid

islets of Langerhans

stomach
small intestine
right kidney

adrenal body

right testis

**Fig 16.1** Sites of the main endocrine glands

The cells of the body that are affected by a particular hormone are called the hormone's **target cells**. Target cells respond to a hormone because they possess **receptors** which are sensitive to the hormone. Receptors are protein. They are found in target cell membranes, cytoplasm and nucleus. Because hormone–receptor binding is specific, hormones only cause a response in their particular target cells.

One mechanism involves receptors in the target cell membrane. When a hormone binds with a membrane receptor, a substance called cyclic AMP (cyclic adenosine-3′,5′,-monophosphate) is produced from ATP (Chapter 5). Cyclic AMP triggers a chain of reactions which results in the activation of certain enzymes, secretion of cell products, protein synthesis, or a change in membrane permeability (Fig 16.2(a)).

Another mechanism involves intracellular receptors. When the hormone enters the target cell, a hormone–receptor complex is formed. The complex activates certain genes. Specific protein (enzyme) synthesis is thus stimulated and the function of the cells is modified (Fig 16.2(b)).

## 16.1 Thyroid and parathyroid glands

### 16.1.1 Thyroid gland

The **thyroid gland** is in the neck close to the larynx (Figs 16.1 and 16.9). Under the microscope the thyroid gland is seen to be composed of globular groups of cells called **follicles** (Fig 16.3(a)). The closely packed follicles are of cubical epithelium and are held together by connective tissue supplied with blood vessels. The epithelial cells secrete thyroid hormones into the cavities of the follicles where they are temporarily stored before they are taken into the blood.

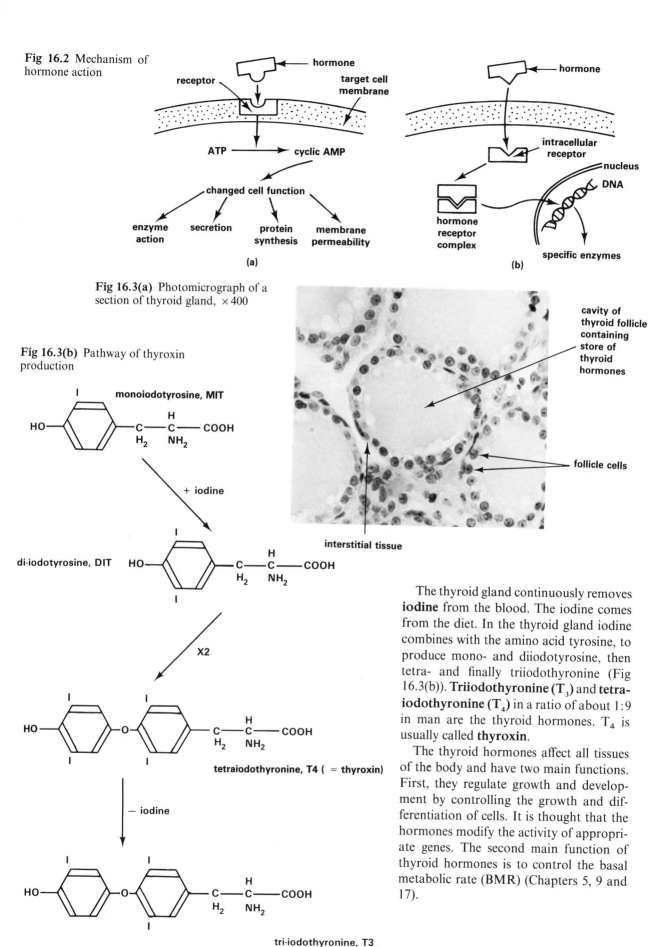

**Fig 16.2** Mechanism of hormone action

**Fig 16.3(a)** Photomicrograph of a section of thyroid gland, ×400

**Fig 16.3(b)** Pathway of thyroxin production

monoiodotyrosine, MIT

di-iodotyrosine, DIT

+ iodine

X2

tetraiodothyronine, T4 ( = thyroxin)

− iodine

tri-iodothyronine, T3

cavity of thyroid follicle containing store of thyroid hormones

follicle cells

interstitial tissue

The thyroid gland continuously removes **iodine** from the blood. The iodine comes from the diet. In the thyroid gland iodine combines with the amino acid tyrosine, to produce mono- and diiodotyrosine, then tetra- and finally triiodothyronine (Fig 16.3(b)). **Triiodothyronine ($T_3$)** and **tetra-iodothyronine ($T_4$)** in a ratio of about 1:9 in man are the thyroid hormones. $T_4$ is usually called **thyroxin**.

The thyroid hormones affect all tissues of the body and have two main functions. First, they regulate growth and development by controlling the growth and differentiation of cells. It is thought that the hormones modify the activity of appropriate genes. The second main function of thyroid hormones is to control the basal metabolic rate (BMR) (Chapters 5, 9 and 17).

**Fig 16.4** Mechanism controlling thyroid activity

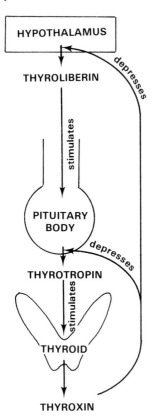

## 16.1.2 Control of thyroid activity

As with several other endocrine glands, growth and secretion of the thyroid gland are controlled by the **pituitary body** (Figs 16.1 and 16.27). The pituitary body secretes a number of **tropins** which affect other endocrine glands (section 16.5.1). The tropins include **thyrotropin**. Growth of the thyroid gland and its output of hormones is stimulated by thyrotropin. Secretion of thyrotropin is in turn suppressed by the thyroid hormones. The interaction is a **negative feedback control** mechanism. Another hormone called **thyroliberin** controls the release of thyrotropin. It is made in the **hypothalamus**, the part of the brain immediately above the pituitary body (Fig 16.4). Such control ensures that activity of the thyroid gland meets the body's needs at any time. It is a homeostatic device which helps maintain the stability of the body's internal environment. Other comparable examples are described in the remainder of the chapter.

## 16.1.3 Measurement of thyroid activity

Thyroid activity can be measured with the use of **radioactive isotopes (radionuclides)** of iodine. Sodium iodide containing $^{131}I$ is given orally in doses which produce harmless levels of radiation to the patient. The radio-iodine circulates in the plasma as **plasma inorganic iodine (PII)**. Much of the iodine in PII is normally absorbed by the thyroid gland and used to make thyroid hormones. The hormones are secreted into the blood where the radio-iodine is carried as **protein bound iodine (PBI)**. When the

**Fig 16.5** Fate of ingested iodine

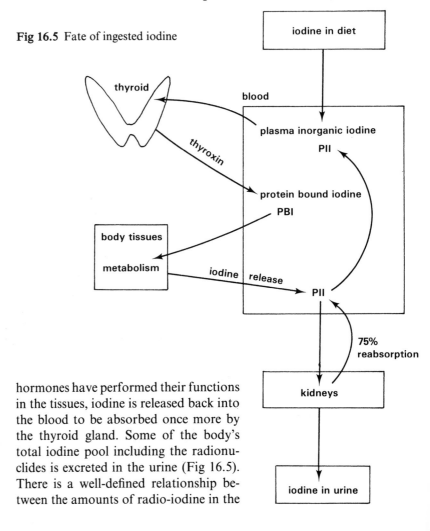

hormones have performed their functions in the tissues, iodine is released back into the blood to be absorbed once more by the thyroid gland. Some of the body's total iodine pool including the radionuclides is excreted in the urine (Fig 16.5). There is a well-defined relationship between the amounts of radio-iodine in the plasma, thyroid gland and urine following a dose of the radionuclide (Fig 16.6(a)). Consequently, measurements of the radiation in the thyroid gland, samples of blood and urine are useful in detecting abnormal activity of the thyroid gland (Fig 16.6(b)).

**Fig 16.6** Relationship between the levels of iodine in the thyroid gland, blood and urine after oral administration of radio-active iodine (after Greig, Boyle and Boyle)

(a) Normal subject

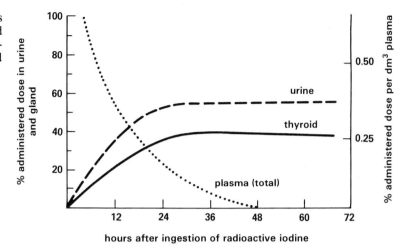

(b) Thyrotoxic subject

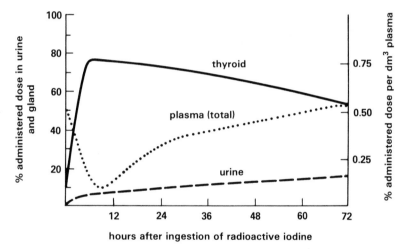

---

How can doctors produce an image of the thyroid gland using radionuclides?

---

An alternative to the iodine uptake studies described above is the use of a radionuclide to produce an image of the thyroid gland. An isotope of technetium, $^{99}Tc^m$, is administered by intravenous injection. The radionuclide is absorbed by the thyroid gland. About 30 minutes later, the gamma radiation emitted from the thyroid gland is used to produce an image with a gamma camera. Images obtained in this way are useful in assessing an overactive thyroid before and after treatment. They can also be used to detect regions of the thyroid, called **nodules**, which secrete thyroid hormones independently of the thyrotropin control mechanism (Fig 16.7).

**Fig 16.7** Thyroid scans, glands shown in outline: (a) normal, (b) cyst in the left lobe results in no uptake of radionuclide; the right lobe is unaffected

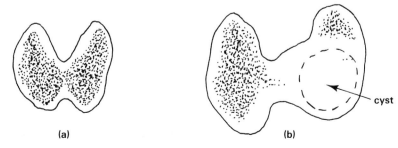

311

### 16.1.4 Malfunction of the thyroid gland

Overactivity of the thyroid gland is called **hyperthyroidism**. The condition can result from a failure of the thyrotropin–thyroid control mechanism, resulting in increased thyroxin output. Another cause of hyperthyroidism is uncontrolled secretion of thyroid hormones by a thyroid tumour. Either way, an increase in basal metabolism occurs with increased heart rate, extreme irritability and loss of body mass. Removal or destruction of part of the thyroid gland may be necessary to check the condition. A common side-effect of an overactive thyroid gland is the secretion by the pituitary body of **exophthalmos-producing substance**. It causes excessive growth of the tissues immediately behind the eyes. Consequently protrusion of the eyes called **exophthalmos** often accompanies hyperthyroidism (Fig 16.8).

Underactivity of the thyroid gland is called **hypothyroidism**. In adults hypothyroidism is the cause of a condition called **myxoedema**. The effects on basal metabolism are the opposite to those of hyperthyroidism. The BMR slows down, a smaller proportion of the energy-rich ingredients in the diet are respired, hence body mass increases. Mental activity also slows down so the patient is less alert than normal. At one time hypothyroidism was common in parts of Derbyshire and the Swiss Alps. Here the local soil and water supply are deficient in iodine and the population was thus deprived of an essential requirement for the synthesis of thyroid hormones. Addition of traces of iodine to table salt has largely overcome the problem.

The abnormal growth and development which accompanies an underactive thyroid gland is particularly distressing in infants. **Cretinism** is the name given to the effects of thyroid deficiency in children. The main symptoms are retardation in mental, physical and sexual development.

Both hyper- and hypothyroidism usually result in excessive growth of the thyroid gland. The enlarged thyroid gland is called a **goitre** which causes swelling of the neck (Fig 16.8).

**Fig 16.8** Facial characteristics of a person with hyperthyroidism. Note the protruding eyes and goitre.

### 16.1.5 Parathyroid glands

The **parathyroid glands** are four small oval bodies embedded in the thyroid gland (Fig 16.9). The parathyroid glands make a polypeptide hormone called **parathyrin** which has a profound effect on the calcium ion

**Fig 16.9** Position of the parathyroid glands

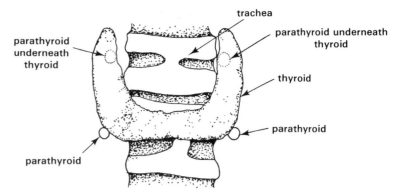

content of the blood and other body fluids. Parathyrin elevates the amount of calcium in the blood which, by negative feedback control, regulates parathyrin output (Fig 16.10).

**Fig 16.10** Mechanism regulating the output of parathyrin

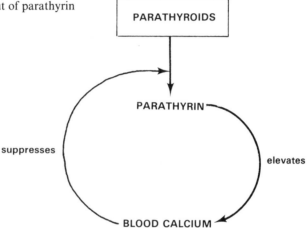

Calcium is an important constituent of body fluids. It participates in many body functions including coagulation of the blood (Chapter 10), nerve and muscle activity (Chapter 14) and teeth and bone formation (Chapters 9 and 7). These functions depend not just on the presence of calcium ions but on a precise concentration of calcium ions. It is therefore important that the amount of calcium is regulated within narrow limits. Calcium is one of the most precisely regulated constituents of the body. The concentration of calcium ions in normal human blood serum ranges between 9 and 11 mg per 100 $cm^3$.

Parathyrin affects blood calcium in several ways, as follows:

### 1 Release of calcium from bone

Bone contains calcium salts, especially calcium phosphate (Chapter 7). Calcium and phosphate(v) ions are continuously released from bones into the tissue fluids and redeposited in bones. Clearly, the composition and structure of bone depends on regulating this reversible process (Fig 16.11). By stimulating the **release of calcium** and phosphate ions from bone, parathyrin helps maintain normal concentrations of calcium and phosphate ions in blood.

**Fig 16.11** Dynamic equilibrium between the calcium and phosphate in bone and that in the tissue fluid and blood

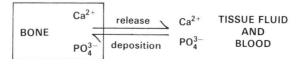

## 2 Reabsorption of calcium from the urine

Calcium and phosphate ions are filtered from the blood in the nephrons. Parathyrin promotes **reabsorption of calcium** ions from the filtrate in the proximal convolutions at the expense of phosphate ions which are excreted (Chapter 13). Parathyrin thus elevates the concentration of calcium ions in the blood while lowering the concentration of blood phosphate ions (Fig 16.12). The inverse relationship between calcium and phosphate ions in blood

$$[Ca^{2+}] \propto \frac{1}{[PO_4^{3-}]}$$

prevents the accumulation of unwanted calcium phosphate in the blood and tissues.

**Fig 16.12** Summary of the effects of parathyrin and calcitonin on blood calcium and phosphate

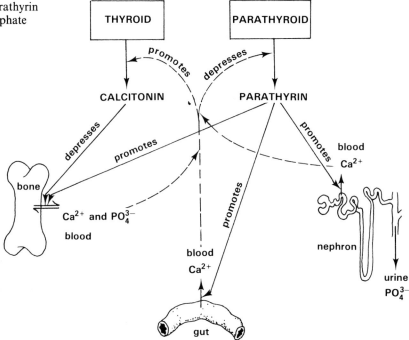

## 3 Absorption of calcium from the gut

Parathyrin is also thought to promote **absorption of calcium** ions from the gut (Chapter 9), thus helping to maintain the normal concentration of calcium ions in the blood.

### 16.1.6 Role of the thyroid gland in calcium control

Another hormone called **calcitonin** also affects the calcium ion concentration of blood. Calcitonin is secreted from **C cells** in the thyroid gland. Its effect is to lower the concentration of calcium ions in blood by causing the deposition of calcium phosphate in bone. Thus it is the combined effects of parathyrin and calcitonin which regulate calcium and phosphate ions to the concentrations necessary for many physiological functions to take place normally (Fig 16.12).

### 16.1.7 Malfunction of the parathyroid glands

Underactivity of the parathyroid glands, **hypoparathyroidism**, results in a lowering of the concentration of calcium ions in blood, **hypocalcaemia**.

If the concentration of calcium ions in blood serum drops below 7 mg per 100 cm³ a condition called **tetany** results. Tetany is characterised by increased excitability of the nervous system. Muscular activity becomes spasmodic and uncontrolled. Surgical removal of the parathyroid glands results in tetany within a few days. Subsequent injection of parathyroid extract quickly corrects the condition but the effect is short-lived. Subsequent maintenance of normal concentrations of calcium ions in the blood requires continued treatment with parathyrin (Fig 16.13).

**Fig 16.13** Effects of removing the parathyroids on the concentration of blood calcium

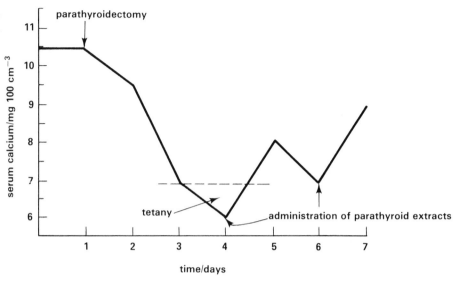

Overactivity of the parathyroid glands, **hyperparathyroidism**, can result in breakdown of bone structure and consequent elevation of the concentration of calcium ions in blood, **hypercalcaemia**. Recent work indicates that the decalcifying activity of parathyrin increases in women after the menopause. Before the menopause, output of parathyrin is inhibited by oestrogen hormones secreted by the ovaries (Chapter 18). After the menopause oestrogen production stops. It explains why the bones of older women are often fragile and prone to fracture.

Why are skeletal fractures and breaks more likely in older women than in men of the same age?

## 16.2 Control of blood glucose

Carbohydrate is the main source of energy released in respiration (Chapter 5). Nearly all the carbohydrate absorbed from the gut circulates in the blood as **glucose** and it is this sugar which is the main respiratory substrate. It is therefore vital that the blood maintains a constant and adequate supply of glucose to all the tissues. It is equally important, however, that the concentration of glucose in blood and tissue fluids is not excessive. Abnormally high concentrations of glucose in the tissue fluid would draw water from cells by osmosis (Chapter 1). Furthermore, the water content of the body as a whole can be affected by the concentration of glucose in the blood. Normally the concentration of glucose in blood is sufficiently low for all the glucose in the renal filtrate to be reabsorbed into the blood in the kidneys (Chapter 13). Thus glucose is not usually excreted in urine. Since reabsorption of water from the nephrons depends on the water potential gradient between the renal filtrate and the blood, any glucose remaining in the filtrate would lower the gradient and reduce water reabsorption. Thus as well as tissue dehydration, a high concentration of glucose in blood could cause total body dehydration.

Mechanisms exist which regulate the concentration of glucose in the blood so that the tissues' metabolic needs are met without adversely affecting the body's water content. A number of hormones help to control the concentration of glucose in blood.

## 16.2.1 Insulin and glucagon

The pancreas is an exocrine gland which secretes pancreatic juice into the gut (Chapter 9). It is situated in a loop of the small intestine just below the stomach (Fig 16.1). Embedded in the pancreas, and sometimes also in the wall of the small intestine, is a large number of microscopic patches of endocrine tissue called the **islets of Langerhans** (Fig 16.14). Histochemical studies show that the islets contain two types of cells called $\alpha$- and $\beta$-cells. They are responsible for the production and secretion of two hormones called **glucagon** and **insulin** respectively. The hormones are discharged directly into the islet blood capillaries. The rest of the pancreas and its exocrine ducts are not involved in hormone production or secretion. Experimentally tying off the pancreatic ducts prevents secretion of pancreatic digestive enzymes but has no effect on hormone secretion.

**Fig 16.14** Photomicrograph of two islets of Langerhans seen in a thin section of pancreas, × 100

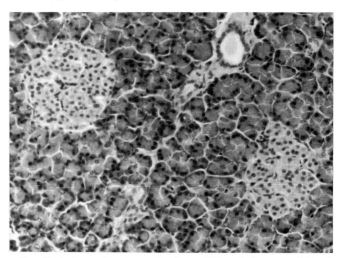

Insulin lowers the concentration of blood glucose while glucagon raises it. Output of the pancreatic hormones is regulated by negative feedback control based on the concentration of glucose in blood (Fig 16.15).

**1 The functions of insulin**

Insulin performs two main functions. The first is to regulate the concentration of glucose in the blood. Glucose is used by the tissues at varying rates dependent on the rate of metabolic activity. More glucose is used during exercise than when the body is at rest. The body's input of glucose varies depending on what has been eaten. The concentration of glucose in the blood draining the gut can double after a meal. The problem then is to reconcile the variable input and respiration of glucose with a relatively constant concentration of glucose in the blood. In dealing with this problem the liver, insulin and glucagon play vital roles.

Blood leaving the gut contains the absorbed products of digestion and passes through the hepatic portal system into the liver (Chapter 9). The liver cells contain enzymes which, under the control of insulin, promote the synthesis of **glycogen**, a polymer of glucose. It is as glycogen that much of the glucose absorbed from the gut is stored in the liver. Some glycogen is also stored in the skeletal muscles. Insulin therefore prevents any undue rise in the concentration of glucose in the blood. Glucose removed from the

**Fig 16.15** Effects of insulin and glucagon on the levels of blood glucose

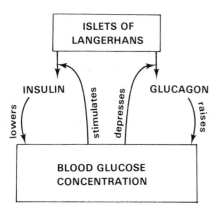

blood for respiration is replaced from the glycogen stores. Glucagon has the opposite effect to insulin. It stimulates the conversion of glycogen to glucose. The balance between the effects of the two hormones results in regulation of the concentration of glucose in blood (Fig 16.16). The negative feedback relationship between the hormones and glucose in blood ensures that glucose is released from the glycogen stores at a rate sufficient to match its uptake from the blood by respiring tissues.

**Fig 16.16** Summary of the fate of glucose in the body

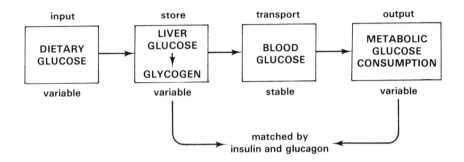

The second main function of insulin is to speed up the rate of entry of glucose into respiring cells. Glucose is taken into living cells by **active absorption**. Insulin greatly increases the rate of glucose absorption, possibly by triggering a membrane carrier mechanism or by acting as a carrier itself. In a normal healthy man a concentration of about 3 to 5 mmol glucose per $dm^3$ of blood is sufficient to meet the requirements of respiring cells. In the absence of insulin, this concentration would have to increase by between ten and twenty times for glucose to enter the cells at the same rate by diffusion only.

### 2 Malfunction of the islets of Langerhans

Underactivity of the islets of Langerhans results in reduced secretion of insulin and is the cause of **diabetes mellitus**. This should not be confused with diabetes insipidus (Chapter 13). In diabetes mellitus the glucose concentration in blood rises, **hyperglycaemia**, and glucose exceeds the maximum concentration which can be totally reabsorbed from the renal filtrate in the kidneys. Consquently glucose is excreted in the urine, a condition called **glycosuria**. The presence of glucose in the urine disturbs the water potential gradient which normally results in water reabsorption from the nephrons. Large volumes of dilute urine are thus produced, a condition called **diuresis**. Diuresis is dangerous because it may bring about dehydration of the body. Since the breakdown of glycogen is uninhibited in diabetes mellitus the stores of glycogen in the liver and muscles are quickly used up. Body fats and proteins are then used as respiratory substrates, causing a rapid loss of body mass.

The condition can be rectified by regular doses of insulin and by eating a carefully controlled diet. A clinical test often used in hospitals to assess whether insulin production is normal involves measurement of **glucose tolerance**. Patients are made to fast for several hours before ingesting 50 g of glucose in 150 $cm^3$ water. The concentration of glucose in the patient's blood is measured immediately and at 30-minute intervals over a period of two to three hours. If necessary the urine is also analysed for glucose. The concentration of glucose in the blood is plotted against time and a glucose tolerance curve is obtained (Fig 16.17). In a healthy individual there is a slight rise in blood glucose concentration after the glucose drink. Insulin then brings the concentration of glucose in the blood down to its original value within about two hours and no glucose is excreted in the urine.

**Fig 16.17** Glucose tolerance curves

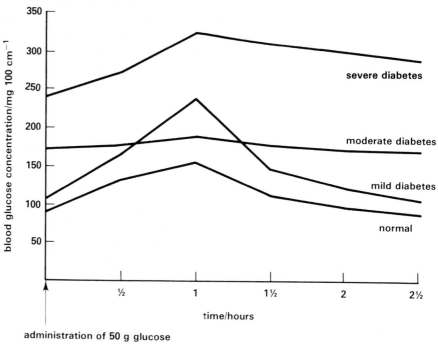

administration of 50 g glucose

**Fig 16.17** Glucose tolerance curves

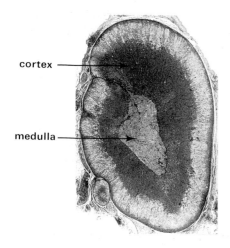

**Fig 16.18** Photomicrograph of a transverse section of an adrenal body from a cat, × 11

cortex

medulla

**Fig 16.19** Molecular structure of cortisol

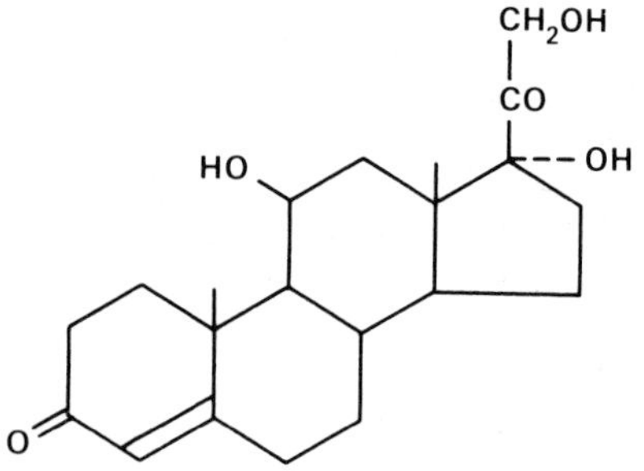

During late pregnancy the renal threshold for glucose may be lower than normal. Thus, even though the concentration of glucose in the blood is normal and only slightly rises after a meal, the rise may be sufficient for some glucose to be excreted in the urine. Glycosuria during pregnancy therefore does not necessarily mean that the mother has diabetes mellitus.

The type of diabetes described above is called **insulin dependent diabetes**. It results from a deficiency of insulin production and must be treated by periodic insulin injection. **Non-insulin dependent diabetes** is a commoner type of diabetes. Insulin secretion is not inhibited, but the membrane receptors in the target cells are insensitive to insulin. Consequently, glucose entry into the cells is less and it accumulates in the blood; hypoglycaemia. Non-insulin dependent diabetes is usually treated by a careful selection of diet in which glucose intake is strictly limited.

### 16.2.2 Cortisol

The reserves of glycogen in the body are limited and in certain conditions fats and proteins can be converted to glucose to supply metabolic demands. The conversion is influenced by a number of steroid hormones called **glucocorticoids** secreted by the **adrenal bodies**, a pair of glands lying close to the kidneys (Fig 16.1). Each adrenal body is divided into an inner **medulla** and an outer **cortex** (Fig 16.18). It is in the cortex that glucocorticoids are made, the most abundant of which is **cortisol** (Fig 16.19).

Cortisol stimulates the conversion of fats and proteins to glucose and is thus involved in regulating the glucose concentration in blood. Output of cortisol is controlled by a tropin called **corticotropin** from the pituitary body. Secretion of corticotropin is affected by the concentration of cortisol

Fig 16.20 Mechanism regulating cortisol production

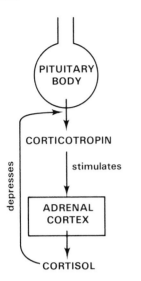

in the blood, another example of a negative feedback control mechanism (Fig 16.20).

Another important function of cortisol is its ability to suppress autolysis of damaged body cells by lysosomes (Chapter 6). Cortisol-like drugs are valuable in promoting the repair of body organs damaged in degenerative diseases such as arthritis. One of the effects of adrenalin, secreted by the medulla of the adrenal glands, is to increase output of corticotropin (section 16.4.1). This partly explains the increased secretion of cortisol which normally accompanies mental and physical stress.

## 16.3 Control of blood sodium and potassium

Sodium and potassium salts, especially chlorides, participate in a variety of body functions, particularly nerve and muscle activity. Consequently the maintenance of stable concentrations of sodium and potassium ions in the blood is vital for normal physiological activity (Chapter 14). Furthermore, because sodium and potassium salts are soluble in water they affect the water potential of the body fluids. The concentrations of sodium and potassium ions in the body are regulated by **mineralocorticoid** hormones made in the adrenal cortex (Fig 16.18).

### 16.3.1 Aldosterone

The most important mineralocorticoid hormone is **aldosterone** (Fig 16.21). This steroid is strikingly similar in molecular structure to cortisol (Fig 16.19). Nevertheless their functions are very different. Aldosterone promotes reabsorption of sodium ions ($Na^+$) into the blood from the filtrate in the nephrons (Chapter 13). Chloride ions ($Cl^-$) usually accompany the sodium ions, probably because of electrostatic attraction. Uptake of sodium ions suppresses reabsorption of potassium ions ($K^+$). Output of aldosterone is inhibited by a high concentration of sodium ions in the blood, yet another example of a feedback mechanism which assists homeostasis (Fig 16.23).

Suppressed output of aldosterone can cause excessive excretion of water by the kidneys resulting in a fall in the blood volume. One effect of the fall is the secretion into the blood of the enzyme **renin** from juxtaglomerular cells in the kidney (Fig 16.22). Renin activates the conversion of a plasma protein called proangiotensin into an active derivative **angiotensin**.

How can glucose appear in the urine of a pregnant woman who does not suffer from diabetes mellitus?

Fig 16.21 Molecular structure of aldosterone

Fig 16.22 Juxtaglomerular cells form part of the afferent arterioles in the kidneys (see Fig 13.4)

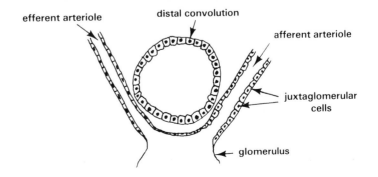

What makes us feel thirsty when our bodies are short of water?

Angiotensin stimulates secretion of aldosterone, thereby aiding water reabsorption by the kidneys. Angiotensin also stimulates the **thirst centre** in the hypothalamus of the brain to trigger off impulses which create the sensation of thirst (Fig 16.23). These are appropriate responses to too little water in the body fluids, and provided water is consumed, the body's state of hydration is rectified.

**Fig 16.23** Effect of aldosterone on water reabsorption by the nephrons. Angiotensin stimulates the thirst centre in the brain as well as aldosterone secretion

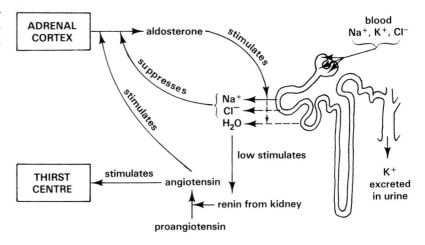

## 16.4 The adrenal medulla

The centre of each adrenal gland, the **medulla**, is derived from the same embryonic tissue which gives rise to sympathetic nerve ganglia. For this reason the medulla of the adrenal glands can be regarded as a modified part of the sympathetic nervous system. Sympathetic nerves, as part of the autonomic system, play an important role in regulating many body functions (Chapter 14). Medullary hormones have similar effects to stimulation of sympathetic nerves.

### 16.4.1 The medullary hormones

Two hormones are made in the adrenal medulla, **adrenalin** and **noradrenalin** (Fig 16.24). Noradrenalin is the neuroeffector transmitter secreted by sympathetic nerve endings. However, adrenalin is the main secretion of the medulla. The effects of both hormones are similar and, like sympathetic activity generally, they prepare the body to expend a lot of energy quickly. In fact, stimulation of the sympathetic nerves activates the medulla and promotes secretion of adrenalin. Normally the adrenal medulla secretes small amounts of adrenalin and noradrenalin into the blood. However, following increased activity of the sympathetic nerves which usually accompanies physical and mental stress, larger quantities are released. Because the medullary hormones prepare mammals to run away from or to face an enemy, they are sometimes called the **flight or fight** hormones. In

**Fig 16.24** Structure of (a) noradrenalin and (b) adrenalin. These hormones belong to a group of chemicals called catecholamines

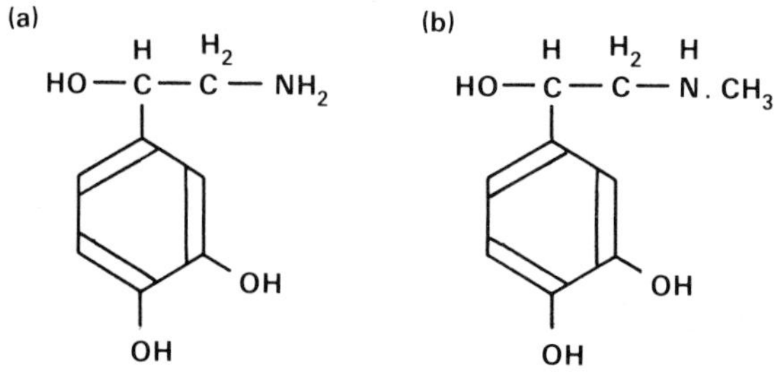

320

How do the 'flight or fight' hormones prepare our bodies to expend a lot of energy quickly in order to respond to a sudden danger?

many respects this is an appropriate description. The effects of adrenalin and noradrenalin can be summarised as follows.

## 1 Effects on the gut and respiratory system

The smooth muscle of the gastro-intestinal tract relaxes and the bronchi become dilated. The thorax is enlarged because the diphragm can now be pushed down further into the abdomen. The net result is that volumes of air larger than normal can be drawn in and out of the lungs, thereby increasing the rate of oxygen uptake by the blood.

## 2 Effects on the cardiovascular system

The heart rate and the power of cardiac contractions increase, with consequent rise of blood pressure. Arterioles in the skin and gut become constricted while the vessels supplying the skeletal muscles dilate. These effects, together with increased ventilation of the lungs, ensure an adequate supply of oxygenated blood to organs such as muscles which produce energy for movement.

## 3 Effects on blood glucose

Glycogen stored in liver and skeletal muscles is converted to glucose, thus providing the source of energy required for increased muscular activity. Adrenalin stimulates secretion of corticotropin and hence cortisol. Some of the extra glucose may therefore come from fats and proteins (section 16.2.2).

## 4 Effects on the nervous system

Adrenalin increases sensitivity of the nervous system, thereby increasing the speed with which the body may react to environmental stimuli. This is of obvious advantage in flight or fight. The various effects of adrenalin are illustrated in Fig 16.25.

**Fig 16.25** Summary of the effects of adrenalin

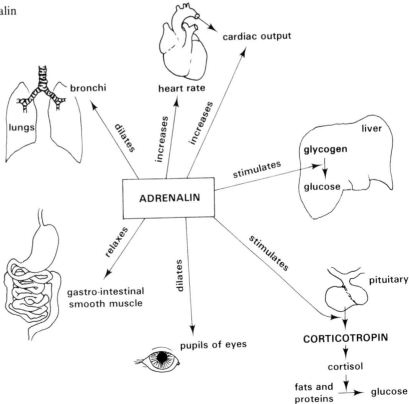

Why is the pituitary body sometimes called the 'master gland' of the endocrine system?

## 16.5 The pituitary body

Projecting downward from the base of the forebrain and almost totally enclosed by bone is the **pituitary body**, also called the **hypophysis**. It consists of two functional glands called the **anterior pituitary**, derived from the embryonic pharynx, and the **posterior pituitary**, derived from the hypothalamus in the forebrain (Figs 16.26 and 14.36).

**Fig 16.26** Photomicrograph of a vertical section through the pituitary body, × 10

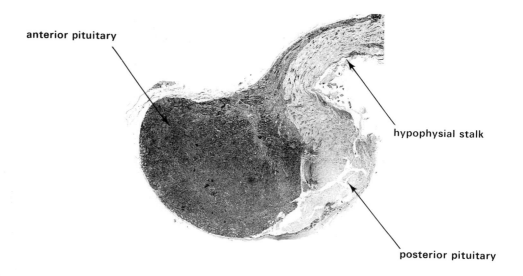

anterior pituitary

hypophysial stalk

posterior pituitary

### 16.5.1 The anterior pituitary

The anterior pituitary secretes several hormones including the various **tropins** referred to earlier in the chapter. **Thyrotropin** controls the secretion of thyroid hormones while **corticotropin** controls the output of cortisol from the adrenal cortex. Other tropins affecting mainly the testes or ovaries are called **gonadotropins** and are also produced by the anterior pituitary. The activities of **follitropin** and **lutropin** are described in Chapter 18.

Another hormone made in the anterior pituitary is **somatotropin**. It causes the release of the growth hormone **somatomedin** from the liver. The main function of somatomedin is to promote the growth of the body's tissues and organs by stimulating the synthesis of macromolecules, especially proteins. The precise mechanism of the stimulation is unknown but it may involve promoting the absorption of vital nutrients, especially amino acids by body cells from tissue fluid, and activation of the genes which control growth.

Secretion of somatotropin occurs throughout life but diminishes after the growing period. The way in which its secretion is controlled is unknown.

Oversecretion of somatotropin in early life leads to **gigantism**, while undersecretion causes **dwarfism** (Fig 16.27). Abnormally high output of somatotropin during adulthood when normal growth is complete leads to **acromegaly**. Many of the internal organs become enlarged, as do the hands and feet. The most striking characteristic is the lower jaw which often grows to protrude forward rather noticeably (Fig 16.28). Body height does not usually increase since after adolescence the limb bones do not normally grow further.

**Fig 16.27** Diagram showing gigantism (left) and dwarfism (right) contrasted with normal growth (centre)

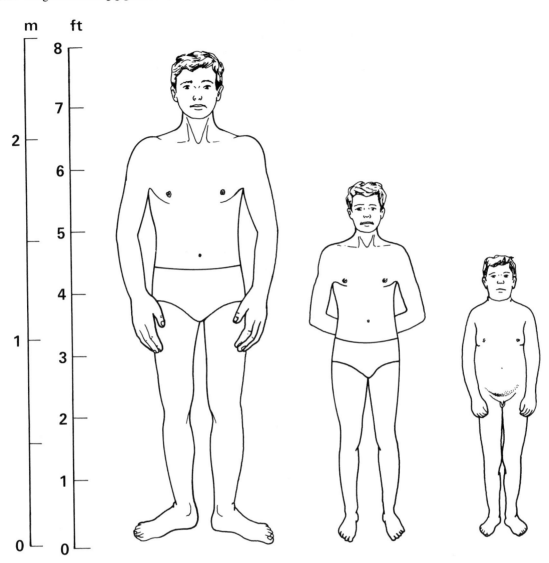

**Fig 16.28** Facial characteristics of acromegaly. Note the protruding jaw and forehead

Secretion by the anterior pituitary is controlled partly by a variety of substances from the hypothalamus. The secretions are carried from the hypothalamus into the anterior pituitary in blood vessels (Fig 16.29). One secretion is **thyroliberin**. Thyroliberin stimulates the release of **thyrotropin** which in turn stimulates secretion of the thyroid hormones (Fig 16.4). Other liberin hormones made in the hypothalamus control the release of the rest of the hormones made in the anterior pituitary.

**Fig 16.29** The hypothalamic-hypophysial portal system responsible for transporting trophins, made in the hypothalamus, into the anterior pituitary

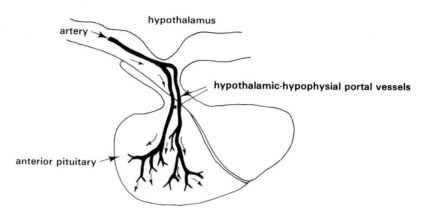

## 16.5.2 The posterior pituitary

The posterior pituitary produces two polypeptide hormones, **vasopressin** and **ocytocin**. Both are made by neurons in the hypothalamus and migrate in the axons of neurons to the posterior pituitary where they are stored. When the neurons connecting the hypothalamus and posterior pituitary are stimulated, vasopressin and ocytocin are released into the blood (Fig 16.30).

**Fig 16.30** Hypothalamic-hypophysial neurons responsible for transporting vasopressin and ocytocin from the hypothalamus into the posterior pituitary

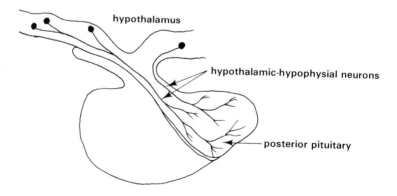

Ocytocin affects smooth muscle and is particularly important in females at the end of pregnancy when it brings about rhythmical contraction of the uterine wall during birth. This function and its role in controlling the ejection of milk from the mammary glands are described in Chapter 18. Vasopressin plays an important role in the regulation of water reabsorption by the renal nephrons (Chapter 13).

# 17 Thermoregulation

One of the many aspects of homeostasis is regulation of body temperature. It is a characteristic feature of birds and mammals, including man. These animals are called **endotherms**. They maintain a relatively high and constant body temperature which enables them to lead active lives even when the temperature of their surroundings is low. All other animals are called **ectotherms**. Their body temperature varies with temperature fluctuation of the environment (Fig 17.1). Regulation of body temperature is called **thermoregulation** and is brought about by balancing heat production in the body with heat loss to the environment.

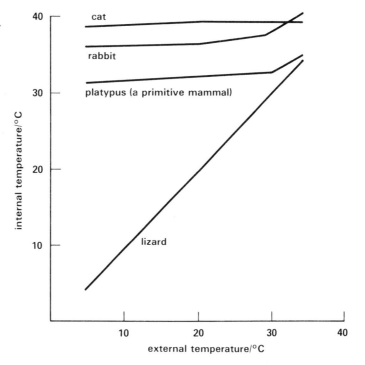

**Fig 17.1** Effects of environmental temperature on the body temperatures of three mammals (endotherms) and a lizard (ectotherm)

## 17.1 Heat production

The main source of heat in the body is tissue respiration (Chapter 5). Over 50 % of the energy released in respiration is heat energy. The rate at which heat is produced is proportional to the metabolic rate which increases greatly during exercise. An adult human produces about $250 \, kJ \, h^{-1}$ of heat energy when resting. Heat production increases to about $1000 \, kJ \, h^{-1}$ during moderate exercise while the rate may go up to $8000 \, kJ \, h^{-1}$ in a few minutes of intense exercise.

### 17.1.1 Measuring metabolic rate

If we fast, energy is released when food reserves in the body such as fat and glycogen are respired. At rest, nearly all the energy released in respiration is eventually given off as heat. Measurement of the heat given off by a resting, fasting individual gives an indication of the energy required to maintain the body's vital functions.

The rate at which the body respires to provide this amount of energy is called the **basal metabolic rate (BMR)**. The BMR of a human can be measured using a human calorimeter which is an insulated room containing water-filled pipes. The temperature of the water entering and leaving the room through the pipes for a given time is recorded. So is the temperature of the air in the room. Heat lost from the body causes a rise in temperature of the water and air. From the data the BMR can be calculated.

A much quicker way of calculating the BMR involves using a **spirometer** (Figs 12.9 and 12.10). A spirometer allows the volume of oxygen taken in by the subject in a given period of time to be measured. For every $1\,dm^3$ oxygen used in respiration, approximately $20.17\,kJ$ of heat energy are released. This is the average of the amount of heat energy released from the oxidation of carbohydrates, lipids and proteins. If a subject takes in $1.5\,dm^3$ oxygen in 5 minutes, which is equivalent to $18\,dm^3\,h^{-1}$, the rate at which heat energy is released is:

$$18 \times 20.17 = 363.06\,kJ\,h^{-1}$$

$$\text{Thus the BMR} = 363.06\,kJ\,h^{-1}$$

What is the basal metabolic rate of a person who consumes $1.48\,dm^3$ oxygen in 4 minutes? Express the answer in units of $kJ\,m^{-2}\,h^{-1}$.

The figure is generally adjusted to take into account differences in surface area of the body (section 17.2.4).

### 17.1.2 Distribution of body heat

Heat is produced unevenly in the body. Skeletal muscle generates a lot of heat during exercise. Another important producer of heat is the liver (Chapter 9). As blood flows through the skeletal muscles and liver it absorbs heat and distributes it to parts of the body where little heat is produced (Fig 17.2). Blood moving through the circulatory system has kinetic energy which is converted to heat energy when the blood meets resistance, mainly in the arterioles. When blood flows near the body surface, heat is lost through the skin.

**Fig 17.2** Distribution of heat on the surface of the body (a) in a cold, and (b) in a hot environment. Figures are in °C

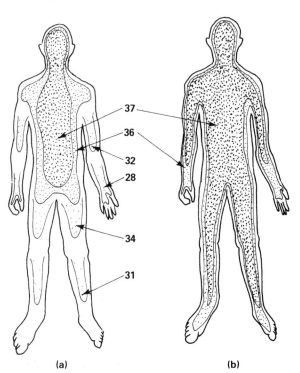

(a)          (b)

## 17.2 Heat loss

We lose body heat in three main ways, **radiation, evaporation** and **conduction**.

### 17.2.1 Radiation

Radiation is the emission of heat from a body to the air. Except in tropical regions, body temperature is generally higher than the air temperature. Heat is therefore radiated to the surrounding air more rapidly than it is gained by the same means. It has been estimated that 60 % of the total heat loss from a naked man sitting in a room kept at 33 °C is by radiation (Fig 17.3).

**Fig 17.3** Approximate relative heat loss by radiation (wavy line), evaporation (broken line) and conduction (straight line)

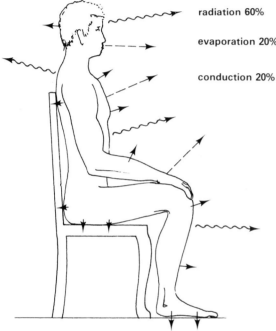

radiation 60%

evaporation 20%

conduction 20%

Measurements of heat coming from the body as infra-red radiation have been used in the detection of certain types of tumour. The patient is kept in a room of constant temperature, generally between 18 and 20 °C, to minimise fluctuations in skin temperature. Photographs are taken with a special camera which is sensitive to infra-red radiation coming from the body. Hot areas of the skin show up as light patches on the picture. Cool areas show up as dark patches. Hot spots occur in areas where blood flow and metabolism are increased, which may be diseased areas of the body. The technique is called **thermography**.

### 17.2.2 Evaporation

When water evaporates from a moist surface, heat energy is taken from the surface. The energy used is the **latent heat of vaporisation**. Our skin contains **sudorific glands** which secrete **sweat** onto the body surface when the body becomes overheated. Sweat is mainly water and when it evaporates the skin is cooled. Evaporation of body water also occurs from the moist linings of our nasal cavities, mouth, trachea and the extensive internal surface of the lungs. In very hot climates the surroundings often have a temperature higher than body temperature. In such conditions the body gains heat from the environment. Profuse sweating and subsequent evaporation of water from the skin's surface prevents overheating. A person can secrete up to $4 \, dm^3$ of sweat per hour in very hot, dry conditions. Dehydration of the body can then become a critical problem.

Even in a moderately warm room when sweating is virtually nil, about 20 % of a person's total heat loss is due to evaporation. This is because, even without sweating, water vapour escapes through the skin. Some of the heat loss is also due to evaporation from the linings of the mouth, nose, trachea and lungs (Fig 17.3).

Many environmental factors affect the rate of evaporation and hence heat loss from the skin. Most notable are the relative humidity of the air in contact with the skin, the air temperature and air movement. They control the rate of evaporation from the body.

### 17.2.3 Conduction

Heat passes from a warm object to a cooler one in direct contact with it, by conduction. You have probably noticed how warm a seat becomes after sitting on it for some time. The greater the temperature difference between the body and other objects touching it, the greater is the rate of conductive heat loss. A cool breeze passing over the skin removes heat by conduction to the air. About 20 % of the total heat lost by the person illustrated in Fig 17.3 can be due to conduction.

### 17.2.4 Surface area and body volume

The amount of heat generated in tissue respiration depends mainly on the **volume** of the body. Because most of the heat is lost through the skin, the amount of heat lost is determined by the **surface area** of the skin.

As we grow, our volume increases in three dimensions whereas our surface area increases only in two dimensions. Consequently, relative to volume our surface area increases at a slower rate. In terms of heat exchange, a bulky person has a larger volume of tissues in which heat is generated, but relative to this, a smaller surface through which heat is lost to the environment. The **surface-area-to-volume (SA/V) ratio** is the area of skin per unit of body volume. A person of volume $56\,dm^3$ and $1.53\,m^2$ surface area has a SA/V ratio of:

$$\frac{1.53}{56} = 0.0273\,m^2\,dm^{-3}$$

In comparison, a person of $80\,dm^3$ volume and $1.97\,m^2$ surface area has a SA/V ratio of:

$$\frac{1.97}{80} = 0.0246\,m^2\,dm^{-3}$$

In other words, each $dm^3$ volume of the smaller person has about 10 % more skin area available for heat loss compared with the bulkier person.

Such considerations highlight one of the factors which limit the extremes of size in terrestrial mammals. Bulky animals face overheating, while small ones lose heat rapidly. Experimenting with small mammals, Pearson, in 1957, obtained data for the metabolic rate of a variety of shrews. He related the metabolic rate to body mass (Fig 17.4). The rate of metabolism controls the rate of heat production. It is, therefore, not surprising that smaller shrews which have a greater relative surface area through which heat is lost show the highest metabolic rate. However, there is a limit to which increased metabolism can compensate for heat loss. It is estimated that a mammal smaller in size than the smallest species of shrew would be unable to obtain energy quickly enough from its food to make good the rate at which heat is lost through its body surface.

Since metabolic rate and surface area are related, measurements of BMR are usually expressed in units which take account of surface area. Expressing BMR in this way allows comparison to be made between animals of different sizes. In humans the BMR is expressed in units of $kJ\,m^{-2}\,h^{-1}$. The surface area of a human is difficult to measure directly but it can be calculated from measurements of body height and mass (Fig 17.5).

Why are there no free-living mammals smaller than shrews?

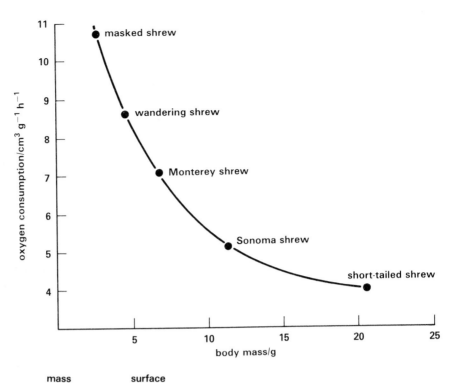

Fig 17.4 Relationship of metabolic rate (oxygen consumption) and body size in a variety of shrews (after Pearson)

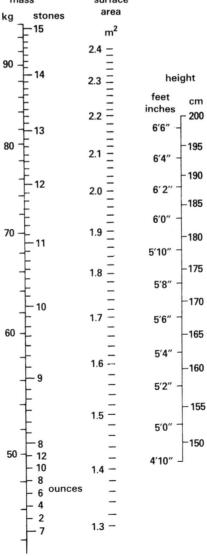

Fig 17.5 Nomogram for determining the surface (skin) area from body mass and height

## 17.3 Thermoregulation

Regulation of body temperature is brought about by balancing heat production and heat loss.

### 17.3.1 The thermoregulation centres

**Thermoregulation** takes place in response to changes in body temperature, but the changes must first be detected. This function is performed by specific parts of the hypothalamus in the forebrain (Chapter 14). In the hypothalamus there are two **thermoregulation centres**. The **cold centre** responds to blood which has a temperature less than normal by triggering off responses which increase heat production and decrease heat loss. The **heat centre** responds to blood which has a temperature higher than normal by triggering off responses which reduce heat production and increase heat loss (Fig 17.6).

**Fig 17.6** Action of thermoregulation centres in the hypothalamus

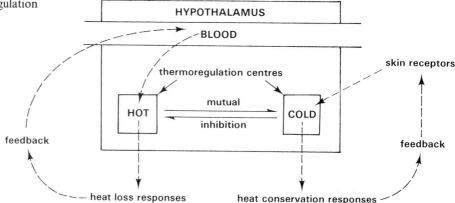

The heat centre acts like a thermostat. It switches on heat loss mechanisms when the temperature of the blood is higher than normal. Conversely it switches on heat conservation mechanisms when the temperature of the blood is lower than normal. The cold centre is inhibited by the heat centre, but becomes active when receptors in the skin signal that the environment is getting cooler.

In this way the body temperature is kept constant, even though the environmental temperature varies. The cold centre also receives information about potential body temperature changes. The information travels along sensory neurons from receptors in the skin which are sensitive to temperature changes outside the body. It is an early warning system which enables the thermoregulation centres to trigger appropriate responses before external changes alter the internal body temperature too much.

### 17.3.2 Regulating heat production

Heat is generated when respiratory substrates taken in as food are oxidised in respiration (Chapter 5). Heat generation in the body is thus ultimately limited by the intake of food. Assuming this to be adequate, other factors become important. Among these is production of the hormone **thyroxin** which controls the BMR (Chapter 16). Increased thyroxin output can double BMR, but the response takes several days to come into effect. A similar but more immediate response is brought about by **adrenalin** made in the adrenal glands and **noradrenalin** secreted by sympathetic nerve endings (Chapters 14 and 16). When the environmental temperature is very low, considerable heat can be generated by **shivering**. Shivering is very rapid alternate contraction and relaxation of the skeletal muscles.

### 17.3.3 Regulating heat loss

There are several means by which we can alter the rate at which heat is lost from the body. Nearly all involve the **skin**. The skin acts as a physical barrier which prevents excessive loss of heat and water from the body and stops foreign matter and pathogenic microbes gaining access to the underlying tissues and organs. The skin has a thin outer layer called the **epidermis**, one of the functions of which is to replace cells which are constantly lost from the surface. Below the epidermis is a much thicker layer, the **dermis**. The dermis contains blood vessels and a variety of receptors sensitive to heat, cold, pressure, touch and pain. Also in the dermis are the **sudorific glands** which produce **sweat**. The sudorific glands are coiled structures, each with a duct leading through the epidermis to the surface.

**Hairs** originate in the dermis and grow in pits called **follicles**. The bases of the follicles are attached to small **erector-pili muscles**, the contraction of which can raise the hairs projecting from the surface. Beneath the dermis are fat deposits contained in **adipose tissue** (Fig 17.7).

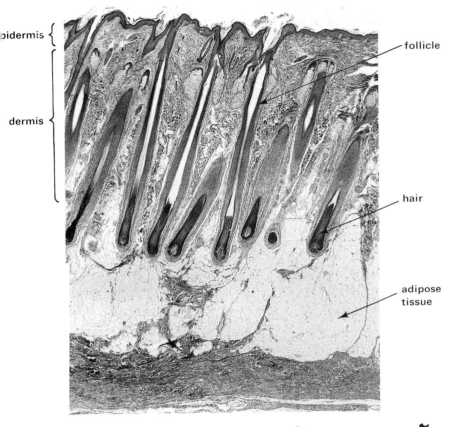

**Fig 17.7** (a) Photomicrograph of a vertical section through the skin from the scalp, × 20

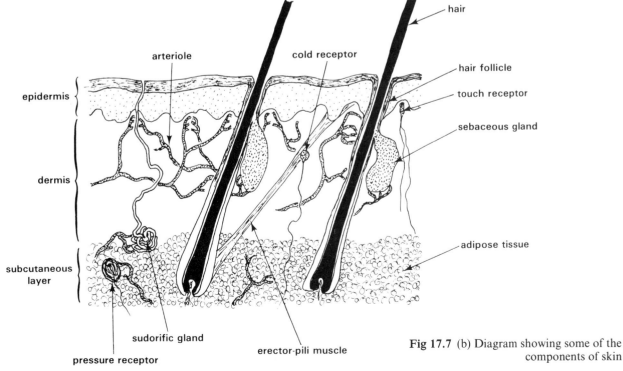

**Fig 17.7** (b) Diagram showing some of the components of skin

### 1 Adipose tissue

Fat is a poor conductor of heat and adipose tissue (Chapter 7) is an effective thermal insulator. The adipose tissue in the skin which insulates against most heat loss makes up **white fat**. Some adipose tissue also makes up **brown fat**. Its cells contain many mitochondria and have a high metabolic rate. Brown fat can be an important generator of body heat. Thin people often have relatively large amounts of brown fat. People who have less brown fat are often overweight and have extensive reserves of white fat. It may be that the body deposits more heat-insulating white fat to compensate for the lack of heat production if little brown fat is present.

### 2 Cutaneous blood vessels

Heat is taken into the dermis by blood in numerous arterioles and capillaries. When the volume of blood flowing through the dermis is high, much heat is lost through the epidermis. This happens when the arterioles in the skin are in a state of **vasodilation** (Figs 17.8(a) and 17.10).

**Fig 17.8** (a) Vasodilation in the skin leading to heat loss through the epidermis

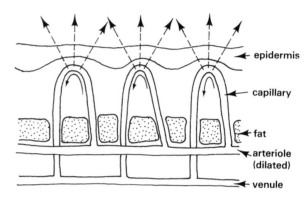

**Fig 17.8** (b) Vasoconstriction in the skin reducing epidermal heat loss to a minimum

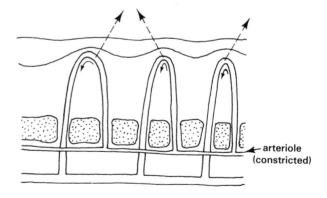

Sympathetic stimulation of the arterioles by nerve impulses from the cold centre of the brain causes **vasoconstriction**, thereby reducing heat loss. Blood flow in human skin can be reduced by vasoconstriction to about one hundredth of the volume which flows when the arterioles are dilated (Fig 17.8(b)). Prolonged constriction of the arterioles resulting from long exposure to intense cold deprives the dermis of the oxygen and nutrients it requires to maintain its metabolic functions. The tissues in the skin may then die or degenerate. This is the cause of **frost bite**.

### 3 Hair

Most mammals are covered with **hair** which helps thermoregulation by keeping a layer of air next to the skin. Air is a relatively poor conductor of heat. Since it is kept in contact with the skin and is not readily replaced by cold air from the surroundings, the temperature gradient between the skin and the trapped air is relatively small. Consequently, although some of the heat in the trapped air is lost to the atmosphere, the total heat loss from the skin in reduced (Fig 17.9(a)). The volume of warm air trapped by the hair depends on whether the hairs are erect or flat. In cold conditions the hairs stand on end, providing maximum heat conservation. In warm conditions the hairs lie flat on the skin, allowing maximum heat loss (Fig 17.9(b)). When the atmosphere is hotter than the body, the air trapped under the hair prevents excessive inward heat transfer (Fig 17.9(c)). This is particularly important to mammals living in hot climates.

**Fig 17.9** Variation in heat exchange with the atmosphere caused by elevation and flattening of hair on the skin in:

(a) a relatively cold environment

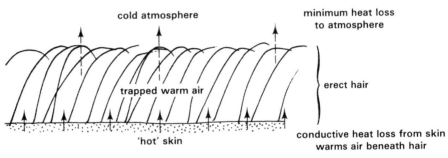

(b) a relatively warm environment and

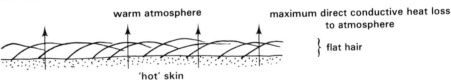

(c) a relatively hot environment

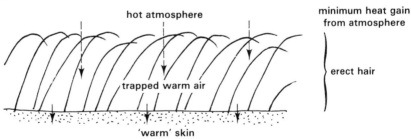

Our bodies do not have abundant hair. We rely on clothes to trap air above our skin, thus keeping us warm in cold weather. Nonetheless, the erector-pili muscles still pull on the hair follicles in our skin. This causes **goose-pimples** in cold weather.

### 4 Sweat

Evaporation of **sweat** from the skin's surface is a means of increasing heat loss. When it is necessary to conserve heat, sweating stops. Activity of the sweat glands is controlled by autonomic nerves. The impulses come from the thermoregulation centres of the brain.

**Fig 17.10** Relationship between evaporative heat loss, skin and hypothalamus temperatures in man following experimental ice meals (after Benzinger)

Benzinger in 1961, experimenting on humans, found that after his subjects had ingested ice, there were strong correlations between fluctuations in evaporation from the body, skin temperature and body temperature (Fig 17.10). Ice in the gut removes heat from the blood. Within a few minutes, a lower blood temperature is detected in the cold centre of the

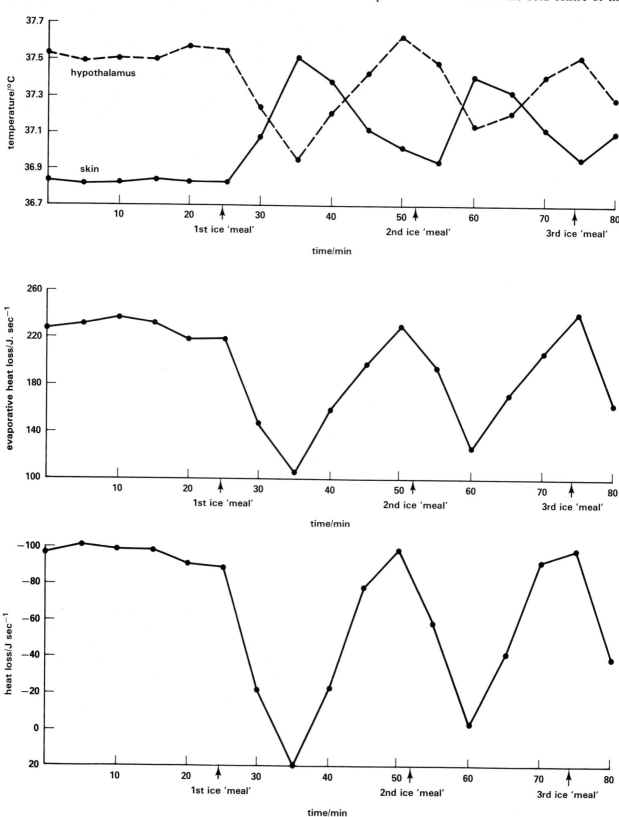

334

hypothalamus. Evaporation from the skin decreases almost at once, hence the skin temperature rises. Within minutes the internal body temperature begins to rise, having initially dropped by only 0.35 °C. After about twenty-five minutes the internal body temperature returns to normal.

Camels and many other mammals living in the desert conserve water at the expense of sweating. Because it sweats so little a camel's temperature can fluctuate by as much as 6 °C.

The temperature of a human does not fluctuate as much as this (Fig 17.11). However, body temperature often rises following an infection. In these circumstances the heat centre responds to blood which is much hotter than normal, by triggering off thermoregulatory mechanisms at about 40 °C compared with 37 °C, the normal body temperature. The maximum upper limit which the human body can survive is about 45 °C and the lower limit is about 24 °C. Above 45 °C, metabolic reactions occur even more quickly, contributing further to body heat. The temperature of the body thus continues to increase indefinitely, causing enzyme denaturation and

**Fig 17.11** Average daily fluctuations in body temperature

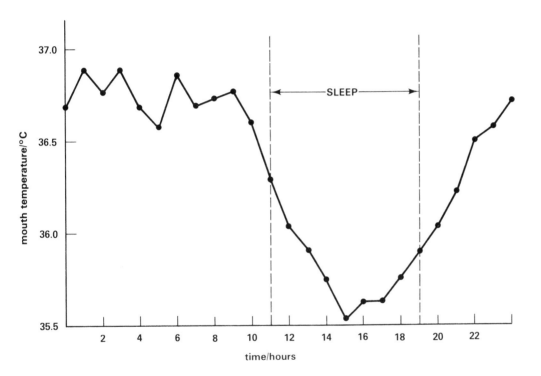

**Fig 17.12** Critical body temperatures

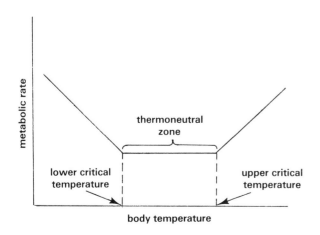

permanent tissue damage. This is an example of a **positive feedback** mechanism which clearly does not contribute to homeostasis. Below 24 °C the body's heat-generating mechanisms fail to work, causing **hypothermia**, a lowering of body temperature (Fig 17.12). Hypothermia is a common cause of loss of consciousness and death in old people who cannot keep themselves warm during cold weather.

Fig 17.13 summarises the mechanisms of thermoregulation in humans.

**Fig 17.13** (a) Summary of the main thermoregulatory mechanisms (A = hot centre, B = cold centre)

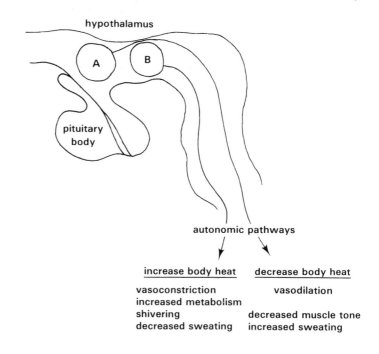

**Fig 17.13** (b) Pathways involved in responses to cold. Compare with Fig 14.1

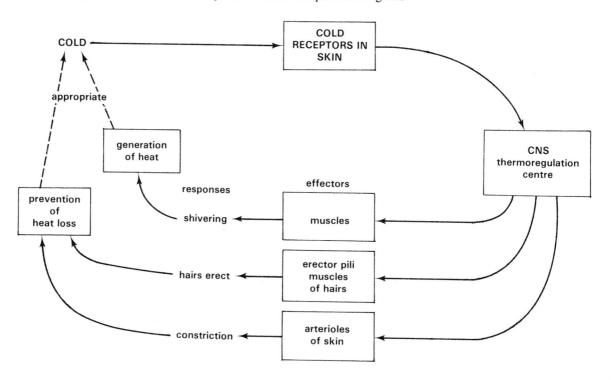

# 18 Reproductive systems

Humans reproduce sexually. For sexual reproduction to occur successfully, gametes have to be mature at the right time. Males and females have to be in the same locality and attracted to each other. Fertilisation must occur and the resulting embryo must be protected and provided with food for it to survive and develop properly.

**Gametes** are made in sex organs called **gonads**, the **testes** in males, the **ovaries** in females. **Gametogenesis** is the process which gives rise to male and female gametes.

## 18.1 The male reproductive system

The production and maturation of **spermatozoa** (sperm), the male gametes, takes place in a pair of compact testes. The testes originate in the abdomen from embryonic tissue which also gives rise to the urinary system. Indeed the urethra, through which urine passes to the outside, also provides the route for sperm to leave the male reproductive system (Fig 18.1). As well as making sperm, the testes produce male sex hormones.

**Fig 18.1** Male reproductive organs (from the side)

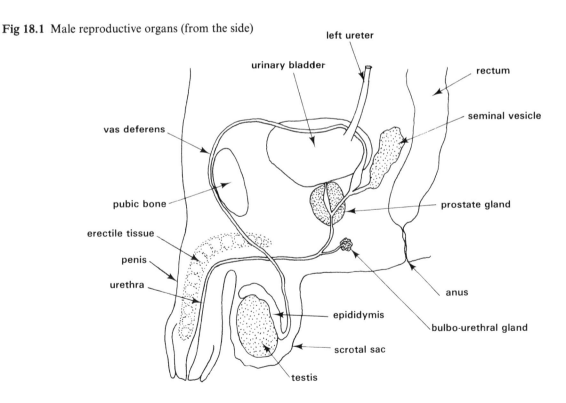

## 18.1.1 Structure of the testes

As they mature the testes descend from the abdomen into two sacs of skin called **scrotal sacs**. Under the microscope a thin section of a testis is seen as thousands of very fine, highly coiled **seminiferous tubules** in which sperm are produced. The seminiferous tubules are continuous with other tubules called the **vasa efferentia, epididymis** and **vasa deferentia** (Fig 18.2). The sperm travel through the system of tubules to the **seminal vesicles** where they are stored before passing out of the body through the urethra. Cutting the vasa deferentia prevents the passage of sperm to the outside. The operation, which is called **vasectomy**, is used to sterilise men and is an important contribution to birth control.

The walls of the seminiferous tubules consist of a layer of cells called the **germinal epithelium** from which sperm originate (Fig 18.3). Sperm at different stages of development are found inside the seminiferous tubules.

**Fig 18.2** Vertical section through testis. The seminiferous tubules have been shown in only one lobule

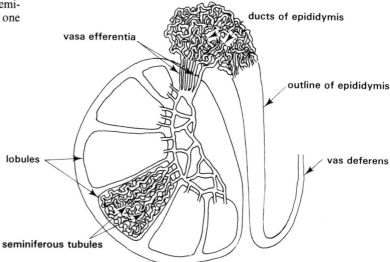

**Fig 18.3** Photomicrograph of a seminiferous tubule seen in transverse section, × 400

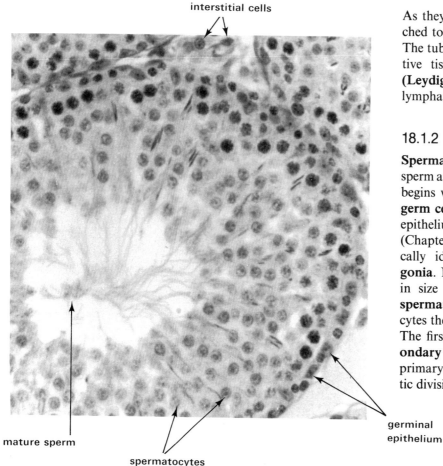

As they mature the sperm become attached to the relatively large **Sertoli cells**. The tubules are held together by connective tissue which contains **interstitial (Leydig) cells**, nerve fibres, blood and lymphatic capillaries.

### 18.1.2 Spermatogenesis

**Spermatogenesis** is the process whereby sperm are made. The production of sperm begins with divisions of the **primordial germ cells** which make up the germinal epithelium. The first divisions are mitotic (Chapter 6) and produce many genetically identical cells called **spermatogonia**. Next the spermatogonia increase in size and mass to become **primary spermatocytes**. The primary spermatocytes then divide by meiosis (Chapter 6). The first meiotic division produces **secondary spermatocytes**, two from each primary spermatocyte. The second meiotic division gives rise to **spermatids**, two from each secondary spermatocyte (Fig 18.4). The final stage of spermatogenesis is differentiation of spermatozoa from the spermatids. Maturing spermatozoa cluster around the Sertoli cells from which they are thought to obtain materials essential for differentiation.

**Fig 18.4** Spermatogenesis

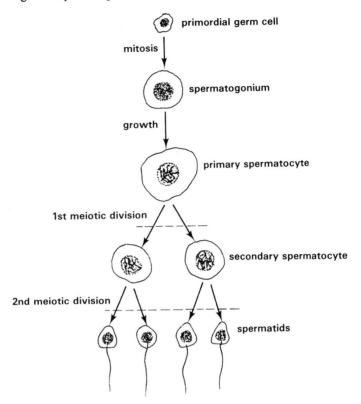

**Fig 18.5** (a) Diagram of a sperm

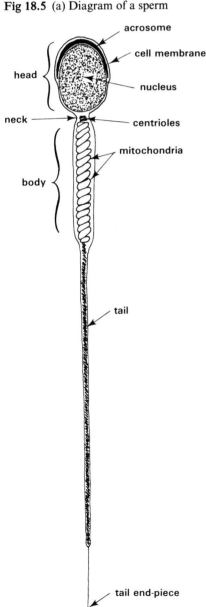

A mature spermatozoon has a **head** which contains a nucleus. Attached to the head is a long **tail** (flagellum). At the base of the tail where it joins the sperm head, there are many mitochondria (Fig 18.5). They provide energy for contraction of protein filaments in the flagellum which consequently undulates. It provides the propulsive force for spermatozoa to swim. Sperm can swim at a rate of 1–4 mm a minute.

Spermatogenesis is inhibited by heat. Because the testes are located in the scrotal sacs where the temperature is lower than in the abdomen, production of sperm is normally encouraged.

**Fig 18.5** (b) Photomicrograph of some human sperm, × 1000

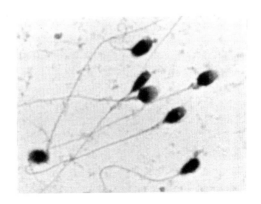

339

**Fig 18.6** Mechanism controlling testosterone secretion

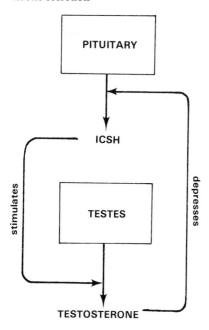

**Fig 18.7** Structure of testosterone

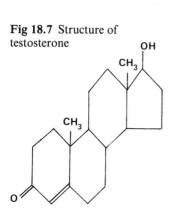

**Fig 18.8** Female reproductive organs (ventral view)

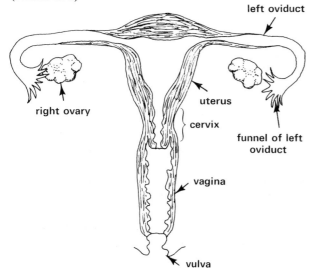

### 18.1.3 Male hormones

Spermatogenesis is stimulated by a gonadotropin hormone called **follitropin** produced by the anterior pituitary (Chapter 16). In both male and female follitropin stimulates gametogenesis.

Another gonadotropin hormone made in the anterior pituitary is **lutropin**. In the male, lutropin is alternatively called **interstitial cell stimulating hormone (ICSH)**. As the name suggests ICSH stimulates the interstitial cells to secrete a male sex hormone called **testosterone**. ICSH and testosterone interact in a negative feedback control mechanism so that the output of testosterone is kept relatively constant (Fig 18.6).

Male sex hormones are collectively called **androgens**. The functions of androgens are twofold. First, they regulate the development of the male **accessory sex organs**. They include the vasa efferentia, epididymes, vasa deferentia, penis and scrotal sacs. The second main function of androgens is to control the development and maintenance of the **secondary sexual characteristics** which in man include growth of facial and pubic hair, a deep voice and general muscular development. Androgens may also be partly responsible for a number of behavioural characteristics. Removal of the testes, castration, does not usually lead to loss of secondary sexual characteristics as androgens are steroids (Fig 18.7) and following castration there is increased output from the adrenal cortex of steroids similar to androgens (Chapter 16).

The development of male secondary sexual characteristics and the production of sperm begin in the human at a period of life called **adolescence**. The point at which sexual maturity is reached is called **puberty**. The testes also produce female sex hormones; their function in males is not clear.

## 18.2 The female reproductive system

Female gametes are called **ova**. Their production and maturation take place in the **ovaries**. The ovaries also produce female sex hormones. A pair of ovaries lies in the abdominal cavity. The ova they release are sucked by ciliary action into funnels from where they move into the **oviducts** and **uterus**. If an ovum is not fertilised it passes into the **vagina** and to the exterior through the **vulva** (Fig 18.8).

### 18.2.1 Structure of the ovaries and oogenesis

Spermatogenesis in the testes (section 18.1.2) is paralleled in the ovaries by **oogenesis**, the production of ova. Unlike the testes, ovaries do not contain tubules but consist mainly of connective tissue. Like the testes, however, ovaries have a germinal epithelium. Ovarian germinal epithelium is on the outside of the ovaries. **Primordial germ cells** in the germinal epithelium divide by mitosis to form many **oogonia**. The oogonia become surrounded by **follicle cells**, also derived by mitosis from the germinal epithelium. The follicle cells and enclosed oogonia are called **primary follicles**. They migrate to the centre of the ovary.

The development of the secondary sexual characteristics is brought about by the sex hormones. How are the characteristics maintained if the gonads are removed?

At birth there are up to 400 000 primary follicles in a human ovary, where they remain dormant until puberty. Hormones from the pituitary gland then start the process of oogenesis. In a woman only about 400 primary follicles normally nature over a reproductive lifetime of about 30–40 years. The end of this period, when follicle development stops, is called the **menopause**. The remaining primary follicles degenerate into cyst-like **atretic follicles** and remain in the ovaries.

Ova usually develop one at a time. Each oogonium grows into a **primary oocyte** which by this stage is surrounded by several layers of follicle cells. Between the dividing follicle cells, cavities appear, filled with follicular liquid. The follicle becomes surrounded by two layers derived from the ovarian connective tissue. An outer fibrous **theca externa** encloses an inner vascular **theca interna**. The mature follicle is called a **Graafian follicle** and can usually be seen bulging at the surface of the ovary (Fig 18.9). During

**Fig 18.9** Photomicrograph of a thin section of ovary, from a rat, × 50

**Fig 18.10** Oogenesis. Compare with Fig 18.4

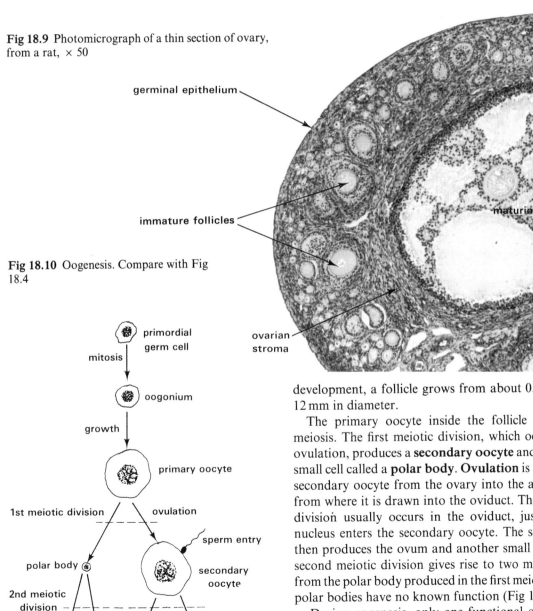

germinal epithelium

immature follicles

maturing follicle

ovarian stroma

primordial germ cell

mitosis

oogonium

growth

primary oocyte

1st meiotic division    ovulation

sperm entry

polar body

2nd meiotic division

secondary oocyte

polar bodies

mature ovum

development, a follicle grows from about 0.05 mm to about 12 mm in diameter.

The primary oocyte inside the follicle now divides by meiosis. The first meiotic division, which occurs just before ovulation, produces a **secondary oocyte** and attached to it, a small cell called a **polar body**. **Ovulation** is the release of the secondary oocyte from the ovary into the abdominal cavity from where it is drawn into the oviduct. The second meiotic division usually occurs in the oviduct, just after a sperm nucleus enters the secondary oocyte. The secondary oocyte then produces the ovum and another small polar body. The second meiotic division gives rise to two more polar bodies from the polar body produced in the first meiotic division. The polar bodies have no known function (Fig 18.10).

During oogenesis, only one functional ovum arises from each primary oocyte, whereas spermatogenesis produces four functional sperm from each primary spermatocyte (section 18.1.2).

Fig 18.11 Mechanism controlling estradiol secretion

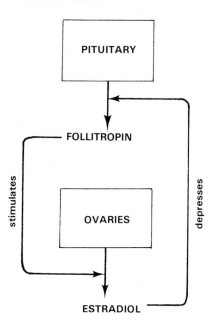

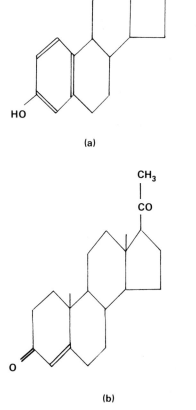

Fig 18.12 Structures of (a) estradiol and (b) progesterone. Compare with Fig 18.7

## 18.2.2 Hormonal control of ovulation

As in the male, gametogenesis in the female is influenced by hormones from the anterior pituitary, particularly follitropin. Unlike spermatogenesis however, oogenesis is not a continuous process. As in other mammals the embryo develops, prior to birth, inside the mother's reproductive tract during the **gestation period**. Since there is limited space in the mother for embryo development, there is a limit to the number of embryos which can develop at a time. The number of embryos carried by a mother during the gestation period is usually determined by the number of ova released at a time from the ovaries. In a woman, only one ovum normally leaves one of the ovaries each month at ovulation. Ova are regularly produced by a mechanism in which follitropin interacts with ovarian hormones to produce a rhythm of activity called the **ovarian** or **menstrual cycle**.

Since the ovarian cycle is a sequence of events which repeats itself in a cyclical fashion, there is no fixed beginning or ending. For convenience of description we shall begin with the onset of follicle development.

Follicle development is stimulated by follitropin from the anterior pituitary. Follitropin also causes the ovary to secrete a female sex hormone, an **oestrogen**. The most prominent oestrogen is called **estradiol**. Estradiol has several effects in the female body, some of which are described later (section 18.2.3). Among these is a negative feedback effect on follitropin (Fig 18.11).

Another effect of estradiol is to cause the anterior pituitary to secrete **lutropin** which brings about ovulation. Ovulation usually occurs 14 days after a follicle starts to form. Lutropin has a second function which is to stimulate the development of the **corpus luteum** from the remains of the ovarian follicle. As it grows, the corpus luteum secretes a hormone called **progesterone** which is similar in structure to estradiol and testosterone (Figs 18.12 and 18.7). Progesterone stimulates growth of the **endometrium**, the lining of the uterus, and its blood supply. Progesterone also inhibits the release of follitropin and lutropin from the anterior pituitary (Fig 18.13). What happens next depends on whether or not the ovum has been fertilised and implanted in the uterine wall.

If implantation occurs, then a new source of progesterone develops, the **placenta** (Chapter 19). Continued production of progesterone inhibits secretion of follitropin, so that the development of new follicles during gestation is prevented. This is an important aspect of pregnancy. If follicles continued to appear and ovulation occurred after the first conception, it might happen that other embryos would be conceived later on. A number of embryos at different stages of development would then be present in the uterus, and a severe physiological burden would be placed on the mother. In addition, relatively violent muscular contractions of the uterine wall occur at birth. Embryos which are not ready to be born are unlikely to survive the contractions.

Maintaining progesterone secretion during pregnancy is important for other reason. Progesterone stimulates growth and improves the blood supply of the endometrium. Maintenance of the

Fig 18.13 Mechanism controlling progesterone secretion

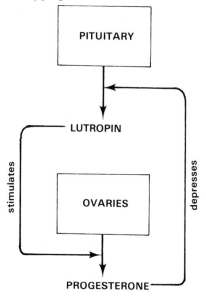

What prevents the continuing production of ova in the ovaries if an ovum is fertilised? Why is this important? What enables ova production to restart in the mother after the birth of her child?

endometrium is necessary for successful implantation of a fertilised ovum and subsequent attachment of the placenta.

If fertilisation does not occur, the lack of lutropin from the anterior pituitary causes the corpus luteum to degenerate, and progesterone production by the ovary then stops. Growth of the endometrium which is promoted by progesterone cannot now be maintained. Consequently, the endometrium breaks down and passes through the vagina to the outside as the **menstrual flow**. Menstruation normally lasts for about four or five days.

Lack of progesterone in the blood also means that secretion of follitropin from the anterior pituitary can begin once more. The cycle restarts with the development of a new follicle. About 14 days elapse between ovulation and the start of a new cycle. With ovulation marking the mid-point, the ovarian cycle takes about 28 days to complete. The length of the cycle varies from woman to woman, and may be different for the same person at different times.

Unwanted pregnancies can be prevented if a woman takes the contraceptive pill. The pill contains chemicals which have effects similar to estradiol and progesterone. In particular, they prevent secretion of follitropin from the anterior pituitary. Consequently no follicles develop in the ovaries and fertilisation is impossible.

Fig 18.14 summarises the hormonal interactions which regulate the ovarian cycle and their effects on the ovary and uterus.

**Fig 18.14** Summary of the main factors involved in the ovarian cycle

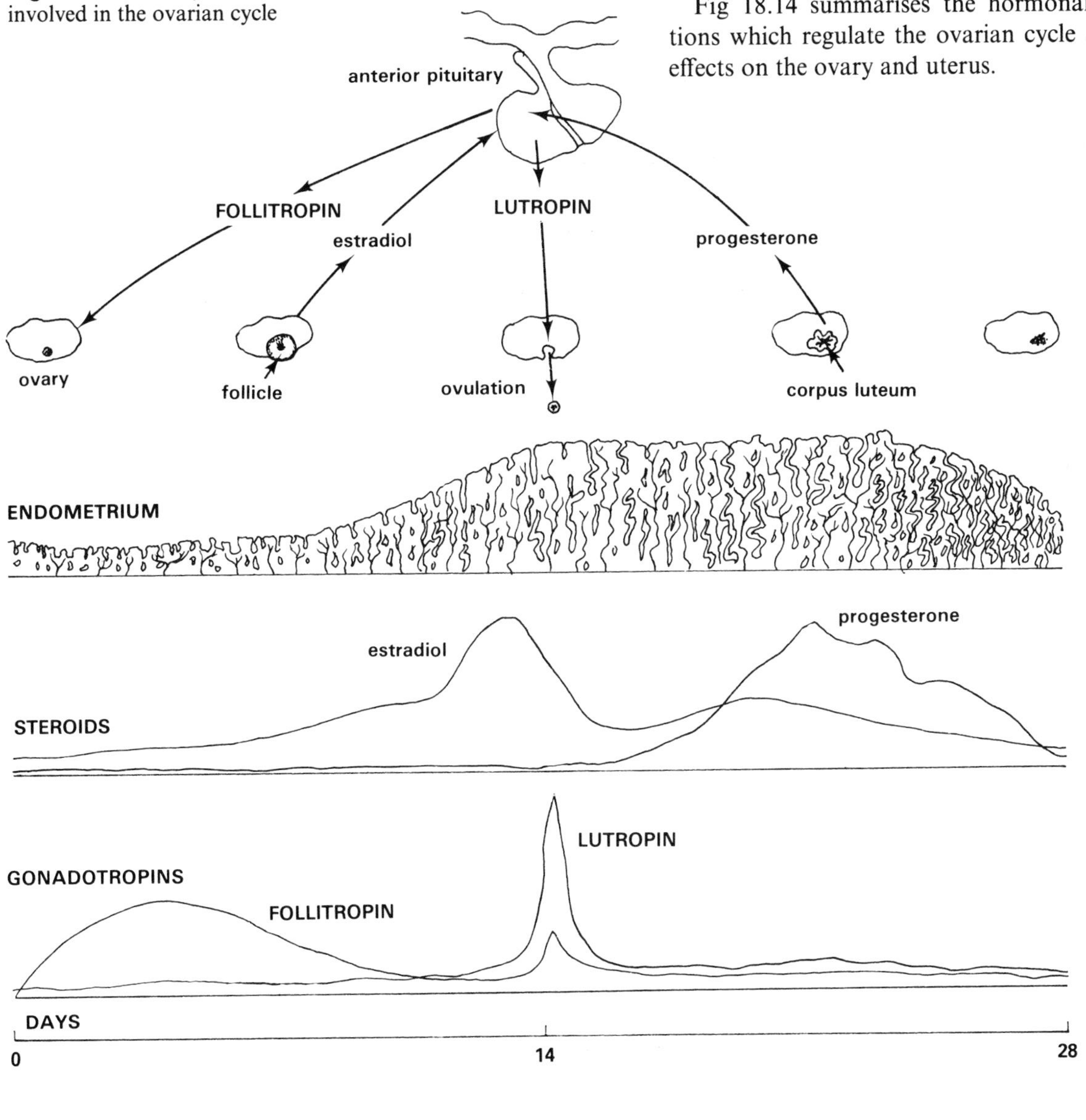

### 18.2.3 Other effects of female hormones

Like androgens in males, oestrogens in females regulate the development of the accessory sex organs. These organs include the **oviducts, uterus, vagina** and **clitoris**. Oestrogens also control the development and maintenance of the secondary sexual characteristics which in women include growth of pubic hair, development of the mammary glands and a relatively broad pelvic girdle.

It has been known for many years that the ovaries secrete androgens as well as oestrogens and that testes secrete oestrogens as well as androgens. The roles of androgens in females and oestrogens in males remain a mystery. However, an important effect of oestrogens in both sexes has been discovered. After the menopause, the skeleton of women can diminish in mass by as much as 2% each year. Oestrogens inhibit the action of **parathyrin** which stimulates the absorption of calcium phosphate from bone into blood and tissue fluid (Chapter 16). After the menopause, when oestrogen secretion stops, output of parathyrin is uninhibited and bone deterioration sets in. Women who have oestrogen treatment during and after the menopause have significantly fewer bone fractures compared with women of a similar age who have not received oestrogen. Men have no such problem of bone deterioration because their testes secrete oestrogens as well as androgens throughout life.

## 18.3 Reproduction

Fertilisation and subsequent development of the embryo occur inside the mother's body. Internal fertilisation requires the introduction of sperm from the male into the female's reproductive tract where the ova are found after ovulation. The transfer of sperm from male to female takes place in an elaborate pattern of behaviour called **copulation**.

### 18.3.1 Copulation

Before copulation the male and female usually interact so that each partner becomes sexually aroused. Sexual excitation of the male results in erection of the penis. The penis is made mainly of spongy **erectile tissue**. Entry of blood into the cavities of the tissue causes **erection**. Appropriate parts of the male's parasympathetic nervous system are activated in response to stimuli from the female, triggering off the sequence of events leading to erection. When erect, the penis can be pushed into the vagina, the wall of which also contains erectile tissue. Appropriate excitation of the female causes this tissue to become gorged with blood, enabling the penis to be gripped as it is pushed back and forth during copulation. The to and fro motion of the penis in the vagina is helped by mucus secretions. In the male the secretions come from the **bulbo-urethral gland** at the upper end of the urethra and from mucus glands along the length of the urethra (Fig 18.1). In the female mucus comes from glands inside the vagina.

When sexual excitation reaches a peak in the male, sympathetic nervous stimulation of the genitalia brings about **ejaculation**. Ejaculation is caused by rhythmic constriction of the genital ducts, beginning in the testes and progressing to the penis. The peristaltic wave pushes sperm from the testes to the urethra of the penis and into the vagina. The ejaculated fluid is called **semen**. Semen consists of sperm suspended in mucus and a milky fluid from the **prostate gland**. The prostate fluid stimulates sperm to swim. Most of the ejaculated fluid comes from the seminal vesicles. In man, approximately 350 000 000 sperm are normally present in each ejaculation.

During copulation the female also becomes excited. The to and fro motions of the penis rub against the erect clitoris. When sexual excitement reaches a peak in the female, peristalsis occurs in her reproductive tract. The direction of peristalsis is opposite to that in the male and it may help to carry sperm into the oviducts. In this way sperm are propelled efficiently to the regions where ova are likely to be found.

### 18.3.2 Fertilisation and implantation of the embryo

Following ovulation, secondary oocytes are sucked from the abdominal cavity into the oviducts by the action of cilia lining the ducts. Oocytes die if they are not fertilised between 8 and 24 hours after ovulation. By this time they have been wafted only a short distance down the oviducts. Consequently fertilisation normally takes place in the upper regions of the oviducts. In some women the tubes are blocked and cannot be opened. Until recently such women had no hope of having their own children. However, research has developed a technique whereby an ovum is removed from the woman, fertilised externally and the resulting embryo placed in its mother's uterus. The technique, which was first performed successfully with humans in 1978 in Britain, produces what are popularly called **test-tube babies**. The procedure is now so well perfected that it is possible to determine the sex of the embryo before it is placed in the mother. Consequently, such mothers can avoid having children who would probably inherit serious defects.

What are 'test tube babies'?

### 1 Fertilisation

**Fertilisation** involves entry of the sperm nucleus into the secondary oocyte. The oocyte is usually surrounded by many small cells called the **corona radiata**. The corona is dispersed by the enzyme **hyaluronidase**, secreted by the sperm when they arrive near an oocyte. Although an oocyte is approached by many thousands of sperm, only one usually enters it. The head of the successful sperm penetrates the oocyte's outer membrane, and the sperm tail is left outside. The oocyte's membrane now becomes impermeable to the entry of other sperm. Entry of additional sperm is undesirable as it could produce a zygote with more chromosomes than normal. This could have adverse consequences for the embryo (Chapter 20).

When the sperm head enters, the secondary oocyte undergoes the second meiotic division to become the ovum proper. Only at this point are both gametes haploid. Fusion of the sperm and ovum nuclei produces a **diploid zygote**. It is from the zygote that the embryo develops. Sometimes the zygote divides into two separate cells from which two individuals develop. These individuals are called **monozygotic twins**. Monozygotic twins, coming from the same sperm and ovum, are genetically identical. Small differences appear as the twins develop. **Dizygotic twins** result from two different zygotes produced from two ova fertilised at the same time. Consequently dizygotic twins are genetically different from one another.

### 2 Implantation

The zygote undergoes several divisions, producing a ball of 16 to 32 cells by the time it reaches the uterus. It is in this form that the embryo is **implanted** in the endometrium (Fig 18.15). Peptidase enzymes from **trophoblast cells** on the outside of the embryo digest part of the endometrium, making space for the embryo. The products of digestion are absorbed by the trophoblast and are used as nutrients by the embryo. Meanwhile the trophoblast, along with surrounding endometrial cells, divide to form the placenta and foetal membranes (Chapter 19).

**Fig 18.15** Photomicrograph of a 15-day human embryo embedded in the uterine wall, × 30

# 19 Growth and development

**Growth** may be described as an increase in the size of the body. We may express size in various ways, height, girth, mass or volume of the body. Growth is **quantitative**. **Development** may be described as changes which occur in the structure and functioning of cells and tissues as they grow. Growth and development occur throughout our lifetime, but especially when we are young.

## 19.1 Foetal growth and development

### 19.1.1 Factors affecting growth and development

Growth and development in cells and tissues include the following processes: **hypertrophy**, an increase in cell size; **hyperplasia**, an increase in cell numbers by mitosis (Chapter 6); **differentiation** into different types of cells which perform different functions (Chapter 7); **organisation** of tissues into organs and organ systems (section 19.1.2).

The exact pattern of specialised development followed by any cell is determined by **genetic** and **environmental** factors. With the exception of gametes (Chapter 18), all cells in the body of an individual normally have the same combination of genes. However, the genotypes of individuals differ, and this is one reason why the rate and pattern of growth and development varies among humans. Diet is one of the most influential environmental factors affecting growth and development. Substances which influence growth and development are called **growth regulators** and include **hormones** (Chapter 16) as well as a variety of other substances called **organisers**.

The **growth rate** varies in different organs and tissues and at different times throughout our lives. The fastest growth rate occurs first in the central nervous system (Figs 19.1 and 19.2). In the generalised growth curve, the fastest rate after birth occurs in the **self-accelerating phase**.

**Fig 19.1** Disproportionate growth of different parts of the body throughout life (from Batt after Stratz)

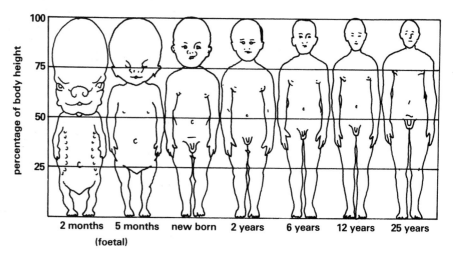

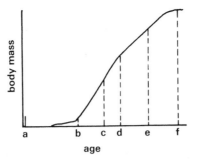

**Fig 19.2** Generalised growth curve, a – conception, b – birth, c – self accelerating phase, d – inflexion point, e – self retarding phase, f – growth plateau (from Batt after Wallace, Palsson and Verges)

Later the growth rate slows down, beginning at the **inflexion point** of the **self-retarding phase**. At the **growth plateau** the growth rate becomes very slow. Fig 19.3 illustrates growth curves for boys and girls based on increases in height and mass.

**Fig 19.3** Growth curve for boys and girls relating age to (a) height and (b) mass

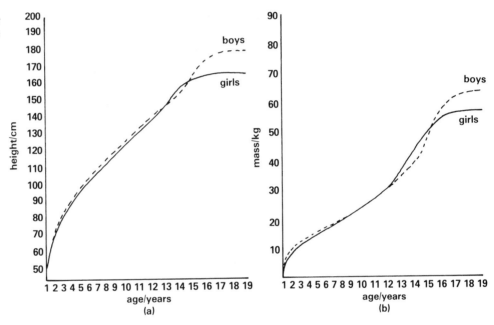

### 19.1.2 Early embryonic development

Following fertilisation (Chapter 18), the zygote divides by mitosis into two cells of unequal size called **blastomeres**. Further **cleavage** divisions form a cluster of 16 to 32 cells called a **morula**. A hollow ball of cells called a **blastocyst** develops from the morula as its cells divide further (Fig 19.4). The **zona pellucida**, which once covered the unfertilised ovum, disappears and the blastocyst becomes attached to the endometrial wall, a process called **implantation** (Fig 18.15).

**Fig 19.4** Fertilisation to implantation during the first week of development

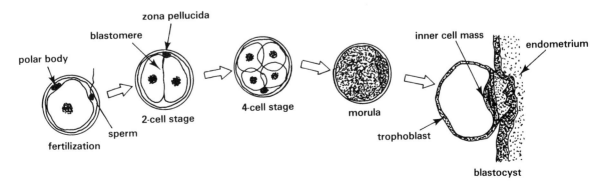

347

The embryo develops from an inner cell mass. The remainder, called the **trophoblast**, gives rise to the extra-embryonic membranes. Part of the inner cell mass becomes organised into a single layer of cells called the **endoderm**, the remainder forms the **ectoderm**. Spaces appear in the ectoderm and join to form the **amniotic cavity**. The now circular embryo consists of a double layer of cells called the **embryonic disc** (Fig 19.5). The endoderm grows around the inside of the trophoblast to form the **yolk sac**.

**Fig 19.5** Embryonic disc

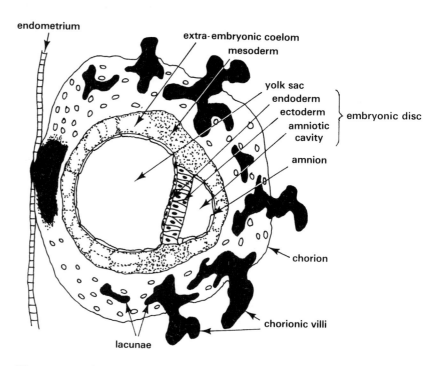

The sac contains yolk in other vertebrate embryos but not in mammals. Nevertheless the name is still used (Fig 19.8). Near the edge of the embryonic disc the endoderm develops into a pouch-like **allantois**. It is far less important in mammalian embryos than in those of other vertebrates such as reptiles and birds where it stores nitrogenous wastes during development (Fig 19.8). The trophoblastic shell grows rapidly to produce a layer of **extra-embryonic mesoderm**. It develops around the inside of the trophoblast and covers the amnion and yolk sac. A **body stalk** joins the embryo to the surrounding trophoblast (Fig 19.6).

**Fig 19.6** Embryo showing body stalk and extra-embryonic mesoderm

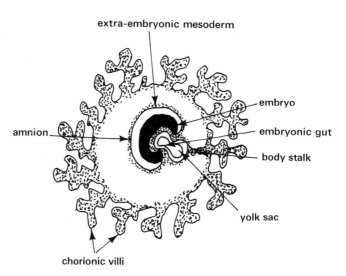

By the fifteenth day the ectodermal cells on two sides of the disc start to move inwards towards each other. The movement causes a thickening, called the **primitive streak** to appear on the disc (Fig 19.7). Ectodermal cells which meet at the streak move downwards into the space between the ectoderm and the endoderm. They become the third germ layer of the embryo, the **mesoderm**. At the head and tail ends of the embryo there are two patches where the mesoderm does not intervene between ectoderm and endoderm. They are called the **buccopharyngeal** and **cloacal membranes**.

**Fig 19.7** Formation of the primitive streak and mesoderm (modified from Dryden)

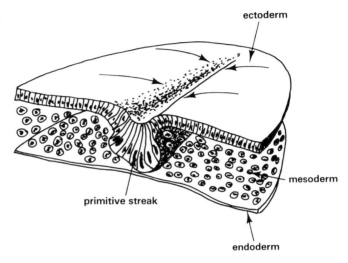

The first mesodermal structure to appear is the **notochord**. It develops later into the intervertebral discs and vertebral centra (Chapter 8).

### 19.1.3 Extra-embryonic membranes

The **amnion** is a layer of cells surrounding the amniotic cavity and originates in the embryonic ectoderm (Fig 19.5). As growth occurs, the amnion surrounds the embryo and joins it ventrally. The stalk connecting the embryo to the trophoblastic shell becomes totally enclosed by amnion. Later the stalk becomes the **umbilical cord** and joins the embryo to the placenta (Fig 19.8).

**Fig 19.8** Developing embryonic membranes

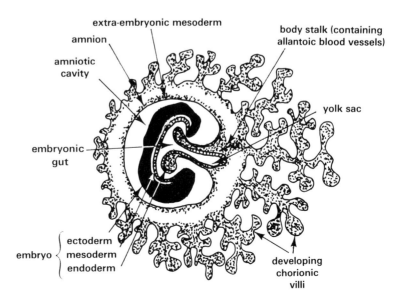

349

The mature amnion consists of two layers derived from embryonic ectoderm and extra-embryonic mesoderm. The **amniotic fluid** increases in volume to about a litre just before birth. Amniotic fluid is produced by the amnion early on. Later, foetal urine accumulates in the amnion. Unusual substances excreted by the foetus can be detected when amniotic fluid is chemically analysed. In this way some inherited foetal abnormalities can be detected before birth (Chapter 20).

Amniotic fluid is swallowed by the foetus from about the fourth month onwards. It is absorbed by the foetal gut and is transported in the blood to the placenta where excess fluid and wastes are transferred to the maternal blood. Amniotic fluid also acts as a shock absorber and allows the foetus to move freely within the uterus. It also helps to maintain a stable foetal body temperature. Microscopic examination of the chromosomes of foetal cells in amniotic fluid is useful in determining certain genetic abnormalities (Chapter 20).

The trophoblastic shell and its extra-embryonic mesodermal lining constitute the **chorion**. During implantation, cavities called **lacunae** appear in the trophoblast and in the endometrium of the uterus (Fig 19.5). The endometrial cells store glycogen and lipids. The lacunae join to form large cavities which fill with maternal blood. Cords of cells from the trophoblastic shell move outwards to form **villi** which push in between the blood-filled lacunae (Fig 19.8). The cells at the tips of the villi divide rapidly so that each lacuna becomes lined with trophoblastic tissue. Blood capillaries appear inside the villi and the extra-embryonic mesoderm lining the trophoblastic shell. At the same time, blood vessels form in the embryo and body stalk. Eventually the embryonic and extra-embryonic vessels join to form a single circulatory system. It is at this stage, when the embryo is about three weeks old, that the heart beings to develop (section 19.1.5).

The chorion and the endometrium in which it is embedded, develop into the **placenta**. The chorionic villi grow and branch into the lakes of maternal blood (Fig 19.9). Foetal blood is delivered to and drained from the placenta by two **umbilical arteries** and an **umbilical vein** (Figs 19.9 and 19.18). Normally foetal and maternal blood do not mix but are separated by a very thin membrane about $2 \mu$m across. Microvilli are found on the surfaces of the cells, thus greatly increasing the area for rapid exchanges of metabolites between the two circulations. Transport across the placental membranes includes diffusion of respiratory gases and active transport of other substances such as nutrients.

**Fig 19.9** Structure of the placenta (from Wendell-Smith)

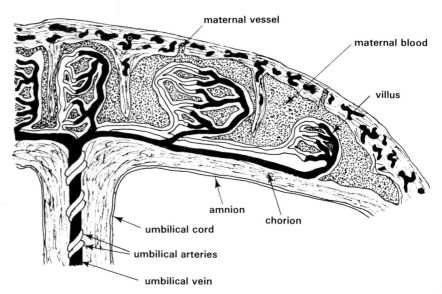

The placenta performs a number of vital functions. They include the exchange of respiratory gases and the transfer of nutrients and waste substances such as urea. The placenta also helps protect the foetus from the mother's immune system. The foetus is not rejected as a transplant would be.

Immediately after implantation, some of the placental cells secrete **human chorionic gonadotropin, HCG**. The hormone helps to maintain the corpus luteum. Continued secretion of estradiol and progesterone from the corpus luteum prevents menstruation. HCG appears in the urine and may be detected after about two weeks of pregnancy. Modern pregnancy tests are based on the detection of HCG in urine (section 11.1.3). **Progesterone** released from the placenta replaces ovarian progesterone when the corpus luteum regresses after about three months of pregnancy.

### 19.1.4 Development of the central nervous system

The ectoderm above the notochord becomes thickened to form a **neural plate** along the length of the embryo. A longitudinal groove appears in the neural plate and the sides grow up to form a **neural tube**. It eventually becomes the central nervous system by a process called **neurulation** (Fig 19.10). Closure of the neural tube begins in the region which will become the midbrain. When the anterior portion of the neural tube is completely closed, it swells producing distinct cranial and caudal portions of the neural tube. In the caudal portion, primitive neurons called **neurocytes** develop into a **mantle layer**. It later becomes the grey matter of the spinal cord (section 14.2.2). The neurocytes produce fibres which extend to surround the mantle as a **marginal layer**. Myelination (section 7.3.3) of the fibres leads to a white appearance. The marginal layer thus becomes the white matter of the spinal cord (Fig. 19.11). The various non-neuronal cells of the spinal cord (section 7.3.4) develop in the early embryo.

**Fig 19.10** Neurulation (modified from Dryden)

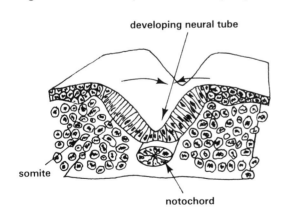

**Fig 19.11** Embryonic spinal cord (modified from Dryden)

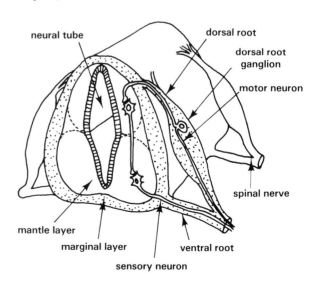

Three vesicles develop in the early brain (Fig 19.12). The **forebrain vesicle** subdivides into two **cerebral vesicles** and a region between them called the **diencephalon**. The **hindbrain vesicle** divides into a cranial portion which gives rise to the **pons varolii** and the **cerebellum**. The caudal portion gives rise to the **medulla oblongata** (section 14.2.4). The lumen of the brain vesicles becomes the brain ventricles. They fill with **cerebrospinal fluid, CSF**.

**Fig 19.12** Main stages in the development of the embryonic brain (modified from Wendell-Smith)

## 19.1.5 Development of the heart and blood vessels

The heart develops from two vessels called **endocardial tubes**, one on each side of the embryonic head (Fig 19.13(a)). When the neural tube closes at the cranial end, the embryonic disc moves ventrally beneath the body. The developing heart thus comes to lie under the neural tube and the foetal gut. The right and left endocardial tubes now fuse (Fig 19.13(b)). Anteriorly the heart is linked to vessels which will become arteries, posteriorly the vessels

**Fig 19.13** Early development of the heart. The plane of section is shown for each diagram. (a) Endocardial tubes. (b) Fusion of endocardial tubes after neurulation (from Dryden)

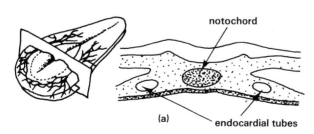

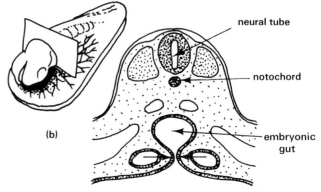

become veins. The single endocardial tube develops four chambers in line. They are the **sinus venosus, atrium, ventricle** and **bulbus cordis** (Fig 19.14(a)). The sinus venosus receives blood from the body of the embryo, yolk sac and the developing placenta through the **common cardinal vein, vitelline veins** and **umbilical vein** respectively.

**Fig 19.14** Main stages in the development of the heart: (a) four chambers 'in line'; (b) and (c) atria rise up behind ventricles; (d) four chambers in final positions

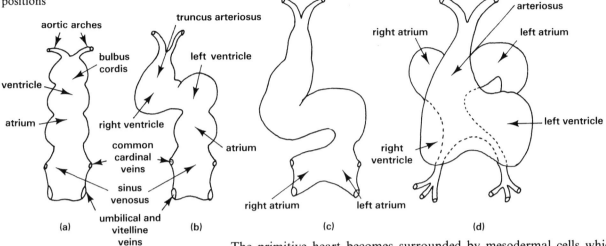

The primitive heart becomes surrounded by mesodermal cells which develop into cardiac muscle. Beating begins between the second and fourth weeks of gestation. The heart is attached at each end to an enclosing membrane called the **pericardium**, in which it is freely suspended. During development the heart twists into an S-shape and the ventricle and atrium each become two chambers. The atria rise up behind the ventricles (Figs 19.14(b) and (c)) and the bulbus cordis becomes a large vessel called the **truncus arteriosus** which fits into a groove between the ventricle (Fig 19.14(d)).

Internally the heart wall begins to grow inwards to separate the four chambers. A thin plate of tissue called the **septum primum** grows down from the top between the two atria, while the **interventricular septum** grows upwards. Two lateral septa separate the atria above from the ventricles beneath. All four septa meet and fuse in the centre. A perforation called the **foramen ovale** appears in the septum primum, enabling blood to pass freely between the atria (Fig 19.15). Another septum called the **septum secundum** grows down from the top. An internal septum grows to divide the truncus arteriosus into two vessels. They twist over each other so that the blood from the left ventricle enters the **aorta** and blood from the right ventricle enters the **pulmonary artery**.

**Fig 19.15** Development of cardiac septa: (a) and (b) septum primum between the atria and interventricular septum; (c) and (d) septum secundum

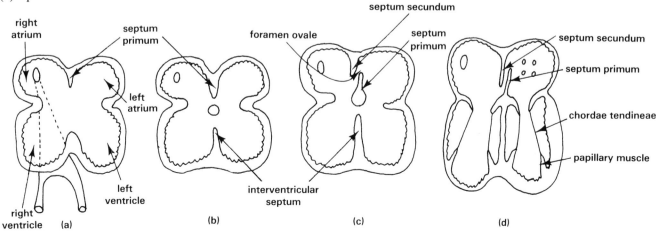

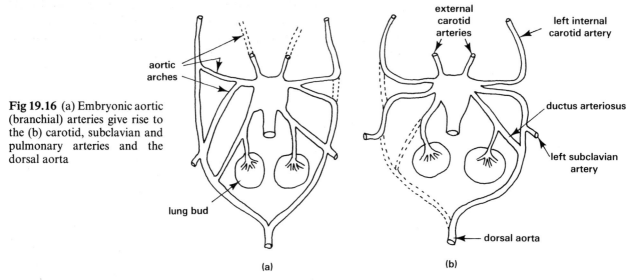

**Fig 19.16** (a) Embryonic aortic (branchial) arteries give rise to the (b) carotid, subclavian and pulmonary arteries and the dorsal aorta

The aorta divides into a series of **branchial arteries** supplying gill clefts in the neck (Fig 19.16). Gill clefts develop in the embryos of many vertebrate groups including fish where they are retained in adult life for gas exchange. Embryological evidence of this kind suggests a common ancestry of all vertebrates and supports the theory of evolution (Chapter 21). In mammalian embryos, the gill clefts and branchial arteries soon disappear and the head and neck receive blood from the aorta in the carotid and subclavian arteries respectively. The veins of the embryo gradually merge into their adult trunks. Most of the veins on the embryo's left side disappear or become attached to those on the right side (Fig 19.17).

**Fig 19.17** (a) Embryonic veins give rise to (b) final venous drainage from the heart

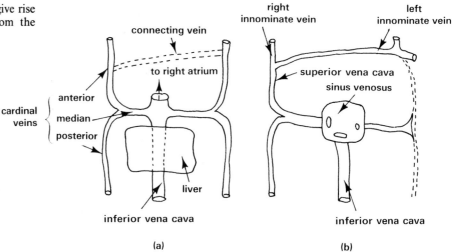

There are several major structural differences between the circulatory system in a mature foetus and an adult (section 10.2.6). The differences are accounted for because the lungs, kidneys and gut are largely non-functional in the foetus. It obtains its oxygen and nutrients and eliminates most of its wastes across the placenta. Foetal blood is delivered to the placenta by two umbilical arteries, contained in the umbilical cord which arises from the abdomen of the foetus. The umbilical arteries are branches of the left and right foetal iliac arteries (Fig 19.18). In the placenta the foetal blood flows in capillaries that project into intervillous spaces containing the mother's blood (Fig 19.9). The foetal blood then drains along a single umbilical vein inside the umbilical cord and back to the foetus. Some of the returned blood is diverted into the foetal liver along the hepatic vein. It is in

354

the liver that most of the foetal blood cells are produced (section 10.1.1). Most of the blood from the umbilical vein enters the inferior vena cava through a vessel called the **ductus venosus**. From here the blood returns to the right atrium of the foetal heart (Fig 19.18).

**Fig 19.18** Diagram of (a) foetal circulation compared with that of (b) an adult

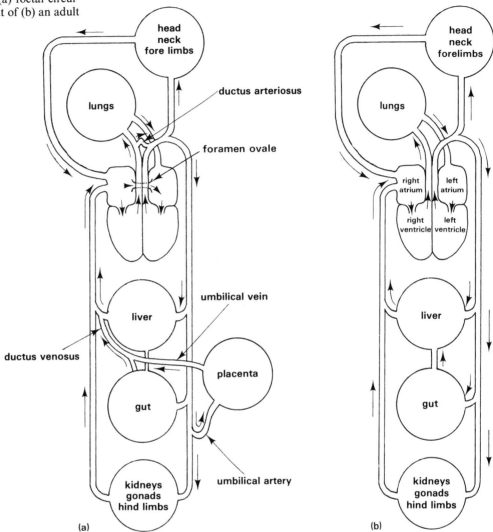

In the foetus, little blood flows to the lungs. Much of the blood in the right atrium, including oxygenated blood from the placenta, flows directly into the left atrium through the foramen ovale in the interatrial septum. A valve in the inferior vena cava directs about 30 % of the blood into the foramen ovale. The rest enters the right ventricle and is pumped into the pulmonary trunk. It is connected to the aorta by a vessel called the **ductus arteriosus**. Consequently, nearly all the blood from the right ventricle enters the systemic circulation and bypasses the lungs (Fig 19.18).

At birth, several major changes transform the foetal circulation to the adult pattern. They coincide with the functioning of the lungs, kidneys and gut and include:

1 Atrophy (degeneration) of the umbilical arteries and vein.
2 Removal of the placenta as the afterbirth.
3 Atrophy of the ductus venosus.
4 Closure of the foramen ovale.
5 Closure and atrophy of the ductus arteriosus.

What major changes occur in the circulatory system at birth? What are the consequences for the baby if the changes do not occur properly?

Some of the changes do not always take place properly. If the ductus arteriosus fails to close then blood continues to pass between the pulmonary trunk and the aorta. The condition is called **patent ductus arteriosus** and can be corrected surgically.

**Tetralogy of Fallot** is a combination of four heart defects causing a condition commonly called **blue baby**. Poorly oxygenated blood gives the skin a blue colour. The main features are as follows:

1  A hole in the interventricular septum.
2  The aorta collects blood from both ventricles and not just the left one.
3  Stenosis, narrowing of the pulmonary valve.
4  An enlarged left ventricle (Fig 19.19).

**Fig 19.19** Main cardiac abnormalities associated with tetralogy of Fallot

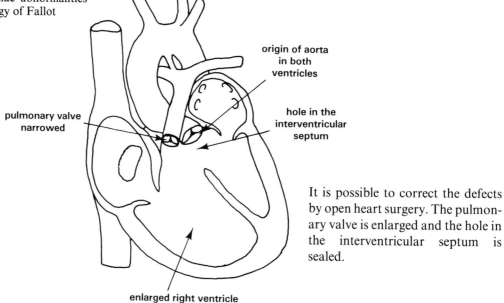

origin of aorta in both ventricles

hole in the interventricular septum

pulmonary valve narrowed

It is possible to correct the defects by open heart surgery. The pulmonary valve is enlarged and the hole in the interventricular septum is sealed.

enlarged right ventricle

### 19.1.6 Development of the skeleton

While the neural plate is forming and closing (Fig 19.10), the mesoderm becomes organised into separate zones including two columns of **somites**. They are solid structures arranged in about 43 pairs either side of the nerve cord (Fig 19.20). Some cells of the somites later contribute to the

**Fig 19.20** Somites (modified from Dryden)

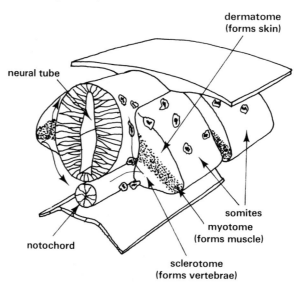

dermatome (forms skin)

neural tube

somites

myotome (forms muscle)

notochord

sclerotome (forms vertebrae)

356

development of muscle and skin. A portion of each somite is called the **sclerotome**. Its cells migrate to surround the nerve cord and the notochord which later develop into the vertebral column. At this stage the notochord extends into a short tail. The development of a tail in embryonic life again points to mammals having a common evolutionary ancestry with other vertebrates.

Each vertebra develops from fusion of the anterior part of one somite and the posterior part of the one in front of it. Vertebrae are thus intersegmental in origin. Spinal nerves project between the vertebrae on each side of the nerve cord (Fig 14.29). The notochord becomes the centrum of each vertebra. Between vertebrae it contributes to the intervertebral discs. Several **primary ossification centres** arise in each vertebra at about nine weeks of age. The primary centres do not fuse until several years after birth. Secondary ossification centres fuse with the rest of the vertebrae at about 25 years of age.

**Fig 19.21** Ossification in a 4-month foetus (modified from Dryden)

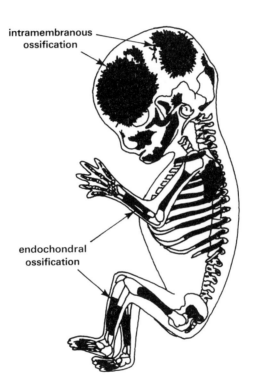

intramembranous ossification

endochondral ossification

Bone develops either by **intramembranous** or **endochondral ossification**. The processes are described in section 8.2.2. Both types of ossification are evident in a four-month-old foetus (Fig 19.21). Intramembranous ossification occurs mostly in the skull. The skull bones are not fused at birth but are separated by spaces called **fontanelles**. They allow the skull to change shape during birth when the head is squeezed through the birth canal (Figs 19.27 and 8.3). Endochondral ossification accounts for the development of the long bones (Fig 8.14) and the pelvic and pectoral girdles (Chapter 8). The clavicle, however, develops by intramembranous ossification. In the limbs the primary ossification centres appear at about eight weeks of age. The secondary ossification centres appear between birth and about twenty years of age (Fig 8.14). Skeletal development must begin in the early embryo to support and protect the soft organs and tissues and to provide anchorage points for muscles to allow movement. However, complete fusion and rigidity of individual bones are delayed until growth and development of the soft organs is completed.

It is not easy to assess the growth of a foetus. Not only is it small and enclosed within the uterus but its growing tissues are very delicate. The use of X-rays presents severe risks of radiation damage, and surgical techniques may cause mechanical injury. In recent years, however, **ultrasound** has been used successfully to assess certain aspects of foetal growth. Ultrasound is sound of a very high frequency, 1 MHz (megahertz) or more. It is far above the upper limit of frequency which we can hear, about 20 kHz (kilohertz) (section 15.2.1). Ultrasound of about 5 MHz is used to investigate foetal structures. In one technique called A-scan, a narrow beam of ultrasound is directed towards the structure being examined, such as the foetal head. Echoes are produced when the beam passes through different features such as the sides of the cranium. The probe used to generate the ultrasound can also detect the echoes. They are converted into vertical deflections of the horizontal electron beam on an oscilloscope screen (Fig 19.22). When suitably calibrated, the oscilloscope trace can be used to measure the distances between foetal structures responsible for the echoes. The distance across the cranium, the **biparietal diameter, BPD** is important. After ten weeks of pregnancy there is a definite relationship between BPD and foetal age (Fig 19.23).

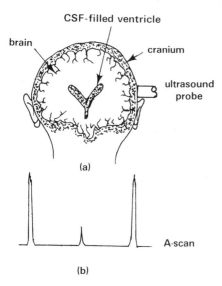

**Fig 19.22** (a) Intracranial structures producing (b) vertical deflections in the ultrasound A-scan

CSF-filled ventricle

brain

cranium

ultrasound probe

(a)

A-scan

(b)

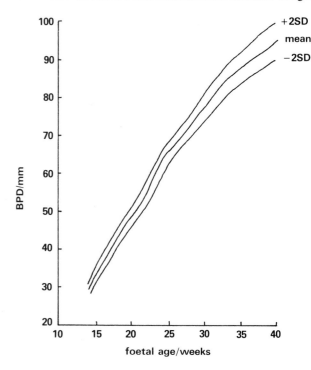

**Fig 19.23** Relationship between biparietal diameter, BPD and foetal age expressed as a range of values $\pm 2$ standard deviations. The range of variation widens as the foetal age increases. Later measurements are thus a less accurate indicator of age.

**Fig 19.24** Ultrasound B-scan showing the foetal head. The two crosses are electronic calipers and are positioned to measure the biparietal diameter. It is 25 mm in this case. Use Fig 19.23 to determine the approximate age of the foetus

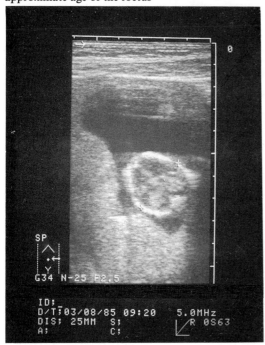

In another technique called B-scan, the ultrasound beam is moved in a fan shape through the foetus. The echoes are detected by the probe and displayed as a two-dimensional picture made of bright dots on an oscilloscope screen. It enables the shape and position of the foetus to be seen.

Modern machines enable doctors to make BPD measurements directly from moving two-dimensional pictures created from ultrasound echoes from the foetus (Fig 19.24).

How can doctors assess the growth of a foetus inside its mother's uterus?

### 19.1.8 Effects of pregnancy on the mother

Considerable physiological as well as anatomical changes occur in the mother during pregnancy. The uterus enlarges greatly, initially because of growth of its muscle fibres and connective tissue. Uterine growth is stimulated in early pregnancy by estradiol. Later uterine enlargement is due to foetal growth (Fig 19.25).

**Fig 19.25** Contours of the uterus at different stages of pregnancy

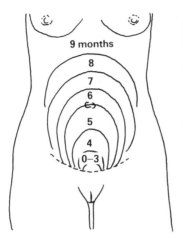

The mother's blood volume goes up by as much as 30 % to accommodate the increased supply required by the enlarged uterus. Elevation of both heart rate and stroke volume (section 10.2.1) causes a rise of up to 33 % in cardiac output. Blood flow increases in the arms and legs, encouraging loss of the excess heat generated by increased metabolism as the foetus grows (Chapter 17).

The enlarged uterus pushes upwards on the diaphragm. However, the thoracic cavity widens, so the tidal volume and pulmonary ventilation of the mother's lungs also increase (section 12.1.4). Delivery of extra oxygen to the placenta is thus ensured by the combined increases in lung and heart activity.

Water usually accumulates in the mother's tissues during later pregnancy. There is a reduction in the concentration of plasma proteins in the mother's blood as its volume increases. Colloid osmotic pressure of the plasma is thus lowered, favouring retention of water in tissue fluids (section 10.3.1). Blood flow and filtration in the kidneys increase. Hence greater quantities of plasma solutes are filtered, some of which the nephrons may be unable to reabsorb. It may account for periodic glucose excretion, glycosuria, during pregnancy (section 16.2.1).

Parathyroid secretion heightens during pregnancy, leading to release of calcium and phosphate from the mother's skeleton. They are transferred to the foetus for growth of its bones (section 16.1.5). About half the calcium phosphate made available in this way is used by the mother in milk production. Iron is required by the foetus and the mother for haemoglobin and myoglobin formation. The total iron content of the adult body is about 3–5 g. A new-born baby has drained about 700 mg iron from the mother and a further 200 mg is required for her to produce milk. Extra dietary intake of iron is required for these reasons (Chapter 9).

Concentrations of cholesterol, phospholipid and fat rise in the mother's blood during pregnancy. Extra adipose tissue (section 7.1.1) is laid down to provide a source of energy for late pregnancy and lactation. Morning sickness often occurs in early pregnancy. Pregnant women often have mild cravings for unusual foods. The causes of these effects are unknown.

The hormone **relaxin** secreted from the placenta causes relaxation of the symphysis pubis between the two halves of the pelvis (Fig 8.7). It permits the cervix to dilate as the uterus grows. **Lactation** is the production of milk by the mammary glands. It is stimulated by the hormone **prolactin** from the anterior pituitary body (section 16.5.1) under the influence of **prolactin**

**releasing factor, PRF**, from the hypothalamus. During pregnancy PRF output rises. However, lactation is inhibited by another hypothalamic hormone called **prolactin inhibiting factor, PIF**. Estradiol and progesterone, initially from the corpus luteum and later from the placenta, stimulate PIF output. After birth when the placenta is removed, PIF secretion subsides and PRF allows prolactin release. During breast-feeding the sucking action of the infant's mouth stimulates the nipple area. Sensory nerve impulses are transmitted to the hypothalamus, resulting in PRF secretion and hence prolactin release. The impulses also cause release of **ocytocin** from the posterior pituitary body (section 16.5.2). Ocytocin encourages contraction of smooth muscles lining the mammary ducts. Milk is thus forcibly expelled into the infant's mouth (Fig 19.26).

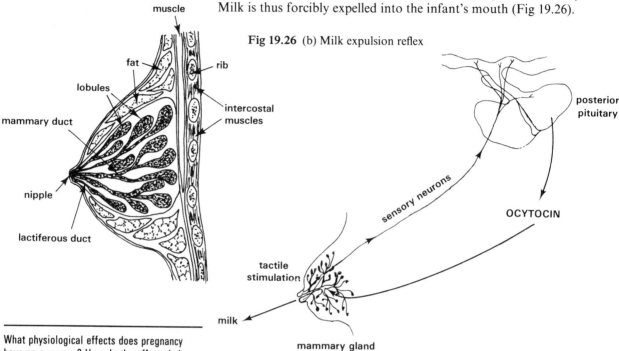

**Fig 19.26** (a) Mammary gland, internal structure

muscle

fat

lobules

mammary duct

rib

intercostal muscles

nipple

lactiferous duct

**Fig 19.26** (b) Milk expulsion reflex

posterior pituitary

sensory neurons

OCYTOCIN

tactile stimulation

milk

mammary gland

---

What physiological effects does pregnancy have on a woman? How do the effects help the foetus to grow?

True **milk** appears a few days after birth. Just before then a fluid called **colostrum** is produced by the mammary glands. Colostrum contains no fat and little lactose compared with milk. It is in colostrum, however, that some kinds of maternal antibodies are transmitted to the baby. They provide important passive immunity against a variety of diseases in early infancy.

### 19.1.9 Birth

The time during which the foetus grows and develops inside the uterus is called **gestation** and usually lasts about nine months. At the end of this period a series of events called **labour** result in **parturition,** that is, **birth** of the baby.

The onset of labour is brought about by several hormones. Among them is ocytocin from the posterior lobe of the pituitary gland. Artificial induction of labour may be brought about by administration of ocytocin into the mother's blood. Ocytocin stimulates rhythmic contractions of the uterine wall. Contractions begin at the top of the uterus and travel downwards in waves. They become regular and produce **labour pains**. The amniotic sac usually ruptures early in labour, releasing amniotic fluid, called 'breaking of the waters'. As labour progresses, the contractions become more and more frequent and intense. The uterine cervix dilates.

**Fig 19.27** Stages of birth viewed from the side. (a) Dilation of cervix, (b) expulsion, (c) delivery of the placenta (afterbirth) (modified from Tortora and Anagnostakos)

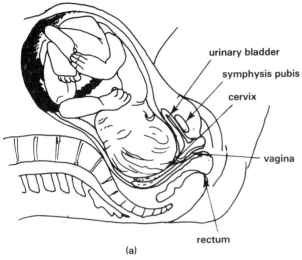

(a)

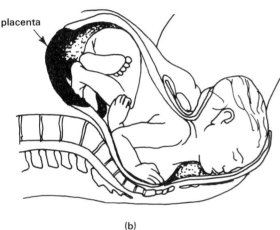

(b)

The baby is pushed downwards, usually head first (Figs 19.27(a) and (b)). Complete expulsion of the baby from the mother is called **delivery** (Fig 19.28).

Finally, the detached placenta, called the **afterbirth**, is expelled by uterine contractions (Fig 19.27(c)). Many damaged blood vessels in the uterine wall are constricted by the final contractions, thus reducing blood loss.

One of the most important changes which occurs in the baby at birth is the establishment of breathing. The lungs must fill with air during the first minute or so of life outside the uterus. An important factor is the presence of a surface active agent (surfactant) in the baby's lungs. It is a lipoprotein which lowers the surface tension of the fluid lining the alveoli, thus making it easier to expand the lungs at inhalation (section 12.1.3). Failure of the lungs of very premature new-born babies to inflate properly is sometimes due to insufficient surfactant production. Such babies suffer from respiratory distress syndrome. They are kept in intensive care until they produce enough surfactant, when their breathing becomes normal.

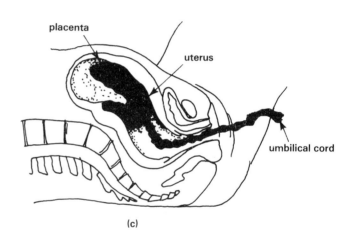

(c)

**Fig 19.28** Delivery (courtesy Camilla Jessel, PAPS)

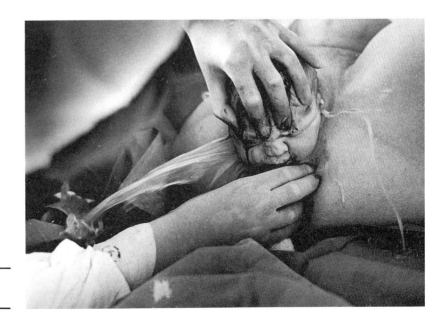

What mechanisms in the body of a pregnant woman bring about the birth of her child?

361

## 19.2 Post-natal growth and development

All the body systems of a new-born baby are immature. The baby is incapable of independent survival and requires post-natal care. As it grows it must continue to be protected from the hazards of its surroundings, provided with food and clothing, and presented with opportunities for learning basic motor and intellectual skills.

### 19.2.1 Childhood

In developed countries the average length of a new-born baby is about 50 cm, the average weight about 3.4 kg (7.5 pounds). Most infants double their birth weight by 4 to 6 months, treble it by twelve months and quadruple it by two years of age. A mature adult is often twenty times or more the birth weight. The relationship to age of body length and mass is different for males and females (Fig 19.3).

The most dramatic changes of infancy coincide with the development of the central nervous system. Several reflexes are usually observed at birth or shortly after, and persist for different periods of time (Table 19.1 and section 14.2.2). Undue persistence of the reflexes, or their absence, may indicate some abnormality in the central nervous system. At three months infants can usually direct their eyes accurately towards objects of interest.

**Table 19.1 Primitive reflexes of the newborn**

| Reflex | Ages at time of appearance | disappearance |
|---|---|---|
| grasp | birth | 4 months |
| head turning (to sound) | birth | — |
| blinking (to visual threat) | 6–7 months | — |
| sucking | birth | 12 months |
| stepping | birth | — |

They recognise their parents and respond differently to strangers, sometimes displaying anxiety and stress. An infant is soon able to manipulate objects with the hands. Unaided walking is usually possible by one year. The emergence of several developmental 'milestones' is illustrated by Fig 19.29.

**Fig 19.29** Some developmental milestones in the first year of life (from Smith, Bierman, Robinson after Johnson, Moore and Ross)

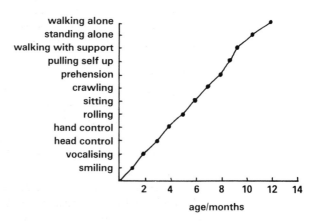

Babies first communicate verbally by crying. Other means of communication, such as smiling, appear at about six weeks. At about three months, broad smiling and chuckling are important means of interaction with parents. Vocal skills begin with simple vowel sounds and, later, single and double syllables such as 'ma', 'da' and 'mama'. An infant is normally able to recognise its spoken name by about nine months, and pronounce a few words by twelve months. At two years the child's vocabulary may include about fifty words. Simple sentences then begin to emerge as the infant learns some rules of language.

The ten-year period from the age of three to early adolescence is marked by rapid physical, mental and emotional growth and development.

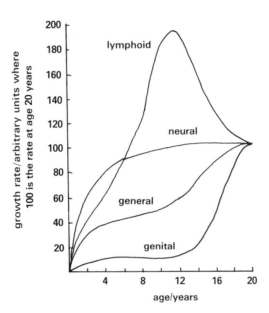

**Fig 19.30** Differential growth rates in some tissues (from Smith, Bierman, Robinson)

Different tissues grow at different rates (Fig 19.30). Body proportions change as the limbs grow faster than the trunk. The skull and brain normally reach approximately adult size by the fifth year. Facial appearance alters as the maxilla and mandible grow rapidly and the milk teeth are replaced by permanent teeth. Further development of the central nervous system leads to increased ability to perform complex tasks requiring a high degree of motor skill and intelligent thought. Most three-year-old children can ride a tricycle. Many four- and five-year-olds can ride a bicycle, dress themselves, button up their coats, select television pro-grammes and even play simple computer games. The eyes of a young child become longer, hence light may be refracted too much before impinging on the retina. For this reason children sometimes go through a phase of myopia. Visual acuity normally improves to 20/20 by the age of five (Chapter 15).

Erikson describes certain **developmental tasks** in children of different ages. In the first year the main task is to establish trust in the world. During the second year the child receives socialising influences such as toilet training and regular sleep and feeding times. An important task is to maintain the child's sense of **self-direction** and to prevent any feelings of doubt or shame. The four-year-old requires a sense of **initiative**. Later, **discipline** and **industry** are important qualities. Children develop reason, the ability to follow logical argument and predict the consequences of their actions. They gain **confidence** and **self-esteem** by performing tasks well and gaining the approval of their parents, teachers and peers. Persistent failure can lead to a poor view of one's self and feelings of inferiority and under-achievement.

**Play** is an important aspect of behavioural and social development. Up to about two years of age, children engage mainly in solitary play. Parallel play follows in which several children may play with the same toys but not interact much with each other. By the age of three, cooperative play appears in which two or more children act complementary roles, such as 'mothers and fathers' or 'doctors and nurses'. Children readily conform to conventions in dress, behaviour, likes and dislikes.

### 19.2.2 Adolescence

**Adolescence** is a period of life in which a **second growth spurt** occurs and **sexual maturity** is achieved. The onset of development of the **secondary sexual characteristics** occurs at **puberty**. In developed countries the average age for the start of puberty is about ten years for girls and about twelve years for boys. The changes are usually complete by the late 'teens. The mechanism that triggers puberty is not known. The gonadotropins **follitropin** and **lutropin** (Chapter 18) increase in output one or two years before puberty. Consequent increase in sex hormone secretion from the gonads and adrenal cortex brings about the changes characteristic of puberty. The growth spurt of puberty is illustrated in Fig 19.31.

**Fig 19.31** Changes in growth rate with age. Note the second growth spurt between 11 and 12 years in girls and 13 to 15 years in boys (from Smith, Bierman, Robinson after Tanner)

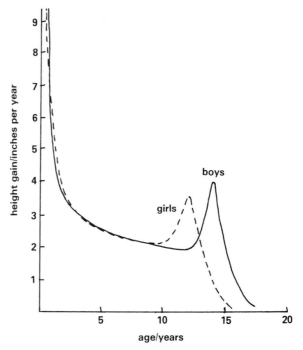

Early signs of puberty in boys include axillary sweating, enlargement of the testes and lengthening of the penis. Later, axillary, pubic and facial hair appear and the voice deepens. In girls the breasts start to develop at the beginning of puberty. About a year after the peak of the growth spurt, menstruation begins, **menarche**. Ovarian cycles do not usually produce ova until about a year later.

Many of the problems of adolescence are psychosocial in nature. Adolescents seek to establish their sense of identity. Ideas are developed about morality and social and personal responsibility. Peer groups form an important base for adolescents who often question generally-accepted values. Sexual awareness and establishment of intimate relationships begin at a time when some adolescents do not have a sound moral awareness or a developed sense of responsibility. Sadly this can lead to unwanted pregnancies and much unhappiness.

### 19.2.3 Adulthood and ageing

**Adulthood** is the period after adolescence when the body is physically and sexually mature. During adulthood most people reach the peak of their life performance in terms of career achievement, material wealth and family responsibility. However, physical and physiological changes continue to occur in different body systems. **Ageing** is accompanied by a deterioration of body functions which can be offset by following a healthy life-style, keeping active and fit and eating a balanced diet. Nonetheless, ageing and eventual **death** are inevitable.

In ageing, the mass of the skeleton declines with loss of mineral content of the bones. The skeleton becomes more fragile and susceptible to fractures and breakages. Skeletal changes are more pronounced in women, probably because of changes in female sex hormone levels in the blood after **menopause** (section 16.1.7). Vertebral compression and loss of muscle tone gradually lead to a shortening of the body and a stooping posture in many old people. The body mass often enlarges due to an increase in adipose tissue, sometimes called 'middle-aged spread'. It may be caused partly by an increasingly sedentary way of life. As we grow older our energy requirement falls. However, food intake does not usually decline to the same extent and fat accumulates. The ability to engage in physically active work declines, due mainly to a reduction in cardiovascular, respiratory and musculo-skeletal efficiency. Many of the physiological changes associated with ageing are illustrated by Fig 19.32.

**Fig 19.32** Decline in several body functions with age (from Smith, Bierman, Robinson after Shock)

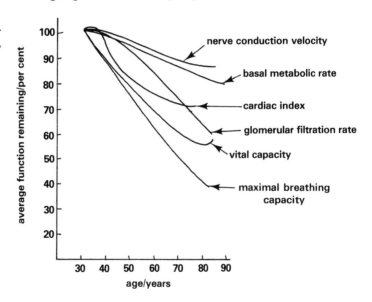

Some of the changes which occur with age can be explained in terms of changes in our issues. The extracellular matrix in connective tissues loses its elasticity and becomes relatively stiff. Elasticity of connective tissues is vital to several body functions. They include lung expansion during breathing and enlargement of the bore of arteries when blood is pumped into them by the heart. Movement of metabolites between the blood and the cells of old connective tissues is also less efficient. Articular cartilage (section 8.2.1) gradually loses its ability to regenerate and may give rise to arthritic problems in middle and old age.

Many cells in the body such as neurons and muscle cells are **non-renewable**. When they die they are lost for ever. Cell loss may contribute to a decline in brain and musculo-skeletal activity with age. Cells may also accumulate a variety of toxic substances as they grow older. Among them are **lipofuscin** and **amyloid**. Their build-up may be due to a decline in lysosome function (section 6.2.4).

**Renewable cells** seem to have a predetermined capacity for division. When they have multiplied a finite number of times, division ceases. In tissue culture embryonic cells divide about fifty times. Cells from a twenty-year-old person divide about thirty times.

Some of the changes which occur with age can be explained in terms of changes at a molecular level. Cell growth and division is regulated by the activities of DNA (Chapter 4). Several factors can cause damage to DNA, leading to the synthesis of abnormal proteins. Excessive body heat may cause disruption of bonds in DNA molecules. Several forms of radiation such as ultraviolet light, gamma-rays and X-rays also damage DNA, as do various chemicals. Such changes cause **gene mutations**.

Cumulative mutations with age account partly for impairment of body activities which depend on protein structure and functions. There is also some evidence that proteins in old cells are less thermostable than those in young cells. Enzymes involved in protein synthesis may also become less specific with age. Consequently, incorrect amino acids may be inserted in the protein chain as it is made.

Improved health care in developed countries has increased the life expectancy of their populations in recent years. In Britain, about 12.5 % of the population were over the age of 65 in 1985. No doubt the quality of life for older people will improve further in future years.

Death may result from any of a variety of possible causes including accident, disease and old age. In all cases, one or more vital organ systems fails and life cannot be sustained.

## 19.3 Measurement of growth

Growth can be measured in a number of ways. We can measure increases in the overall mass of the body as time proceeds (Fig 19.3(b)). Alternatively, we can measure changes in size such as height (Fig 19.3(a)). Neither of these methods is entirely satisfactory since each provides only part of the growth picture. Also, many body systems grow and develop at different rates (Fig 19.30). Body mass may be affected by temporary factors such as obesity and hence may not reflect true growth patterns.

Whatever methods are used to measure growth, we must be able to refer to what is accepted as 'normal'. People differ genetically from each other, even their close relatives. Consequently there is no fixed **growth norm**. We must take account of **variation** in establishing what is 'normal growth' for the population.

One way to determine growth norms is to measure regularly the height and mass of individuals from birth to early adulthood. Such a **longitudinal study** is time-consuming and difficult to complete. It is also difficult to include enough individuals in the study to obtain data representative of the entire population.

Another way to determine growth norms is to measure the height and mass of a large number of children of different ages. The data from such a **cross-sectional study** can then be put together to form a composite growth curve. The only advantage of this approach is that it can be completed in a much shorter time than a longitudinal study. It suffers from the problem of choosing representative samples. Also, various factors such as nutrition which affect growth may change from one generation to another. Because of improved nutrition the average height and mass of children is greater now than it was, say, twenty years ago.

The construction of a growth curve involves recording growth measurements made over an individual's lifetime. How can we construct growth curves without having to wait that long?

Embryonic growth in the uterus is more difficult to measure because the foetus is inaccessible. The recent introduction of ultrasound techniques has made it possible to measure growth of certain body structures. They include the biparietal diameter across the embryonic cranium and the crown–rump length along the foetal back. The growth curves illustrated in Figs 19.2, 19.3 and 19.23 show average data for the UK population.

# SECTION 3

# Heredity and Evolution

# 20 Genetics

Genetics is the science of **inheritance**, the way in which characteristics or traits are passed from parents to their offspring. Because humans reproduce sexually, meiotic nuclear division is an integral part of the life cycle. Meiosis produces gametes with variable combinations of genes (Chapter 6). For this reason children are never identical to their parents.

It is desirable to be acquainted with some commonly used genetic terms before studying the inheritance of characteristics. A **gene** is a part of a molecule of DNA which codes the synthesis of a polypeptide. Protein molecules consist of one or more polypeptides (Chapter 2). The characteristics of organisms are largely determined by the types of protein, especially enzymes, they are able to synthesise. The combination of genes an organism has is called its **genotype** or **genome**. The **phenotype** is the characteristics an organism has. Interaction of the genotype and environment determine the phenotype. Each of the characteristics of an organism is usually determined by two **alleles**. An individual in which both alleles are similar is described as **homozygous** or pure-breeding for that characteristic. Alleles of **heterozygous** individuals are dissimilar. Organisms can be homozygous for some traits and heterozygous for others. Alleles are thus alternative forms of a gene.

## 20.1 Autosomal inheritance

The chromosomes of somatic cells exist as homologous pairs (Chapter 6). In many organisms, including humans, one pair are the sex chromosomes, the rest are called **autosomes**. Gregor Johann Mendel (Fig 20.1), an Austrian monk, was the first to discover how genes carried by autosomes are inherited. He collected the seeds from several varieties of the garden pea and devised a procedure for following the passage of characteristics from parents to their offspring. In 1865 Mendel presented the results of his studies to the local Natural History society. A year later his findings were published in the Annual Proceedings of the Society. They lay neglected until the beginning of this century when several geneticists extended his

**Fig 20.1** (a) Gregor Johann Mendel (Ann Ronan Picture Library)

**Fig 20.1** (b) The monastery at Brno where Mendel worked

methods to other species and confirmed his conclusions. At the time Mendel was alive, little was known of human inheritance. For instance, it was not until 1901 that Garrod worked out the way in which the human characteristic alkaptonuria (Chapter 4) is inherited. A few years later, Landsteiner's discovery of ABO blood groups led to an explanation of how we inherit our blood group.

### 20.1.1 Monohybrid inheritance

Mendel's first experiments were designed to follow the inheritance of one well-defined characteristic. His procedure was to cross-pollinate pure-breeding varieties of the garden pea, collect the seeds and sow them the following year. The plants which grew were then allowed to self-pollinate, and the seeds they produced were again collected. The characteristics of the plants which grew from the seeds were carefully noted. One of the traits Mendel worked with was the height of the plants. In one investigation he crossed a pure-breeding tall variety with a pure-breeding dwarf variety. All the progeny ($F_1$) were tall but when self-pollinated they produced $F_2$ plants in which there was a ratio of 3 tall : 1 dwarf (Fig 20.2). This is known as a **monohybrid ratio**. A third of the tall $F_2$ plants were pure-breeding. The remainder, on selfing, produced a mixture of tall and dwarf progeny, showing that they were heterozygous. The tall characteristic, which the $F_1$ plants showed, Mendel termed **dominant**. Its contrasting trait, dwarfness, was termed **recessive**.

**Fig 20.2** Inheritance of height in the garden pea

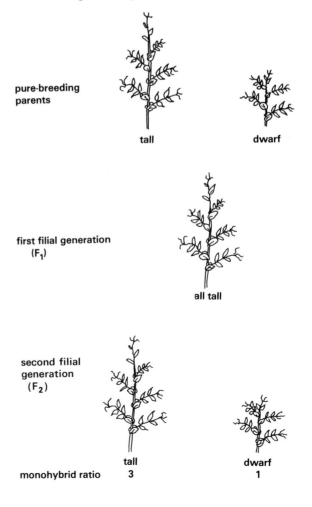

pure-breeding
parents

tall       dwarf

first filial generation
($F_1$)

all tall

second filial
generation
($F_2$)

tall       dwarf

monohybrid ratio    3       1

The observations led Mendel to formulate his First Law of Inheritance which states:

**Of a pair of contrasting characteristics, only one can be represented in the gametes.**

The law can be more readily understood if symbols are used to represent the alleles which determine the characteristics. Using **T** as the allele for tallness and **t** for dwarfness the experiment may be depicted as shown in Fig 20.3. The fact that plants of genotype **Tt** are hybrids can be detected by crossing them with a pure-breeding recessive. This is called a **back-cross** or **test-cross** (Fig 20.4). The 1:1 ratio obtained is termed a **back-cross ratio**.

**Fig 20.3** Genetic explanation of inheritance of height in the garden pea

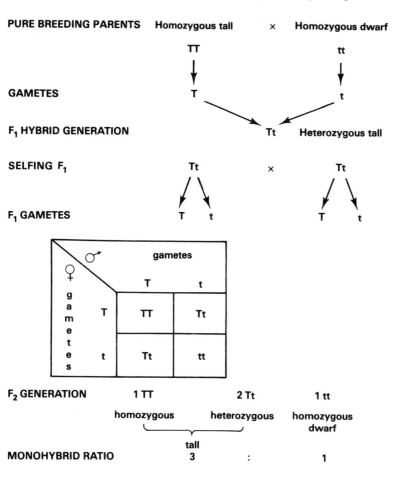

**Fig 20.4** A back-cross

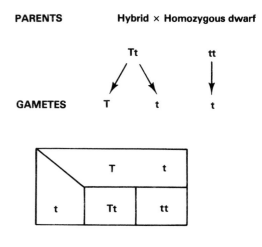

---

What ratio would result in a backcross if the tall plants were homozygous?

Mendel observed that seven other characteristics of the garden pea were inherited in a comparable way.

An example of a human trait which follows the same pattern of inheritance is **Huntingdon's chorea (HC)**. The condition was first described by a Dr Huntingdon in the USA in 1872. People with HC gradually lose muscular coordination and develop severe mental deterioration. The symptoms do not usually begin to appear until the affected person is about thirty-five years of age. By this time many people, unaware that they have the disease, will have passed the dominant allele for HC to their children (Fig 20.5). At the moment there is no way of detecting carriers of the allele.

**Fig 20.5** Family pedigree showing inheritance of Huntingdon's chorea (after Clarke 1970)

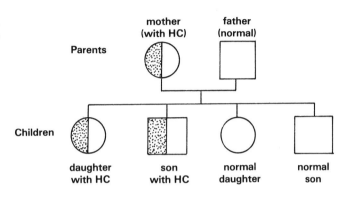

Throughout this chapter the symbols ○ and □ are used to denote females and males respectively in family pedigrees.

Using appropriate symbols, produce a genetic diagram similar to that shown in Fig 20.3 to explain the inheritance of HC in this family.

Other human characteristics determined by dominant alleles are **brachydactyly** and **polydactyly**. People with brachydactyly have a very short bone in the middle of each finger. Consequently each finger is only a little longer than the thumb. In the homozygous state the gene causes severe defects of many bones, and death occurs shortly before or after birth. Thus all surviving individuals with brachydactyly are heterozygous. Polydactylous people have six fingers on each hand and six toes on each foot. The extra toes and fingers of individuals who are homozygous for polydactyly are of normal size. Heterozygotes have much smaller extra digits and occasionally they are not externally visible (Fig 20.6).

**Fig 20.6** A polydactylous hand and foot

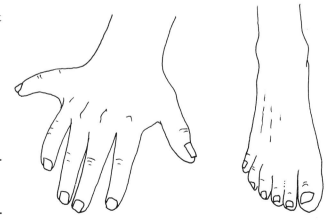

Is the individual whose hand and foot are shown in Fig 20.6 likely to be homozygous or heterozygous for polydactyly?

371

**Albinism** and **fibrocystic disease of the pancreas (FCD)** are caused by recessive alleles. Hence people with these conditions are homozygous recessive. The family pedigree in Fig 20.7 shows how FCD was inherited by a child whose parents were carriers.

**Fig 20.7** Family pedigree showing inheritance of FCD (after Clarke 1970)

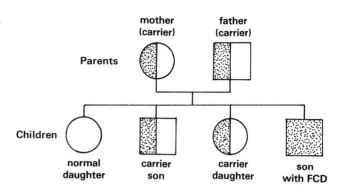

## Chromosome behaviour and monohybrid inheritance

Organisms which reproduce sexually form gametes. The gametes of humans are sperm and ova (Chapter 18). They are formed by meiotic nuclear division which cause them to have a haploid chromosome number (Chapter 6).

Mendel's First Law of Inheritance is also called the **Law of Segregation** because alleles become segregated in meiosis (Fig 20.8).

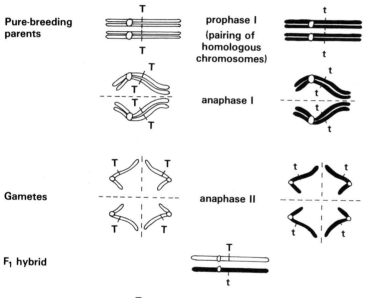

**Fig 20.8** Segregation of alleles in meiosis

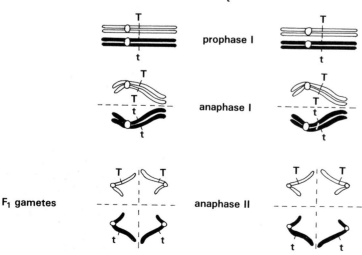

T = allele for tallness
t = allele for dwarfness

372

## 20.1.2 Dihybrid inheritance

Mendel also studied the simultaneous inheritance of two characteristics. In one experiment he traced the inheritance of seed colour and texture. He first crossed a pure-breeding variety having round and yellow seeds with another pure-breeding variety having wrinkled and green seeds. All the $F_1$ plants had round, yellow seeds, showing these to be dominant traits. When self-pollinated, the plants which grew from the $F_1$ seeds produced four kinds of seeds: round and yellow, round and green, wrinkled and yellow, wrinkled and green in a ratio of 9:3:3:1 respectively (Fig 20.9). This is a **dihybrid ratio**. The results led Mendel to formulate a Second Law of Inheritance which states:

> **Each of a pair of contrasting characteristics segregates independently of those of any other pair.**

Once more, it is easier to understand the law using symbols to represent the alleles (Fig 20.10). Mendel observed that several other pairs of contrasting characteristics of the garden pea were inherited in the same way.

**Fig 20.9** Inheritance of seed colour and texture in the garden pea

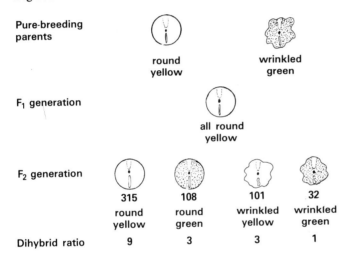

| | | | | |
|---|---|---|---|---|
| Pure-breeding parents | round yellow | wrinkled green | | |
| $F_1$ generation | all round yellow | | | |
| $F_2$ generation | 315 round yellow | 108 round green | 101 wrinkled yellow | 32 wrinkled green |
| Dihybrid ratio | 9 | 3 | 3 | 1 |

**Fig 20.10** Genetic explanation of inheritance of seed colour and texture

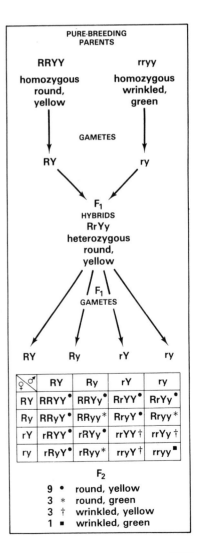

PURE-BREEDING PARENTS

RRYY homozygous round, yellow     rryy homozygous wrinkled, green

GAMETES

RY     ry

$F_1$ HYBRIDS RrYy heterozygous round, yellow

$F_1$ GAMETES

RY   Ry   rY   ry

| ♀\♂ | RY | Ry | rY | ry |
|---|---|---|---|---|
| RY | RRYY • | RRYy • | RrYY • | RrYy • |
| Ry | RRyY • | RRyy * | RryY • | Rryy * |
| rY | rRYY • | rRYy • | rrYY † | rrYy † |
| ry | rRyY • | rRyy * | rryY † | rryy ▪ |

$F_2$

| 9 | • | round, yellow |
|---|---|---|
| 3 | * | round, green |
| 3 | † | wrinkled, yellow |
| 1 | ▪ | wrinkled, green |

## Chromosome behaviour and dihybrid inheritance

A study of the behaviour of chromosomes in meiosis reveals how independent segregation occurs (Fig 20.11). The alleles which determine the two pairs of contrasting characteristics are located on different pairs of homologous autosomes. Because the chromosomes of one pair separate independently of the other pair, the alleles segregate independently. Mendel's Second Law is also known as the **Law of Independent Assortment**.

**Fig 20.11** Independent segregation of two pairs of alleles in meiosis

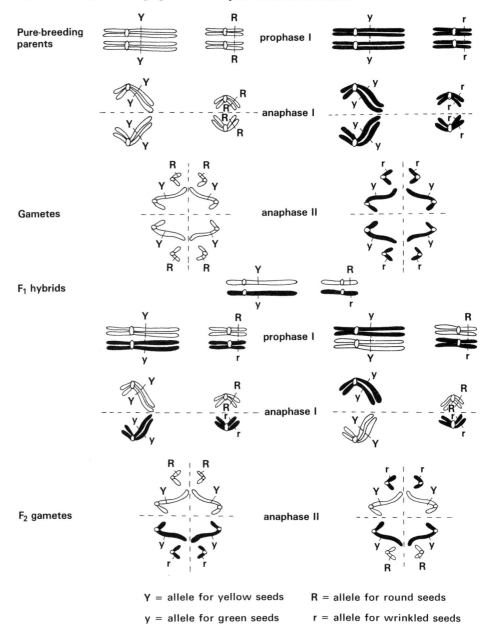

Y = allele for yellow seeds    R = allele for round seeds

y = allele for green seeds    r = allele for wrinkled seeds

### 20.1.3 Polyhybrid inheritance

Tracing the simultaneous inheritance of more than two pairs of contrasting characteristics is more difficult. Nevertheless, the principles of segregation and independent assortment still apply. Thus there is an increase in the variety of combinations of genes in the gametes produced by hybrid $F_1$

individuals, giving rise to an even greater variety of genotypes in the $F_2$ generation (Table 20.1). Many quantitative characteristics show this pattern of inheritance (section 20.3).

**Table 20.1 Relationship between number of alleles and $F_2$ genotypes**

| Pairs of alleles | No. of gene combinations in $F_1$ gametes | No. of genotypes in $F_2$ generation |
|---|---|---|
| 1 | $2^1 = 2$ | $3^1 = 3$ |
| 2 | $2^2 = 4$ | $3^2 = 9$ |
| 3 | $2^3 = 8$ | $3^3 = 27$ |
| $n$ | $2^n$ | $3^n$ |

### 20.1.4 Co-dominance and incomplete dominance

Complete dominance is not always observed in an allelic pair. For example, there is a pair of alleles which express themselves equally in people of blood group AB (section 20.1.5). Such alleles are **co-dominant**.

In other instances an allele may show **incomplete dominance** over its partner. A typical example is seen in people who have the **sickle cell trait** (Chapter 12). The condition is caused by an allele which codes the production of haemoglobin S. Individuals who are heterozygous produce equal amounts of haemoglobin S and normal haemoglobin A. They have a relatively mild form of anaemia. Homozygotes produce only haemoglobin S and have a severe form of anaemia. It is therefore apparent that the allele for producing normal haemoglobin A does not completely suppress the allele which regulates production of haemoglobin S. Fig 20.12 shows how children with sickle cell disease could be among the progeny of parents who have sickle cell trait. Another example of incomplete dominance is seen in people who suffer from **thalassaemia** (Mediterranean anaemia). The condition is caused by an allele **Hb$^B$** which is only partly dominated by the allele **Hb$^A$** which controls production of normal haemoglobin A. It means that people who are heterozygous for the condition **Hb$^A$Hb$^B$** have a mild form of anaemia (thalassaemia minor). Homozygotes **Hb$^B$Hb$^B$** suffer from very severe anaemia (thalassaemia major) and usually die in childhood.

**Fig 20.12** Inheritance of sickle cell disease

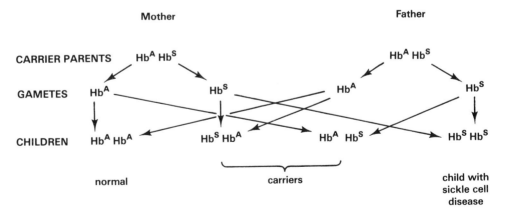

### 20.1.5 Multiple alleles

The patterns of inheritance described so far involve two alternative forms of a gene. Many of the characteristics of most organisms are determined by such alleles. However, some traits are determined by several forms of an allele known as **multiple alleles**. A good example is seen in the inheritance of blood groups of the **ABO system**. As described in Chapter 11 the grouping is based on the presence or absence of A and B antigens on red blood cells. The alleles $I^A$ and $I^B$ which code production of A and B antigens respectively are co-dominant. Both are dominant to the allele **i** which does not code for antigen production (Table 20.2).

**Table 20.2  Genotypes of A, B, AB and O blood groups**

| Group | Antigens | Genotypes |
|-------|----------|-----------|
| A | AA or AO | $I^AI^A$ or $I^Ai$ |
| B | BB or BO | $I^BI^B$ or $I^Bi$ |
| AB | AB | $I^AI^B$ |
| O | none | ii |

If a mother's blood group is B and her child's is AB, which blood group(s) could the father have?

If a mother and her baby belong to group O, the father could be of group O, A or B but not AB (Fig 20.13). Evidence of this kind is sometimes used in legal cases where the paternity of a child is in question.

**Fig 20.13** Inheritance of ABO blood groups

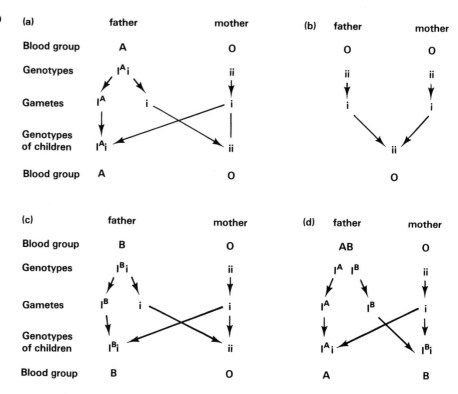

### 20.1.6 Epistasis and gene interaction

**Epistatic genes** prevent the action of other genes. The recessive gene for albinism inhibits the activity of several genes which determine the amount of melanin in the eyes, hair and skin. In a similar way the recessive gene for blue eye colour is epistatic to genes which control the amount of melanin in the iris.

Genes may also **interact** to produce unusual phenotypes. A pair of alleles determines the type of haptoglobin, a plasma protein, we produce. People who are homozygous have the genotype $\mathbf{Hp^1Hp^1}$ or $\mathbf{Hp^2Hp^2}$ and produce different varieties of haptoglobin. In heterozygotes the alleles $\mathbf{Hp^1Hp^2}$ interact to form a third variety of the protein.

### 20.1.7 Linkage

Up to now we have considered patterns of inheritance in which the genes are located on separate autosomes and thus segregate independently. However, each chromosome carries a number of genes. Genes which are on the same chromosomes are said to be **linked** and are inherited *en bloc*.

A well-known case of linkage in humans involves the genes for **ABO blood groups** and the **nail-patella syndrome (NP)**. People with the syndrome have small, discoloured nails especially on the thumbs, index fingers and the first and second toes. The patella is missing or is small and pushed to one side of the knee. The syndrome is caused by a dominant allele which is on the same chromosome as the allele for ABO blood groups (Fig 20.14). Most people with NP syndrome belong to the two most common blood groups, A and O (Table 20.3). Examination of the pedigrees of families in which the syndrome has occurred clearly reveals that the ABO and NP genes are linked (Fig 20.15).

**Table 20.3 Relationship between frequencies of ABO blood groups and NP syndrome**

| Blood group | A | O | B | AB |
|---|---|---|---|---|
| % normal | 49.4 | 38.4 | 8.5 | 3.7 |
| % with NP | 38.1 | 44.6 | 14.4 | 2.9 |

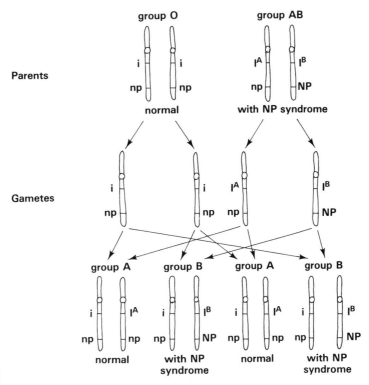

**Fig 20.14** Inheritance of linked genes for ABO blood group and NP syndrome

**Fig 20.15** Pedigree of a family showing linkage between ABO blood group and NP syndrome

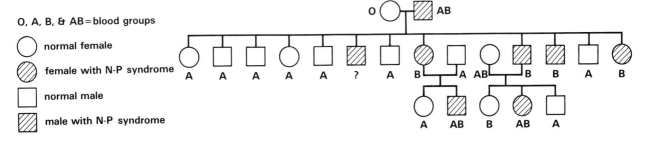

O, A, B, & AB = blood groups

○ normal female

◧ female with N-P syndrome

□ normal male

▨ male with N-P syndrome

**Fig 20.16** Separation of linked genes during crossing-over

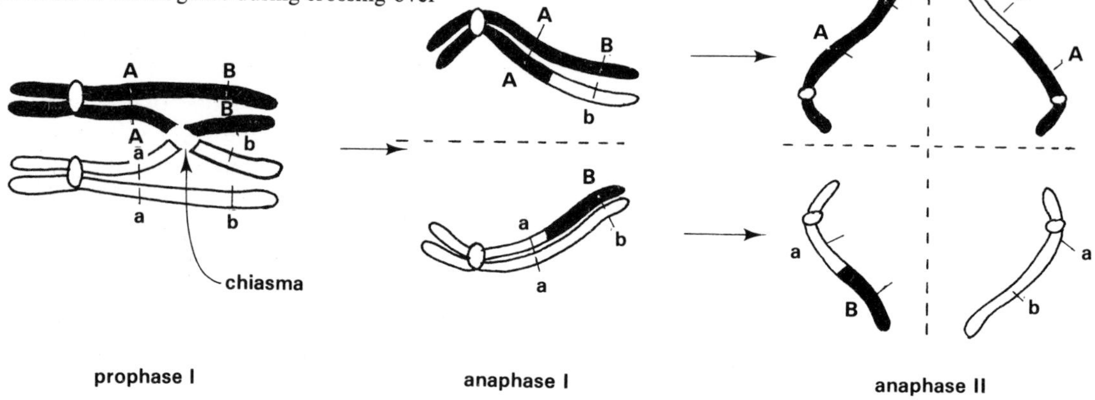

prophase I       anaphase I       anaphase II

**Fig 20.17** Inheritance of closely linked genes

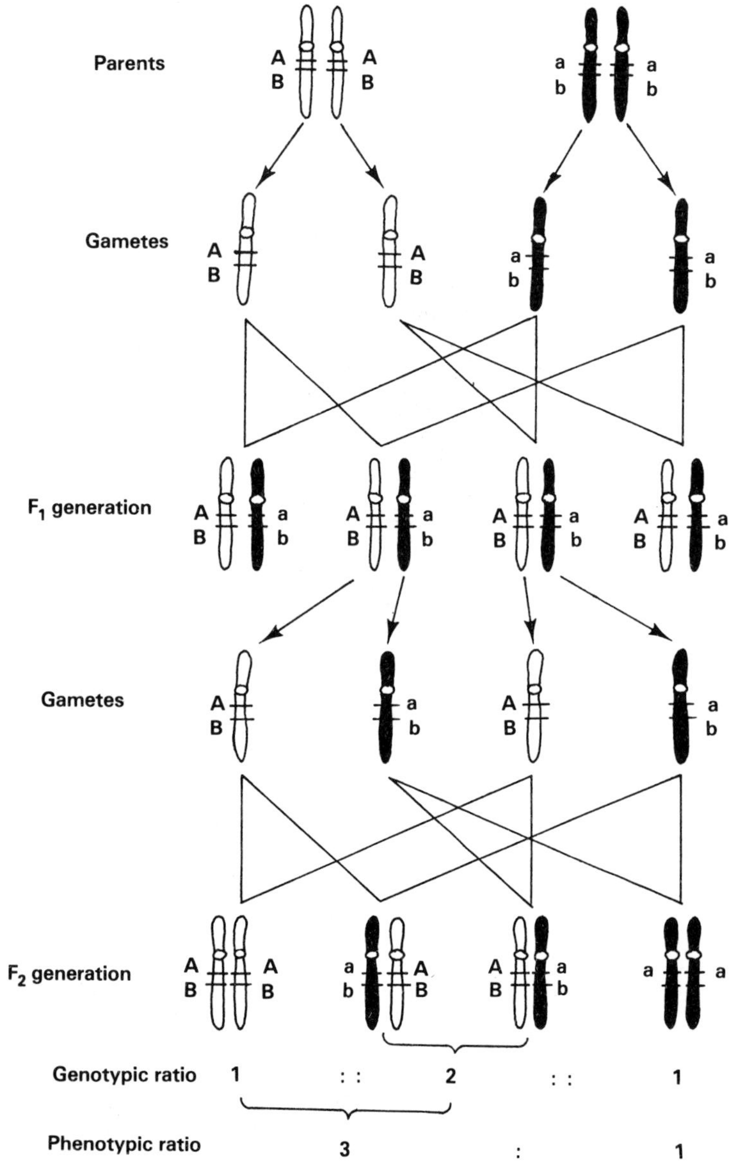

Parents

Gametes

F₁ generation

Gametes

F₂ generation

Genotypic ratio   1    : :    2    : :    1

Phenotypic ratio      3      :      1

Linked genes can be separated when crossing-over occurs at metaphase I of meiosis (Fig 20.16). Crossing-over produces a relatively small number of gametes containing **recombinant genes**. Linkage contrasts markedly with independent assortment which gives rise to equal numbers of each combination of genes in the gametes. Consequently most of the $F_2$ progeny arising in the inheritance of linked genes have the characteristics of one or other of their parents. The few which have traits from both parents are produced from gametes containing recombinant genes.

In general terms, the closer a pair of linked genes are on a chromosome, the less is the probability that a chiasma will form between them. Thus a very small proportion of recombinants indicates close linkage. The absence of recombinants tells us that the genes are so close that they are not separated by crossing-over. When this is the case, the ratio of progeny in the $F_2$ generation is 3:1 (Fig 20.17). Compare this with the 9:3:3:1 dihybrid ratio which occurs when the genes are not linked.

Studies of linkage have made it possible to locate the positions of some genes on several of the chromosomes of humans (Fig 20.18).

## 20.1.8 Autosome abnormalities

Fig 20.19(a) shows the chromosomes of a human female. They can be arranged in homologous pairs. The chromosomes of the first twenty-two pairs are called **autosomes**, the twenty-third pair are the sex chromosomes. People are sometimes born with autosomes which are unusual in number or structure. Failure of a pair of homologous autosomes to separate at anaphase of the first meiotic nuclear division occurs

378

**Fig 20.18** Map of human chromosome 1 (after Emery 1977)

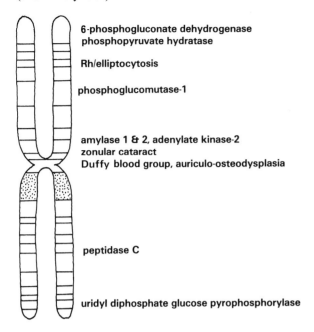

6-phosphogluconate dehydrogenase
phosphopyruvate hydratase

Rh/elliptocytosis

phosphoglucomutase-1

amylase 1 & 2, adenylate kinase-2
zonular cataract
Duffy blood group, auriculo-osteodysplasia

peptidase C

uridyl diphosphate glucose pyrophosphorylase

fairly often. It gives rise to gametes having one less or one more autosome than normal. If fertilised by a normal gamete, the offspring will have 45 or 47 chromosomes respectively, rather than the usual number of 46. Such a condition is called **polysomy**. Loss of an autosome is lethal in humans. **Down's syndrome** is a condition in which individuals usually have an extra twenty-first autosome (Fig 20.19(b)). The extra autosome often becomes joined to one of the fourteenth pair of autosomes. The chromosome number then appears normal. However, the translocation can be detected from the unusual structure of one of the fourteenth autosomes.

People with Down's syndrome have characteristic features (Fig 20.19(c)). They are mentally retarded and have a relatively short life expectancy. Although some live into their forties, others die much younger from congenital heart disease or from common infections. Even so, they are loving, trusting individuals and often serve to strengthen the bonds within families.

**Fig 20.19** (b) Chromosomes of a female with Down's syndrome

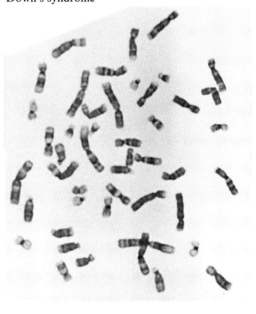

**Fig 20.19** (a) Chromosomes of a normal female

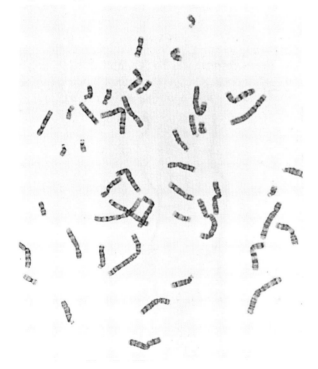

How many pairs of chromosomes are there in Fig 20.19(a)? Is there any difference in the number of chromosomes shown in Fig 20.19(b)?

**Fig 20.19** (c) A child with Down's syndrome (courtesy A. Bezear and Heinemann Medical Ltd)

Very occasionally all the homologous pairs of chromosomes fail to separate in meiosis. This produces gametes with a diploid chromosome number. If fertilised by a normal gamete, a triploid offspring with an extra set of chromosomes arises. The condition is called **polyploidy**. Only a few instances of triploid humans have been recorded and they all died shortly after birth. However, polyploidy is quite common among plants. Polyploid crop plants are usually more vigorous than their diploid parents and therefore produce higher yields. For example, bread wheat is a hexaploid with a diploid number of 42 chromosomes whereas its ancestors had a diploid number of 14.

## 20.2 Non-autosomal inheritance

### 20.2.1 Inheritance of sex

In a male the sex chromosomes are different in size and shape and are called **X** and **Y chromosomes**. Females have two X chromosomes (Fig 20.20). Note that the longer arm of the X chromosome has no counterpart in the Y chromosome.

In the first nuclear division of meiosis, the sex chromosomes pair and then segregate, just as homologous pairs of autosomes do (Chapter 6). As a result, half of the sperm normally have 22 autosomes and an X chromosome; the other half have 22 autosomes and a Y chromosome. For this reason the male is known as the **heterogametic** sex. All the ova normally have 22 autosomes and an X chromosome. This is why the female is called the **homogametic** sex.

There is a 50 % chance that an ovum is fertilised by a sperm carrying an X chromosome. The zygote so formed will be XX and develop into a female. There is also a 50 % chance that the sperm carries a Y chromosome, and the XY zygote so produced will develop into a male (Fig 20.21).

**Fig 20.20** Sex chromosomes

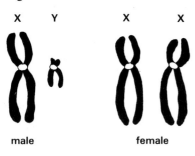

male          female

**Fig 20.21** Inheritance of sex

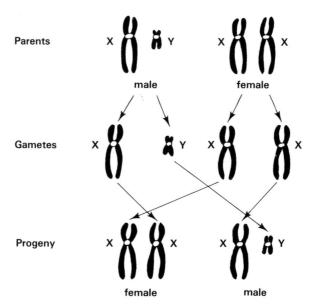

Sometimes the sex chromosomes fail to separate at anaphase of meiosis. When this happens, some ova develop with an excess of X chromosomes and others with none, or some sperm develop with an excess of X or Y chromosomes. Possible outcomes are as follows.

380

**i. An XX ovum is fertilised by a Y sperm.** The XXY zygote produced develops into a male with **Klinefelter's syndrome**. During early life the child may be normal. At puberty, however, the testes will have grown to only about half the expected size and fail to produce sperm. Though obviously infertile, males with the syndrome are capable of sexual intercourse, their ejaculate coming from the prostate and other accessory sex glands (Chapter 18). They also tend to grow taller than average, have high-pitched voices and develop small breasts. About 75 % of individuals with Klinefelter's syndrome are XXY. Others are usually XXXXY, having developed from an ovum containing four X chromosomes fertilised by a Y sperm. Such an ovum would arise if the pair of X chromosomes failed to separate at anaphase in both meiotic divisions.

**ii. A normal ovum is fertilised by a sperm containing two Y chromosomes.** A sperm of this kind is produced if the Y chromosome failed to split at anaphase II of meiosis. The zygote is XYY, and often gives rise to tall aggresive male offspring. Recent studies have shown that men in this category are frequently inmates of institutions for the criminally insane.

**iii. An XX ovum is fertilised by an X sperm.** The XXX zygote would develop into a female with **triple-X syndrome**. Many individuals with this condition are physically and mentally normal and fertile. Others do not menstruate and are sterile.

**iv. An ovum with no X chromosomes is fertilised by a Y sperm.** Zygotes of the YO type have never survived. It indicates that the X chromosomes carry genes which are essential for life.

**v. An ovum lacking X chromosomes is fertilised by an X sperm.** The XO zygote produced develops into a female with **Turner's syndrome**. At birth the child has a thick fold of skin on both sides of the neck causing the neck to appear very wide when viewed from the front or back. Other than this the girl appears normal until puberty when the secondary sexual characteristics fail to appear (Chapter 18). Consequently the sex organs remain child-like, the ovaries do not produce ova and there is no menstruation or breast development.

### 20.2.2 Sex linkage

As well as carrying genes which regulate sexual development, the sex chromosomes, especially the X chromosome, carry genes which determine other traits. Such genes are said to be **sex-linked**. If they occur on the longer arms of the X chromosome, their presence will be apparent in males even when the alleles are recessive. This is because there are no corresponding alleles on the Y chromosome. In females the other X chromosome may carry dominant alleles which would mask recessive alleles on this part of the X chromosome (Fig 20.22).

**Fig 20.22** Sex-linked genes

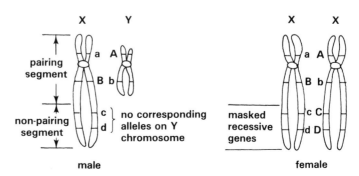

Two well-known examples of sex-linked traits in humans are **haemophilia** and **red–green colour-blindness**. Both conditions are caused by recessive alleles located on the longer arms of the X chromosome. For females to have haemophilia or red–green colour-blindness the alleles must be present on both X chromosomes. Males who have these traits carry the alleles only on their single X chromosome. For this reason haemophiliac and colour-blind males are more common than females. Women who have the recessive alleles on one X chromosome, and the dominant alleles on the other for blood clotting and colour vision, are called **carriers**. They do not have the diseases but can transmit the diseases to their children (Fig 20.23).

**Fig 20.23** Inheritance of haemophilia (after Pedder and Wynne 1972)

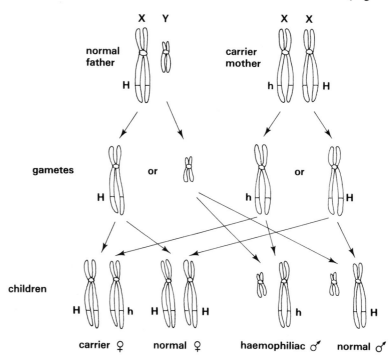

H = gene for normal clotting of blood
h = gene for haemophilia

**Fig 20.24** Pedigree of descendants of Queen Victoria (after Pedder and Wynne 1972)

Perhaps the most publicised pedigree showing the inheritance of haemophilia is that of the descendants of Queen Victoria (Fig 20.24).

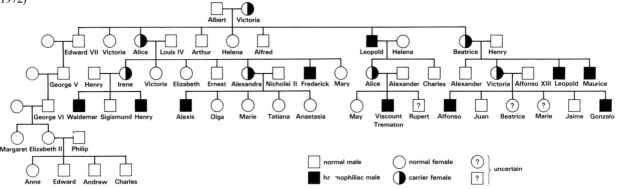

## 20.3 Inheritance of quantitative characteristics

The examples of inheritance dealt with so far have been concerned with **discontinuous variation**. These are clear-cut alternatives of a given trait such as pigmentation in contrast to albinism. Humans also have traits

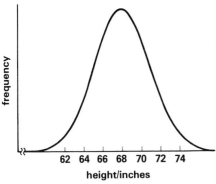

**Fig 20.25** A frequency distribution curve for height of men in South East England (from Carter 1962)

**Table 20.4 Polygenic inheritance**

| Gametes | ABC | ABc | AbC | aBC | Abc | aBc | abC | abc |
|---------|-----|-----|-----|-----|-----|-----|-----|-----|
| ABC | 12 | 11 | 11 | 11 | 10 | 10 | 10 | 9 |
| ABc | 11 | 10 | 10 | 10 | 9 | 9 | 9 | 8 |
| AbC | 11 | 10 | 10 | 10 | 9 | 9 | 9 | 8 |
| aBC | 11 | 10 | 10 | 10 | 9 | 9 | 9 | 8 |
| Abc | 10 | 9 | 9 | 9 | 8 | 8 | 8 | 7 |
| aBc | 10 | 9 | 9 | 9 | 8 | 8 | 8 | 7 |
| abC | 10 | 9 | 9 | 9 | 8 | 8 | 8 | 7 |
| abc | 9 | 8 | 8 | 8 | 7 | 7 | 7 | 6 |

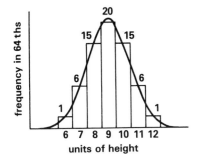

**Fig 20.26** A normal distribution curve for stature, assuming that the alleles with the properties stated in the text determine this characteristic

where there is a range of variation. A typical example of a characteristic showing **continuous variation** is **height**. Within a population of humans there is a range of height. Quantitative characteristics of this kind are usually determined by several genes.

If the frequency of height of a large number of humans of a given age group is plotted graphically, a **frequency distribution curve** is obtained (Fig 20.25).

The observations can be explained as follows. Let us assume that three pairs of alleles **Aa**, **Bb** and **Cc** control height. Let us also assume that each dominant allele allows for two units of height, while each recessive allele allows for only one unit. The tallest individuals will have the genotype **AABBCC** (12 units) and the shortest **aabbcc** (6 units). Heterozygotes for all three genes **AaBbCc** will have a height of nine units, exactly half way between the extremes. The distribution of height among the progeny of heterozygotes can be worked out as in Table 20.4. In reality, height in humans is controlled by more genes than shown in this model.

When displayed as a graph, the data appear as in Fig 20.26. It is a **normal distribution curve**. Note the similarity between this curve and the frequency distribution curve. The resemblance supports the notion that continuously variable characteristics are mainly determined by the additive effects of several genes. This is called **polygenic inheritance**.

**Pigmentation** of the skin and eyes is also determined by polygenes. Skin and eye colour depend on the amount of melanin produced in the Malpighian layer and iris respectively. Melanin is synthesised only if a person has a dominant allele to code its production. However, the amount of melanin produced is regulated by polygenes. As a consequence, a range of skin colour from black to almost white is possible, whereas eye colour can be dark, medium or light brown, hazel, green or blue. People with blue eyes have no pigment in the iris but there is melanin in the choroid layer of their eyes. Albinos are double recessive and are thus unable to produce any melanin. For this reason their skin is white. The eyes of albinos are pink because of the blood capillaries in the iris.

Another characteristic determined by polygenes is the **Rhesus blood group**. The Rhesus group to which a person belongs depends on three dominant alleles **C, D, E** and their recessive alternatives **c, d** and **e**. With the exception of **d**, the alleles can regulate production of red cell antigens. However, only the **D** antigen is usually important in blood transfusion and pregnancy (Chapter 11). The most common genotype for Rhesus group is **CcDdee** (Table 20.5). Double dominant (**DD**) and heterozygous (**Dd**) individuals are Rhesus positive, whereas double recessives (**dd**) are Rhesus negative.

**Table 20.5 Frequency of some genotypes for Rhesus group (after Emery 1975)**

| Genotype | Frequency/% | Phenotype |
|----------|-------------|-----------|
| CcDdee | 33 | Rhesus positive |
| CCDDee | 17 | Rhesus positive |
| ccddee | 15 | Rhesus negative |

## 20.4 Population genetics

A population is a group of individuals living in the same locality. No two individuals are exactly alike. Even identical twins are not the same, despite the fact that they are derived from the same zygote and thus have the same genotype. They are different because they did not develop in identical environments. One of the twins may have been in a more favourable position to obtain nourishment from the placenta. Consequently it could grow more rapidly and would be larger at birth. Hence the full genetic potential of an individual is realised only in an ideal environment. **Environmental differences** are the cause of some of the variation observed among humans. However, much of the variation is due to **genetic differences**.

### 20.4.1 Inbreeding and outbreeding

The genes of humans are passed in their gametes to the next generation. Thus all the genes of children come from a common pool of parental genes. Many of the individuals in a population have many alleles in common. They also have some alleles which are different from those of other individuals. Gametes are formed by meiotic nuclear division which brings about recombination of parental genes. Provided mating is at random, genetic variation constantly occurs among the individuals in each generation. This is called **outbreeding**. Inbreeding limits the amount of genetic variation because fertilisation is not at random. It also increases the proportion of homozygous individuals in a population (Fig 20.27).

**Fig 20.27** Effect of successive inbreeding on the proportion of homozygotes in a population

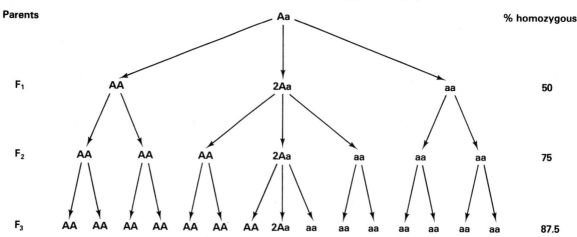

At one time it was quite common for close relatives to marry and have children. In ancient Egypt, for example, it was the custom for cousins and even brothers and sisters to get married. Provided the parents are healthy and are not carriers of harmful recessive alleles, there is a high probability that their offspring will be normal. However, the probability that they will have children with physical and mental abnormalities is high if there is a genetic disorder in the family. For instance, if two cousins have an uncle with phenylketonuria, the chances of their having a child with this defect is 1 in 36. In contrast, the probability of unrelated parents having such a child is 1 in 10 000. For this reason, **inbreeding** among humans is normally avoided these days. Religious, cultural and legal barriers discourage close relatives from marrying.

## 20.4.2 The Hardy–Weinberg equation

Each characteristic of an individual is usually determined by a gene pair, the alleles of which often occur in dominant and recessive forms. If the genotypes of two parents for a given characteristic are known, it is possible to calculate the probability of their having a child with a particular genotype. For example, if the father is of blood group A with a genotype $I^A i$, and the mother is of blood group B with a genotype $I^B i$, the probability that they will have a child of blood group O is 0.25 or 25 % (Fig 20.28). It is worked out by multiplying the probability of sperm having the allele i by the probability of ova having the same allele.

**Fig 20.28** Determining the probability of the genotypes of children

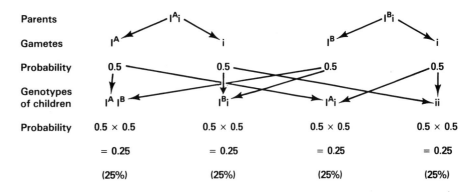

In a population the probability that a sperm may contain such an allele may be different from the probability for an ovum. Nevertheless, the same principle applies. When a population breeds at random it contains homozygous dominant, homozygous recessive and heterozygous individuals. Consider an allele **A** and its recessive form **a**. Individuals of genotypes **AA**, **Aa** and **aa** are present in the population and contribute their genes to a large pool of gametes. If the probability that a gamete contains **A** is $p$ and the probability that it contains **a** is $q$, the proportions of genotypes among the progeny can be calculated as shown in Table 20.6.

**Table 20.6  Determination of frequencies of genotypes of progeny in a population producing gametes of known frequencies**

| | ♂ gametes | |
|---|---|---|
| | A($p$) | a($q$) |
| ♀ gametes | | |
| A($p$) | AA($p^2$) | Aa($pq$) |
| a($q$) | Aa($pq$) | aa($q^2$) |

Because there is 100 % probability that a gamete contains **A** or **a** then $p + q$ = 100 % or 1.0. There is also a 100 % probability that the progeny are of the genotypes **AA + 2Aa + aa**. Thus:

$$p^2 + 2pq + q^2 = 1.0 \text{ or } 100\%$$

This is called the **Hardy–Weinberg equation**. It has been verified with a number of human traits, including our ability or otherwise to **taste**

**phenylthiocarbamate (PTC)**. Some people are unable to taste PTC, others experience a bitter taste when a small amount of it is placed on the tip of the tongue. The allele for tasting (**T**) is dominant to that for non-tasting (**t**). Non-tasters must therefore have the genotype **tt**, whereas tasters could be **TT** or **Tt**. In the early 1930s Snyder tested a large number of North American parents for their ability to taste PTC. His results are summarised in Table 20.7. Applying the Hardy–Weinberg equation to the findings:

$$p^2 + 2pq = 0.712$$
$$\text{and } q^2 = 0.288$$
$$\text{Therefore } q = \sqrt{0.288}$$
$$= 0.537$$

$$\text{Now } p + q = 1.0$$
$$\text{Thus } p + 0.537 = 1.0$$
$$\text{and } p = 1.0 - 0.537$$
$$= 0.463$$

**Table 20.7 Frequencies of tasters and non-tasters of PTC among parents in Ohio, USA (after Snyder 1932)**

| Phenotype | Percentage | Frequency |
|---|---|---|
| taster | 71.2 | 0.712 |
| non-taster | 28.8 | 0.288 |
| total | 100 | 1.000 |

The probability of someone having the genotype **TT** ($p^2$) is therefore:

$$(0.463)^2 = 0.214 \text{ or } 21.4\%$$

The probability of someone having the genotype **Tt** ($2pq$) is:

$$2(0.463 \times 0.537) = 0.497 \text{ or } 49.7\%$$

In this way Snyder calculated the proportions of the various genotypes among the parents. He also detected the same ratio of tasters and non-tasters among their children. It was therefore apparent that there had been no change in the proportions of individuals of the three genotypes in successive generations. This is known as the **Hardy–Weinberg equilibrium**. It is disturbed when:

  **i.** mating is non-random,
  **ii.** the alleles in question mutate,
  **iii.** migration occurs between populations, and
  **iv.** there is natural selection against one of the genotypes (section 20.4.3).

The gene pool of many populations is constantly changing because people frequently move within their own country or they emigrate to another country. Such movement eventually leads to intermarriage of people of different cultures and races. The resulting exchange of genes between such populations is called **gene flow**. Population genetics studies also reveal that the probability of someone carrying undesirable recessive alleles is often quite high (Table 20.8).

**Table 20.8 Frequencies of some recessive alleles in humans (after Clark 1970)**

| Trait | Approx. frequency of persons affected | Probabilities of carriers |
|---|---|---|
| albinism | 1 in 20 000 | 1 in 70 |
| phenylketonuria | 1 in 25 000 | 1 in 80 |
| alkaptonuria | 1 in 1 000 000 | 1 in 500 |

However, in many parts of the world small communities are isolated from the general population for cultural, political or religious reasons.

Should a gene mutation arise in an isolated population, it is likely to increase in frequency because of inbreeding. This is known as the **founder principle** and obviously disturbs the Hardy–Weinberg equilibrium. The equilibrium may also be changed in isolated populations as a consequence of **random genetic drift**. If for example allele **X** is at high frequency in the general population, whereas allele **x** is rare, the recessive allele is still retained because of random mating. In a small isolated population, only a few people would carry the **x** allele. If they do not have children the recessive allele will disappear from the gene pool of the next generation. Such genetic drift is thought to account for the high frequency of blood group A among Blackfeet Indians in contrast to the majority of North American Indians who are mainly blood group O.

### 20.4.3 Natural selection

Instances occur when a person with a particular genotype is unable to reproduce. For example, the genotypes for haemoglobin production in people of African origin are **SS** (normal), **Ss** (sickle-cell trait) and **ss** (sickle-cell disease). Those with **sickle cell disease** do not normally survive to reproductive age. Hence their genes are not passed on to the next generation. On the contrary, people who are homozygous dominant are more susceptible to some forms of malaria than are heterozygotes. The loss of **s** alleles to the next generation is thus to some extent counterbalanced by the loss of **S** alleles in people dying of malaria before they produce children. For these reasons the heterozygotes are the main contributors of genes to each new generation. As a consequence, the harmful **s** allele is retained at a high frequency in such populations.

New genes also constantly arise by mutation (Chapter 4). However, a mutated gene is only retained in a population if it is advantageous or neutral in terms of survival. An increase in frequency of a harmful gene is usually prevented because individuals having such genes may not live to produce children. Preserving two or more forms of a gene in a population by means other than mutation is called **genetic polymorphism**. Sickle-cell anaemia and ABO blood groups are examples of how alternative forms of a gene may be maintained at relatively high frequencies. The allele for sickle-cell anaemia is advantageous in helping heterozygotes to resist malaria. The multiple alleles for the ABO blood groups appear to be neutral so far as survival is concerned.

## 20.5 Detection of carriers

**Carriers** are people who are heterozygous for an undesirable recessive allele. Parents are often unaware that they carry a defective allele until they have a child which has a physiological, biochemical, physical or mental impairment because the child is homozygous recessive.

Simple biochemical tests can sometimes be used to detect carriers. Blood can be tested to determine whether or not parents carry the gene for **acatalasia**. The red blood cells of normal people contain the enzyme catalase which breaks down hydrogen peroxide into water and oxygen. However, homozygotes have twice as much catalase activity as heterozygotes. The activity is determined by mixing a known volume of blood with a known amount of hydrogen peroxide and measuring the frothing (oxygen production) which occurs in a given time. Homozygous recessives lack catalase altogether. They are often healthy but are prone to sepsis of the mouth. Biochemical tests on samples of blood can also be used to detect carriers of **phenylketonuria** and **galactosaemia** (Chapter 4). People who carry the gene for **haemophilia** can be identified by testing the clotting of their blood.

In some instances it is possible to detect whether or not a foetus will develop into a defective child. The technique most often used for this purpose is **amniocentesis** (Fig 20.29). A small sample of amniotic fluid is withdrawn from around the developing foetus through a fine hollow needle which is pushed through the mother's abdominal wall and uterus. The fluid contains urine which the foetus has produced. Amniotic fluid can be analysed to see if it contains unusual constituents such as homogentisic acid, a sign that the foetus has **alkaptonuria** (Chapter 4). Also in the fluid are foetal skin cells. They can be grown in a cuture medium and tested to see if important enzymes are missing. An alternative technique called **chorionic villus sampling** is currently being tried out. A sample of embryonic tissue is taken from the chorionic villi (Chapter 18) by suction using a narrow tube inserted into the womb via the vagina. This procedure can be carried out at 10–12 weeks of pregnancy compared with 16–18 weeks for amniocentesis.

**Fig 20.29** Amniocentesis

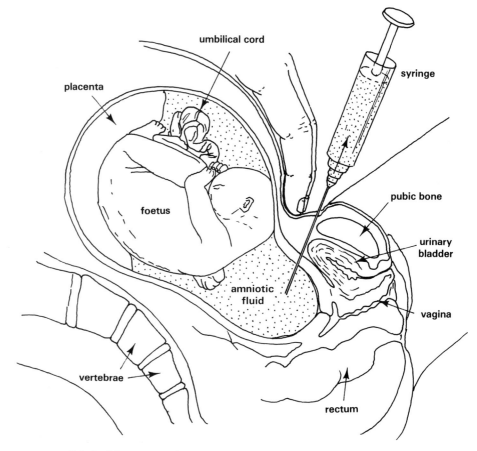

## 20.6 Karyotyping

Unusual numbers of chromosomes or abnormal chromosomes may be detected by allowing cells to divide in culture and examining the chromosomes microscopically. The technique is called **karyotyping**. In the 1950s Tijo and Levan developed a technique which clearly demonstrated that the diploid chromosome number in humans is 46. The technique which is still universally used in **cytogenetic laboratories**, is as follows.

A sample of amniotic fluid is centrifuged to sediment the foetal cells (Fig

20.30). The cells are sucked into a pipette and are then aseptically transferred to a nutrient solution containing **phytohaemagglutinin** which stimulates the cells to divide by mitosis. After about three days at 37 °C, several mitotic divisions will have occurred. Some **colchicine** is now added which prevents the formation of the spindle fibres. Cell division is thus stopped at metaphase when the chromosomes are shortest and thickest. Next, some weak saline solution is added to the medium, causing the cells to absorb water by osmosis. They swell up and the chromosomes spread out. Some of the cells are placed on a glass slide and viewed with a light microscope. The chromosomes are photographed and the images of individual chromosomes cut out of the picture. They are now paired in decreasing order of size and numbered 1–23, the last pair being the sex chromosomes. The arrangement of chromosomes is a **karyotype** (Fig 20.31).

**Fig 20.30** Preparation of a karyotype

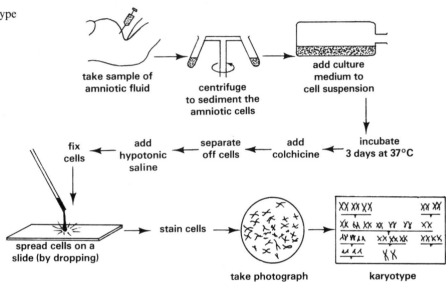

The technique of karyotyping makes it possible to detect foetuses with abnormal chromosome numbers at a relatively early stage in gestation. A decision can then be taken whether or not to terminate the pregnancy. Karyotyping of foetal cells also makes it possible to tell the sex of a child well before it is born. The same technique can be applied to white blood cells to diagnose chromosome abnormalities after birth.

**Fig 20.31** Karyotype of a male with Down's syndrome

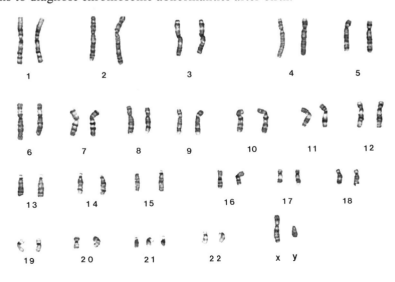

389

## 20.7 Barr bodies, Y chromosomes and white cell drumsticks

In 1949 Murray Barr reported the presence of a small semicircular body lying just inside the nuclear membrane of normal female cells to which a stain specific for DNA had been applied (Fig 20.32). The **Barr body**, as it is now called, does not occur in the cells of normal males. This is because at least two X chromosomes must be present for the body to appear. Barr body studies can be used to determine the sex of a foetus, applying the stain to foetal cells obtained by amniocentesis. It is another way of finding out the sex of a child before it is born. Smears of buccal epithelium can also be examined using this technique when there is doubt as to the sex of athletes participating in major events such as the Olympic Games.

**Fig 20.32** Nucleus of a cell of a normal female

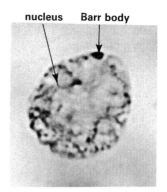

These days a piece of amnion is also routinely examined for Barr bodies at birth. The reason for doing so is to identify children who may have an unusual complement of sex chromosomes. The number of Barr bodies in a cell is one fewer than the number of X chromosomes. Thus normal females have one Barr body and normal males have none. However, triple-X females or XXXY males will have two Barr bodies. If the baby is male and a Barr body is present then Klinefelter's syndrome is suspected. The absence of a Barr body in a female child would indicate that it may have Turner's syndrome.

**Fluorescent microscopical techniques** can be used to examine cells for Y chromosomes. The cells are stained at interphase with acridine dyes and viewed with an ultraviolet microscope. Y chromosomes take up the dye and emit a green light (fluorescence), enabling their number to be determined. The test is now performed routinely at birth on cells from the umbilical cord. It is also performed on buccal smears of female athletes where their sex is in doubt.

Polymorphonuclear granulocytes of normal females have a body shaped like a **drumstick** protruding from one of the lobes of the nucleus (Chapter 10). Drumsticks do not occur in such cells of normal males. The number of drumsticks is one less than the number of X chromosomes. Examining blood smears microscopically for drumsticks is therefore another way of finding out the sex of a child and of detecting unusual numbers of X chromosomes.

## 20.8 Genetic counselling

With the aid of the techniques described in the previous three sections it is now possible to counsel parents about the possibility of genetic defects in their unborn children. The parents are then able to decide whether or not to have a pregnancy terminated. Decisions of this kind are not easy to arrive at. Some people believe it ethically and morally wrong to prevent any foetus from reaching maturity. Others are of the opinion that children with very serious abnormalities should not be born because of the suffering they may have to endure.

**Fig 20.33** Predicting the probability of fibrocystic disease

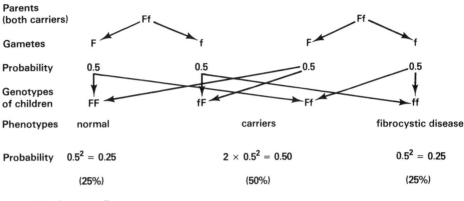

F = allele for normality
f = fibrocystic disease

**Genetic counselling** is available to families with a history of genetic abnormalities and those without. For example, both parents may be carriers of a defective allele. The first child they produce may be a homozygous recessive for a condition such as fibrocystic disease of the pancreas. It is then possible to determine the probability of further children having the defect. In this instance it would be 25 % (Fig 20.33). In a similar way it is possible to predict that of the sons born to a normal man and a woman who is carrier for haemophilia, half will be normal and half will be haemophiliacs (Fig 20.34).

Some abnormalities in babies are not caused genetically. The embryo may be affected by **teratogenic agents** while inside the uterus. For example, it is now well known that a woman who has rubella (German measles) in early pregnancy is very likely to have a child with defective sight, hearing or heart function. Drugs taken by pregnant women, as thalidomide was in the 1960s, may also seriously impair foetal development. These and other related factors are discussed more fully in Chapter 19.

**Fig 20.34** Predicting the probability of haemophilia

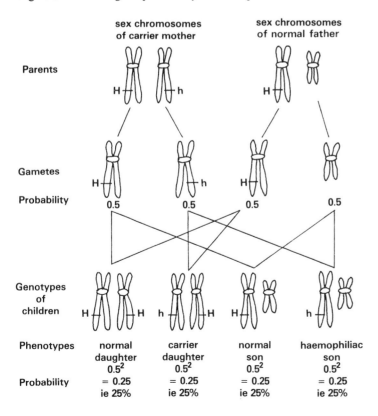

391

# 21 The origin and evolution of life

The earth is populated by an enormous variety of plant and animal life. Some organisms clearly resemble others and there is little difficulty in accepting the fact that they are closely related. Others are so different in appearance that it would seem that they are unrelated. Evolution implies that all living organisms, however unalike, arose from common ancestors. They are all the descendants of previous forms of life, and in this sense they are related to each other.

For centuries people have wondered how life originated and how the diversity of living organisms evolved. Scientists can still only speculate about the origin of life. Much more is known about how life evolved, mainly due to the work of Charles Darwin and Alfred Russel Wallace (section 21.4.2) in the nineteenth century, and to the findings of twentieth century geneticists and biochemists.

## 21.1 The origin of life

Radionuclide methods of dating (section 21.2.1) indicate that the earth was formed about $4.6 \times 10^3$ million years ago, whereas the oldest fossils are only $3.2 \times 10^3$ million years old. The first $1.4 \times 10^3$ million years of the earth's history are thus something of a mystery. Assuming that the ancestors of today's living organisms originated on earth, the way in which it happened has to be the subject of informed guesswork, and there are many theories as to how life first began.

J B S Haldane appears to have been the first to propose that life originated when the earth's atmosphere was devoid of oxygen gas. In 1929 he suggested that an atmosphere lacking oxygen would have no ozone layer which today stops the penetration of most of the ultraviolet radiation from the sun. Ultraviolet light is assumed to have provided energy for organic molecules to be made from atmospheric gases such as carbon dioxide, ammonia and water vapour. The organic molecules gradually accumulated in the oceans, in Haldane's words, as a 'dilute soup'. One of the difficulties of accepting Haldane's theory is that ultraviolet light can disrupt chemical bonds in organic molecules as well as providing energy for synthesis.

A few years later, the Russian biochemist A I Oparin came up with similar ideas as to how life began. There was little interest at the time in Haldane's and Oparin's suggestions. However, with present-day knowledge, it is reasonable to speculate that the main steps in the early evolution of life were probably as follows:

1 **Synthesis of organic monomer molecules** such as amino acids, nitrogenous bases and sugars.
2 **Formation of organic polymers** such as proteins and nucleic acids.
3 **Isolation of the polymers** from the non-living environment, followed by their **replication**

### 21.1.1 Monomer synthesis

An indication of the kinds of organic monomer which can be synthesised from simple gases came in the early 1950s. Harold Urey and Stanley Miller in the USA carried out a series of laboratory experiments in which they subjected a variety of gas mixtures to ultraviolet light and an electric spark (Fig 21.1). The electric spark simulated lightning flashes. These may have been an additional or alternative source of energy for the synthesis of organic molecules in the earth's early atmosphere. Urey and Miller observed that gas mixtures such as hydrogen, methane, ammonia and water vapour gave rise to, among other things, a variety of amino acids.

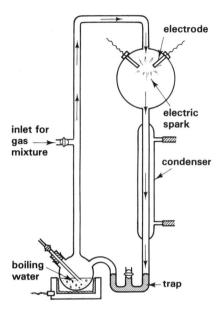

**Fig 21.1** Apparatus used by Miller and Urey to simulate monomer synthesis

electrode

electric spark

inlet for gas mixture

condenser

boiling water

trap

Other mixtures for example, carbon dioxide, carbon monoxide, nitrogen and water vapour, gave similar results. Several of the amino acids formed in the experiments performed by Urey and Miller are found in the proteins of living organisms.

In the early 1960s using other gas mixtures, Oró and Kimball made purine and pyrimidine bases as well as ribose and deoxyribose. These are found in the monomers of which nucleic acids are synthesised (Chapter 4).

### 21.1.2 Polymer synthesis

When amino acids join by peptide linkages to form proteins, water is one of the products of the reaction (Chapter 2). The same happens when nitrogenous bases and phosphorylated pentose sugars join to make nucleic acids. They are both examples of **polymerisation** by condensation reactions. It is difficult to see how condensation reactions could readily occur in the earth's primaeval oceans. Here water was abundant and would have kept the monomers apart. Bernal has suggested that the amount of water in the vicinity of the monomers may have been reduced by their adsorption on to clay or mica particles. Freezing and drying are alternative methods which may have concentrated the monomers prior to polymerisation. However, polymers formed in such a way would have their monomers arranged at random. In contrast, the monomers of proteins and nucleic acids formed by living organisms are in non-random sequences (Chapters 2 and 4). Scientists do not yet know how such orderliness may have come about. Presumably simple polynucleotides able to replicate and code the synthesis of simple enzyme molecules would have to be formed to achieve specific sequencing.

### 21.1.3 Polymer isolation and replication

In living cells, the energy required to build polymers from monomers comes from ATP (Chapter 5). In prebiological systems, it may have come from ultraviolet light and possibly lightning.

Oparin in the Soviet Union and Fox in the USA have closely studied some of the ways in which natural polymers may have become separated from their surroundings. The tendency of biological polymers to form polymer-rich droplets called **coacervates** has been examined by Oparin.

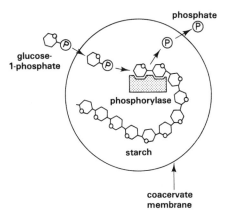

**Fig 21.2** Polymerisation in a coacervate

phosphate

glucose-
1-phosphate

phosphorylase

starch

coacervate
membrane

He reports that coacervates are readily formed from aqueous suspensions of proteins and polysaccharides, and proteins and nucleic acids. Some coacervates are unstable and quickly sediment from suspension. Oparin has shown that coacervates can be made stable by adding enzymes to the suspensions. The enzyme molecules concentrate in the coacervates, where they are able to catalyse biochemical reactions. One such enzyme is phosphorylase which catalyses the polymerisation of glucose phosphate to form starch (Fig 21.2). As starch accumulates, the coacervates grow and eventually break up into smaller coacervates which continue to grow in the same way. Could this be the way in which biological polymers were first replicated?

In addition to condensation reactions, Oparin has shown that coacervates can also bring about electron transport. For example, coacervates containing dehydrogenase enzymes reduce redox dyes such as methyl red when an electron carrier such as NADH is added to the suspension. Furthermore, he has succeeded in getting coacervates containing chlorophyll to develop a reducing ability when exposed to visible light. This is comparable to the way in which chloroplasts reduce $NADP^+$ in photosynthesis (Chapter 5).

Fox has worked mainly with protein coacervates. When heated in water at 130–180 °C, they aggregate spontaneously into spheres 1–2 $\mu$m in diameter. The spheres have a boundary which resembles the structure of a unit membrane, though it lacks the double phospholipid layer (Chapter 6). They grow by absorbing protein dissolved in the surrounding water and split in much the same way that bacteria multiply by fission. In addition, they are able to catalyse a variety of biochemical reactions, for example glucose oxidation. The experiments of Oparin and Fox show that biological polymers may form stable systems capable of catalysing biochemical reactions. Such systems may well have been the forerunners of living cells.

### 21.1.4 The first cells

The first cells to evolve were possibly **unicellular organisms**. Because oxygen gas was absent in the earth's oceans at the time, they probably acquired energy by fermenting some of the organic monomers synthesised from atmospheric gases. In many ways they were probably similar to anaerobic bacteria living today. The capacity of the earth to support **consumer organisms** of this kind was limited by the rate at which monomers were formed. When photosynthetic organisms evolved, the earth's capacity to support life became greatly enhanced. Such **primary producers** probably used hydrogen sulphide as a source of hydrogen for reducing carbon dioxide. This form of photosynthesis is used by some bacteria today. At a later stage, organisms able to use water instead of hydrogen sulphide in photosynthesis appeared. They were probably similar to blue-green algae. It is these organisms which are found as microfossils in rocks at least $3.2 \times 10^3$ million years old.

The release of oxygen as a photosynthetic product resulted in a gradual but dramatic change in the composition of the earth's atmosphere. An important consequence of the change was the formation of the ozone layer in the upper atmosphere which reduced the penetration of ultraviolet light from the sun. It ended the non-biological production of organic matter. From this time on, synthesis of organic molecules was brought about mainly by photosynthesis. Another considerable effect was that organisms began to evolve aerobic methods of respiration. Aerobic respiration is more efficient at energy release than anaerobic respiration (Chapter 5).

## 21.2 Evidence of evolution

The reason why biologists believe that the enormous variety of life has evolved from simple ancestors is based on three main sources of evidence. They are the evidence from **fossils**, from **affinities** between organisms and from the **geographical distribution** of organisms.

### 21.2.1 Fossils

A **fossil** is any buried object which points to the existence of prehistoric life. Fossils are formed in various ways. One of the commonest is where a hard part of an animal, such as the shell or skeleton, immersed in water, has become infiltrated or replaced by minerals causing it to become **petrified** – changed to stone. In other instances, the body of an organism has become embedded in mud which hardened to form a **mould** when the body subsequently decayed. When the mould later filled with minerals brought in by water, a stony **cast** was formed, similar in shape to the organism. Casts of the burrows of mud-dwelling invertebrates have been formed in this way. Alternatively, an **impression** of the shape of the body or its surface structure may be left in material which later changes into rock (Fig 21.3).

**Fig 21.3** Fossils

cast of an ammonite

shell of a gastropod

tooth of a shark

impression of fern leaves

cast of a trilobite

**Fig 21.4** Rock strata at Aberystwyth, Wales

Most fossils are found in **sedimentary rocks** formed from mud and silt deposited in lakes and on the sea bed millions of years ago. Earth movements later raised many sedimentary rocks above sea level. Often the rocks are in **strata**, the oldest at the bottom and the most recently formed at the top (Fig 21.4). Some indication of the course of evolution can be obtained by comparing the fossils in the strata.

It is also possible to age the rocks by radionuclide dating. Nearly all elements exist as several nuclides, that is, as atoms having slightly different atomic masses, e.g. $^{12}C$ and $^{14}C$. Some nuclides are unstable and decay by releasing sub-atomic $\alpha$ and $\beta$ particles until they become more stable. Some examples are as follows:

$$^{238}uranium \rightarrow\,^{206}lead$$
$$^{40}potassium \rightarrow\,^{40}calcium +\,^{39}argon$$
$$^{14}carbon \rightarrow\,^{12}carbon$$

The time required for half a given quantity of an unstable nuclide to break down into a stable form is its **half-life**. The half-lives of many unstable nuclides have been determined. It is thus possible to estimate accurately the age of a fossil by measuring the ratio of unstable : stable nuclides, for example $^{238}U :\,^{206}Pb$ in the rock in which the fossil was found. If the ratios of several nuclides are measured at the same time, the dating is cross-checked and reliable. The order in which living organism were evolved can thus be determined. It usually coincides with the order suggested from rock strata studies.

Other information of the kinds of living organism existing in the past comes from **amber**, the hardened resin of primitive seed-bearing trees, in which whole insects were trapped. The **frozen bodies** of mammoths and woolly rhinoceroses found in Siberia, and the **pollen-grains** trapped in peat bogs and sediments, have also provided valuable information.

It is important to realise that fossils provide only a sketchy history of previous life. Many fossils have been found purely by chance when rocks have been disturbed in mining, quarrying and in the construction of roads, bridges and buildings. Many more lie undiscovered. Fossils of soft-bodied organisms are not plentiful. This is presumably because the organisms quickly decayed after death. For some forms of life such as fungi there is hardly any fossil evidence.

## 1 The first fossils

Rocks over $6 \times 10^2$ million years old belong to the **Precambrian** geological period. Until quite recently there was no fossil evidence of life as far back as this. In the 1950s two American geologists, Barghoorn and Tyler, reported the presence of fossil microorganisms called **stromatolites** in Precambrian rocks near Lake Superior, Ontario. The fossils, which are $3 \times 10^3$ million years old, resemble bacteria and blue-green algae. Living stromatolites have since been found in the coastal waters off Western Australia. They are layered communities of bacteria and blue-green algae (Fig 21.5).

Living blue-green algae are able to photosynthesise and some can fix atmospheric nitrogen. If the fossil stromatolite microorganisms were also able to do this, they may have had a profound influence on the course of evolution. By releasing oxygen gas, a product of photosynthesis, they may have created the conditions in which aerobic organisms first evolved. Before this time the earth's atmosphere is thought to have contained very little oxygen gas.

Fig 21.5 (a) General view of living stromatolites (Western Australian Tourism Commission)

Fig 21.5 (b) Section of a fossil stromatolite (B&B)

## 2 Animal and plant fossils

The oldest animal and plant fossils suggest that the earliest forms of life were simple and probably aquatic. Fossil burrows have been found in rocks of the late Precambrian period, less than $7 \times 10^2$ million years old. Living invertebrate animals which form similar burrows are worm-like creatures. They bore through mud and soil. It is likely that the fossil burrows were formed by animals of this kind. Precambrian rocks dated as $6.8$–$5.8 \times 10^2$ million years old contain fossils which resemble **jellyfish** and **annelids**. Fragments of fossil invertebrate skeletons have also been found in rocks of the late Precambrian period. However, we do not know to which group of invertebrates they belong. These earliest animals probably lived near the bottom of the oceans where they fed on particles of organic debris suspended in the water or deposited on the mud.

By the end of the **Cambrian** period, a wide variety of invertebrate groups had evolved. Some later became extinct, others have living descendants. One of the groups resembled **sponges**, another looked like **molluscs**. Another of the groups were **agnathans**, jawless fishes which fed by ingesting particles filtered from the water through their gills. The plant life which existed at the time resembled **algae**. Some were single cells, others multicellular. They enriched the oceans with oxygen and detritus which may have stimulated evolution of the range of consumer organisms. During the **Ordovician** period, the first **bryophytes** appeared.

In the **Silurian** period, the first **jawed fishes** evolved, their jaws enabling them to use a wider choice of food which may have been the reason for the greater variety of fish in the succeeding **Devonian** period. At this time the first vascular land plants had appeared. They were **psilophytes**, a group of extinct pteridophytes (Fig 21.6) which most likely grew in soil covered with a shallow layer of water. As plants began to colonise the land, invertebrates such as annelid worms and arthropods probably followed.

**Amphibians** were well established by the **Carboniferous** period. Many were fairly large omnivorous animals which fed on the evolving terrestrial plant and animal life. They used water for fertilisation and embryo development, as do living members of this group. Later amphibians were smaller and consequently less noticeable to predatory vertebrates which were beginning to emerge. It was from the smaller amphibians that present-day amphibians probably descended. At this time, vascular land plants were abundant. They included many of the larger extinct **pteridophytes** (Fig 21.7) such as the giant horsetails and tree ferns. Primitive **spermatophytes** called the pteridosperms had also evolved by this time. Some of them secreted resin which attracted terrestrial insects. Some of the insects have become fossilised in amber formed from the resin.

**Fig 21.6** *Rhynia*, a psilophyte

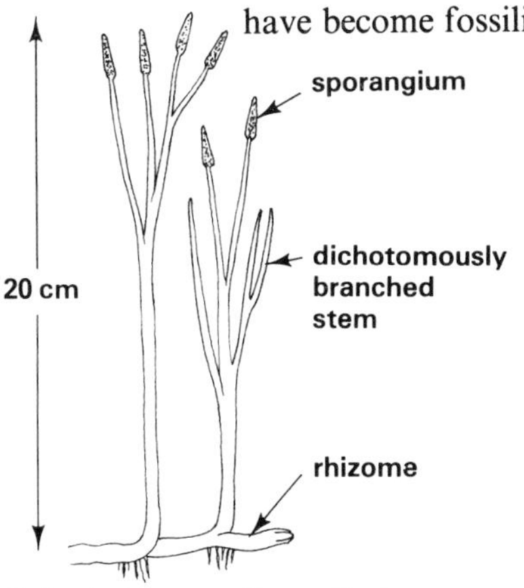

sporangium

20 cm

dichotomously branched stem

rhizome

Psilophytes were the earliest vascular terrestrial plants. None survive today

**Fig 21.7** Tree-sized pteridophytes, now extinct

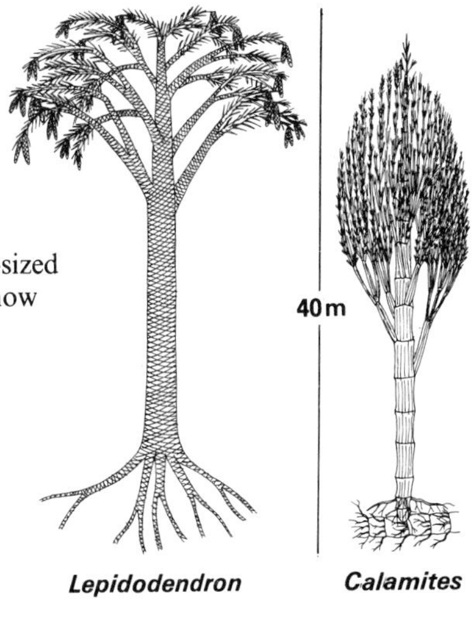

40 m

*Lepidodendron*      *Calamites*

The earliest known **reptiles** appeared in the **Permian** period. Like living reptiles, they probably did not depend on water for reproduction and were able to colonise the land quickly. Here they preyed on and competed with the larger amphibians. Most were lizard-like and some had extremely large dorsal fins (Fig 21.8). In the **Triassic** period, they were replaced by the **dinosaurs** which survived well into the **Cretaceous** period, over $1.5 \times 10^2$ million years later. They were the most successful of terrestrial vertebrates to evolve up to this time and they exploited every ecological niche. Some, such as the pterodactyls were the first known vertebrates to fly (Fig 21.9). Other dinosaurs were awesome predators, while some were herbivores of enormous size (Fig 21.10).

**Fig 21.8** A pelycosaur *Dimetrodon*, one of the earliest known reptiles

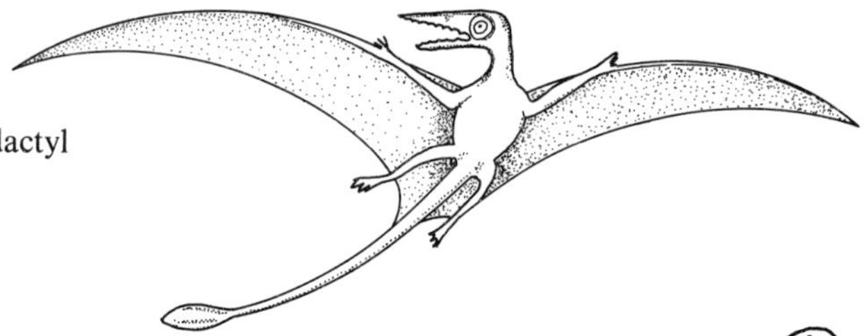

**Fig 21.9** A pterodactyl

**Fig 21.10** (a) *Tyrannosaurus*, a predatory dinosaur. (b) *Brontosaurus*, a herbivore

(a)

(b)

Rocks of the Triassic period also contain fossils of small animals which clearly had mammalian features. During the **Jurassic** period, the terrestrial vegetation included a wide range of **gymnosperms**. It was at this time that the first known **angiosperms** also appeared. Although the dinosaurs still remained dominant among the terrestrial vertebrates, **birds** and **egg-laying mammals** began to emerge.

The dinosaurs became extinct in the Cretaceous period. By then a variety of **placental mammals** had evolved, including what are thought to be man's ancestors. The diversity of angiosperms also increased at this time.

The variety of pollinating insects which evolved in the subsequent **Tertiary** period reflects the fact that spermatophytes became the dominant vegetation on land at the time. Simultaneously there was an increase in the variety of birds and terrestrial mammals.

The major phases of evolution described to date are summarised in Table 21.1, p.400.

### 21.2.2 Affinities

All living organisms have common features. Some have many features which are almost identical with those of another organism. In other words, they have a close **affinity** to each other and for this reason they are classified in the same group. Among the features which indicate that living organisms are related by descent are **anatomical, immunological, biochemical, embryological** and **behavioural affinities**. The relationship is more reliably established when a number of the features are common.

**1 Anatomical affinities**

When the anatomy of various animals, which superficially appear to be unrelated, is compared, similarities may be evident which suggest a common ancestry. During the course of evolution, vertebrate limbs have become adapted to serve a variety of purposes. For example, the forelimbs of a whale are used as paddles, the forelimbs of birds for flight and the hand of humans for grasping and manipulating objects. Even so, the limbs of all vertebrates develop from the same embryonic tissues and fit a common plan known as the **pentadactyl limb** (Fig 21.11). For this reason they are called **homologous structures**.

Fig 21.11 Forelimbs of (a) man, (b) whale and (c) bird

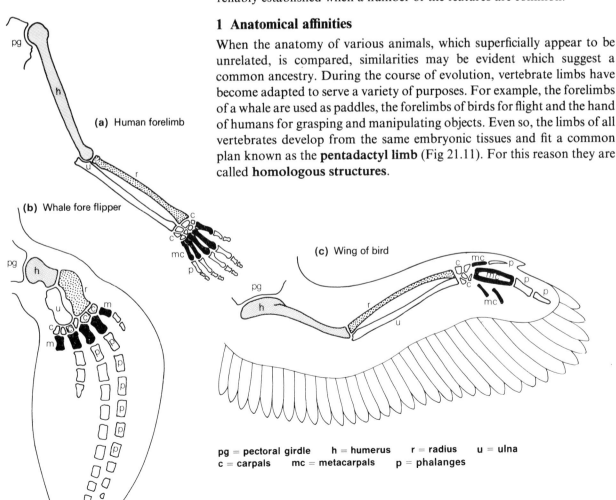

(a) Human forelimb

(b) Whale fore flipper

(c) Wing of bird

pg = pectoral girdle    h = humerus    r = radius    u = ulna
c = carpals    mc = metacarpals    p = phalanges

**Table 21.1**

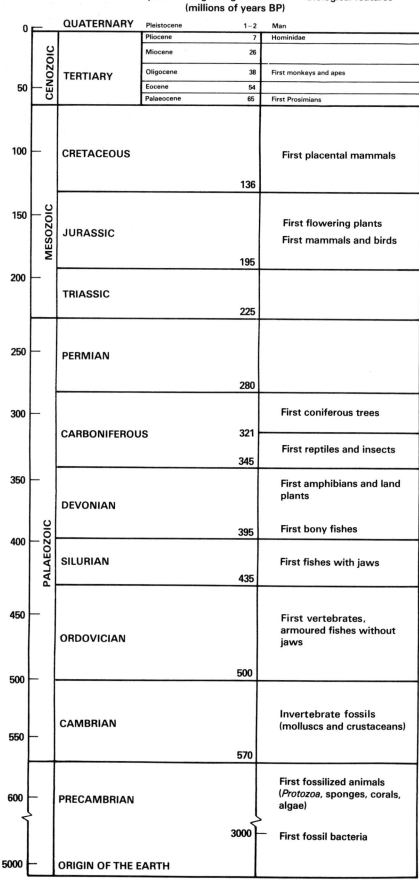

| 10⁶ Years | Era | Period | Epoch | Beginning of Period (millions of years BP) | Biological features |
|---|---|---|---|---|---|
| 0 | CENOZOIC | QUATERNARY | Pleistocene | 1–2 | Man |
| | | TERTIARY | Pliocene | 7 | Hominidae |
| | | | Miocene | 26 | |
| | | | Oligocene | 38 | First monkeys and apes |
| 50 | | | Eocene | 54 | |
| | | | Palaeocene | 65 | First Prosimians |
| 100 | MESOZOIC | CRETACEOUS | | 136 | First placental mammals |
| 150 | | JURASSIC | | 195 | First flowering plants / First mammals and birds |
| 200 | | TRIASSIC | | 225 | |
| 250 | PALAEOZOIC | PERMIAN | | 280 | |
| 300 | | CARBONIFEROUS | | 321 | First coniferous trees |
| | | | | 345 | First reptiles and insects |
| 350 | | DEVONIAN | | 395 | First amphibians and land plants / First bony fishes |
| 400 | | SILURIAN | | 435 | First fishes with jaws |
| 450 | | ORDOVICIAN | | 500 | First vertebrates, armoured fishes without jaws |
| 500 | | CAMBRIAN | | 570 | Invertebrate fossils (molluscs and crustaceans) |
| 600 | | PRECAMBRIAN | | 3000 | First fossilized animals (*Protozoa*, sponges, corals, algae) / First fossil bacteria |
| 5000 | | ORIGIN OF THE EARTH | | | |

Some animals have **vestigial structures** which in related creatures are fully developed. For instance, whales have vestigial pelvic girdles but lack hind limbs. This suggests that whales in common with other vertebrates have evolved from the same ancestors. Humans have a vestigial appendix suggesting that their forebears were herbivores.

The examples given above show how structures of similar origin have become adapted for different roles in various creatures. They illustrate **adaptive radiation** and are examples of **divergent evolution**. There are instances where structures of quite different origin have evolved to carry out similar functions. The wings of both insects and birds are organs used for flight, yet they do not have the same origin. They are **analogous structures**, which illustrates that **convergent evolution** can also occur.

## 2 Immunological affinities

In 1904, George Nuttall devised a procedure based on immunological methods for determining the relationship between animals. When, for example, the serum of a human is injected into the blood of an experimental animal, the latter forms **antibodies** against the proteins in the serum (Chapter 11). If the serum from the experimental animal is then mixed with more human serum, the antibodies combine with those serum proteins which caused their production to form a precipitate. Should the antibody-containing serum from the experimental animal be mixed instead with serum from another animal, the amount of precipitate formed is a measure of the affinity between human and the other animal. The more precipitate, the nearer the relationship (Table 21.2). Such methods suggest that humans are more closely related to the chimpanzee than to other primates.

**Table 21.2 Precipitin reactions with human serum (from Marshall 1978)**

| Serum | Percentage precipitation |
|---|---|
| human | 100 |
| gorilla | 64 |
| baboon | 29 |
| deer | 7 |
| kangaroo | 0 |

## 3 Biochemical affinities

Between man and chimpanzee there is only one difference in the sequence of amino acids in the $\alpha$-globin and two differences in the $\beta$-globin fractions of **haemoglobin**. This contrasts with seventeen differences between human and horse haemoglobin. It points to man and chimpanzee having a more recent common ancestor than other mammals.

Another protein which has been studied in this respect is **cytochrome $c$**, a respiratory co-enzyme (Chapter 5). The amino acid sequence of cytochrome $c$ has been determined for organisms as diverse as bacteria, yeasts, invertebrates and vertebrates. The variations are caused by slight differences in the nucleotide sequence of DNA which codes the production of the co-enzyme (Chapter 4). In this way it is possible to compute the minimum number of nucleotide differences between the genes which code the synthesis of cytochrome $c$ in various organisms. A comparison of the differences shows how closely the organisms are related (Fig 21.12). The results agree well with the fossil record.

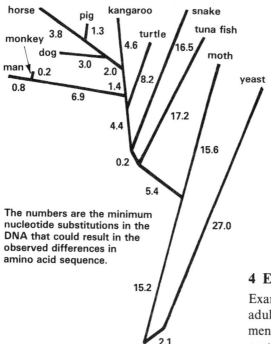

The numbers are the minimum nucleotide substitutions in the DNA that could result in the observed differences in amino acid sequence.

**Fig 21.12** Computer predicted relationship between a variety of organisms based on differences in the amino acid sequence of their cytochrome *c* (after Ayala 1978)

## 4 Embryological affinities

Examination of the **embryos** of vertebrates, which as adults are very different, reveals similarities in development. The shapes and positions of the brain, eyes, nostrils, limbs and tail are comparable in animals as remotely related as fish, amphibians, reptiles, birds and mammals. At some stage in development, each of the embryos has gill clefts to which deoxygenated blood is pumped from the heart (Chapter 19). With the exception of fish, the gill clefts later degenerate.

## 5 Behavioural affinities

Organisms whose patterns of **behaviour** are similar are more likely to be closely related than those whose behaviour is very different.

Jane Goodall in Tanzania has observed many aspects of chimpanzee behaviour. Apart from man, chimpanzees are the most habitual users of simple tools. They use sticks for attacking, exploring, poking and teasing. They collect ants and termites using small twigs and blades of grass. They clean themselves with leaves and use stones to crack nuts. The play behaviour of adult and juvenile chimpanzees resembles that of humans. Before it is weaned, a young chimpanzee is also held by its parents in a way similar to that in which human babies are nursed (Fig 21.13). The observations further substantiate the claim that chimpanzees and humans have common ancestors.

**Fig 21.13** Chimpanzee suckling its young (London Zoological Society)

### 21.2.3 Geographical distribution

Each of the main land masses of the world has its characteristic plant and animal life. Marsupials are confined to Australia and South America, elephants to Africa and India.

Wallace was one of the first naturalists to survey the world-wide distribution of birds and mammals. He noted that there are six major land masses separated from each other by natural barriers such as mountain ranges and oceans. The number of species native to Africa, Australia and South America is about twice as great as in Asia, Europe and North America. It is thought that the first mammals evolved in Asia and migrated into Africa and Europe. From here they spread into Australia and the Americas using bridges of land which at that time connected the continents.

**Fig 21.14** Changes in positions of major land masses with time as a result of plate tectonics

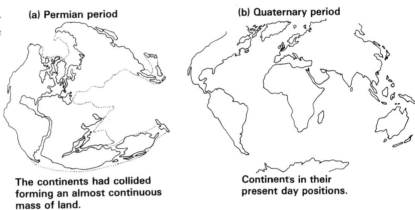

(a) Permian period

The continents had collided forming an almost continuous mass of land.

(b) Quaternary period

Continents in their present day positions.

Movement of the land masses by the processes of **plate tectonics**, caused Australia and South America to become cut off in the Cretaceous period (Fig 21.14). It is assumed that placental mammals were just beginning to evolve at this time, and plate tectonics explains why marsupial mammals are found in these areas. Africa and North America became separated from Europe later on, by which time many species of placental mammals had evolved. They are more successful than marsupials, which explains why they are more prevalent in Asia, Africa, North America and Europe. Because Asia did not move apart from Africa, migration is still possible between these land masses. It is the reason why elephants are found in both Africa and Asia.

**Fig 21.15** Bush baby, a prosimian (after J Z Young 1976)

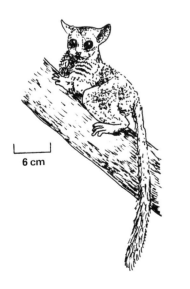

6 cm

## 21.3 Human evolution

Man is a member of the class **Mammalia** which taxonomists divide into about twenty orders including the **Carnivora** (bears, cats and dogs), **Cetacea** (whales and dolphins), **Chiroptera** (bats) and **Insectivora** (shrews and hedgehogs). The order to which man belongs is the **Primates** which has two sub-orders, the **Prosimii** and **Anthropoidea**. The prosimians include the earliest fossil primates as well as a few living representatives such as the lemur, loris and bush baby (Fig 21.15). They are probably similar to what our ancestors were like 60–70 million years ago, small tree-dwelling creatures. At the beginning of the Eocene epoch these early primates disappeared. At that time the earth's climate was generally warmer than it is now. The widespread tropical forests were the habitats of many early prosimians, some of which were similar to lemurs. About 70 distinct species have been described. It is from these creatures that anthropoids probably evolved. The evidence as to how this evolution took place is incomplete.

Fig 21.16 *Ramapithecus* (after RA Leakey 1981)

### 21.3.1 Early anthropoids

Fragments of several kinds of anthropoids have been discovered dating from the middle of the Oligocene epoch, about 35 million years ago. It appears that by then there was a clear distinction between ape and monkey lines of descent. From then on monkeys evolved along a separate path. The stock from which apes and man evolved thus probably diverged from that of monkeys 35–25 million years ago. Little is known of the adaptive features responsible for the divergence, but they were probably very small. Our ancestors were in many ways similar to monkeys.

During the Miocene epoch the climate became generally cooler and drier. Open grassy plains replaced large tracts of forest and stimulated the evolution of horses and other fast-running herbivores. Fossils of the teeth and jaws of Miocene apes have been found. Some were ancestors of gibbons, others of apes. A third group of the genus *Ramapithecus* are more like humans with rounded jawbones, small canines and spade-like incisors. A dentition of this kind would allow the jawbone to be moved from side to side in a grinding action. The diet probably included grass seeds collected from open grassland. Seed collection was probably performed by the hands suggesting that *Ramapithecus* was bipedal (Fig 21.16). This is only speculation because no other parts of the skeleton have been found. Palaeontologists do not agree on when the ape and human lines diverged. Traditionally the lines are thought to have been distinct since Miocene times, ten million years ago. However, some evidence indicates a more recent separation. Various alternatives are shown in Fig 21.17. Most typically human characteristics evolved within the Pleistocene epoch.

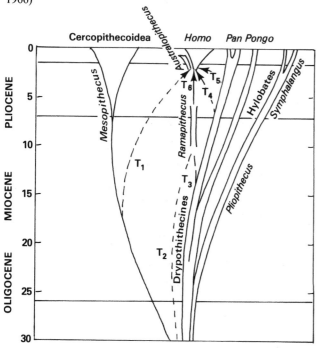

Fig 21.17 Possible course of evolution of old-world *Anthropoidea* $T_1$–$T_6$ show possible origins of the *Hominidae* in increasing order of probability (from J Z Young 1976 after Campbell 1966)

### 21.3.2 *Australopithecus*

Several human-like fossils from the early Pleistocene period have been discovered this century in various parts of Africa. Some taxonomists put them all into the genus *Australopithecus* with two species, *A. africanus* and *A. robustus*. Fossils of *A. africanus* were found at Taungs in the Transvaal, South Africa by Raymond Dart in 1925, and by Louis Leakey at Olduvai Gorge, Tanzania in the early 1960s. *A. africanus* has a flattened face with a forward projecting jawbone (Fig 21.18). The foramen magnum pointed upwards, a feature taken by some to indicate an upright gait. The teeth were generally like those of humans except the premolars and molars which were bigger. Large numbers of bones of various animals including hares, gazelle and wildebeest are frequently present in caves where fossils of *A. africanus* have been unearthed. This suggests that its diet included the flesh of herbivores which lived on nearby open grassland.

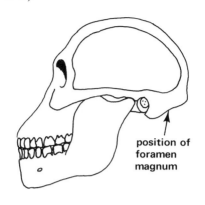

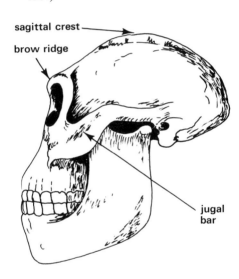

Alternatively it may be that *A. africanus* was among the prey of carnivores such as leopards and hyenas which used the caves as lairs. One of the fossils of *A. africanus* had a cranial volume of about 800 cm³ and was named *Homo habilis* by Leakey. Habiline means skilful and refers to the ability of this early primate to make simple tools. They were nothing more than stones which it may have used to throw at prey.

In 1938 Robert Broom found fragments of fossil crania and teeth of *A. robustus* in a cave at Kroomdraar, South Africa. A complete cranium of this species was discovered by Mary Leakey at Olduvai Gorge in 1959. The popular press christened him 'nutcracker man'. The skull of this species is heavy with a prominent ridge at each brow (Fig 21.19). Its cheek teeth were also larger and the cranium had a prominent sagittal crest to which a large temporalis muscle was probably attached. The jugal bar was also thick indicating that *A. robustus* had a large masseter muscle. The facts strongly suggest that vegetable material figured prominently in its diet. This is supported by scanning electron microscope studies of its teeth. They show that the tooth-wear pattern is similar to that of wild chimpanzees whose food is mainly hard, thick-skinned fruits. A diet of this kind is of low quality and has to be eaten in large quantities. It also has to be thoroughly chewed to help its digestion. The evolution of large grinding cheek teeth and massive jaw muscles in *A. robustus* enabled it to cope with such a diet. Neither species of *Australopithecus* had a diastema, a gap between the incisor and premolar teeth, seen in apes.

Something is also known of the pelvic girdle of *Australopithecus* (Fig 21.20). The ilium is short and broad as in modern man. However, it has only a small area of articulation with the sacrum and its acetabulum is simpler than that of modern man. The ischium is relatively long, hence the hamstring muscle was probably inserted a long way from the acetabulum. The arrangement enabled the hindlimb to be drawn backwards and straightened, making a bipedal gait possible.

**Fig 21.20** Pelvic bones of *Australopithecus* and man (from J Z Young 1976 after Broom and Robinson 1955). The ilium of *Australopithecus* resembles that of man but the ischium is much longer

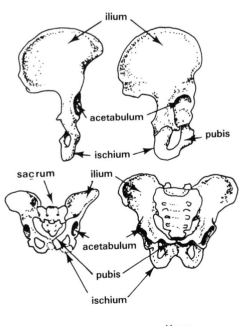

*Australopithecus*　　　　*Homo*

405

A third australopithecine species called *A. afarensis* was unearthed from river sediments in the Afar Triangle, Ethiopia in the early 1970s. The fossils are up to 3 million years old. Nearly 40 % of the bones of one skeleton, nicknamed Lucy, have been pieced together. Owen Lovejoy examined the bones in great detail and commented:

'They look incredibly primitive above the neck and incredibly modern below. The knee looks very much like a modern human joint; the pelvis is fully adapted for upright walking and the foot, although a curious mixture of ancient and modern is adequately structured for bipedalism.'

### 21.3.3 Appearance of *Homo*

The following is an abbreviated form of Le Gros Clark's definition of *Homo*:

'A genus of the family Hominidae, distinguished mainly by a large cranial capacity with a mean value of more than $1100 \, cm^3$, mental eminence variably developed; dental arcade evenly rounded with no diastema, first lower premolar bicuspid, molars variable in size with a relative reduction of the last molar; canines relatively small with no overlapping after the initial stages of wear; limb skeleton adapted for fully erect posture and gait.'

The scientific evidence suggests that man gradually evolved to such a state. Some human characteristics such as bipedalism were evident in *Australopithecus*. Others, for example cranial capacity, were not. The cranial volume of modern man ranges from 1000 to $2000 \, cm^3$, with a mean of $1400 \, cm^3$, compared with an estimated range of $400-600 \, cm^3$ for *Australopithecus*. Hence, various human characteristics evolved at different times in different populations. This phenomenon is called **mosaic evolution**.

Fossils have been found showing that creatures anatomically intermediate between *Australopithecus* and modern man existed. They support the idea that man's evolution was gradual.

### 1 Homo erectus

The first fossils of *H. erectus* were discovered in the early 1890s by Dubois in Java, hence this species is often called Java man. Subsequently other specimens have been located in China (Peking man), Germany, East Africa, South Africa, Algeria and Morocco. The remains are mainly bones of the skull, teeth, a mandible and a femur. The skull was thick with a cranial volume ranging from 775 to $1225 \, cm^3$ (mean $980 \, cm^3$). The forehead was less steep than in modern man and had prominent ridges at the brow (Fig 21.21). *H. erectus* had a heavier jaw and larger teeth than present-day man. However, the teeth were typically human in appearance and had wear patterns similar to those of modern man indicating that *H. erectus* was probably omnivorous. East African fossils of this kind are about 5 hundred thousand years old, those from Java are dated as 7 hundred thousand years old. The Chinese specimens are younger, dating from 5 to 3.5 hundred thousand years ago. They were discovered in caves alongside heaps of charcoal and ashes, suggesting *H. erectus* used fire. Also in the caves were the remains of deer and rhinoceros, together with simple tools. They were mainly sharp-edged stone flakes 2–3 cm long and were probably used to cut up carcasses of large animals. The flakes were knapped from stones, the remnants of which look like crude axeheads. As far as can be judged the rest of the skeleton of *H. erectus* was very much like our own.

With reference to Le Gros Clark's definition, suggest an order in which the characteristic features of *Homo* may have evolved.

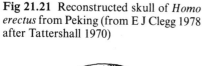

**Fig 21.21** Reconstructed skull of *Homo erectus* from Peking (from E J Clegg 1978 after Tattershall 1970)

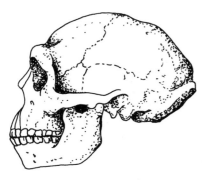

Our limited knowledge of *H. erectus* tells us they lived throughout Africa, Asia and Europe in the early to middle Pleistocene epoch from about a million to 2.5 hundred thousand years ago. Their distribution extended much further north than *Australopithecus*. This may be related to the ability of *H. erectus* to use fire and so keep warm in cooler climates.

### 2 Homo sapiens neanderthalensis

Fossils of *H. sapiens* were discovered in 1856 in the Neander Valley near Dusseldorf, West Germany. They are referred to as Neanderthal man, *H. sapiens neanderthalensis* and are no more than 50 thousand years old. Other sites where similar remains have been found are China, Gibraltar, North and South Africa. An almost complete skull and most of a skeleton was dug up at La Chapelle-aux-Saints, France in 1908.

Neanderthal man had a large skull with a cranial volume ranging from 1300 to 1600 cm$^3$. There is a prominent brow ridge across the slanted forehead. The jawbone and teeth are larger than those of modern man. Points of muscle attachment on the stout limb bones are well marked, suggesting a muscular build. The average height was about 1.67 metres (5 feet 8 inches). Neanderthal features first evolved at a time when the world was in a warm interglacial period. They were fully developed by the last Ice Age about 70 thousand years ago. Bulky bodies and short limbs are suited to cold conditions as there is less surface area per unit volume from which to lose body heat. They also used fire. *H. sapiens neanderthalensis* was thus well adapted to cope with low temperatures.

Neanderthal man made a much wider variety of tools than his predecessors. Many of them were made from stone, including hand axes, knives and scrapers. Others were of bone. Their production required skilful trimming and fine control of the hands. They also point to the Neanderthals as meat eaters, preying on herbivores such as reindeer and horses.

As in all populations, fossils of Neanderthal man show considerable variation. Some palaentologists divide them into 'classical' and 'progressive' types (Fig 21.22). The progressive type is more like modern man. He had a more rounded cranium, a high forehead and smaller brow ridge.

**Fig 21.22** Skulls of (a) a classic, *Homo sapiens neanderthalensis*, and (b) a progressive neanderthal, *H. sapiens steinheimensis* (from J Z Young 1976)

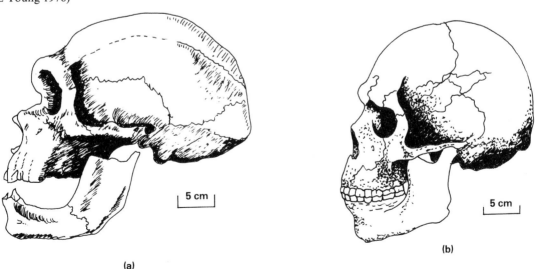

(a)

(b)

Skulls of this description have been found at Swanscombe, Kent and Steinheim in Germany. They are dated at about 250 thousand years old and are thought by some people to be related to progressive Neanderthals. Others believe them to be more like *Homo erectus* who lived at that time.

More recent finds of Neanderthals have been made in caves at Mount Carmel, Israel. In one there are the remains of the classical type, in another nearby are fossils of the progressive type. Carbon dating gives an age of about 45 thousand years. It appears that both types were in existence for many thousands of years. The progressive type is more like modern man, and he is thought to be our ancestor.

### 3 Homo sapiens sapiens

All human remains from about 35 thousand years ago to the present are modern man, *Homo sapiens sapiens*. He first appeared during the last glacial period when the climate placed great stress on man's capacity for survival. Even in southern France and Spain the summer temperature probably averaged only 16 °C. One of the most famous finds of early modern man was made at Cro-Magnon in the Dordogne, France. In all respects he is typical of present-day man. Other similar remains have come to light at Combe-Capelle, France and Lautsch, Czechoslovakia.

Many of the typical features of modern man reflect the arboreal existence of our early ancestors. One of the ways our predecessors probably moved in trees was to swing from branch to branch. This requires great mobility of the upper limb, especially at the shoulder joint, and the ability to flex the phalanges of the hands. Being able to twist the forearm also helps. Moving about in this way places great demands on motor coordination. The ability to judge distances is also important. Having eyes at the front of the head enables fields of vision to overlap to give stereoscopic vision which permits distances to be judged with precision.

Many mammals have scent glands and smell plays a significant role in communication with each other. In trees, communication by smell is not very effective and visual signals such as facial expression and audible signals are more useful. The evolution of such systems of communication may have been instrumental in the development of social life. Another effect of moving among trees is that it would be disadvantageous to carry large numbers of young. Producing one or only a small number of offspring made it possible to carry youngsters for a long time after they were born, giving them a long infancy in which they could learn a lot from their parents. Having mammary glands on the chest would permit the infants to be suckled while the mother is on the move.

The food of our tree-dwelling forebears was probably mainly leaves and fruit. The evolution of a herbivorous dentition enabled our ancestors to deal with a diet of this kind. Our teeth retain many of these features. The incisors have flat, sharp upper edges suited to cutting leaves and nibbling fruit. The canines are reduced in size and the molars and premolars provide grinding surfaces on which plant food is reduced to pieces small enough to be swallowed.

Bipedal locomotion evolved when our ancestors began to colonise the expanding grassy plains during the Miocene period. Our long-armed predecessors probably used their knuckles to support their bodies as they moved on their hind legs. A full striding bipedal gait requires the gluteus maximus muscle to be used as an extensor of the hip, thus permitting the trunk to be raised into an upright position. The propulsive thrust of the legs is created by the quadriceps and calf muscles which extend the knee and ankle respectively. Moving about on hindlimbs gave our ancestors a better view of their surroundings and freed the hands which could thus be used for grasping food.

Colonisation of the plains probably brought with it a change to an omnivorous diet. Lizards, birds and small mammals were killed at first, large mammals later. As the hands evolved, sticks and stones could be used to throw at prey, and simple tools for cutting up carcasses could be made. An important feature in this respect was the lengthening of the thumb so that it could be opposed to the fingers (Fig 21.23). This made it possible for the hands to grip and to manipulate objects with precision. These facilities would help a hunting way of life. To strike prey would require weapons to be held with a power grip. Making and using simple tools such as stone flakes for butchering carcasses could only occur if the hands could be used precisely to manipulate objects.

**Fig 21.23** (a) the power grip, and (b) the precision grip (from J Z Young 1976 after Napier 1956)

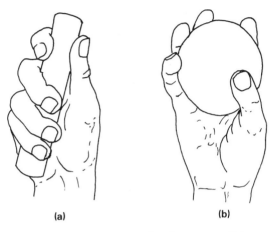

(a)        (b)

Success in hunting animals may well have been the main selective advantage in the evolution of many other human characteristics. Because our ancestors were unable to run as quickly as their prey they had to resort to sustained tracking, often over rough terrain. The human foot allows movement for long periods of time over uneven ground. The forearms can lift heavy loads such as the carcasses of large prey.

Another human feature which may well have evolved at the same time as the hunting habit was loss of body hair. It may be that the exertions involved in hunting, especially in a hot climate, placed considerable stress on the body's cooling system. Loss of body hair would facilitate heat loss. Evaporation of sweat from large numbers of sweat glands in the skin helps maintain an even body temperature during exertion (Chapter 17). Pigmentation of the skin is an advantage in sunny climates. A dark skin absorbs ultraviolet light (Fig 21.24), lowering the risk of sunburn. One of the serious side-effects of sunburn is that it inhibits sweating. In addition, deep

**Fig 21.24** Skin reflectance curves for individuals of four different racial groups (from E J Clegg 1978)

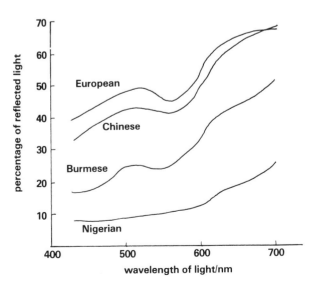

409

pigmentation protects against skin cancer caused by absorption of too much UV light. In cloudy, temperate regions, lack of skin pigment is of some advantage. Light coloured skin permits UV light to penetrate deeper into the skin where ergosterol is converted to vitamin D. Dark skinned individuals living in such conditions may therefore suffer from rickets if their diet is inadequate in vitamin D (Chapter 9). A summary of the main stages of man's evolution is given in Table 21.3.

**Table 21.3 Main stages of man's evolution (from JZ Young 1976 after Campbell 1964)**

| Years BP × $10^3$ | EPOCH | FOSSILS | COUNTRY/AREA |
|---|---|---|---|
| | Upper Pleistocene | H.s. sapiens<br>H.s. neanderthalensis<br>? H.s. steinheimensis | Asia, Europe<br>Africa, Israel<br>Europe |
| 100 | | | |
| 200 | Middle | H. erectus | Morocco |
| 300 | | H.s. steinheimensis<br>H.s. steinheimensis | Germany<br>England |
| 400 | Pleistocene | | |
| 500 | | H. erectus | China, Hungary |
| Lower | | A. robustus | Germany<br>Java<br>Tanzania (Olduvai)<br>S. Africa (Kromdrai) |
| 1000 | Pleistocene | A. africanus | S. Africa (Taungs) |
| | | A. ? robustus | Tanzania (Olduvai) |
| 1500 | | A. ? africanus<br>(= H. habilis) | Tanzania (Olduvai) |

### 21.3.4 Evolution of human populations

Socialisation was a feature of tree life where our ancestors probably moved about in troupes. A way of life which later included hunting needed co-operation among the individuals in trapping large animals and carrying the prey. It is unlikely that females with young or pregnant females would participate in hunting. They could remain at a base camp or cave where the young could be cared for. Hence there was probably some division of duties in family groups of early man. Aggression and non-cooperation among the young could be suppressed during the long period before puberty. In this way a stable social system could evolve.

Traditionally it is thought that early modern man lived in small populations as nomadic hunters and food gatherers. There is evidence that, rather than merely relying on whatever came along, they also manipulated food resources. Many of the sites where early man lived in southern France were strategically placed for driving and corralling herds of reindeer. Many of their tools were made of bone rather than stone and included spear heads, awls and needles. It is thus possible that they also followed and stalked migrating reindeer. At this time in South Africa early man was capturing a variety of large mammals including Cape buffalo. He also collected shellfish such as mussels as well as birds and fish.

About ten thousand years ago most human populations adopted some form of agriculture. The **Agricultural Revolution** moved ahead with great speed and accompanied an enormous increase in world population. It has been estimated that the total number of humans was between five and ten million at the start and within 8000 years it had risen to three hundred million.

The transition to an agricultural way of life occurred independently in several places. The three major centres where agriculture developed were Central America, Northeast China and the Fertile Crescent of the Middle East. The change is linked with an improvement in the climate following the last Ice Age. As the ice sheets began to melt the land became colonised with trees, shrubs and grasses. Hunter-gatherers included grass seeds in their diet. As grasslands enlarged, humans exploited the new rich food resource and this led to a population explosion. Another view is that population growth was the cause rather than the result of the Agricultural Revolution. Hunting and gathering could not provide enough food for the increasing numbers of humans. A change to agriculture provided enough food to meet the needs of the growing population.

**Fig 21.25** The Fertile Crescent shown as the shaded area on the map (from R A Leakey 1981). Settled agriculture was a way of life there 10 000 years ago

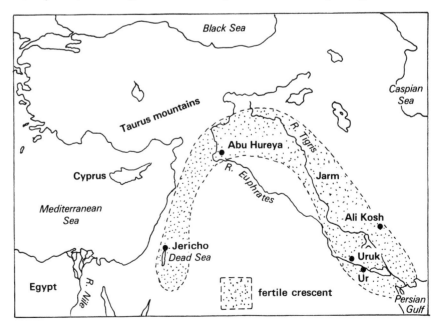

The **Fertile Crescent** (Fig 21.25) has been intensively studied as a model of the probable events of the Agricultural Revolution. It is an area sheltered by the foothills and mountains of Israel, Jordan, Syria, Turkey and Iran.

411

Though it escaped the rigours of the Ice Age its climate at the time was drier and cooler than today. The vegetation in the area was mainly small stands of oak trees among which grew large-seeded wild grasses, the ancestors of cereals. Hunter-gatherers collected the seeds and captured rabbits, goats and gazelle on the surrounding sparse grassland.

When the climate improved after the Ice Age, tree and shrub cover became more extensive and was accompanied by a spread of the large-seeded grasses. They became a substantial part of the diet. Even more important, grain in excess of immediate requirements could be stored for use when food was scarce. This enabled populations to lead a sedentary existence and village life evolved. Supportive evidence comes from an excavation carried out in the early 1970s at Abu Hureya in the Euphrates Valley, northern Syria. This was the site of a small village occupied between 11 500 and 7500 years ago. Among the finds were grinding stones, small flint sickles and cereal grains of three types. One of these was a primitive wheat called Einkorn.

There is evidence that such cereals were even cultivated. To grow crops land has to be broken up. Disturbing the ground leads to the appearance of weeds, and the seeds of weeds were among the items discovered at the village. The transition from gathering wild cereal grains to their systematic cultivation may have happened quite naturally. As the population of a village grew, its inhabitants began to specialise in particular crafts as a contribution to community life. In this way basket-makers, tool makers, potters and weavers began to emerge.

From the Fertile Crescent, agriculture spread as the early farmers moved out and colonised new areas. Using genetic markers in present-day populations throughout the near East and Europe, researchers have recently estimated that the spread of agriculture took place on average at a rate of about 1 km a year. In Central America the practice of agriculture evolved independently about 5000 years ago, in the valleys and foot-hills of Mexico especially. In Peru people living near the sea abandoned hunting and gathering and took to fishing. They turned to agriculture later.

When sedentary communities appeared, trade developed between them. Trading centres such as Jericho evolved. Jericho is thought to be the world's first city, established 10 500 years ago. Organisation of large numbers of people required central control and authority invested in chiefs. Social hierarchies thus became established. The property of such societies needed protection from neighbouring populations. Construction of fortifications such as the huge wall around Jericho required planning, organised cooperation between the inhabitants and the development of new building skills. Many of the world's towns and cities probably grew up in this way.

In the **Industrial Revolution**, an agricultural way of life changed to one based on industry. Between 1730 and 1850 Britain was transformed from a mainly agricultural to a predominantly industrial country, beginning with the invention of simple machines such as the spinning jenny and the steam engine which made it possible to mechanise the textile industry. Major developments in mining and industrial organisation soon followed. New bridges, canals, railways and ships radically improved transport. Industry became concentrated around the coalfields, and large new industrial towns such as Birmingham, Manchester, Newcastle-upon-Tyne and Glasgow emerged as country people flocked in to seek employment. These changes soon followed in many other countries. With them came many of the problems such as pollution and disease control facing us today. They are considered more fully in Chapter 25.

## 21.4 How evolution came about

There have been many attempts to explain how life evolved into the enormous variety of living organisms which now inhabit the earth. The theories of Lamarck, Darwin and Wallace are particularly important.

### 21.4.1 Lamarckism

**Jean-Baptiste Lamarck** (1774–1829), a French naturalist, was the first modern biologist to propose how the variety of living organisms has evolved. His theory was based on two main premises:

**i** That organisms acquired new features because of an inner need.
**ii** That the acquired features are inherited by future generations.

Lamarck's ideas are often illustrated with reference to the evolution of the giraffe's long neck. According to Lamarck, the ancestor of the giraffe was a short-necked creature whose inner need to eat the leaves of trees caused it to stretch higher so that its neck became progressively longer. The offspring of these animals inherited the longer neck so that they could feed in the same way. It is the second part of the theory which is incorrect. Living organisms may acquire new characteristics during their lifetime. For example, a blacksmith acquires large biceps because of the nature of his work. However, the children of a blacksmith have biceps which are normal in size. It is thus evident that acquired characteristics are not inherited.

### 21.4.2 Darwinism

**Fig 21.26** Charles Darwin (Ann Ronan Picture Library)

**Charles Darwin** (1809–1882) (Fig 21.26) was well acquainted with the variety of life. Between 1831 and 1836 he worked as a naturalist on HMS Beagle on its voyage around the world. He made careful records of the diversity and distribution of living organisms, particularly in South America, the Pacific Islands and Australia. Darwin was also a keen geologist. He made an extensive collection of fossils from various parts of the world.

After many years of observation and experiment, Darwin's theory of evolution was presented in a paper to the Linnaean Society of London in 1858. At the same meeting, a paper written by **Alfred Russel Wallace**, an English naturalist who had recently explored the East Indies and Australia, was also read. Darwin and Wallace had arrived independently at the same conclusions. Darwin's theory of evolution, illustrated by numerous examples, was published in 1859 in the ***Origin of Species by Means of Natural Selection***. It was based on the following observations and deductions:

**i.** Most organisms have the potential to produce very large numbers of progeny. For example, in a single season a female cod lays about $85 \times 10^6$ eggs, a frog $1–2 \times 10^3$ eggs, an orchid produces $1.7 \times 10^6$ seeds. There must be intense **competition** because the numbers of adult organisms remain fairly stable. Some degree of differential death occurs either among gametes or zygotes, or during development. Very few sperm produced by man fertilise ova, and relatively few ova are fertilised, hence there is a high death rate among human gametes.

**ii.** All organisms, even members of the same species, **vary**. You have only to look around you to notice that this is so!

**iii.** The progeny which are most likely to survive competition are those which have the combination of features which enable them to cope best with their environment. This is called **survival of the fittest by natural selection**. Fitness can be defined as the capacity of an individual to survive and the probability of it producing viable offspring.

**iv.** The features favoured by natural selection are **inherited**.

Darwin's interpretation as to how the long neck of the giraffe was evolved would go something like this:

Among the ancestors of the giraffe were animals with a range of neck lengths. Those with long necks could browse on the leaves of trees. In this way they could also avoid the competition with other herbivores which ate vegetation growing nearer the ground. The short-necked ancestors succumbed to the competition and became extinct. The long-necked variety survived and gave rise to long-necked progeny which are the ancestors of the modern giraffe.

### 21.4.3 Variation

Darwin had no idea of the origin and causes of variation (Chapters 4 and 6). Nor did he know how characteristics were inherited. Mendel's work (Chapter 20), though published in Darwin's lifetime, was neglected until 1900, well after Darwin's death. Man reproduces sexually, consequently a set of chromosomes from each parent is inherited by a child. Genetic variability in gametes can arise by random orientation and crossing over between chromosomes during meiosis (Chapter 6). This is partly why humans vary in many characteristics such as height, colour of skin, hair and eyes, blood pressure, resistance to disease and intelligence.

Inheritable variation is of two kinds, **continuous** and **discontinuous**. Continuous variation is caused by the random assortment of many genes which affect some characteristics such as height, skin and eye colour. Discontinuous variation occurs where a characteristic is determined usually by a single gene (Chapter 20). The discovery of the nucleic acids and their functions have made it possible to explain the causes of discontinuous variation (Chapter 4). Changes in the genetic code are known as **gene mutations**. All genes mutate, but some do so more frequently than others. **Chromosome mutations** are caused by an alteration in the number of chromosomes or by fragmentation of chromosomes. Mutations may occur in any body cell but are inherited only when they appear in gametes. Somatic mutations may occur in any of our body cells, giving rise to local effects, for example, a white lock in a head of dark hair.

Gradual accumulation of somatic mutations may cause the changes occurring in our bodies as we age. Physical and chemical agents are **mutagenic**. The main physical agent is high energy radiation. Mutagenic chemicals include methanal and mustard gas. If they are to be effective in bringing about evolutionary change, mutations must not affect the viability of individuals before reproductive age. Unsuitable mutations may cause death and may not therefore be inherited.

### 1 Gene mutations

Proteins perform numerous functions in living organisms (Chapter 2), so they determine many of an organism's characteristics. The role played by a protein is determined largely by its primary structure, the sequence of amino acids in its molecule. The primary structure affects the three-dimensional structure, the **conformation** of the protein molecule. It is the conformation which determines the protein's function. For example, the type of substrate which binds to the active centre of an enzyme depends on the conformation of the enzyme molecule. If the primary structure of the enzyme is abnormal, its conformation may be so altered that substrate molecules cannot bind to the active centre.

The sequence of amino acids in proteins is precisely regulated by the genetic code (Chapter 4). Any change in the code, a **gene mutation**, causes abnormal proteins to be made. Sometimes the change is harmless. It may even be beneficial. If this is so, the new gene may confer an advantage on its possessor. As a consequence, it may increase in frequency in a population, especially if it is a dominant gene. Usually gene mutations are harmful, because the abnormal proteins which are synthesised are unable to perform their normal functions. Individuals with such mutations may not survive to reproductive age, so the mutated gene is not passed to the next generation. Gene mutations are the cause of haemophilia and sickle-cell anaemia in humans (Chapter 10). Interestingly, people who are heterozygous for sickle-cell anaemia are resistant to malaria, so it is more likely that they will survive to childbearing age than normal individuals in countries where malaria is endemic. It is for this reason that the gene for sickle-cell anaemia is frequent among people who live in tropical countries.

## 2 Chromosome mutations

During meiosis, one or more chromosomes may break. A variety of things can happen to fragments of broken chromosomes:

i. they may become lost (**deletion**),
ii. they may become attached to the end of another chromosome (**translocation**),
iii. they may become turned round and rejoin the chromosome from which they became detached (**inversion**),
iv. they may become inserted into another chromosome (**duplication**).

Clearly, these changes may alter the complement of genetic material in the gametes.

Exposure to ultraviolet light is one of the causes of chromosome fragmentation. Nucleic acids strongly absorb UV light. New varieties of moulds such as *Penicillium* are produced by irradiation with UV light. Many new strains of crop plants have been developed by exposing their seeds to $\gamma$ radiation. Sometimes a pair of homologous chromosomes fails to separate in meiosis, giving rise to gametes with one chromosome less and one chromosome more than normal. The progeny formed from such gametes are called **polysomics**, having fewer or more chromosomes than normal. Down's syndrome in humans is caused by an extra chromosome (Chapter 20).

In other instances, whole sets of homologous chromosomes do not separate in meiosis, so that diploid gametes are produced. Fusion with a normal haploid gamete gives rise to progeny with a triploid chromosome number. This condition is called polyploidy and is more often observed in plants than in animals. Many modern varieties of crop plants are **polyploids** (section 21.4.4). Polyploids are usually more vigorous than their diploid counterparts.

### 21.4.4 Selection

Darwin saw the important role of selection as an evolutionary force. In *The Origin of Species by Natural Selection* he described many examples of both **natural** and **artificial selection**.

### 1 Natural selection

A number of experimental and field studies carried out this century have provided an insight into how selection works in natural conditions. One of the first studies was of the peppered moth, *Biston betularia*. Collections of preserved peppered moths show that until almost the middle of the nineteenth century the only variety known had the speckled, light colour of pepper dust. In 1845, a black moth of the same species was caught in Manchester. By the end of the eighteen-hundreds the black variety made up 98 % of the population of the moth in the Manchester area.

The change in frequency of the gene for black colour coincides with the spread of heavy industry in the locality. Before the Industrial Revolution the air in and around Manchester was clean. The bark of trees growing in the area was covered with lichens. Against such a background the speckled variety of the moth is camouflaged (Fig 21.27). Insectivorous birds presumably had difficulty in seeing it, so the moth had a good chance of surviving and reproducing. The soot and smoke generated by heavy industry polluted the air and was deposited on the bark of trees. Many of the lichens were killed by the high concentrations of sulphur dioxide in the atmosphere. Against a dark background the speckled moth would be seen easily and taken by insect-eating birds. In contrast, the black form would be less conspicuous. It would be less likely to be taken as part of the diet of insectivorous birds. Consequently its survival was favoured. It was for this reason that the gene for black colour gradually became more frequent in subsequent generations.

To prove that natural selection was the cause of the changed gene frequency, Dr H B D Kettlewell bred large stocks of the two varieties of the peppered moth, which were marked and then released in two areas:

i. in polluted Birmingham where over 90 % of the indigenous peppered moth population was the black variety, and

ii. in an unpolluted rural area of Dorset where none of the peppered moths was black.

He observed and filmed insectivorous birds such as robins, redstarts and hedge sparrows feeding in the two localities. The birds were particularly severe on the variety of moth which was not camouflaged by its background. Kettlewell also recaptured the surviving marked moths (Table 21.4). The experimental fieldwork confirmed that **natural selection** was the cause of the increase in frequency of the gene for blackness among

**Fig 21.27** Non-melanic and melanic varieties of the peppered moth (Heather Angel)

**Table 21.4 Natural selection of the peppered moth in two localities (from Marshall 1978)**

|  | Non-melanic | Melanic |
|---|---|---|
| Dorset, 1955 (unpolluted) |  |  |
| released | 496 | 473 |
| recaptured | 62 | 30 |
| % recaptured | 12.5 | 6.3 |
| Birmingham, 1953 (polluted) |  |  |
| released | 137 | 447 |
| recaptured | 18 | 123 |
| % recaptured | 13.1 | 27.5 |

peppered moths in industrialised areas. This is an example of **industrial melanism**. It has since been observed among many other types of moth in Britain, Europe and the USA.

Many other examples of natural selection have since come to light. A number of them have important implications in terms of human health and survival. For example, strains of malarial mosquitoes resistant to insecticides, especially DDT, have evolved in some parts of the world. For this reason malaria has returned to countries from where it had been eradicated in recent years.

## 2 Artificial selection

For centuries man has selected livestock and crop plants because they have desirable features such as resistance to disease and high yield. Where this happens it is called **artificial selection**. In *The Origin of Species* Charles Darwin quoted many examples of how livestock had been improved by artificial selection.

A well-documented example of the influence of man in selection is seen in the evolution of wheat. In wild wheats the ripe ear is brittle and breaks up to release each seed-containing spikelet separately. It brings about efficient dispersal of the seeds. This type of wheat is of little use for cultivation because the ear breaks off when the grain is harvested. Over the past 10 000 years or so, man has selected non-brittle varieties of wheat for agriculture. One of the first was Einkorn wheat, *Triticum monococcum*, from Iran with a diploid number ($2n$) of 14 (genome AA). It was widely cultivated in Neolithic times. Einkorn wheat is, however, a low-yielding variety, so it was later replaced by Emmer wheat, *Triticum dicoccum*, a tetraploid with a chromosome number of 28 (genome AABB). Half of its chromosomes (AA) are thought to have come from Einkorn wheat. The other half (BB) probably came from a grass *Agropyron*. Emmer wheat was grown for thousands of years by the Ancient Egyptians. It is still cultivated on a small scale in a few parts of India and Russia. One of its disadvantages is that it is difficult to remove its seeds from the chaff in threshing. For this reason 'naked' wheats have been selected in more recent times. Bread wheat, *Triticum aestivum* ($2n = 42$, genome AABBDD) is the most widely grown naked wheat today. It is a hexaploid, probably a hybrid of Emmer wheat and the grass *Aegilops* (Fig 21.28).

**Fig 21.28** Evolution of bread wheat, *Triticum aestivum*

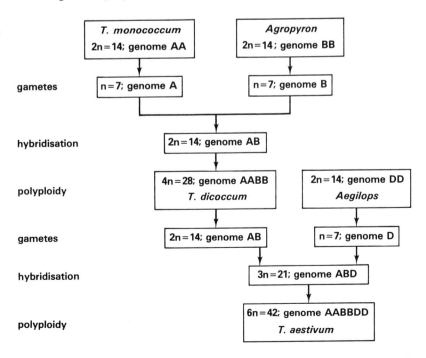

417

## 21.4.5 Isolating mechanisms

A **species** can be defined as:

> **a natural population, the individuals of which are actually or potentially capable of breeding with each other to produce fertile viable progeny, and which do not interbreed with members of other species.**

Not all individuals in a natural population have the same chance of breeding with one another. Every species is thus made up of a number of breeding populations called **demes**. When demes become isolated they become subjected to different selection pressures. Consequently the gene pool of a deme can be so changed that the individuals in the deme may in time lose their ability to breed with other demes of the same species. They may then be thought of as a new species. There are several ways in which isolation can occur.

### 1 Geographical isolation

One of the most striking examples of how new species can arise when demes become spatially separated was observed by Darwin. On his journey in HMS Beagle, Darwin visited the volcanic islands of the Galapagos Archipelago, west of Ecuador. He noticed that there were just twenty-four species of bird on the islands, most of which he had not seen on the South American mainland. Fourteen of the species were finches belonging to four genera. In his account of the journey in *Voyage of the Beagle*, Darwin referred to the finches as follows:

> 'One might really fancy that from the original paucity of birds on this archipelago, one species had been taken and modified to different ends.'

The species of Darwin's finches can be distinguished by their feeding habits and the shape of their beaks. Finches are usually ground-dwelling, seed-eating birds. However, on the Galapagos Islands the finches eat a variety of foods (Fig 21.29). Some are seed-eaters, others feed on cacti. Of the

**Fig 21.29** Galapagos finches

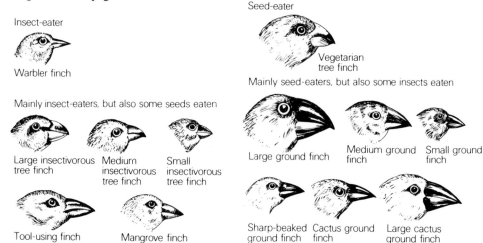

Insect-eater
Warbler finch

Mainly insect-eaters, but also some seeds eaten
Large insectivorous tree finch   Medium insectivorous tree finch   Small insectivorous tree finch
Tool-using finch   Mangrove finch

Seed-eater
Vegetarian tree finch

Mainly seed-eaters, but also some insects eaten
Large ground finch   Medium ground finch   Small ground finch
Sharp-beaked ground finch   Cactus ground finch   Large cactus ground finch

insectivorous finches, one feeds like a woodpecker. Subsequent studies of the anatomy and reproductive behaviour of the finches have revealed the supposed evolutionary relationship between the species.

It is assumed that sometimes after the Galapagos Islands were formed by volcanic eruption, seed-eating finches flew across from the mainland of nearby South America. The vegetation on the newly formed islands was sparse, hence seeds were scarce and the finches adopted a variety of new eating habits. This is another example of **adaptive radiation**. In time the demes of finches became so different in their general habits that they no longer interbred. As a result, fourteen new species gradually evolved from the single species which first colonised the islands.

Comparable patterns of evolution have been observed in island populations in Britain. The mainland, Hebridean and Shetland forms of the British wren are slightly different in colour and wing length and do not normally interbreed. At the moment they are considered to be varieties of the same species. The Orkney vole exists as five sub-species on different islands off the coast of Scotland.

It is possible that **geographical isolation** played a part in the evolution of man by separating races adapted to different climates. With global transport it is unlikely that such isolation will contribute to our future evolution.

## 2 Reproductive isolation

Even when related demes live in the same locality, they do not always breed with each other. The causes of this are twofold:

### A. Pre-fertilisation

*Seasonal barriers* Many flowering plants and mammals reproduce seasonally because of different responses to day length. It is probable that some species have evolved from demes which reproduce at a different time of the year from other demes in a population.

*Behavioural barriers* The plumage of male finches living on the Galapagos Islands is so similar that it is difficult for us to identify the species to which they belong. However, female finches have no problems in identifying a male of the same species. During courtship the male passes food to the female in his beak. It is the shape of the male's beak which enables the female to recognise her mate. In this way interbreeding between demes does not happen.

*Mechanical barriers* The genitalia of male and female animals in related demes may be different in size. It is often assumed that copulation may thus be prevented.

*Physiological barriers* Many related demes of flowering plants are reproductively mature at the same time and there are no spatial, behavioural or mechanical barriers. Pollination may occur, yet they do not interbreed. The reason is that the stigma and style of the recipient flower do not provide suitable physiological conditions for germination of pollen grains from a related deme.

### B. Post-fertilisation

Some demes intermate, but do not produce viable progeny. Among the reasons for this are:

*Hybrid inviability* Fertilisation occurs when cross-mating takes place between many species. However, the embryos do not fully develop, hence the progeny are not born alive. The reasons for this are not fully known.

*Hybrid sterility* Mules are produced by crossing a horse with a donkey. Because the chromosomes from the parents do not pair at meiosis, the gametes of mules are non-viable. It is for this reason that mules are sterile and the genes of horses and donkeys remain isolated.

## 21.4.6 Species formation

Once a deme has been isolated from the rest of its species, natural selection acts on variants within the deme, changing the frequency of genes. The genome of an isolated deme may, as a result, become so different from the genomes of other demes that some degree of reproductive incompatibility exists. After prolonged isolation, total sterility may develop, so that the isolated deme becomes a new species.

The stages in the formation of a new species involve the appearance of:

  i. a **variety** – a recognisable type within a deme,
 ii. a **race** – a partially isolated deme,
iii. a **sub-species** – a deme which is partially fertile with the original species, and finally
 iv. a **species** – a deme which is infertile with the original species.

The process is not irreversible. Where it has not gone to stage iv, interbreeding may take place if the barriers preventing it are removed.

There are many demes of man, differing from each other in gene pool and thus in appearance. Viable offspring are produced when humans of any two demes mate. Hence, there is only one species of man, *Homo sapiens*. The various demes of man are races called *Homo sapiens sapiens*.

Evolution is a controversial subject of which much is still to be learned. Darwin compared it with trying to unravel the story of a novel of which only one page was available with half the lines missing and in each line half the words missing. Shortly after the publication of *The Origin of Species*, Darwin was severely criticised by the clergy who favoured the idea of **Special Creation**. However, in the hundred or so years since Darwin's death many new and important biological and geological discoveries have served to confirm his ideas of how the variety of life we see today has come into being.

---

What are the main races of man and what are their typical features?

---

420

# SECTION 4
# The Environment

# 22 Growth of populations

The number of individuals in a population of any species that can be supported in a given habitat is called the **carrying capacity**. The density of a population is the number of individuals per unit area or volume of the habitat. Some of the factors which determine the size of the population are **density-independent**. They affect the growth of the population whether it is dense or sparse. Others are **density-dependent**, usually affecting the population only when it is dense. To understand how such factors regulate populations it is useful to study a model of population growth. The rate of growth is the number of individuals added to a population in a given time. Studying the growth of human populations is vital as it enables the development and implementation of strategies which allow for population changes. Governments should then be better able to plan resources and direct the activities of farmers, fishermen, engineers, health care personnel and others to the population's advantage.

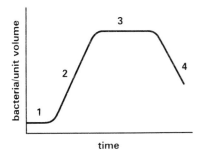

**Fig 22.1** Growth of a population of bacteria

## 22.1 Principles of population growth

A model for examining the principles of population growth is a culture of bacteria. This can be prepared by inoculating 10–15 cm³ of nutrient broth with the species being studied, then incubating the culture at an optimum temperature. At known time intervals the number of bacteria per unit volume of medium is determined. Four phases of population growth are usually seen (Fig 22.1).

### 22.1.1 Lag phase

In this phase the population grows rather slowly. There are several reasons for the slow growth. The bacteria may have been in a dormant state, and time is required before their metabolism begins to work efficiently. They may have been growing previously in a different medium and have to adjust to the new medium.

### 22.1.2 Log (exponential) phase

Here the population increases at its most rapid rate. It doubles its number in a given time. There are no factors limiting growth. Nutrients and oxygen are in plentiful supply and there is ample space. It is called the log phase because a straight line is obtained if the logarithm of the number of bacteria is plotted against time.

### 22.1.3 Stationary phase

By this time the population size is stable and does not increase further. The habitat's carrying capacity has been reached. The key factors controlling growth here are density-dependent factors such as the amount of nutrients and supply of oxygen. Shortage of nutrients and oxygen prevents any further increase in the density of the population. This is known as **environmental resistance**.

**Fig 22.2** Effect of temperature on the growth of a bacterial population

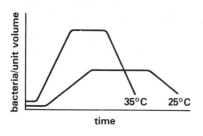

### 22.1.4 Death phase

At this stage the carrying capacity of the environment has declined. It is unable to support such a high density of bacteria and they begin to die. They may die of starvation, a shortage of oxygen, or waste products may be present in toxic amounts.

If the investigation is repeated at 10 °C lower, the growth rate would be about half of what is seen at the optimum temperature. The change in temperature has this effect whatever the density of the bacterial population (Fig 22.2). It is a density-independent factor. Remember that the laboratory model does not necessarily indicate what happens in populations of the same organism outside the laboratory. It has not allowed for the possibility of resource renewal, competition with other species, disease, predation and climatic changes. Nevertheless, there are two very important conclusions to be drawn from such studies. The first is that the density of a population will increase up to the carrying capacity of its environment. Secondly, as the carrying capacity alters, so does the population density (Fig 22.3).

**Fig 22.3** Effect of population size on carrying capacity of the environment (from Owen, *What is Ecology?* published by OUP)

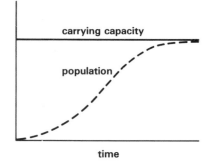

(a) Stability resulting from pressures exerted by carrying capacity

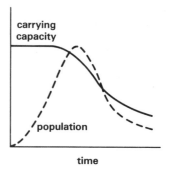

(b) Decrease in population caused by reduction of carrying capacity

**Fig 22.4** Growth of the human population (from J Z Young 1976 after Dorn 1966)

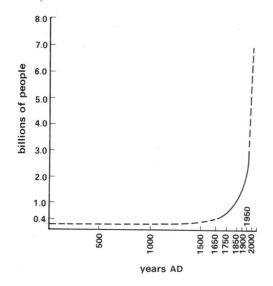

## 22.2 Growth of human populations

How far can the principles outlined above be applied to human populations? We first need to know the size of the human population over a long period of time. In many countries, population censuses have been carried out only in the last 150 years or so. Hence the figures for earlier times are estimates based on burial sites, parish registers and other documents. Populations increase when the number born exceeds the number dying in a given time. The human population appears to have had a very long lag period which ended about 300 years ago (Fig 22.4). Between 4000 BC and about 1659 AD the population size doubled every 2000 years. However, from the middle of the seventeenth century the growth rate began to increase considerably. In less than 200 years the population had doubled again and a further doubling had occurred by 1930. What has been happening for the past 300 years is **hyper-exponential growth**. The interval required for the population to double its size is almost halved as time goes on. This contrasts with exponential growth where the interval of time required for the population to double remains fixed.

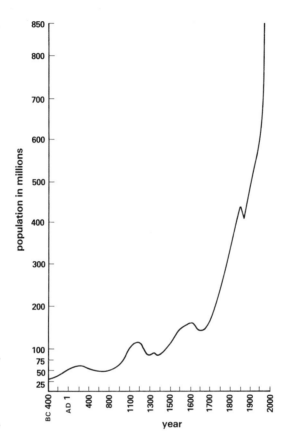

**Fig 22.5** Growth of the Chinese population (from McEvely and Jones 1978)

Use Fig 22.5 to calculate the rate of growth of the population of China over the past century. How does this compare with the average for the world population shown in Fig 22.4?

The explosive increase in the numbers of people has not occurred evenly throughout the world. It first began mainly in western countries and subsequently spread to most other parts of the globe. Another important feature is that growth of the human population in every country has occurred in spurts (Fig 22.5). The rate of growth also differs from one country to another (Table 22.1). Consequently there is an enormous range

**Table 22.1  Population growth in selected countries for 1978 (World Population Data Sheet 1979)**

| Country | Annual birth rate | Annual death rate | Doubling time (y) |
|---|---|---|---|
| | (per 1000 population) | | |
| Kenya | 51 | 12 | 18 |
| India | 34 | 15 | 36 |
| China | 20 | 8 | 58 |
| USSR | 18 | 10 | 87 |
| USA | 15 | 9 | 116 |
| Sweden | 12 | 11 | 693 |
| United Kingdom | 12 | 12 | — |

of **population density** (Table 22.2). Java is the most densely populated country in the world with 426 people km$^{-2}$; in Britain the figure is 218 km$^{-2}$. Considerable variation also occurs between and within countries. For example, in China the population density is 75 people km$^{-2}$ but for Tibet the figure is only 1.2 km$^{-2}$, in the Yang-tse Kiang valley there are over 500 km$^{-2}$ whereas part of Hong Kong has the highest population density with a phenomenal 290 000 km$^{-2}$.

**Table 22.2 Population, area and population density for major areas of the world (from UN and Unesco Statistical Yearbooks 1981 and 1984)**

| | Population $\times 10^6$ | | | | | | | | Area/ km$^2 \times 10^3$ | Density per km$^2$ (1980) |
|---|---|---|---|---|---|---|---|---|---|---|
| | 1950 | 1955 | 1960 | 1965 | 1970 | 1975 | 1980 | 1982 | | |
| World | 2525 | 2757 | 3037 | 3354 | 3696 | 4066 | 4432 | 4586 | 135837 | 33 |
| Africa | 220 | 245 | 275 | 312 | 355 | 407 | 470 | 499 | 30330 | 15 |
| America | 330 | 370 | 415 | 462 | 509 | 558 | 612 | 635 | 42082 | 15 |
| Asia | 1389 | 1524 | 1693 | 1887 | 2111 | 2352 | 2579 | 2676 | 27576 | 94 |
| Europe | 392 | 408 | 425 | 445 | 459 | 474 | 484 | 758 | 4937 | 98 |
| USSR | 180 | 196 | 214 | 231 | 242 | 253 | 265 | | 22402 | 12 |
| Oceania* | 12.6 | 14.2 | 15.8 | 17.5 | 19.3 | 21.2 | 22.8 | 23 | 8510 | 3 |

* Includes Australia, New Zealand,
  Melanesia and Polynesia.

## 22.3 Factors affecting growth of human populations

The irregular pattern of growth of human populations (Fig 22.5) indicates that different factors are at play regulating the population size at different times. Although some of them are closely related to the density of populations and others not, it is likely that some factors fall into both categories.

### 22.3.1 Food production

**Thomas Malthus** (1776–1834) was one of the first writers to emphasise the relationship between population growth and food production.

In 1798 he published *An Essay on the Principle of Population* in which he argued that populations failed to grow in many instances because they were unable to increase the amount of food they required. The power of man to raise his food supply was, Malthus stated, limited by the scarcity of land. However, in the course of human evolution the tendency has been to raise the amount of food needed per head on smaller and smaller areas of land. It may well be that the demand for extra food created by the growing human population has constantly stimulated the development of new ways of coping with the problem. Many of the agricultural practices we see today are immensely more efficient than those of past years (Chapter 24). There is good reason to believe that innovations will continue to keep pace with the increased demand for food.

Early man living as a **hunter-gatherer** required a relatively large area of land per head to provide sufficient food. This is because much of his food came from primary consumers. The density of herbivorous prey was thus limited mainly by the amount of vegetation growing per unit area on which they could graze (Chapter 23). The numbers of humans supported in this way can be judged by population studies of Australian aborigines before the arrival of white colonists towards the end of the eighteenth century. About 30 km$^2$ of land was required to provide sufficient food to maintain each person at a subsistence level.

When humans turned to agriculture as a way of life, many more people could be supported per unit area. A simple 'slash and burn' method of agriculture is still used today in many parts of the world. A patch of forest is cut down and the trees and shrubs burned. Their ash helps to fertilise the cleared ground. Crops are grown on the land until its fertility is exhausted, when a new patch is cleared and the cycle repeated. Agriculture as crude as this barely produces enough food to support 10 people $km^{-2}$.

When humans became **settled agriculturalists**, growing crops on the same land year after year, the output of food was sufficient for up to 100 people $km^{-2}$. Indeed, this density of population would be required to perform the various steps involved in growing crops such as ploughing, tilling, sowing, weeding and harvesting. Remember that only simple tools had been invented by then, so many of the tasks had to be performed by hand.

Mechanisation later greatly increased the output per head of farm workers. Coupled with this have been the development of higher yielding strains of crop plants and the use of fertilisers to stimulate crop growth. Pesticides have greatly reduced crop loss. These improvements now mean that in countries where **advanced agriculture** is practised, a single farm labourer can produce enough food for at least forty families to enjoy a high standard of living. Financial reward is usually sufficient incentive for farmers to increase the yield of their crops, and food production worldwide increased by about 25 % between 1970 and 1980.

There is little doubt that sufficient food can be grown at present to enable everyone to live at more than subsistence level. The food 'mountains' of the European Economic Community (EEC) is evidence of the ability of some countries to produce more food than is required by their populations. In this sense, therefore, we have not yet exceeded the carrying capacity of our environment. Indeed, future technical and agricultural innovations should enable us to increase further the earth's carrying capacity in terms of food production.

One approach is founded on **biotechnology**. For example, work is now being carried out on the production of protein from a variety of micro-organisms including fungi, bacteria and algae. So far, **Single Cell Protein (SCP)** makes only a small contribution to the world's protein supply. In the near future, as the human population grows and the shortage of protein becomes more acute, SCP is likely to become increasingly important. Compared with meat production, the manufacture of SCP is much more rapid and efficient. A bullock weighing 500 kg produces 0.45 kg of protein per day, while the same amount of yeast, for example, yields 50 000 kg of protein per day. Comparable results are obtained with bacteria.

Energy-rich byproducts of the petroleum and natural gas industries are among the raw materials required to grow SCP on a large scale. Supplies of these raw materials will dry up as reserves of oil are exhausted (section 24.7.1). Alternative sources of energy will then have to be sought. Algae use the apparently inexhaustible energy of the sun. For this reason they may become important producers of SCP in the future.

Food production is uneven because some parts of the world have a climate more suited to growing crops than others. In countries where good crops can be grown, farmers are usually efficient, making use of the latest agricultural techniques. Many of the developing countries cannot finance modern methods of crop production. Crops sometimes fail for climatic reasons such as **drought**, as has happened in recent years in Ethiopia, Sudan and other North African countries. In other instances crops have been grown only to be eaten by **pests** such as locusts or ravaged by diseases. For these reasons most people in developed countries are well fed whereas malnutrition and even starvation are common in the Third World (Fig 22.6).

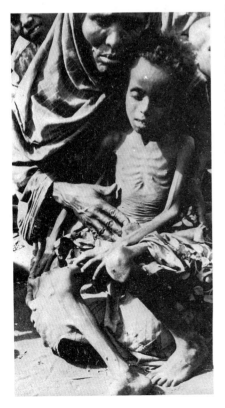

**Fig 22.6** A starving child (The Photo Source)

**Famine** is not new to mankind. Checks in the growth of many populations over the centuries have been due to crop failures. Between 1845 and 1847 the population of Ireland was reduced by over 2.5 million. Many emigrated, especially to the USA, others died of starvation. Most of the deaths were among peasants whose staple diet was potatoes. The potato crop had been devastated by a fungal disease called late blight.

When famine strikes, populations can be helped by importing food from elsewhere. The tragedy is that the countries where famine is rife are least able to pay for such help. Neither can they afford to improve their agriculture because of the high cost of fertilisers, irrigation schemes and mechanisation. Governments of some of the richer nations and voluntary agencies already do what they can to ameliorate the immediate effects of famine. Perhaps the best hope of avoiding famine in Third World countries lies in better training and education. In this way they may eventually become self-sufficient in terms of food production.

### 22.3.2 Fertility

The **fertility** of a woman is defined as the number of live children she produces by the time her reproductive life ends. It differs from **fecundity** which is her capacity to reproduce. Fecundity is determined by her reproductive life-span, the number of live children at each birth, and the length of time between births. Some foetuses abort and children are sometimes stillborn. Nevertheless it should be possible for a woman to have a live birth on average every 2.5 years. Assuming a reproductive life of 30 years each mother could produce at least 12 live offspring. This degree of fertility is achieved by few human populations these days. Women in many developed countries are less fertile now than in past years (Fig 22.7).

**Fig 22.7** Change in fertility with time (from J Z Young 1976 after Clark 1967)

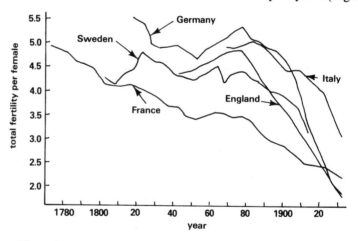

There is insufficient data for the fertility of populations in countries worldwide to be compared. Perhaps the best available index is **birth rate**, the total number of children born each year per 1000 of a population. Based on birth rate, countries can be divided into three categories (Table 22.3). There are many reasons, some cultural, others social or religious for the variation in birth rate. In many rural areas of African countries, infertility is frequently grounds for divorce. Large numbers of children counteract the ravages of infectious diseases and malnutrition, leaving enough to survive to working age. In contrast, infant mortality is now so low in the developed nations that parents can be fairly sure that their children will live to maturity. Reliable, acceptable and readily available methods of **contraception** have contributed immensely to the declining birth rate in these parts of the world.

**Table 22.3 Relative birth rates in various parts of the world**

| Birth rate (live births/1000 of population) | Country or region |
| --- | --- |
| Above 30 | Africa, Latin America, South East Asia |
| 20–30 | Spain, Portugal, Poland, China |
| Under 20 | Japan, USA, USSR, most of Europe |

**Fig 22.8** (a) Survivorship curves (from J Z Young 1976 after Clark 1967)

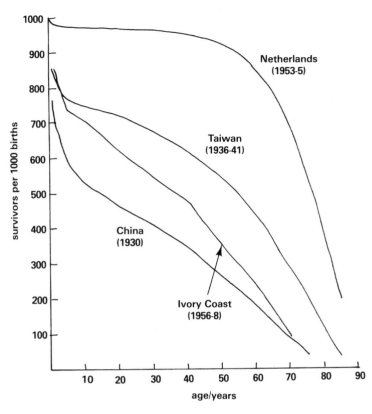

**Fig 22.8** (b) Change in survivorship curves with time for Sweden (from J Z Young 1976, after Clark 1967)

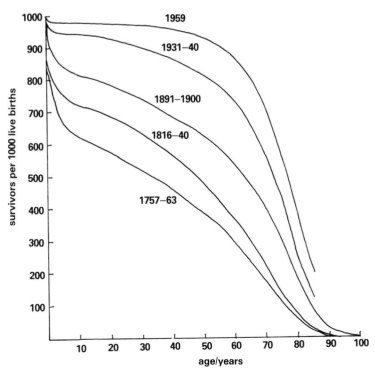

## 22.3.3 Life expectancy

The number of people dying per thousand of a population each year is the **crude mortality rate**. Where the mortality rate is high, the chance of any individual living to an advanced age is very slim. A century ago in Britain, a woman could expect to reach the age of about 45 years on average, a man 40 years. Today is is roughly 75 and 70 years respectively. One of the reasons for the greatly increased life expectancy is the reduced mortality rate among children. These days 29 infants out of every 1000 born in Britain die before their first birthday. A hundred years ago the figure was nearly 400. Remarkable advances, in medicine in particular, have contributed enormously to the increased rate of survival. Many previously fatal infectious diseases, for which there was no cure in the past, can now be cured, or prevented by vaccination (Chapter 11). We now have a much better understanding of how infectious diseases are spread (Chapter 25). People can be advised to avoid passing on hereditary diseases (Chapter 20), and dietary diseases can be avoided as our knowledge of human nutrition has improved (Chapter 9).

How successful such advances have been can be judged from **survivorship curves**. They show that life expectancy is increasing in many countries (Fig 22.8). It means that more and more people of all ages survive longer, thus adding to the existing population and contributing to a population explosion. Wars have intermittently reduced the life expectancy of humans in many parts of the world. During the six years of the Second World War, some twenty-two million people, soldiers and civilians, lost their lives. Fatal accidents constantly lessen life expectancy too. The shorter life span of men in many countries is because they often take on dangerous occupations such as mining.

What are the probable reasons for the change in survivorship of the people of Sweden shown in Fig 22.8(b)?

**Fig 22.9** Population pyramid for USA in 1957 (from J Z Young 1976 after Stamp 1960)

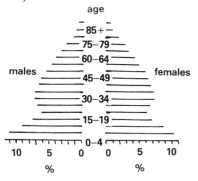

## 22.4 Population structure

Knowing the age structure and sex ratio it is possible to construct a **population pyramid**. The population is divided usually into five-year age groups, the youngest at the bottom and the oldest at the top. The percentage of females is shown on the right and males on the left (Fig 22.9). Three kinds of pyramid have been recognised which match rapidly expanding, stable and declining populations (Fig 22.10). In countries with a 'young' population, fewer old people must support the large number of youngsters. The reverse is so in countries with 'aged' populations. Fig 22.11 shows the age structure of several countries in the 1950s.

Over a period of time populations have undergone changes in structure and will continue to do so. Thus in the 1890s it was fashionable to have large families in Britain. We then had a high proportion of youngsters in our population. By the mid-1940s there was a bulge about the 35–40 age group. This was mainly due to the fall in birth rate and increased life expectancy. The bulge reached the 45–50 age group by the mid-1950s, but at the same time there was also an increase in children within the 5–10 age category (Fig. 22.12). Changes of this kind are called **demographic transition**. Population pyramids also tell us something of the history of a population. Notches within a pyramid for males of age group 15–30 are often due to the heavy loss of life in wars.

**Fig 22.10** Three kinds of population pyramid: (a) stable, (b) declining, (c) rapidly expanding populations (after Sands 1978)

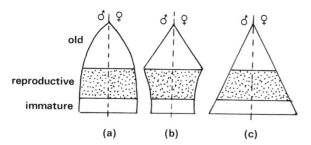

**Fig 22.11** Age composition of selected countries (from J Z Young 1976, after Stamp 1960)

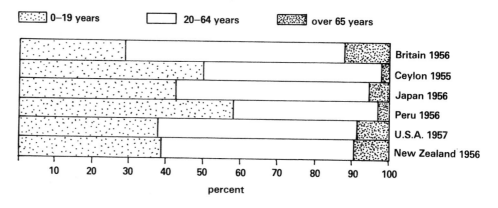

**Fig 22.12** Changes in structure of the British population: (a) 1891, (b) 1947, (c) 1957 (from J Z Young 1976 after Stamp 1960)

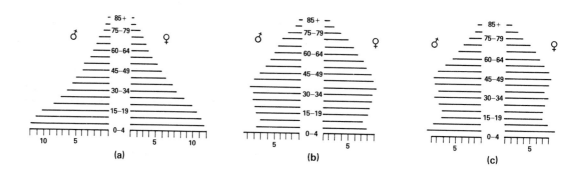

## 22.5 Present and future population growth

### 22.5.1 Present growth

We have already seen that many interacting factors determine the rate at which the human population grows. Foremost are the existing number of people, the number of fertile females, their fertility, and life expectancy. Clearly there are variations in the rate at which populations grow in different parts of the world. On this basis it is possible to identify two main categories.

### 1 Stable populations

Most European countries and Japan have populations whose numbers are more or less stable. The main reasons for this are the decline in birth rate and increased life expectancy. A few countries of central Africa, south America and south-east Asia have stable populations for very different reasons. There the birth rate is high and life expectancy is short.

### 2 Growing populations

These have a birth rate which is higher than the death rate, hence the population is increasing rapidly. Over half the total number of people today live in growing populations. They are mainly in Africa, Latin America and most of south and east Asia.

### 22.5.2 Future growth

Predicting future growth of human populations is a difficult task. The reason is that many factors affect population growth and these factors change with time. Changes in fertility with time illustrate the complexity of the problem. In 1890 each British mother had given birth on average to 7.6 children. In 1975 it was down to 2.3. Knowledge, availability and acceptance of methods of **birth control** by the British population have enabled families to be planned. The same applies to populations in most developed countries.

**Family planning** programmes have started more recently in some developing countries. Getting people who are illiterate and suspicious to use sophisticated contraceptive techniques voluntarily is not easy. At one time the Government of India encouraged males to have **vasectomy** and women of child-bearing age to be fitted with **intra-uterine devices** by offering them rewards of transistor radios. In China it is now illegal for women under 25 years of age to marry. When they do they are limited to having only one child. For these reasons the birth rate in both countries has fallen. If the birth rate declines further, their populations will still continue to grow rapidly for some time. This is because the large numbers of children already produced will start families in about 20 years' time. Between them, India and China have about 40% of the world's population at present. This figure is bound to go up in the foreseeable future.

Improvements in medical care are also likely to continue, and the consequent increased life expectancy in developing countries will add to their population problems. Bearing in mind that human fertility may rapidly change and that people world wide may in future benefit from better health care, the United Nations in 1958 constructed models from which future population growth could be predicted. Based on two

**Table 22.4 Estimates of world population × 10⁶ (from United Nations Population Study No. 28, 1958)**

| Estimate | 1975 | 1980 | 1990 | 2000 |
|----------|------|------|------|------|
| Low      | 3590 | 3850 | 4370 | 4880 |
| Medium   | 3830 | 4220 | 5140 | 6280 |
| High     | 3860 | 4280 | 5360 | 6900 |

important assumptions they came up with three estimates (Table 22.4). The assumptions were:

**i.** the birth rate in countries where fertility is low will not alter, and

**ii.** in countries where there is a high birth rate this will remain at its present level (high estimate), decline from 1975 onwards (medium estimate) or decline from 1958 onwards (low estimate).

The actual world population in 1980 exceeded the high estimate (Table 22.2). It illustrates the great difficulty of predicting the long-term rate of growth of the human population. Another estimate of population growth was made by Unesco in 1984. The subsequent change in distribution of people throughout the world is shown in Table 22.5. Note that the proportion of people living in developing countries is expected to increase.

**Table 22.5 Estimated world population of humans 0–24 years old in major areas of the world (from Unesco Statistical Yearbook 1984)**

| | Population × 10⁶ | | |
|---|---|---|---|
| | 1985 | 1990 | 2000 |
| World | 2550 | 2687 | 2946 |
| Africa | 350 | 408 | 542 |
| America | 343 | 364 | 413 |
| Asia | 1553 | 1616 | 1693 |
| Oceania | 11.3 | 11.7 | 12.5 |
| Europe (inc. USSR) | 293 | 286 | 286 |
| Developed countries | 469 | 466 | 476 |
| Developing countries | 2081 | 2220 | 2470 |

The data must be treated with caution because rapid changes in reproductive behaviour could significantly alter the outcome.

# 23 The environment: ecosystems

The total volume of the earth in which living organisms exist permanently is called the **biosphere**. It is divided into four major **habitats** – marine, estuarine, freshwater and terrestrial. Man is a successful terrestrial animal and occupies all the major biogeographical zones, called **biomes**, such as temperate deciduous forest, tropical rain forest, steppe and tundra (Fig 23.1). Among the reasons for our success as a land animal are our inventiveness, ability to use a wide variety of foods in our diet, efficient reproduction and tolerance of environmental stress.

**Fig 23.1** (a) Distribution of the earth's biomes (from Sands 1978)

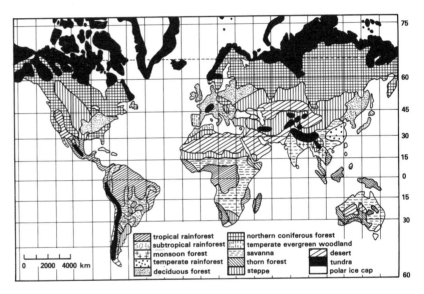

tropical rainforest — northern coniferous forest
subtropical rainforest — temperate evergreen woodland
monsoon forest — savanna — desert
temperate rainforest — thorn forest — tundra
deciduous forest — steppe — polar ice cap

0   2000   4000 km

**Fig 23.1** (b) Tropical rain forest (Brian Rogers/Biofotos)

An **ecosystem** is part of a biome. It may be a pond, stream, woodland or seashore inhabited by a characteristic **community** of living organisms. A **population** is a group of individuals of the same species living in the same place at the same time. Human populations are thus part of the community of life around them. The role an organism performs in a community is called its **niche**. Ecosystems contain more or less self-sufficient communities which are at equilibrium with each other and with their environment. Energy and matter are exchanged between communities and their surroundings. The activities of communities cause energy to flow and matter to be cycled in ecosystems.

## 23.1 Trophic relationships

Green plants are the life-blood of every ecosystem. They make organic molecules by photosynthesis from simple inorganic molecules. For this reason they are called **producers**. Organic substances are eaten mainly by animals which are the **consumers** in ecosystems. **Primary consumers** are mainly herbivorous animals which feed on green plants. Predators, scavengers and many parasites are **secondary consumers**. They feed on primary consumers. **Tertiary consumers** feed on secondary consumers.

431

When producers and consumers die, their bodies are broken down by **decomposers** such as bacteria and fungi, which also dispose of animal faeces. Dead organic matter is also consumed by animals, called **detritivores**. Millipedes, woodlice and earthworms feed mainly on dead vegetation in soil ecosystems. In aquatic ecosystems, particles of dead organic matter suspended in the water are eaten by detritivores such as bivalve molluscs and lugworms.

It is convenient to think of the feeding relationships in ecosystems as a **food chain** with several **trophic levels** (Fig 23.2). In reality, feeding relationships are rarely as simple as this. Omnivores such as man can be primary, secondary and tertiary consumers; so too can decomposers and detritivores. It is therefore more informative to show feeding relationships as **food webs** (Fig 23.3). This gives a much clearer picture of what happens to food in a community.

**Fig 23.2** Trophic levels in a food chain

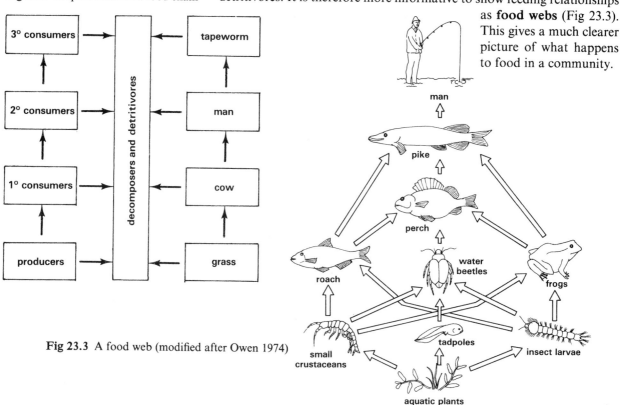

**Fig 23.3** A food web (modified after Owen 1974)

## 23.2 Energy flow

Energy can be transformed from one form to another. Green plants, for example, convert the energy in sunlight into the chemical bond energy of photosynthetic products. When energy is converted from one form to another, some energy always appears as heat. A proportion of the energy in respiratory substrates is released as heat energy.

The energy source for all the earth's ecosystems based on photosynthesis is the sun. About $641 \times 10^4 \, \mathrm{kJ \, m^{-2} \, y^{-1}}$ of solar energy reaches the earth's atmosphere. A good deal of solar energy does not, however, penetrate the atmosphere. It is reflected or absorbed and radiated back into space by the ozone layer, dust particles and clouds. The precise amount reaching the earth's surface depends on geographical location. Britain gets about $105 \times 10^4 \, \mathrm{kJ \, m^{-2} \, y^{-1}}$, which is roughly a third of the energy received in tropical countries. Between 90 and 95 % of the energy getting to the surface of the earth is reflected by vegetation, soil and water or absorbed and radiated to the earth's atmosphere as heat. It means that only between 10 and 5 % is left for producers to make use of.

### 23.2.1 Producers

Let us assume that 100 units of energy per unit time reach the leaves of a crop plant. What exactly happens to the energy? About half is not of appropriate wavelengths to be absorbed by chloroplast pigments. Roughly a quarter of what is absorbed ends up as a chemical bond energy in photosynthetic products such as starch. The rate at which the products are formed is called **gross primary productivity (GPP)**. A substantial amount of the gross production is respired by the plant. What is left, in this example just 5.5 % of the energy reaching the leaf, is net production (Fig 23.4). The rate at which the products of photosynthesis accumulate is known as **net primary productivity (NPP)**.

**Fig 23.4** Fate of solar energy reaching the leaf of a crop plant (data from Hall 1979)

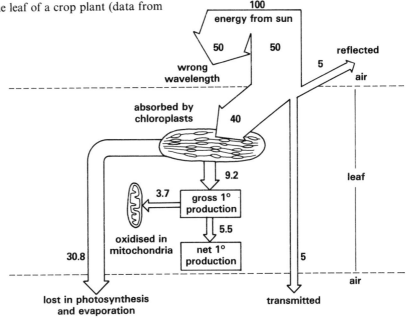

Because they photorespire more, plants which use the $C_3$ photosynthetic pathways have lower NPP values than $C_4$ species (Chapter 5). Productivity in temperate areas could in theory be improved by growing $C_4$ crops. Unfortunately, many $C_4$ species are native to tropical climates and do not grow well in temperate countries. Those which can be cultivated, such as maize, are less productive because of the lower temperature and less intense light (Table 23.1). One approach to improve the productivity of $C_3$ species such as wheat has been the development of early maturing varieties. This ensures the crop is ready for harvesting before the hot summer weather begins to stimulate photorespiration. An alternative strategy, which is being actively researched, is to spray $C_3$ crops with a chemical to block photorespiration. Another possibility is to breed out the biochemical pathway of photorespiration.

**Table 23.1 The growth of maize under different conditions (from Bland and Bland 1983)**

| Plant | Region | Net primary production $(g\,m^{-2}\,day^{-1})$ | Irradiance $(J\,cm^{-2}\,day^{-1})$ | Conversion of photosynthetically active radiation (%) |
|-------|--------|------|------|------|
| maize | California | 52 | 2000 | 9.8 |
| maize | Europe | 17 | 1200 | 4.6 |

In winter, fields in temperate climates often have no crops growing in them. During the early growth of a crop, a good deal of sunlight is absorbed or reflected by the soil. It is only when plants are fully grown that their leaves absorb the maximum amount of sunlight. When the NPP is averaged for the whole year, it is therefore much less than in the growing season. In natural ecosystems where the climate allows plants to grow all year, the NPP is higher. In others, climatic factors limit the rate of photosynthesis, so the NPP is lower (Table 23.2).

The NPP determines the amount of carbohydrate food and hence the energy available to the consumers in the community. Thus the ability of any ecosystem to support consumers is determined by the NPP of its plants. This will be a vital factor in ultimately determining the number of humans the earth can carry.

Table 23.2 Net primary productivity of some of the world's ecosystems (after Whittaker 1975)

| Ecosystem | NPP ($g\,m^{-2}\,y^{-1}$) | |
| --- | --- | --- |
| | Range | Mean |
| open ocean | 2–400 | 125 |
| lakes and streams | 100–1500 | 250 |
| continental shelf | 200–1000 | 360 |
| temperate grassland | 200–1500 | 600 |
| temperate forest | 600–2500 | 1250 |
| tropical forest | 1000–3500 | 2000 |

### 23.2.2 Consumers

Let us consider the transfer of energy from producers to primary consumers. **Secondary productivity** is the rate at which consumers accumulate energy in the form of cells and tissues. The first point to bear in mind is that **herbivores** do not eat all the vegetation in an ecosystem. Cattle grazing a pasture will eat the grasses and palatable weeds such as buttercups. They leave inedible weeds such as nettles and thistles. Even the plants that are grazed are not eaten entirely. The roots are inaccessible and part of the shoot system is also left uneaten. It means that only part of the NPP of the ecosystem is transferred to the primary consumers. The amount depends to some extent on the variety and density of primary consumers. In pasture grazed by cattle it is about 30–45 %.

Fig 23.5 shows the fate of energy consumed by a bullock grazing on pasture. Nearly two-thirds of this energy, 1909/3056 kJ, is excreted in urine and egested in its faeces. The remaining third, 1147/3056 kJ, is absorbed into the circulatory system. Roughly 90 % of the absorbed energy, 1022/1147 × 100, is lost as heat or is belched out as methane gas. Only about 10 % of the absorbed energy, 125/1147 × 100, is used to synthesise new cells and tissues. The secondary production of the bullock is thus only 0.6 %, 125/21 436 × 100, of the NPP of the pasture.

**Fig 23.5** Fate of energy in a year's growth of grass from 1 m² of pasture (data from *Nuffield Study Guide in Advanced Biology* 1974)

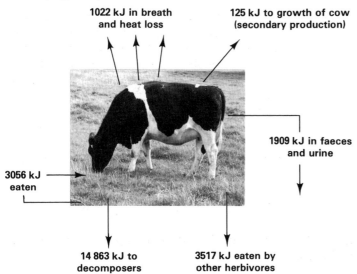

1022 kJ in breath and heat loss

125 kJ to growth of cow (secondary production)

3056 kJ eaten

1909 kJ in faeces and urine

14 863 kJ to decomposers

3517 kJ eaten by other herbivores

**Carnivores** have a much higher secondary productivity. This is because a protein-rich diet is more readily and efficiently digested. They do not have energy-consuming symbiotic microbes in their digestive tracts and their faeces contain much less undigested matter. Only about 20 % of the energy intake is lost in their faeces and urine. They absorb almost twice as much energy per unit mass of food compared with herbivores.

The proportion of absorbed energy released in respiration is determined by the activity of the consumer and whether or not it maintains a constant body temperature. **Endotherms** have a stable temperature. The energy in absorbed food makes good the energy lost as heat to their surroundings. In cold weather, more absorbed energy is used for this purpose. **Ectotherms** do not use energy in this way and direct a higher proportion of absorbed energy into secondary production.

Of the consumers eaten by humans, fish are probably the most productive (Table 23.3). This is one reason why there has been increased interest in fish-farming in recent years. Changes have also been made in the production of farm animals. They involve intensive methods of farming such as raising poultry by the broiler method. Keeping livestock indoors reduces the amount of energy they use for movement and minimises heat loss through the skin.

**Table 23.3 Production of a range of animals (mainly after Open University 1974)**

| | | | % Absorbed | % Respired | % Produced |
|---|---|---|---|---|---|
| **Endotherms** | herbivores | rabbit | 50.0 | 47.4 | 2.6 |
| | | cow | 37.5 | 33.4 | 4.1 |
| **Ectotherms** | carnivores | trout | 86.0 | 55.9 | 30.1 |
| | | spider | 91.0 | 62.0 | 29.0 |
| | herbivores | grasshopper | 37.0 | 24.0 | 13.0 |
| | | caterpillar | 41.0 | 17.6 | 23.4 |

## 23.2.3 Decomposers and detritivores

The proportion of the primary production used by herbivores varies from one ecosystem to another. Up to 90 % of the phytoplankton in the oceans is eaten by microphagous animals. About 50 % of the primary production of pastureland is grazed, whereas in forests as little as 10 % is consumed by herbivores. The faeces of herbivores and most of what is left of the primary production are used as an energy source by **decomposers** and **detritivores**.

One ecosystem in which the energy flow through decomposers has been determined is a salt marsh in Georgia, USA (Table 23.4). The primary production available to the animals and decomposers living in the marsh is $34\,354 - 15\,370 = 18\,984\,kJ\,m^{-2}\,y^{-1}$, of which the decomposers use $16\,287\,kJ$.

The proportion used by the decomposers is thus $\dfrac{16\,287}{18\,984} \times 100 = 85.8\,\%$.

It is thought that this is a reflection of the energy flow through the decomposers in many ecosystems.

**Table 23.4 Energy flow in a Georgia salt-marsh (after Teal 1962)**

| Process | Energy $(kJ\,m^{-2}\,y^{-1})$ |
|---|---|
| gross production | 152 323 |
| producer respiration | 117 969 |
| net production | 34 354 |
| decomposer respiration | 16 287 |
| export from marsh | 15 370 |
| primary consumer respiration | 2495 |
| secondary consumer respiration | 201 |

## 23.2.4 Whole ecosystems

In most ecosystems there are many species, and feeding relationships are often poorly known. The few ecosystems where the flow of energy is known in detail are fairly simple. A freshwater spring is a simple ecosystem. The producers are diatoms, green filamentous algae and small flowering plants such as duckweed. When they die, the producers become detritus and sink to the water bed. Some of the detritus is carried away by the current. Detritus also enters the spring with the incoming water. The detritivores include freshwater lice, bivalve and gastropod molluscs and small crustaceans. Larvae of the caddis fly feed on larger pieces of dead vegetation. Predatory turbellarians and midge larvae are the consumers.

Fig 23.6 Flow of energy in Silver Springs, Florida (after Odum 1957)

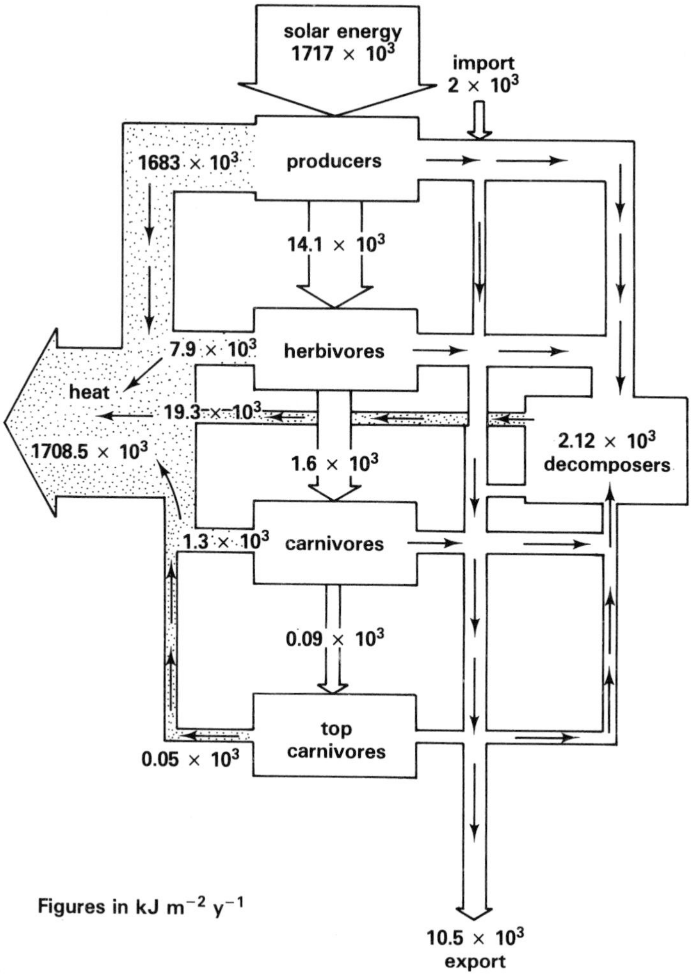

Figures in kJ m$^{-2}$ y$^{-1}$

The flow of energy between the various groups of organisms is shown in Fig 23.6. The first important thing to notice is that the energy leaving the system is the same as the amount of solar energy coming in. The second is that the rate at which energy passes into the animals at each trophic level is about 10 % of that entering the previous level. For example, the secondary consumers in the spring receive $1.6 \times 10^3 \, \text{kJ} \, \text{m}^{-2} \, \text{y}^{-1}$ from the primary consumers which get $14.1 \times 10^3 \, \text{kJ} \, \text{m}^{-2} \, \text{y}^{-1}$ from the producers. The rate at which energy passes to the secondary consumers compared with the rate it enters the primary consumers is thus

$$\frac{1.6}{14.1} \times 100 = 11.4\%.$$

This is called the **gross ecological efficiency**. Because only about 10 % of the energy passes from one level of consumers to the next, the number of trophic levels is limited to two or three in most ecosystems.

## 23.3 Ecological pyramids

The food and energy relationships in ecosystems can be displayed as bar diagrams. The producers are placed at the bottom, the primary and secondary consumers in the middle and tertiary consumers at the top. Decomposers and detritivores are not often shown. Diagrams of this kind often have the shape of a pyramid and are called **ecological pyramids.** They are of three types: of numbers, of biomass and of energy.

### 23.3.1 Pyramids of numbers

These show the number of organisms at each trophic level. The length of each bar is proportional to the number of organisms. A major drawback of number pyramids is that they fail to distinguish between the sizes of organisms. A mature tree in a forest ecosystem is counted the same as a diatom in a lake. A cow in a meadow is equated with a leaf-eating caterpillar in a woodland. For this reason, number pyramids are bulged at the middle when the producers are large and few in number as in forest ecosystems. They are inverted when the consumers carry parasites. For example, a single rose bush may support thousands of aphids, which in turn may be the hosts for millions of microbial parasites (Fig 23.7).

**Fig 23.7** Pyramids of numbers

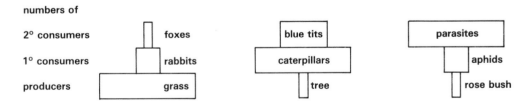

### 23.3.2 Pyramids of biomass

These show the dry mass of the organisms in the trophic levels at a particular point in time. The length of the bars is proportional to the biomass in each level. They tell us what is called the **standing crop**. One of the problems with this approach is that it does not allow for changes in biomass at different times of the year. Deciduous trees, for example, have a larger biomass in summer than in winter when they have shed their leaves. Another difficulty is that the rate at which the biomass accumulates is not taken into account. A mature tree has a large biomass which increases slowly over many years. In comparison, the standing crop of diatoms in a lake is very small, yet production is high because their turnover rate is so rapid. This is why inverted pyramids of biomass are obtained for aquatic ecosystems (Fig 23.8).

**Fig 23.8** Pyramids of biomass: (a) a field in Georgia USA, and (b) the English Channel

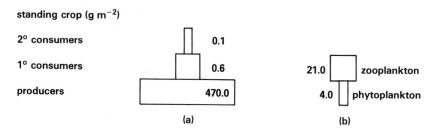

### 23.3.3 Pyramids of energy

Why is the transfer of energy to primary consumers less efficient than between other trophic levels?

These show how much energy passes from one trophic level to the next. The length of the producer bar is proportional to the amount of solar energy used annually in photosynthesis. The other bars show the rate at which energy passes along the food chain (Fig 23.9). Note that the transfer of energy from producers to primary consumers is less efficient than between the other trophic levels. Energy pyramids are more informative than pyramids of numbers and biomass. They tell us how much energy is required to support each trophic level. Because only a proportion of energy in a level is transferred to the next, energy pyramids are never inverted, nor do they have a central bulge.

**Fig 23.9** Pyramid of energy in Silver Springs, Florida

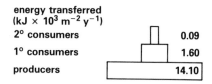

## 23.4 Cycling of matter

Energy comes into ecosystems from the sun and is lost as heat to the surroundings. In contrast, matter is cycled between organisms and their environment. The availability of minerals and water affects the distribution and abundance of plants and hence, to a large extent, man.

### 23.4.1 Nitrogen cycle

Nitrogen is vital to plants and animals for the synthesis of amino acids, proteins and nucleic acids. The earth's atmosphere contains over 78 % by volume of nitrogen gas, yet in this form it is useless to most living organisms. Plants take in nitrogen as ammonium and nitrate ions. On a global scale, farmers add about $75 \times 10^6$ tonnes of ammonium and nitrate fertilisers to cultivated soils each year. Treating fields with nitrogen fertilisers is essential, because most of the nitrogen previously in the soil is removed in the crop at harvesting. In uncultivated soils, the crop is returned each year as dead leaves, roots, animal carcasses and faeces. Nitrogen in the detritus is converted to ammonium and nitrate ions and is absorbed once more by the vegetation. The cycling of nitrogen takes place in several steps (Fig 23.10).

Fig 23.10 The nitrogen cycle

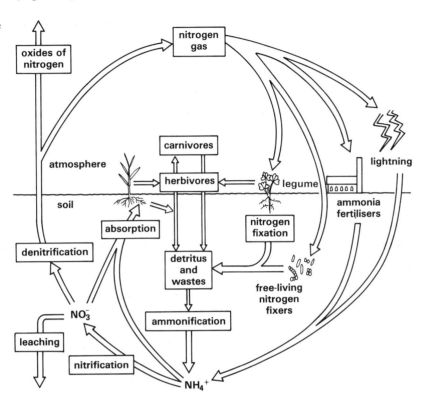

**1 Ammonification** A wide range of saprophytic soil bacteria and fungi decompose detritus. Peptidase enzymes released by the decomposers hydrolyse proteins to amino acids, some of which are absorbed and used to synthesise microbial protein. Unwanted amino acids are deaminated in a way comparable to that which occurs in the mammalian liver, and ammonia is released. As much as $7000–10\,000 \times 10^6$ tonnes of nitrogen enter soil annually in this way. Some ammonia escapes into the atmosphere, but much of it is converted in the soil to nitrate ions by nitrifying bacteria.

**2 Nitrification** Well aerated soil contains chemo-autotrophic **nitrifying bacteria**. They oxidise ammonia to nitrate ions in two stages. The first stage is the conversion of ammonia to nitrite ions, mainly by bacteria of the genus *Nitrosomonas*. Nitrite ions are toxic to plants, but they are oxidised as quickly as they are formed into nitrate ions by bacteria of the genus *Nitrobacter*.

Plants absorb nitrate ions, but these are also rapidly leached from soil by rain water. In anaerobic conditions, as in waterlogged soils, nitrate ions are reduced to nitrite, ammonia, nitrogen and oxides of nitrogen by **dentrifying bacteria** such as *Pseudomonas denitrificans*.

**3 Nitrogen fixation** Something like $200–300 \times 10^6$ tonnes of soil nitrogen are lost each year by leaching, denitrification and ammonification. The loss is made good by **nitrogen-fixing bacteria and blue-green algae**. They are able to reduce nitrogen gas to ammonia which they use to form amino acids. Free-living bacteria such as *Azotobacter* and *Clostridium* and blue-green algae such as *Nostoc* account for $170–270 \times 10^6$ tonnes of nitrogen fixed annually. The rest is fixed by symbiotic bacteria, notably *Rhizobium* which lives in the nodules of legume roots (Fig 23.11).

**Fig 23.11** (a) Root system of a broad bean plant, × 0.75

**Fig 23.11** (b) Section through a root nodule, × 30

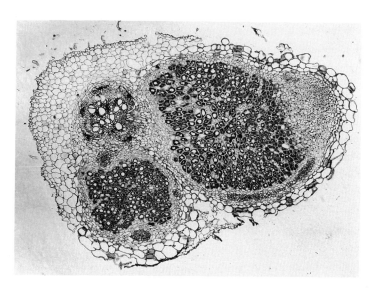

Nitrogen fixation is a complex process and differs in detail from one fixer to another. Fig 23.12 summarises what probably happens in *Rhizobium*. The enzyme **nitrogenase** catalyses the transfer of hydrogen from NADH to nitrogen gas, reducing it to ammonia. The ammonia combines with carboxylic-acids formed in Krebs' cycle (Chapter 5) to make a range of amino acids. These can be used in transamination, giving rise to all the amino acids required by the nitrogen fixer. Any surplus to requirement is used by the leguminous host for protein synthesis, thereby stimulating its growth.

**Fig 23.12** Mechanism of nitrogen fixation in *Rhizobium*

nitrogenase enzymes

keto-glutaric acid

nitrogen gas $\longrightarrow$ ammonia $\longrightarrow$ glutamic acid

NADH  NAD$^+$

transamination

proteins $\longleftarrow$ other amino acids

### 23.4.2 Carbon cycle

The earth's atmosphere contains $2.5 \times 10^6$ million tonnes of carbon dioxide. The oceans contain 150 times more. Terrestrial and aquatic organisms exchange about $100-200 \times 10^6$ tonnes of carbon dioxide annually with their surroundings. The exchange is mainly brought about by three activities (Fig 23.13).

**Fig 23.13** The carbon cycle

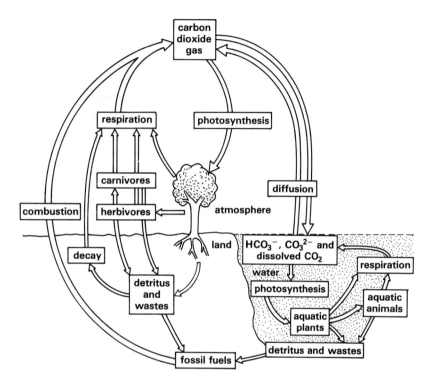

**1 Photosynthesis** Producers use carbon dioxide in photosynthesis. Each year, terrestrial plants remove between 50 and $100 \times 10^6$ tonnes of carbon dioxide from the atmosphere. Marine algae use about the same amount from the oceans. The standing crop of plants is an enormous reservoir of carbon. They hold about twice as much carbon as is contained in the atmosphere. Carbon fixed in photosynthesis millions of years ago, especially during the Carboniferous period, is preserved in vast reserves of fossil fuels such as coal, oil and natural gas. Peat is a recently-formed fossil fuel.

**2 Respiration** Living organs make carbon dioxide as a respiratory product. The amount of respiratory carbon dioxide released by producers is over half the amount they fix in photosynthesis. Consumers contribute about 10 % and decomposers and detritivores roughly 40 % of the carbon dioxide returned to the environment as a product of respiration.

**3 Combustion** Carbon dioxide is released when fossil fuels are burned. It is estimated that about $1-10 \times 10^6$ tonnes of carbon dioxide enter the atmosphere annually in this way. The amount of fossil fuels burned each year has increased over the past century. It is thought to be one of the reasons why the mean concentration of carbon dioxide in the earth's atmosphere is slowly rising (Fig 24.1). The cutting down of vast tracts of tropical forest may be another cause, because there is now less vegetation to remove atmospheric carbon dioxide in photosynthesis.

### 23.4.3 Other elements

Sulphur is cycled in the atmosphere, soil and living organisms in much the same way as nitrogen. Other important elements such as phosphorus and potassium do not have gaseous forms and cycle in a different way (Fig 23.14).

**Fig 23.14** (a) The sulphur cycle. (b) The phosphorus cycle

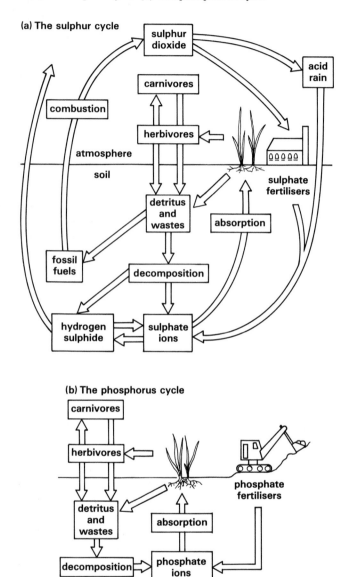

### 23.4.4 Water cycle

Water has many roles in living organisms and life on earth is impossible without it (Chapter 1). The total mass of water vapour in the atmosphere is equivalent to a mean rainfall of 2.5 cm a year over the earth's surface. The average rainfall is 90 cm a year, so water in the atmosphere is cycled 90/2.5 = 36 times a year. Rain does not fall evenly throughout the world. The amount falling on land determines to a large extent the abundance and distribution of terrestrial plants. Two processes are the cause of the cycling of water (Fig 23.15).

**Fig 23.15** The water cycle

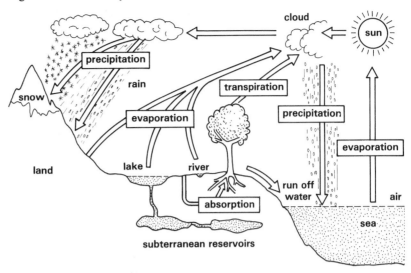

**1 Evaporation** The earth's atmosphere has a lower water potential than living organisms, aquatic habitats and all but the driest of environments. Consequently, water constantly **evaporates** into the air. Most evaporation occurs over the oceans, from which about $1500 \times 10^6 \, \text{m}^3$ of water vapour pass into the atmosphere each day. This compares with approximately $200 \times 10^6 \, \text{m}^3$ each day from the land, roughly half from the soil and half from vegetation. Water vapour is less dense than air, so it rises to the upper atmosphere where the temperature is lower. Here it condenses, forming clouds of water droplets. Most clouds form over the oceans, where evaporation is greatest. Clouds sometimes evaporate when they absorb heat from the sun. Alternatively they are blown elsewhere, often over land where water precipitates from them.

**2 Precipitation** Water precipitates from clouds as rain, hail or snow. The mean daily **precipitation** over land is about $100 \times 10^6 \, \text{m}^3$ more than that lost by evaporation. The difference is because this amount of water runs off the land into the oceans in streams and rivers every day. The mean water vapour content of the atmosphere remains constant because evaporation over land and sea balances precipitation. It is important to realise that evaporation and precipitation are uneven in different parts of the world. It explains why some terrestrial habitats are constantly wet, whereas others are arid. In many areas, precipitation is also seasonal and wet spells alternate with dry ones.

The rate at which water is cycled causes the rapid removal of a variety of atmospheric pollutants. The clearing of a haze after a shower is because dust, soot, smoke and other particles suspended in the air have been washed out. Rain also dissolves other atmospheric pollutants, such as sulphur dioxide (Chapter 24).

The cloud cover in the atmosphere affects the earth's **heat balance**. Rain clouds reflect or absorb about 20 % of the radiation from the sun, mainly infra-red rays. Clouds also reflect heat radiating from the earth back to the earth's surface. It helps prevent the loss of too much heat to outer space.

To what extent can man influence the water cycle for his own advantage?

443

# 24 Effects of human activity on the environment

There are many toxic substances in the biosphere. Some are part of the natural cycling of matter. For example, when volcanoes erupt they discharge poisonous gases and ash is scattered for miles around. However a good many toxic substances in the environment result from human activity and are of recent origin. Some are the products of industry and our attempts to eradicate or control pests on a large scale. Others are the results of people living close to each other in towns and cities where the sheer mass of domestic waste poses disposal problems.

Substances we introduce into the environment and which adversely affect living organisms, humans included, are called **pollutants**. Growth of the human population, industry and agriculture are at the heart of many aspects of pollution.

## 24.1 Pollution of the atmosphere

The earth's **atmosphere** is a mixture of gases which is about 2000 km thick. The lowermost layer, about 10 km thick, is called the **troposphere**. It is in this layer that gas exchange occurs between the atmosphere and living organisms. Above it is the **stratosphere**, which contains ozone, a gas which absorbs ultraviolet and infra-red radiation from the sun. In this way terrestrial organisms are protected from excessive exposure to such forms of radiation. Ultraviolet radiation can cause mutations, while infra-red radiation raises the temperature of living organisms and can cause sunstroke and sunburn.

Table 24.1 shows the composition of the troposphere at sea level in a clean-air area. Variable amounts of other gases such as neon, helium, krypton, xenon, methane, carbon monoxide, hydrogen, ammonia, sulphur dioxide, hydrogen sulphide, nitric and nitrous oxides are also present. They are there as part of the natural cycling of matter. Industrial activities have altered the proportions of the atmosphere's natural constituents. In recent times, human technology has also caused new ingredients to enter the atmosphere. There is growing evidence that such changes are harmful to life.

**Table 24.1 Composition of the troposphere at sea level**

| Gas | Volume (%) |
| --- | --- |
| nitrogen | 78.08 |
| oxygen | 20.96 |
| argon | 0.93 |
| carbon dioxide | 0.03 |

The amount of water vapour varies from 0 to 4%.

### 24.1.1 Carbon dioxide

The burning of fossil fuels such as coal, oil and natural gas provides people with most of the energy they require for domestic, industrial and agricultural purposes. When such fuels are burned, **carbon dioxide** is released into the air. Data obtained from widely spaced sampling points indicate that the amount of carbon dioxide in the air is increasing (Fig 24.1). Two major effects of increasing atmospheric carbon dioxide are possible:

1 It may stimulate photosynthesis by plants. This is unlikely however as other factors such as light intensity and scarcity of minerals limit photosynthesis.

**2** The earth's surface reflects some of the heat from the sun, thus warming the atmosphere. When air is enriched with carbon dioxide it reduces the amount of heat escaping into space just as the glass in a greenhouse does. For this reason it is called the **greenhouse effect**.

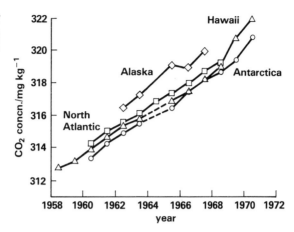

**Fig 24.1** Annual mean concentration of $CO_2$ in the air at three widely spaced sites (from Spedding *Air Pollution* published by OUP)

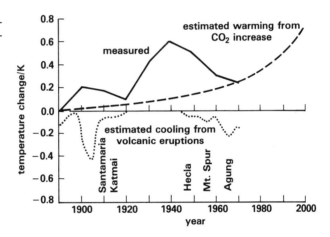

**Fig 24.2** Temperature changes, measured and estimated in the northern hemisphere (from Spedding *Air Pollution*)

Fig 24.2 shows the predicted change in world temperature up to the year AD 2000, based on present emissions of carbon dioxide. Of course, the use of alternative sources of energy such as nuclear fuels may in the near future bring about a reduction in the emission rate. The falls in measured temperature are thought to be due to increases in particulate matter in the atmosphere from industry and volcanic eruptions. Modern industries have reduced the dust, soot and smoke they release, so any future falls in the measured temperature are more likely to be caused by natural events. If, however, the overall result is a gradual increase in temperature, it will bring about melting of the polar ice-caps, thus raising the level of the oceans. One of the consequences of this would be flooding of large areas of low land on which many modern cities have been built. Vast areas of agricultural land would also be inundated.

### 24.1.2 Sulphur dioxide

**Sulphur dioxide** is released into the air when fossil fuels are burned and in the smelting of ores (Table 24.2). The background concentration is between 0.3 and 1.0 $\mu g\,m^{-3}$, but in areas near heavy industry the concentration is as high as 3000 $\mu g\,m^{-3}$.

**Table 24.2 Annual sulphur dioxide emissions for Britain in 1965 (after Robinson and Robbins 1970)**

| Source | Emission $\times 10^6$ (tonnes) |
|---|---|
| coal | 102.0 |
| petroleum | 28.5 |
| copper smelting | 12.9 |
| lead smelting | 1.5 |
| zinc smelting | 1.3 |

445

Sulphur dioxide is oxidised to sulphate ions in the air. When precipitated in rainfall, sulphate can be absorbed from the soil by plant roots. Sulphur is an essential nutrient for plant growth, so its absorption as sulphate stimulates the growth of crops. The effect could be important in poor countries where chemical fertilisers are not used for economic reasons.

Human life is endangered when people are exposed to high concentrations of sulphur dioxide in combination with smoke rising from the combustion of fossil fuels (Fig 24.3). The increased number of deaths in such conditions is mainly due to bronchitis, pneumonia and heart failure. Laboratory experiments indicate that sulphur dioxide slows ciliary activity in the respiratory tract. Particulate matter is then not cleared in the usual way and reaches the alveoli, where it causes irritation and interferes with gas exchange. Bronchitis is one of the main causes of death in men over 45 years of age in Britain. About 30 million working days a year are lost by people who suffer from the condition.

Fig 24.3 Human death rates and pollution of the air with soot and $SO_2$ in London (from Mellanby, *The Biology of Pollution*, Inst. Biol: Studies in Biology, published by Edward Arnold)

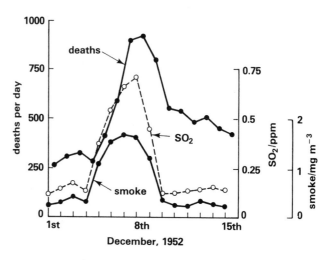

Plants absorb sulphur dioxide through their stomata. The leaves of some important crop plants such as barley, wheat and cotton are damaged by high concentrations of sulphur dioxide in the air. Others such as potatoes, maize and onions are resistant. Lichens are especially sensitive and have been used as indicators of aerial pollution (Fig 24.4).

The implementation of the *Clean Air Act* in 1956 has done much to reduce the amounts of sulphur dioxide and smoke in the air around and in British towns and cities in recent years (Fig 24.5). Such improvements can only contribute to a reduction of some of the harmful effects mentioned. Even so, something like 30 million tonnes of sulphur dioxide are still released into the air in western Europe each year. It is thought that much of the sulphur dioxide is blown by the prevailing winds to Scandinavia, where it falls as sulphuric acid in **acid rain** (Fig 24.6). A high concentration of sulphur dioxide in the air is toxic to plants. Acid rain also lowers the pH of soil, bringing toxic elements such as aluminium into solution. The extensive damage to plantations of conifers in Sweden and Germany in recent years is probably due to a combination of such factors. The high concentration of aluminium leached from soils receiving acid rain is thought to be the reason for the massive decline in fish populations in Norwegian lakes in the past decade. So serious is the threat from acid rain that it was the sole topic at the 1982 *Conference on the Environment* held in Stockholm.

Fig 24.4 Distribution of lichens in the Newcastle-on-Tyne area (after Gilbert 1970)

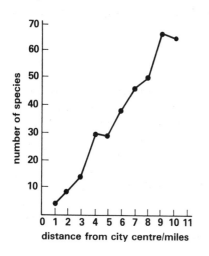

446

**Fig 24.5** Soot and SO$_2$ emissions in London air after the passing of the Clean Air Act (from Mellanby *The Biology of Pollution*, Inst. Biol: Studies in Biology, published by Edward Arnold)

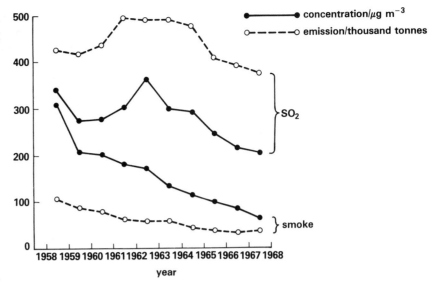

**Fig 24.6** Accumulation of acid rain over Southern Norway (from Ottaway *The Biochemistry of Pollution*, Inst. Biol: Studies in Biology, published by Edward Arnold)

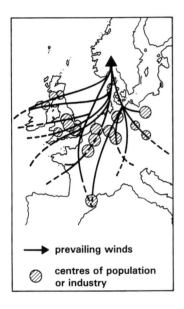

### 24.1.3 Particulate matter

The most obvious effect of the presence of particles in the air is a reduction in visibility. The **smogs** of the 1950s contained large amounts of smoke particles. Smoke, soot and ash particles are released into the air when fossil fuels are burned and ores refined. Ash and soot particles are heavy and are deposited near to their source of release. The smaller smoke particles remain suspended in the air for much longer. They are breathed in and may permanently blacken the alveoli, where they cause irritation and interfere with gas exchange. It was this which contributed to the increase in human deaths in the London smogs of the 1950s. Smoke particles may also contribute to the various forms of pulmonary ailments which are common among smokers.

### 24.1.4 Carbon monoxide

Haemoglobin has a much higher affinity for **carbon monoxide** than oxygen. Breathing in high concentrations of carbon monoxide reduces the blood's ability to carry oxygen to respiring tissues. Many people suffer from oxygen deficiency in their tissues when the carbon monoxide haemoglobin content in their blood reaches 5 %. Cigarette smokers often have up to 10 % carbon monoxide haemoglobin in their blood. This is probably why babies born to mothers who smoke heavily are smaller than those of non-smokers.

### 24.1.5 Other pollutants of the air

**Nitrogen oxides** play an important role in the formation of **photochemical smog** which was first recognised in Los Angeles, USA. Both nitric oxide and nitrogen dioxide, as well as hydrocarbons, are present in the exhaust gases of motor vehicles. In sunny weather they become trapped near the ground by thermal inversion, causing a severe reduction in visibility. Ozone and peroxyacyl nitrates (PANs) are formed in the smog and cause eye and lung irritation, in addition to damage to vegetation and motor-vehicle tyres.

At present **lead** is considered to be the most serious heavy metal pollutant in the atmosphere (section 24.2.1).

447

## 24.2 Pollution by heavy metals

Heavy metals are those with relative atomic masses above 100. **Lead, mercury** and **cadmium** are in this category. They accumulate in living organisms, reaching concentrations which inhibit enzyme activity.

### 24.2.1 Lead

Not all environmentalists agree that the present concentration of lead in our surroundings is harmful to life in general. However, there are many pieces of evidence which show that the amount of lead in the environment is increasing. There is lead in the air we breathe, in the water we drink and in the food we eat.

### 1 Lead in the air

Combustion of petrol in motor vehicles accounts for over 80 % of the lead in air. Lead is added to petrol as **tetraethyl lead (TEL)** which serves as an anti-knock compound. It improves the efficiency with which petrol is burned in internal combustion engines. Most of the lead comes out in the exhaust gases as **lead chloro-bromide** ($PbClBr$). In rural areas, about 20 % of lead taken in by humans comes from the air. It is over 50 % in towns and cities. People who work close to motor vehicles have higher concentrations of lead in their blood than those who do not (Table 24.3).

**Table 24.3  Concentrations of lead in the blood of people in Cincinnati, US (after Lagerwerff 1972)**

| Occupation | Lead concentration ($\mu$g per 100 cm$^3$ blood) |
| --- | --- |
| office worker | 19 |
| policeman | 21 |
| postman | 23 |
| parking attendant | 34 |
| garage mechanic | 38 |

### 2 Lead in water

Tributaries of some Welsh rivers such as the Rheidol and Ystwyth are devoid of fish life. They receive water from disued **lead mines**. In places, the concentration of lead is more than 0.5 ppm, which is toxic to freshwater fish. Only a few species of algae and insects survive.

Lead poisoning was common among Ancient Romans. It happened because lead tanks and pipes were used to supply drinking water. The water pipes in some older houses today are made of lead. They contribute substantially to the intake of lead by some people.

### 3 Lead in food

Lead in the air is washed out by falling rain. Soil near to busy roads is more likely to be polluted in this way than elsewhere. Composts made from **sewage sludge** may also contain relatively high concentrations of lead (Chapter 25). Plant roots absorb lead from polluted soil. Foliage may also take in lead from the air. Either way, the amount of lead in plants may impair human health if the contaminated plants are eaten. The amount of lead in **canned foods** is higher than in fresh foods. This is because lead solders are used to seal cans.

## 4 Lead in humans

Whether inhaled or ingested, lead is absorbed into the blood and **accumulates** in the liver, kidneys, and bones. About 90 % of lead in the body is in the bones and teeth. The amounts of lead absorbed from air, food and water have been measured in the USA (Table 24.4). Food is the main source, wherever people live. In urban areas, lead from air makes up a large proportion of absorbed lead. There is evidence that the **mental development** of children in urban areas may be affected by the lead they breathe in. For this reason, it is policy in Britain to reduce lead in petrol by two-thirds in 1985 and to have lead-free petrol on sale by 1987. Tobacco smokers take in more lead than non-smokers, because nicotine is thought to stimulate the absorption of lead by the lungs.

**Table 24.4 Average daily intake of lead by people in USA (after Walker 1975)**

| Source | Lead absorbed per day ($\mu$g) |
| --- | --- |
| food | 17.0 |
| water | 1.0 |
| urban air | 10.4 |
| rural air | 0.4 |
| tobacco smoke | 9.6 |

### 24.2.2 Mercury

An estimated 230 tonnes per year of mercury reach the oceans from natural sources, mainly the weathering of rocks. Worldwide production of mercury for industrial and agricultural uses amounts to more than 7000 tonnes annually, of which 25 to 50 % is believed to be discharged into the environment. Mercury is used extensively as a floating electrode in the production of chlorine and caustic soda from brine. It is also used as a catalyst in the manufacture of vinyl plastics. Organic mercury fungicides are used to preserve wood pulp and as seed dressings to ensure good germination.

### 1 Mercury in the air

When coal and oil are burned, mercury is released into the air. **Ore-refining** is another source of aerial mercury. Mercury also evaporates from the earth's crust. High concentrations of atmospheric mercury detected by sniffer aircraft are valuable indicators of where ores of mercury can be found.

### 2 Mercury in water

Rain picks up mercury as it falls through the air. Waste water from paper mills and factories making vinyl plastics also contains mercury. Most of the mercury accumulates in mud at the bottom of rivers, lakes and seas. There it is converted to **methyl mercury** by methane-producing bacteria. Methyl mercury is volatile and very toxic. It is absorbed by aquatic organisms which may be part of a food web involving humans.

### 3 Mercury in food

In the recent past, **sprays** and **seed dressings** containing mercury were used a great deal to prevent and to control many plant diseases caused by pathogenic fungi. The annual usage of mercury worldwide for this purpose was 140 tonnes in 1967. It fell to about half this amount by 1970 and the trend is still downward.

Mercury is retained in the upper layers of well drained soils. From here it is absorbed by the roots of plants. Mercury in sprays is taken in through the leaves. Movement to the fruit and other edible parts of plants ensues. Mercury poisoning has occurred in wood pigeons consuming dressed seed. In Sweden, Canada and the USA, freshwater fish have been shown to contain much higher concentrations of mercury than marine fish (Table 24.5). It is thought to have entered lakes in the waste water from **wood pulp mills**. The pollution is so bad in parts of North America that commercial fishing in some lakes has been banned in the past few years.

**Table 24.5 Concentrations of mercury in Swedish freshwater and marine fish (after Walker 1975)**

| Species | Habitat | Mercury concentration (ppm) |
|---|---|---|
| mackerel | | 0.06–0.08 |
| herring | marine | 0.03–0.11 |
| cod | | 0.15–0.19 |
| pike | freshwater | 5.00 |

### 4 Mercury in humans

The appalling events at Minamata and Niigata, Japan, in the 1950s underlined the danger to man of pollution by mercury. At Minamata 111 people were seriously affected by **mercury poisoning** and 46 of them died. Shellfish caught in Minamata Bay made up a large proportion of their diet. It was discovered that a vinyl plastics factory had for some time discharged waste water into the bay. Table 24.6 shows the concentrations of mercury in the seawater and in marine organisms from Minamata Bay in 1960. Note how the mercury had **accumulated** higher up the food chain.

Post-mortem examinations revealed high concentrations of mercury in the kidneys, liver and brain of those poisoned (Table 24.7). At Niigata, 120 people were poisoned by mercury in similar circumstances. The victims had eaten up to three fish meals a day. At both places the survivors were partially or totally paralysed and had lost most forms of sensation. Many had become blind. Mothers who three years earlier had shown no signs of mercury poisoning bore infants who suffered from mental retardation and cerebral palsy. The developing foetuses had evidently accumulated mercury from their mothers' blood.

**Table 24.6 Concentrations of mercury in seawater and marine organisms from Minamata Bay in 1960**

| Source | Mercury concentration |
|---|---|
| seawater | 1.6–3.6 ppb |
| plankton | 3.5–19.0 ppm |
| shellfish | 30.0–102.0 ppm |

**Table 24.7 Concentrations of mercury in body organs of Minamata victims, 1, and a control group, 2, of humans**

| Organs | 1 | 2 |
|---|---|---|
| kidneys | 106 ppm | 1.0 ppb |
| liver | 42 ppm | 3.0 ppb |
| brain | 21 ppm | 0.1 ppb |

### 24.2.3 Other heavy metals

**Copper** enters streams and rivers in drainage water from **copper mines** and in the liquid effluent from copper-plating works. It is a very toxic metal. Freshwater plants and fish are killed by less than 1 ppm of copper. Fish absorb it through their gills. The rate at which water passes over the gills therefore determines how much copper is absorbed in a given time. In water containing a low concentration of dissolved oxygen, gill ventilation is rapid. In such conditions fish are more quickly killed by copper polluting the water.

Copper also pollutes the spoil heaps of copper mines. Its presence limits the vegetation to copper-tolerant species of plants called **metallophytes**. Pollution of soil by copper has occurred in some English apple orchards. It arrived there from copper-containing **fungicides**, such as Bordeaux mixture, which have been sprayed on the trees for over a hundred years. The polluted soil is devoid of most kinds of animal life. The absence of earthworms is thought to be the reason why a thick layer of poorly decayed detritus accumulates on the soil surface.

Zinc appears in streams and rivers in the effluent from **zinc mines**. Water boatmen and the larvae of caddis and stone flies tolerate up to 500 ppm. Freshwater fish are killed by less than 1 ppm. As with copper, fish are more quickly killed in poorly aerated water. Most terrestrial plants can tolerate up to 7 ppm of zinc in soil. The spoil heaps near zinc mines contain much more than this and are inhabited by zinc-tolerant metallophytes.

The air in the vicinity of zinc smelters is polluted with oxides of zinc and **cadmium.** Humans absorb cadmium from the air through their lungs, and there is some evidence that people working and living near smelters suffer from acute **cadmium poisoning**. Cadmium concentrates in the kidneys, causing renal damage and high blood pressure. The drainage water from zinc mines also contains cadmium. In the Jintsu Basin of Japan, where river water containing such effluent is used to irrigate paddy fields, people suffer from a chronic form of cadmium poisoning called *itai-itai*. Minerals are withdrawn from the bones and there is severe pain. Cadmium is taken up by the rice crop, which is a staple part of the diet of the Japanese. They therefore ingest much more cadmium than people living in unpolluted areas.

## 24.3 Pesticides and fertilisers

A wide range of toxic chemicals called **pesticides** is used by man to control or eradicate pests. **Fertilisers** promote the growth of crops, making more food available for mankind.

### 24.3.1 Herbicides

**Herbidices** make up about 40 % of the world's production of pesticides. They are used to control or eliminate weeds. Gardeners use herbicides to get rid of daisies, dandelions, plantains and other weeds growing in lawns. Farmers use them to eliminate weeds such as thistles, docks and poppies which grow among cereal crops and in pastures. Herbicides are also used in forestry to kill forest weeds such as rhododendrons and gorse. Railways are kept free of weeds by spraying herbicides onto the track.

Many herbicides are similar to plant hormones called auxins. Auxin-like herbicides are biodegradable and are soon broken down to harmless products by soil bacteria. For this reason they seldom cause environmental problems. However, an impurity which is present in one group of herbicides has recently given cause for concern. Several widely-used herbicides are derivatives of trinitrophenol, for example 2,4,5-trichlorophenoxyacetic acid (2,4,5T). In manufacturing trinitrophenol, an impurity called **dioxin** is formed and this is usually present in 2,4,5T. Dioxin has been shown to be harmful to humans. It is so potent that half a gram of it can kill 3000 people. Although it is normally present in only small amounts in 2,4,5T, dioxin is a potential threat because a good deal of the herbicide is released into the environment. In the USA over 3000 tonnes of 2,4,5T are used annually, mainly in forestry. In the UK, about 3 tonnes are used each year, chiefly by British Rail and the Forestry Commission.

The effects of dioxin on humans and its potential effects have been studied in the following three main ways.

### 1 Laboratory studies

When small amounts of dioxin are painted on to the ears of rabbits, the treated area develops a very severe form of acne known as **chloracne**. When given to animals in their food and water, dioxin has been shown to be one of the most potent of poisons (Table 24.8). The animals which died in these experiments showed general wastage of body tissues and organs.

**Table 24.8 Minimum dose of dioxin required to kill 50% of a batch of animals (LD50 values)**

| Species | LD50 (ppm) |
|---|---|
| rat | 0.0005 |
| mouse | 0.1000 |
| guinea pig | 0.0006 |

Tumours and cancer were frequent. A number of serious **teratological defects** (defects of the foetuses) were observed when dioxin was given to pregnant animals. They included cleft palate, kidney malfunction, enlarged livers and internal haemorrhage. In monkeys, whose anatomy and physiology are very similar to those of humans, spontaneous abortion, haemorrhage, gangrene of the fingers and toes and loss of hair occurred, even at the lowest dosage tested.

### 2 Vietnamese and American forests

Eleven million gallons of 'Agent Orange' (50 % 2,4,5T) were sprayed by the Americans onto the forests of Vietnam in the 1970s. The herbicide **defoliated** the trees, making it easier to observe enemy movements. It has since been shown that defects among children born in the contaminated areas are much higher than normal. USA servicemen working in the area later fathered children with physical and mental handicaps. Dioxin was found in the bodies of the fathers. More recently, miscarriages in women living in the forests of Oregon, USA, where 2,4,5T has been used for several years, have been greater than expected. It has led to a temporary ban on the use of the herbicide in this part of the USA. For these reasons there has also been a campaign to stop the use of 2,4,5T in Britain.

### 3 Factory accidents

There have been several major explosions at factories where trinitrophenol is produced. One of the most recent was at Seveso, Italy, in 1976. Cats and dogs died in large numbers. There were 400 cases of chloracne in children alone. Miscarriages and cancer rates were higher than normal. Chloracne among workers and their families, abnormal liver function and higher incidences of skin and heart disease followed an explosion at Bolsover, England in 1968. The full effects of an accident at Nitro, West Virginia, USA in 1949 are still unknown. Recently, a comprehensive survey of the health of the affected people has been carried out. The results are awaited and could be vitally important because some conditions such as cancer often have a long latent period. There is little doubt that some dioxin was released in the explosions and was taken into the bodies of workers and people living nearby through the skin and lungs and by swallowing.

### 24.3.2 Fungicides

Compounds of mercury and copper are the most serious pollutants of the fungicides. Their environmental effects are discussed in sections 24.2.2 and 24.2.3.

### 24.3.3 Insecticides

Chemicals which kill insects have been known for a long time. Before 1940, the **insecticides** most widely used were natural products such as pyrethrum and nicotine. Since then, two groups of synthetic insecticides, **organochlorine** and **organophosphorus** compounds, have replaced the older insecticides. Like pyrethrum and nicotine, organophosphorus insecticides are biodegradable and are quickly broken down into harmless products in the environment. Although very toxic, they are not pollutants if used in a responsible way. In contrast, organochlorine insecticides persist for many years in the environment. Their persistence has been the cause of pollution.

The best known of the organochlorine insecticides is **DDT** (dichlorodiphenyl trichloroethane). It was first made nearly a century ago, but its insecticidal powers were not realised until the Second World War. Its non-toxicity to man and its effectiveness in killing insects led people to believe that DDT would soon eradicate insect pests. Malaria, yellow fever and other insect-borne diseases would become a thing of the past. Other organochlorine insecticides such as aldrin, dieldrin, heptachlor and lindane were just as effective in eradicating insect pests of crops. Hopes were high that the agricultural use of such compounds would eliminate starvation in developing countries. Today the widespread use of DDT and some of the other organochlorine insecticides has been banned in many developed countries. People die in their millions each year of insect-borne diseases and crops are still ravaged by insect pests. What has gone wrong?

### 1 Overuse

Rachel Carson's aptly titled book, *Silent Spring* (1963) publicised the ecological effects of the **overuse** of organochlorine insecticides in the USA. In places where farmers were grossly exceeding the recommended dose rate, there were hardly any songbirds. Some had been killed by breathing in air containing large amounts of organochlorine insecticides or by eating contaminated food. The massive reduction of the songbird population was, however, mainly caused by the insecticides reducing their fertility. Many of the eggs which were laid failed to hatch. Enough of the insecticides had also been leached from the land to kill fish in nearby rivers and lakes.

### 2 Persistence

Organochlorine insecticides are very stable substances. They persist unchanged for a long time in the environment (Table 24.9). **Persistence** increases the chances of poisoning non-target organisms. In the 1950s thousands of seed-eating birds were poisoned after eating cereal grain dressed with aldrin and dieldrin. Endotherms convert DDT to less toxic DDE and aldrin to dieldrin. The amounts of DDE and dieldrin in the flesh and eggs of birds provide an estimate of the extent of pollution of the environment by organochlorine insecticides. Such a survey was carried out between 1960 and 1965 in Britain after reports of a severe decline in the numbers of predatory birds such as the peregrine falcon and golden eagle

**Table 24.9 Persistence of some organochlorine insecticides (after Walker 1975)**

| Compound | Time for 95% loss (years) |
|----------|---------------------------|
| Dieldrin | 4–30 |
| DDT | 5–25 |
| Lindane | 3–10 |

**Fig 24.7** (a) Distribution of peregrine falcons in Britain in 1961. The first number shows the percentage of territories occupied, the second (in box) the percentage in which young were reared (from Mellanby 1967 after Ratcliffe)

(Fig 24.7). The survey showed that predatory birds had higher concentrations of DDE and dieldrin in their bodies than other species (Fig 24.8). The insecticides had probably entered the flesh-eating species along a food chain.

**Fig 24.7** (b) A peregrine falcon (RSPB/William S. Paton)

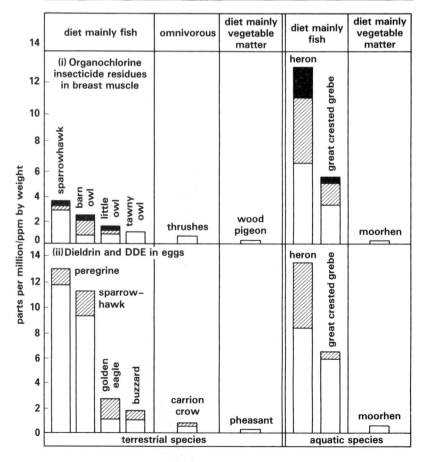

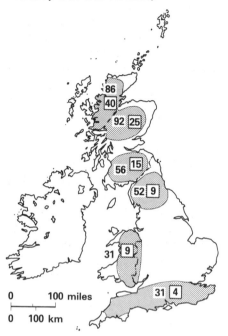

**Fig 24.8** Residues of organochlorine insecticides in British birds (from Walker 1975)

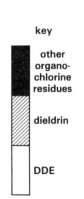

key

other organo-chlorine residues

dieldrin

DDE

| | diet mainly fish | | | | omnivorous | diet mainly vegetable matter | diet mainly fish | | diet mainly vegetable matter |

(i) Organochlorine insecticide residues in breast muscle

sparrowhawk, barn owl, little owl, tawny owl, thrushes, wood pigeon, heron, great crested grebe, moorhen

parts per million/ppm by weight

(ii) Dieldrin and DDE in eggs

peregrine, sparrow-hawk, golden eagle, buzzard, carrion crow, pheasant, heron, great crested grebe, moorhen

terrestrial species     aquatic species

454

Because they are insoluble in water, the organochlorine compounds are stored in fatty tissues instead of being excreted. In this way they **accumulate** at each trophic level. Toxic amounts of DDE and dieldrin are probably released into the bloodstream when fat reserves are used at times of food shortage and at egg-laying. Laboratory experiments show that sublethal amounts of organochlorine insecticides cause birds to lay eggs with thin, fragile shells. When incubated by the parents, they are likely to be broken. A study of the eggshells of peregrine falcons showed that they had become significantly thinner after organochlorine insecticides came into general use in Britain in the 1940s. Circumstantial evidence of this kind led to a restriction of the use of organochlorine insecticides for agricultural purposes in the USA. Britain and other western countries adopted similar measures in the late 1960s. Subsequently there has been a marked improvement in the breeding success of predatory birds.

### 3 Resistance

Soon after DDT was introduced, houseflies in Italy were found to be unaffected by it. They had acquired a means of converting DDT to less toxic DDE. It is now known that over 100 species of insect vectors have developed a **resistance** to organochlorine insecticides (Fig 24.9). It is for this reason that any hope of eradicating malaria in the near future has subsided. More than 200 species of insects which are pests of crops are also resistant. Whilst this is not a pollution problem, it does give cause for concern to those who have to control insect pests and insect-borne diseases.

The adverse environmental effects of pesticides have often caused them to be seen in a rather bad light. However, it is important to take an objective view. There is little doubt that pesticides have saved millions of people from disease and starvation. Much of the publicity centred around them has stimulated research into producing safer pesticides and alternative ways of pest control.

**Fig 24.9** The number of insects of public health importance which have developed resistance to insecticides (from Ottaway *The Biochemistry of Pollution*, Inst. Biol: Studies in Biology, published by Edward Arnold)

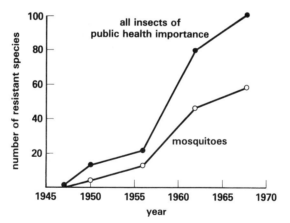

### 24.3.4 Fertilisers

Before **fertilisers** became available the increasing demand for food by the rapidly expanding human population was achieved by using more and more land for growing crops. The amount of new land which can be used for agricultural purposes is getting less. In some countries land which was farmed has been used for building roads, houses and factories. It has been estimated that at least an acre of agricultural land is lost every day in the UK for these reasons. One alternative is to bring into cultivation land which has not previously been farmed. Recent legislation such as the *Protection of Wildlife Act* has been designed to conserve habitats in which wildlife flourishes. The EEC currently produces more food than it requires and farmers are being paid not to cultivate some of their land. It reflects the fact that high yields can be produced from less land. Fertilisers have helped farmers to achieve such yields in developed countries. Without fertilisers Britain would produce about half the food that is presently grown. The Food and Agricultural Organisation of the United Nations estimates that roughly three-quarters of the extra food required to feed the human population by the end of this century will come from intensification of agriculture. The remainder can only be achieved by destroying large tracts of wild forest and draining wetlands.

Doubts have been cast by some people as to the wisdom of using intensive methods for producing crops. It is therefore necessary to assess objectively the disadvantages as well as the benefits of applying fertilisers to stimulate crop growth.

### 1 What are fertilisers?

Three elements, nitrogen, potassium and phosphorus, are required in large quantities by growing crops. Fertilisers consist of these elements mainly in the form of **ammonia, potash** and **rock phosphate** respectively (Fig 24.10). Most of the ammonia used to make fertilisers in Britain comes from atmospheric nitrogen and North Sea gas. In soil, ammonia is rapidly oxidised to nitrate. A significant amount of potash is mined in Britain but much is imported, as is all of the rock phosphate used in fertiliser production.

**Fig 24.10** (a) A bag of fertiliser (ICI Agricultural Division)

**Fig 24.10** (b) Applying fertilizer to a field (ICI Agricultural Division)

Nitrogen is used to synthesise amino acids, proteins, nucleic acids and chlorophyll. Phosphorus is required for plants to make nucleic acids, phospholipids, ATP and $NADP^+$. Potassium activates a wide range of plant enzymes. Hence adding fertilisers to land stimulates crop growth (Fig 24.11).

**Fig 24.11** Effect of nitrogen concentration on the growth of rice (after Bland & Bland 1983)

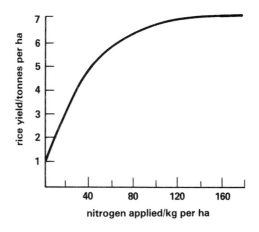

456

## 2 Alternative sources of plant nutrients

Some nitrogen, phosphorus and potassium are present in organic materials such as **guano** (bird droppings), **hoof and horn, dried blood** and **municipal compost**. However, supplies of these are fairly small and they make up less than 0.1 % of the UK fertiliser market. An alternative source of plant nutrients is **sewage** and **animal manure**. The amounts of nutrients which sewage could provide is relatively small compared with agricultural needs. For example, 1 200 000 tonnes of nitrogen are currently applied to farmland annually in Britain. It has been calculated that sewage could provide only 60 000 tonnes of nitrogen a year. Animal manure is used widely on the land and provides about 600 000 tonnes of nitrogen annually. The main difficulties of using sewage sludge and municipal compost is that they often contain high concentrations of heavy metals such as lead, zinc and cadmium which are toxic to humans and livestock. It is also costly to treat and transport them to farms.

Another way of increasing the nitrogen content of soil is to grow **leguminous crops** such as clover, peas and beans in **rotation** with other crops. Nitrogen from the air is fixed in the root nodules of legumes and is left in the soil for subsequent crops to absorb. One of the aims of genetic engineering (Chapter 4) is to transfer the genes for nitrogen fixation from bacteria to non-leguminous crop plants, thus eliminating the necessity for crop rotation. Even so, fertilisers would have to be applied to obtain maximum yields.

Some people will not eat foodstuffs grown on land which has been treated with fertilisers and pesticides. Their diet is made up of foods which have been grown in soils whose fertility is maintained by applying organic materials. **Organic farming** is usually labour intensive and produces crops which are more expensive than those grown by methods using pesticides and fertilisers.

## 3 Effects of fertilisers on the soil

The amount of fertilisers added to soil is insignificant compared with the total mass of soil. Fertilisers simply increase the amounts of chemical elements which are found naturally in all soils. At Rothamsted Experimental Station, fertilisers have been used on plots of land for over a hundred years. Yields of wheat and grass grown on the plots are bigger than ever. It appears that fertilisers have no detrimental effects on soil in the long run. The reason why fertilisers improve soil fertility is that they stimulate the growth of bigger root and shoot systems of the crop. When the crop is harvested the roots and often much of the shoot system are ploughed back into the ground. Here they are decomposed to form **humus** which improves the structure and fertility of the soil. Humus promotes the formation of aggregates of soil particles called **crumbs**. A soil with good crumb structure is usually well drained and well aerated. It is easier to plough and harrow and provides a good tilth in which seeds germinate well. Humus also slowly releases nutrients for plant roots to absorb.

The risk of soil erosion by heavy rain is reduced in well drained soils, especially in tropical countries. The well developed root systems of crops prevent rain washing away the soil.

Why are crops raised by organic farming usually more expensive than those grown by conventional methods?

## 4 Effect of fertilisers on water supplies

Potash and phosphate are firmly held in the soil, whereas nitrate is readily leached out by rain water. About 30 % of the nitrate entering rivers in England and Wales comes from land drainage. The remainder comes mainly from sewage effluent. Roughly a third of the nitrate in the water draining from land originates from fertilisers. The other two-thirds comes from the decomposition of organic matter.

457

There are two causes for concern attached to nitrate in water:

**i. Eutrophication** The enrichment of rivers, lakes and reservoirs with nutrients stimulates algal growth. The eventual death and decay of the algae increases the biochemical oxygen demand (BOD) of the water and depletes the dissolved oxygen, so that freshwater fish and other aquatic life are killed. Large quantities of algae in the water may also block the filters of water-treatment works. Although fertilisers contribute to these undesirable changes, the main cause is sewage. In general, **eutrophication** is not a serious problem in the UK, though it causes concern elsewhere in the world.

**ii. Human health** Nitrate is reduced to nitrite in the gut of young mammals including human babies. Nitrite ions combine with haemoglobin to form methaemoglobin which cannot carry oxygen in the blood. Babies under twelve months of age are mostly at risk. Since 1945 at least ten fatal cases of **methaemoglobinaemia** have been diagnosed in the UK, mainly in East Anglia. Here agriculture is the main industry and some drinking water comes from wells into which nitrate is leached. For this reason the World Health Organisation has indicated that water containing more than 100 ppm of nitrate is unfit for consumption. Water Authorities now constantly monitor the nitrate content of drinking water to ensure that it does not go above this limit.

## 5 Fertilisers and energy

Fertilisers are relatively little used in developing countries because they are too expensive. One reason for this is that energy is required to make fertilisers and energy is costly. Even so, less than 1 % of the total energy used in Britain goes into the manufacture of fertilisers. As plants grow they trap solar energy in photosynthesis, thus increasing the amount of energy-rich foods for humans to eat. For each unit of energy required to make fertilisers, several units of energy are made available in plant foods for people to consume (Table 24.10).

**Table 24.10 Energy ratios for various crops (from *Fertilisers and our Environment*, Fertiliser Manufacturer's Association 1981)**

| Crop | Energy input:output ratio |
| --- | --- |
| potatoes | 1:1.3 |
| barley | 1:2.6 |
| wheat | 1:3.1 |
| sugar beet | 1:3.3 |

Promoting the growth of crops is the most efficient way of making more energy available in human food. In general, fertilisers are not thought of as a threat to our environment. On the contrary, the careful use of fertilisers has done much to increase crop yields, making more food available for the growing human population. In effect fertilisers have provided a substitute for more land. In 1980 the production of crops in the USA took up 120 million hectares of land and 24 million tonnes of fertilisers. If a third of the fertiliser had been used, another 50 % more land would have been needed to yield the same amount of crop.

## 24.4 Radiation

Ever since life evolved, living organisms have been subjected to various forms of **radiation**. The most obvious is visible light from the sun. The sun also radiates ultraviolet and infra-red waves. Radiation comes from outer space too as cosmic rays, and from radioactive minerals in rocks. Natural forms of radiation are not thought of as pollutants. But radiation introduced into the environment by human activity is a form of pollution.

### 24.4.1 Forms of radiation

The forms of radiation which pose a threat to the well-being of living organisms fall into two categories:

#### 1 Electromagnetic waves

**Gamma rays** ($\gamma$-rays) are electromagnetic waves with a wavelength of about $10^{-12}$ m. Their very short wavelength gives them considerably more energy than visible light and consequently much better powers of penetration. Gamma rays are emitted when radioactive atoms (radionuclides) disintegrate or when atomic nuclei are bombarded with sub-atomic particles.

#### 2 Sub-atomic particles

**Alpha particles** ($\alpha$-particles) are given off when radionuclides break down. They are positively charged, consisting of two protons and two neutrons. When they collide with anything solid, $\alpha$-particles lose a good deal of energy. **Beta-particles** ($\beta$-particles) are electrons and have a negative charge. They are not very penetrating. Like $\alpha$-particles, they soon lose energy when they hit something solid. Because of this energy transfer, $\alpha$- and $\beta$-particles are biologically among the most damaging forms of radiation. However, they are unlikely to do very much damage unless they arise inside a living organism. **Neutrons** have no net charge. Each neutron consists of one proton and one electron. They are emitted when certain elements are bombarded with $\alpha$-particles or with $\gamma$-rays. Neutrons are more penetrating than $\alpha$- and $\beta$-particles. Like $\gamma$-rays, they can do a great deal of harm, even when originating outside living organisms.

Some radionuclides give off radiation for only a short time. Others do so for many years. The time required for a radionuclide to emit half its radiation is called its **half-life** (Table 24.11). The half-life and type of radiation emitted determine the degree of danger.

**Table 24.11 Half-lives of some radionuclides (after Ottaway 1980)**

| Nuclide | Name | Radiation | Half-life |
|---------|------|-----------|-----------|
| $^{90}$Sr | Strontium | $\beta$-particles | 28 y |
| $^{129}$I | Iodine | $\beta$-particles $\gamma$-rays | $16 \times 10^6$ y |
| $^{131}$I | Iodine | $\beta$-particles $\gamma$-rays | 8 days |
| $^{137}$Cs | Caesium | $\beta$-particles $\gamma$-rays | 30 y |
| $^{237}$Np | Neptunium | $\alpha$-particles | $2.2 \times 10^6$ y |
| $^{239}$Pu | Plutonium | $\alpha$-particles $\gamma$-rays | $2.4 \times 10^4$ y |

The unit of radiation most meaningful to biologists is the gray (Gy). 1 Gy is equal to 1 J of radiation energy absorbed by 1 kg of body mass.

### 24.4.2 Sources of radiation

Natural sources account for more than 80 % of the radiation absorbed by humans (Table 24.12). There is little that can be done about natural radiation. Living organisms have had to cope with it since life began. What is of concern is the radiation released by human activities.

**Table 24.12 Radiation absorbed by people in Britain (after Mellanby 1980)**

| Source | Annual dose ($\mu$Gy) | Percentage |
| --- | --- | --- |
| natural background | 870 | 83.8 |
| medical | 140 | 13.5 |
| nuclear fallout | 21 | 2.02 |
| miscellaneous | 7 | 0.67 |
| radioactive waste | 0.1 | 0.01 |

Though non-natural radiation presently accounts for a relatively small amount of the total, it is on the increase and adds to the risks created by natural sources. Two human activities which release radiation into the environment are the use and testing of **nuclear weapons** and the use of **nuclear fuels**. Both are emotive subjects to people the world over.

In the early 1960s there was a dramatic increase in the amount of $^{90}$Sr in the milk consumed in Britain. It followed the testing of nuclear weapons by the USA and USSR. Soon after the Nuclear Test Ban Treaty, contamination by $^{90}$Sr reverted to normal and has stayed there since (Fig 24.12).

**Fig 24.12** Average ratio of $^{90}$Sr to Ca in milk in Britain (from Mellanby *The Biology of Pollution*)

**Thermal nuclear reactors** use natural uranium, a mixture of $^{235}$U and $^{238}$U. When bombarded with neutrons, uranium decays into a variety of **fission products**, including $^{90}$Sr, $^{129}$I, $^{131}$I, $^{137}$Cs and $^{239}$Pu. The fission products which are potentially most dangerous to living organisms are $^{90}$Sr, $^{129}$I, $^{131}$I and $^{137}$Cs. Radioactive iodine absorbed by grasses and eaten by cows appears in their milk. Humans drinking contaminated milk concentrate radio-iodine in the thyroid gland. In this way the thyroid can be exposed to high doses of radiation. $^{90}$Sr accumulates in bone adjacent to the marrow. Its long half-life means that it irradiates dividing bone marrow cells for years. Caesium is taken up by soft tissues, especially muscle.

Accidents have happened at nuclear power plants, spilling fission products into the surrounding countryside. The chief risk is failure of the cooling system in a reactor. The build-up of heat could be enough to melt a reactor. Partial failure of the cooling system in a fast reactor caused an explosion at Three Mile Island near Pittsburgh, USA, in 1979. The effects are still being assessed. The worst nuclear accident in Britain happened at

Sellafield, Cumbria, in 1957 when $^{131}$I and $^{137}$Cs were released. As a precautionary measure, milk produced for several weeks within $5\,km^2$ of the nuclear station was declared unfit for consumption.

Without doubt the most serious accident in a nuclear power plant occurred in 1986 at Chernobyl in the Ukraine, USSR. On 26 April one of the four reactors exploded, releasing a huge cloud of radioactive dust. Within days, over 100 000 inhabitants were evacuated from Chernobyl and the surrounding area. Crops in a zone of diameter 29 km were immediately designated as unfit for harvesting. A reserve supply of drinking water was hurriedly arranged for Kiev, the Ukrainian capital. Before the accident, water for about half of Kiev's population came from the River Dnieper which has tributaries in the Chernobyl area.

The radioactive cloud drifted across Europe, and in less than a week radiation levels in Scandinavia had increased five-fold. In the UK the highest fallout occurred in the wettest areas, the uplands of Scotland, Cumbria and Wales. Within a month the concentration of $^{137}$Cs in the muscles of sheep was so high that the animals were declared unfit for human consumption. Slaughter was permitted about two months later when the amount of radioactivity in livestock had fallen to acceptable levels.

The full consequences of the Chernobyl catastrophe may never be known. Medical experts have predicted an inevitable increase in the incidence of cancers such as leukaemia in Europe as a whole as well as the USSR. At a news conference in August 1986 the Soviet government disclosed that 31 Russians had been killed in the accident and that 203 were hospitalised with acute radiation sickness.

Another problem faced by the nuclear fuel industry is what to do with spent reactor cores which are still radioactive. In Britain, high and medium radioactive waste is sealed in concrete containers. If stored indefinitely, the fission products decay and become harmless. The difficulty is in deciding where they should be kept. Storage in underground caverns is the disposal method used mainly at present. Prior to 1983, Britain dumped hundreds of tonnes of radioactive waste at sea. In 1984 the London Dumping Convention stopped the disposal of such waste in the Atlantic Ocean. Hence the government agency for the disposal of radioactive waste is presently seeking dumping sites on land.

### 24.4.3 Biological effects of radiation

The effects of radiation on living organisms are two-fold.

**1 Genetic effects**

Living cells are very sensitive to radiation at interphase. It is at this stage in the cell cycle that replication of DNA occurs. Radiation alters the genetic code by breaking hydrogen bonds between the pairs of nitrogenous bases in the DNA molecules. Bases may be lost or reassembled in a way different from that in the original DNA. Such changes are called **gene mutations** (Chapter 4). The faulty DNA then codes the synthesis of abnormal proteins which may be defective in function. Gene mutations may be passed to future generations in gametes. Intense doses of radiation cause chromosomes to fragment. Chromosome fragments do not behave as normal chromosomes in nuclear division. They often fail to become attached to the spindle and are lost when the cytoplasm cleaves. The ends of large chromosome fragments may join up. It is then impossible for the chromatids to separate at anaphase (Fig 24.13). Such changes are **chromosome mutations**.

Where are the proposed sites earmarked by the British government for dumping nuclear waste? What precautions will be taken to prevent leakage to the surrounding area?

**Fig 24.13** Possible behaviour of fragmented chromosomes during nuclear division

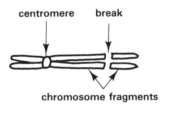

centromere    break

chromosome fragments

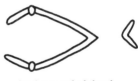

broken ends joined

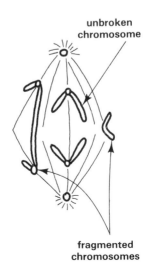

unbroken chromosome

fragmented chromosomes

461

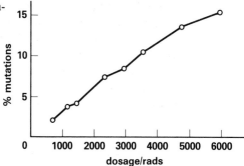

**Fig 24.14** Effect of X-rays on the mutation rate of *Drosophila*

Muller, in 1927, using the fruit-fly *Drosophila*, was the first to demonstrate that radiation was a cause of gene mutations. He also showed that the mutation rate was proportional to the dosage of radiation (Fig 24.14).

The probability of a mutation occurring increases with each exposure to radiation. The critical exposure period for any organism is when it is producing offspring. It is then that mutations may occur in cells which give rise to progeny formed asexually or in gamete-forming cells. For women the critical period extends over thirty child-bearing years. It has been estimated that the rate of mutation among humans would double if women absorbed an additional 0.5 Gy over their reproductive life span. If the increase was only 0.01 Gy, in each generation the mutation rate would increase by 1.5 %.

### 2 Somatic effects

Table 24.13 summarises the observations made on Japanese people exposed to very high doses of whole-body radiation following the nuclear bomb explosions at Hiroshima and Nagasaki in 1946, and on fishermen from the Marshall Islands exposed to radioactive fall-out near a nuclear-weapons testing station in the Pacific Ocean in 1954. The children of pregnant women exposed to radiation at Hiroshima and Nagasaki weighed much less than normal, had smaller brains, and many suffered from severe mental and physical handicaps. Survivors at Hiroshima were more likely to have **leukaemia** if they were near the centre of the explosion (Fig 24.15). Fall-out of $^{90}$Sr from the testing of nuclear weapons may well have been the cause of the increased number of cases of leukaemia in Britain in the 1960s.

**Fig 24.15** Relationship between the incidence of leukaemia and distance from centre of the atomic explosion at Hiroshima, Japan

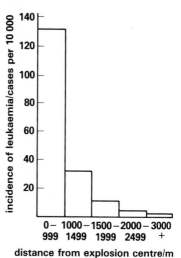

**Table 24.13 Somatic effects of radiation**

| Dosage (Gy) | Death within | General effect | Specific effects |
| --- | --- | --- | --- |
| >100 | 1 day | CNS syndrome | Vomiting, diarrhoea, convulsions, brain damage, coma |
| 10–100 | 2 weeks | GI syndrome | Gut lining not replaced, inability to digest food, vomiting, diarrhoea, emaciation |
| 2–10 | 3–4 weeks | BM syndrome | Blood cells not replaced, anaemia, fever, vomiting, haemorrhages, loss of hair |
| <2 | unknown | life-shortening | Inability to make antibodies, small wounds remain septic for long periods |

CNS = central nervous system. GI = gastro-intestinal. BM = bone marrow.

As with pollution by pesticides, it is important to try to see the threat from radiation in perspective. Reserves of fossil fuels are being used at such a rate that some experts predict they will be exhausted in the foreseeable future. At present, nuclear fuel offers the main alternative source of energy for industrial and domestic use. As more nuclear power plants are built, the risk of radiation pollution is likely to increase. So far, rigorous safeguards have been taken. Although accidents have happened and some radiation has escaped from nuclear power stations, there has been no widespread danger to human life. However, the amount of radioactive waste to be disposed of is growing rapidly. There is concern about the safety of dumping at sea where the containers may eventually corrode. Nuclear waste buried underground may escape if earth tremors occur. The very long half-lives of some fission products increase the possibility of contamination. Future generations could then inherit the hazard of pollution by the radioactive wastes of today's nuclear fuel industry.

A worse hazard would come from the wide-scale use of nuclear weapons. The radiation released by explosions of nuclear bombs worldwide would kill most of us immediately and many parts of the earth could be uninhabitable for a very long time.

## 24.5 Pollution of aquatic habitats

The more obvious of society's waste products, such as plastic containers, glass bottles and tin cans which are often dumped in streams, rivers, lakes and the oceans, are eyesores. There is little evidence that they harm organisms. Agricultural, domestic and industrial wastes are the main pollutants of aquatic habitats.

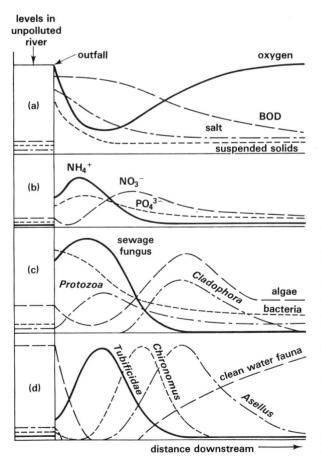

**Fig 24.16** Effects of organic effluent on (a) physical changes, (b) chemical changes, (c) flora, (d) fauna of a river (from Hynes *The Biology of Polluted Waters* published by Liverpool University Press, 6th imp. 1978)

### 24.5.1 Freshwater pollution

**1 Rivers**

The pollution of British rivers by heavy metals is described in section 24.2. Other industrial wastes such as ammonia and cyanide are also toxic to aquatic life. Particulate matter released by industry can pollute rivers too. Coal washings and china clay settle on the beds of rivers, fouling the spawning places of salmon and trout. When suspended in water, the particles reduce penetration of light for aquatic plants and asphyxiate fish by blocking their gills.

**Sewage** is the biggest pollutant of fresh water. It is the waterborne waste of society (Chapter 25). The effects of discharging untreated sewage into a river have been studied by Hynes (Fig 24.16). One of the most striking features is a substantial and immediate drop in the amount of oxygen dissolved in the water. It happens because organic matter stimulates decomposer organisms, especially bacteria, which break down suspended solids in the sewage. As they respire, the decomposers use up dissolved oxygen. The amount of oxygen used in breaking down the organic matter into simple soluble products, such as ammonia, sulphate and phosphate, is the **biochemical oxygen demand (BOD)** of the sewage. The ammonia is quickly oxidised to nitrate by nitrifying bacteria. Some of the phosphate may have come from detergents in the sewage. Downstream from the sewage outfall, the dissolved oxygen concentration returns to normal as oxygen is absorbed by the river from the air. A gradual fall in BOD occurs simultaneously as the decomposers are dispersed by the flowing water.

Changes in the flora and fauna of the river accompany the chemical changes. Filamentous bacteria called sewage fungus thrive near the outfall. They are anaerobic and tolerate the high concentration of ammonia which makes the water alkaline. Algae, *Cladophora* in particular, are sensitive to pollution by sewage. Downstream they flourish, stimulated by the minerals released from the decomposed organic matter. Their numbers then return to normal as the minerals are used up or diluted by water from tributaries. The fauna most tolerant of sewage are annelids of the family *Tubificidae* (tubifex worms). Their blood contains a form of haemoglobin with an exceptional affinity for oxygen. It enables them to extract oxygen even when the concentration of dissolved oxygen in water is very low. Larvae of the midge *Chironomus* are the next most tolerant. They too have a type of haemoglobin which binds very efficiently with dissolved oxygen. The water louse *Asellus* begins to appear when the dissolved oxygen begins to recover. Clean-water fauna such as the freshwater shrimp *Gammarus* and larvae of the caddis fly, stonefly and mayfly are wiped out by sewage. They reappear when the amount of dissolved oxygen is back to normal (Fig 24.17).

**Fig 24.17** Changes in the fauna of a polluted river (from Sands 1978)

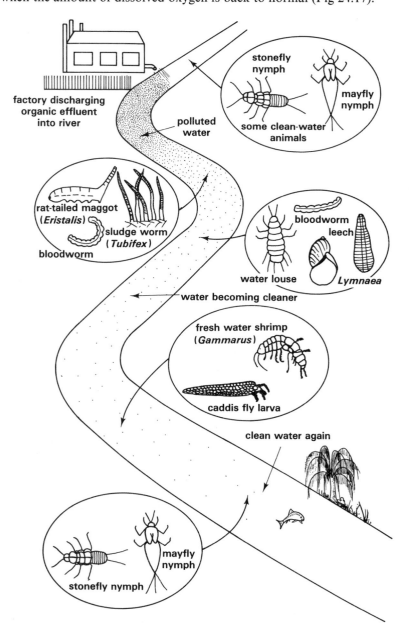

464

The numbers and types of animals at different sampling points in a river provide a useful guide to the extent and type of pollution. Given time, a river cleans itself in the way described. However, if it receives successive outfalls of sewage close to each other, the stretches of water between the outfalls do not have time to recover. In these circumstances the water becomes deoxygenated. Anaerobic bacteria partly decompose the sewage, releasing toxic ammonia and hydrogen sulphide. Rivers as badly polluted as this have an obnoxious smell and contain little flora or fauna.

What is popularly called **thermal pollution** has occurred in some of our rivers. A good deal of water is taken from rivers to cool industrial machinery. The biggest user of water for cooling purposes in Britain is the Central Electricity Generating Board. When returned, the water is hotter than when it was removed. In an unpolluted river, the increase in temperature lowers the amount of dissolved oxygen. At the same time, it increases the metabolic rate of aquatic organisms and therefore raises their oxygen demand. Up to now no gross changes in the flora and fauna have been observed in Britain, because the increase in temperature is slight. In other parts of the world, salmon and trout have disappeared from stretches of clean rivers which have been warmed to a greater extent. Ironically, water from cooling towers that is returned to polluted rivers accelerates their return to normal. While cascading down cooling towers (Fig 24.18) water takes up a good deal of oxygen from the air, so the water returned to polluted rivers often contains more oxygen than the water extracted. The increase in temperature and oxygen also encourages decomposers to break down organic matter more quickly. In this way the BOD returns to normal sooner.

**Fig 24.18** Cooling towers of the electricity generation station at Farndon, Notts. with River Trent in foreground

## 2 Lakes

Lakes are fairly still bodies of water often with no outlet. Any pollutants they receive will thus accumulate gradually. Reference has already been made to pollution of lakes by acid water draining from land polluted with sulphur dioxide and by mercury in industrial effluent (sections 24.1 and 24.2). In unpolluted lakes, the water contains only small amounts of dissolved minerals. The low mineral content limits the growth of aquatic plants. In such **oligotrophic lakes** there is consequently little organic matter to be decayed. For this reason, the water has a low BOD and remains well oxygenated. When polluted with sewage or with drainage water from nearby arable land, the water is enriched with minerals. The water is now **eutrophic** and growth of phytoplankton is stimulated. Blooms of algae, blue-greens especially, quickly appear. Some of them release toxins which kill fish. When the algae die, they contribute substantially to the dead organic matter which sinks to the bed of the lake. Here it creates a high BOD, and what oxygen there is in the water is soon used by decomposers. When this point is reached, anaerobic bacteria release ammonia, methane and hydrogen sulphide, making the water unfit for most forms of aquatic life. If there is a large amount of organic matter to be decayed, anaerobism may extend to the upper layers of water.

There are two ways in which the effects of such pollution can be reversed. One is to dredge the lake bed. Clearly this is too large a task for all but the smallest lakes. The other is to oxygenate the water so that decomposers break down the organic matter. This expensive approach is presently being tried in an attempt to clean up Lake Geneva in Switzerland. For some of the world's largest lakes it seems that the effects of eutrophication are irreversible. The flora and fauna of large parts of Lake Erie in Canada are pollution-tolerant species. Masses of decaying algae and dead fish are regularly washed up on the lake's shores. The water is unfit for bathing and the local fishing industry has virtually collapsed.

## 24.5.2 Marine pollution

Unlike the water in a lake, sea water moves continuously. In some seas the movement brings about a rapid turnover of water. Turnover of all the water in the North Sea takes two years. For this reason, pollutants are soon carried away and dispersed. The effects of pollution are more likely to be serious in shallow, enclosed seas. Here the turnover may take hundreds of years and marine organisms soon concentrate toxic pollutants. This was a major contributory factor in the poisoning of people with **mercury** at Minamata, Japan (section 24.2.2).

**Eutrophication** has been another cause of pollution in enclosed seas. Within the past 30 years, parts of the southern Baltic Sea have been made anaerobic in the same way as an enriched lake. Enrichment with phosphate fertiliser coming from surrounding agricultural land is mainly to blame. Eutrophication also occurs when raw or partly-treated sewage is discharged at sea, a common feature of many coastal towns on the Mediterranean Sea. Bathing in sewage-tainted sea water is unpleasant and likely to be a health risk. Humans have also contracted typhoid fever after eating shellfish from sewage-contaminated water.

Shellfish also concentrate **organochlorine insecticides** in their tissues. Oysters, for example, accumulate DDT to a concentration 70 000 times that of DDT in seawater, and as little as 0.001 ppm reduces their growth. Organochlorine insecticides pass from one trophic level to another in the sea, as in other ecosystems, and with comparable effects. The decline of predatory marine birds such as the herring gull, the osprey and brown pelican in various parts of North America in the 1960s is thought to have occurred because they accumulated high concentrations of DDT residues in their bodies.

Marine organisms concentrate radionuclides in their tissues too. They are often used as a measure of the amounts of radionuclides in the environment near coastal nuclear power plants. So far the evidence indicates little danger to marine life or to humans eating seafood.

**Oil** has captured the headlines as one of the most serious forms of marine pollution. When the *Torrey Canyon* went aground off Land's End in 1967, about 60 000 tonnes of crude oil were spilled. Much of it was washed up on Cornish beaches. The rest was blown across the English Channel and ended up on the French coast. The foundering of the *Amoco Cadiz* off the coast of France in 1978 caused 200 000 tonnes of crude oil to be released. Much of the oil came up on the beaches of Brittany. There have been many other similar incidents elsewhere. Large spillages have also occurred more recently from oil rigs in the North Sea and at oil terminals in the Shetland Islands. It has been estimated that up to 10 million tonnes of crude oil are released annually into the world's seas and oceans.

**Fig 24.19** Guillemot polluted with oil (RSPB/Michael W. Richards)

The worst casualties of oil pollution are fish-eating birds. From the air, oil slicks resemble patches of plankton where fish are likely to be feeding. Attracted by what looks like a good source of food, the birds settle on the slick or dive through it. Their feathers become clogged with oil, stopping the birds from flying (Fig 24.19). Oil-soaked feathers are poor insulators of body heat, so many of the birds die of hypothermia. The birds attempt to rid themselves of the oil by preening, but in doing so they swallow crude oil and are poisoned. Something like 100 000 sea birds, especially guillemots, were killed by the oil spillage from the *Torrey Canyon* alone.

Crude oil also kills seaweeds, molluscs and crustaceans when it is washed up on rocky shores. The seaweeds grow again quite quickly, often better than before. Browsing molluscs which keep them in check are slower to return, but within two years the shore is more or less back to normal. The oil disappears because it is broken down by aerobic bacteria. Concerned by the possible harm to their tourist industries, British and French authorities tried various measures to deal with the oil from the sunken tankers. In Britain the polluted shoreline was sprayed with detergents to disperse the oil. Because detergents are toxic to marine life, recovery of populations of tidal organisms was slower when this was done. The French sprinkled chalk and sawdust on the slicks to sink the oil before it could be washed up on the shore. Molluscs on the seabed were poisoned by the oil and by the mid 1980s their numbers had still not returned to normal.

The world's oceans contain over $14 \times 10^{18} \, m^3$ of water. The earth's atmosphere extends up to 2000 km. It is tempting to believe that such huge volumes can absorb all the wastes of human activities without any ill effects. Indeed, there is a catchphrase which says '*dilution is the solution to pollution*'. Unfortunately, aerial pollutants are concentrated by natural forces such as rain or snowfall. Marine pollutants do not become evenly dispersed. Some of the world's largest freshwater lakes have been unable to take the wastes of human activities without irreversible and undesirable effects. Living organisms have the knack of accumulating toxic materials and passing them on to others. The unwanted products of the activities of some people are the problems of others.

## 24.6 Manipulation of the environment by man

Man is the only creature able to manipulate the environment on a large scale. In this context his role as predator, agriculturalist and constructor are of particular importance.

### 24.6.1 Man as a predator

Humans kill animals of many species and they contribute to our omnivorous diet. Domesticated stock such as pigs, sheep, cattle and fowl are farmed to maintain a stable supply for market. Wild prey also make up part of our food supply. Marine animals, notably fish and whales, figure prominently in the diets of many humans. The seas are generally thought of as a common resource which anyone may fish. Improved fishing methods have meant that more and more fish have been taken in recent years from the world's seas and oceans (Table 24.14). Fishing vessels now travel over a thousand miles from their home ports. Echo-sounders, radar equipment and helicopters are used to spot shoals of fish. Factory ships stay at sea for months, processing and refrigerating the catch from fishing fleets. Intensive fishing sooner or later results in a fall in the catch. The fleet then moves to other fishing grounds. It often takes many years for **overfished** areas to recover. There is the fear that, in the not-too-distant future, fish stocks

**Table 24.14** Amounts of fish taken from the world's oceans in recent years (after Lucas and Critch 1974)

| Year | Global catch (tonnes) |
| --- | --- |
| 1940 | $20 \times 10^6$ |
| 1963 | $46 \times 10^6$ |
| 1970 | $70 \times 10^6$ |

467

Fig 24.20 (a) Numbers of five species of whale caught over a 25 year period (from Lucas & Critch 1974)

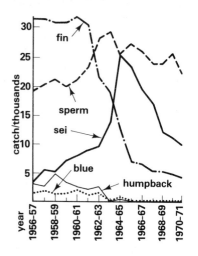

Fig 24.20 (b) Fin whales (Eric and David Hoskins)

generally will become so depleted that the oceans will be unable to meet the demands placed on them.

Fluctuations in fish numbers are sometimes due to natural causes. Many food fish eat plankton. Their numbers depend to some extent on the amount of plankton available to them. It has been suggested that the decline of the herring shoals in the North Sea during the 1930s may have been caused partly by overfishing and partly by a fall in phosphate concentration. Phosphate stimulates growth of plankton. Recovery of the herring population in the 1960s was probably due to the fish overcoming the intensive fishing of earlier days and also because of increased phosphate concentration. The phosphate came from domestic detergents which became popular at that time.

There is no doubt that **predation** by humans has been the main cause of the dramatic fall in numbers of many species of whale this century. Following the severe decline in the numbers of blue whales in the 1940s, by which time they had been hunted almost to the point of extinction, whaling fleets switched to catching fin whales. When the fin whale population slumped in the 1950s from overhunting, whalers took to capturing sei whales. In a little more than a decade the sei whale population fell drastically (Fig 24.20).

Studies on wild populations have made it possible to find out how the sizes of natural populations are regulated. The findings should enable us to control the size of wild populations to our advantage. To keep the population size stable, the rate at which mature stock is harvested must be equalled by replacement from immature stock (Fig 24.21). This keeps the population in the log phase of growth, and the carrying capacity of the environment is not overstretched. If the population reaches the stationary phase before it is harvested, it may take a long time to restore its optimum number.

Fig 24.21 Model of population regulation

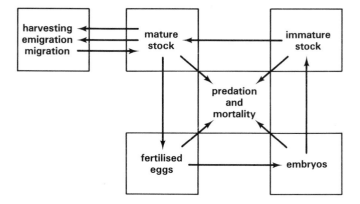

Maintaining stable fish stocks in a massive habitat such as the sea is easier said than done. Up to now there has been little serious attempt worldwide to regulate populations of marine fish. Shoals are usually a mixture of mature and young stock. If a net with a small mesh is used, many young fish are caught, leaving too few to allow populations to recover rapidly. International agreements limit the catches of fish which can be taken from territorial waters. However, the open seas belong to nobody. Their stock is fished with little regard for the future. This is a great shame as the oceans are potentially the richest source of protein in the world.

Various measures have been taken to try to conserve whale and fish stocks. A worldwide ban on the hunting of humpback whales came into being in 1966. A similar measure was taken to protect the blue whale in 1967. A ten-year ban on the capture of any whales was proposed at the International Whaling Commission (IWC) in 1972 but was turned down. In 1986 the IWC agreed to a cessation of whaling but several countries continue to catch whales for what they describe as scientific purposes. Catch quotas of over-exploited species of fish have been set by international agreement. Much still needs to be done to rectify the mistakes of the past.

### 24.6.2 Man as an agriculturalist

Humans first began to mould the environment when they adopted **settled agriculture** as a way of life. Forest, woodland and scrub were cleared to provide the land required to grow enough food for the growing population. **Deforestation** still goes on today in developing countries. Vast tracts of tropical forest in the Amazon basin have been cleared in recent years, and poor Brazilian families have been given financial inducements to settle as farmers in the cleared areas. The plan has backfired because the soil over much of the area is infertile and the settlers cannot afford fertilisers. Heavy tropical storms have also washed away a lot of the bare soil which had previously been held in place by tree roots. **Soil erosion** is not a new problem encountered in our attempts to turn more land to food production. The Hwang Ho river of China is better known as the Yellow River in the West. Its colour comes from soil washed into it from the surrounding land. In drier areas wind has eroded the soil from farmland, creating huge dust storms, the most publicised of which occurred in the USA in the 1930s. Erosion by wind and rain has occurred on a much larger scale for hundreds of years in North Africa where goats and cattle have overgrazed the vegetation.

Another way of increasing the area of land for agriculture is to drain wet areas. Wetlands such as bogs and fens are the homes of unique plants and animals. Ecologists are concerned that some species may become extinct as their habitats are destroyed. **Habitat impoverishment** for wildlife has also come about where scrub, forest and woodland have been put to the plough. The intensive methods of modern farming also pose a threat to wildlife. Farmers have removed hedgerows to allow large machines such as combine harvesters to work efficiently. Many species of wild birds, insects and flowering plants have hedgerows as their habitats. The prairie-like fields created by farmers are used to raise **monocultures**, crops of a single species such as wheat. Herbicides are sprayed on land to discourage weeds, insecticides on the crops to destroy insect pests. Weeds are the food plants for many desirable insects such as butterflies. Insects are an important part of the diets of many species of wild birds.

### 24.6.3 Man as a constructor

The constructions of humans have made a great impact on the environment. Vast areas of land are now taken up by towns and cities connected by intricate transport systems. Though we are unable to control the weather, we manipulate the climate immediately around us with air-conditioning and central heating.

Human populations require a constant supply of drinkable water (Chapter 25). Water is often provided by building huge dams and flooding enormous areas of moorland in upland regions. Again, there is a loss of habitat to wildlife. Our ability to store water has contributed to our energy needs from hydroelectric schemes. Stored water is also used for irrigation purposes, opening up deserts for agriculture. Thus, in 1985 over 40 000 hectares of desert land in south Peru was provided with water. It will make a substantial contribution to the country's attempts to become self-sufficient in food production.

## 24.7 Depletion of natural resources

Some of the natural resources we use are **renewable**. We grow more trees to replace the timber we have used. Crops are cultivated to replenish our food stores. Water and air are constantly available. Other resources are **non-renewable**: they cannot be replaced. There is anxiety that we may be using non-renewable resources at such a rate that reserves will soon be depleted. Two resources in this category are fossil fuels and metal ores.

### 24.7.1 Fossil fuels

There are many different sources of energy and a shortage would occur only if all sources were near to exhaustion. The demand for energy is rising continually. One estimate indicates an average global increase of 5 % a year. It means that we will have to double the world supply of energy every 20 years to maintain present standards of living. In countries where the standard of living is going up and where industry is growing the figures are significant. An increase of 300 to 400 % over the next 20 years has been estimated for the USA.

Energy sources which are exhaustible are mainly **fossil fuels**, coal, oil and natural gas. In the 1920s **coal** provided as much as 80 % of man's energy, mainly as steam power or electricity. These days it has taken second place to **natural gas** and **oil**. World reserves of coal are believed to be more than $5 \times 10^{12}$ tonnes. At the present rate of consumption they can be expected to last for at least another 100 years. Oil supplies are unlikely to last that long as more motor-driven vehicles are used. The figure most often quoted for oil reserves is $2.5 \times 10^9$ barrels. It should last for at least 50 years. Natural gas has the smallest reserves, and it is difficult to judge how long it may last. It may be possible to provide additional oil from alternative sources such as coal and shale. Further reserves of all fossil fuels may also be discovered in the future. Nevertheless, if we are to rely exclusively on fossil fuels it is evident that the demand for energy cannot be met in the foreseeable future.

**Nuclear energy** is an option which may be used more as our reserves of fossil fuels dwindle. However, people everywhere are suspicious of nuclear power plants (section 24.4.2). Alternative forms of energy have to be considered. **Hydro-electricity** has been used for some time and is renewable. Its main drawback is that it requires massive capital expenditure to construct dams. The number of sites suitable for dam construction is limited, and there is also a limit to the amount of land we can eventually flood. On a smaller scale, **windmills** and **watermills** can provide renewable power. Energy in **tidal movement** has also been tapped in parts of France and the USSR where the rise and fall of the sea is sufficiently great. **Animals**, especially the horse and ox, are another renewable source of energy for work which man may have to turn to on a wide scale.

One source of energy not yet successfully drawn on directly on a large scale is **sunlight**. **Solar panels** can be built to absorb heat from the sun.

Water in the panels is warmed up, so that less gas and electricity is required to provide heat. Although solar panels are unlikely to provide much of the energy required in cloudy, temperate areas, they clearly make an important contribution in relatively cloudless parts of the world.

One approach to extending our energy reserves is to use energy more efficiently. Even in developed countries a lot of energy could be saved by raising the standard of insulation of homes and workplaces by installing double glazing and cavity wall filling. At the moment many people are reluctant or unable to find the money required. As energy costs rise, however, more of us may be prepared to save energy.

### 24.7.2 Metal ores

**Iron ore** is used more than any other type of ore. It is refined to provide iron and steel which are used in countless ways, notably in ship and bridge construction and the manufacture of motor vehicles and machinery. Since 1950 the world demand for iron ore has increased at least four times, mostly in the developed countries. At this rate, known high-grade iron ore reserves would be exhausted by about 2050 AD. This is probably a pessimistic forecast because as reserves run down several factors could extend their life. New reserves of high-grade ore may be discovered. If not, known reserves of low-grade ore could be tapped. Alternatively, used iron and steel could be recycled more efficiently. At present about half of the iron and steel produced comes from scrap collected by dealers and from waste produced by the iron and steel foundries. Substantial quantities of waste iron and steel are used in making cans and end up on domestic refuse dumps. Some municipal authorities have invested in metal extraction facilities as part of their refuse disposal programme. Finance raised from selling metal waste offsets the capital expenditure involved. Another possibility is that other metals may replace iron and steel in some of their traditional uses. A present day example is aluminium, and other metals may be used in future.

There is no easy way of telling how long our non-renewable resources will last. What is clear is that one day each of the resources will come to an end. We can either look for ways of delaying the inevitable or we can seek alternatives.

# 25 Diseases and parasites of humans

Diseases fall into three main categories.

1 Those arising from the genotype of the individual. Examples of **hereditary diseases** such as inborn errors of metabolism and sickle cell disease are described in Chapter 4 and 12 respectively.
2 Those caused by parasites such as viruses, bacteria, fungi and protozoa, flatworms, roundworms and arthropods. **Parasitic diseases** are the main subject of this chapter.
3 Those which are caused by harmful environments. Several examples of **environmental diseases** such as malnutrition, and lead and mercury poisoning have already been described (Chapters 9 and 24 respectively). This chapter deals with some important occupational diseases caused by hazardous working environments. We shall also look at some of the ways in which diseases can be prevented and controlled.

A **parasite** is a living organism which feeds on another, called the **host**, to the host's detriment. Some parasites cause us little harm, others can kill us. Among the most harmful parasites are microbes. Many of the flatworm, roundworm and arthropod parasites are macroscopic and big enough to be seen by the naked eye. **Ectoparasites** live on our skin and hair, **endoparasites** live inside us.

## 25.1 Microbial parasites

Disease-causing micro-organisms are called **pathogens**. Pathogenic micro-organisms include viruses, bacteria, fungi and protozoans.

### 25.1.1 Viruses

**Viruses** were first discovered in the 1890s when it was shown that the sap extracted from diseased tobacco plants was capable of causing the same disease in healthy plants even though it had been passed through a filter capable of trapping bacteria, the smallest organisms known until then. Hence viruses are smaller than bacteria. Indeed, viruses are the smallest organisms known to this day. The largest are no more than 300 nm in diameter, about a third the diameter of a typical bacterium. The smallest virus is only 10 nm diameter, one hundred times smaller than a bacterium (Table 25.1). Because they are so small, no-one ever saw a virus until the electron microscope was introduced in the 1930s.

**Table 25.1 Range of size of some viruses**

| Virus | Diameter (nm) | Disease |
|---|---|---|
| vaccinia | 250–300 | smallpox |
| myxovirus | 100–300 | measles, mumps, influenza |
| varicella | 180–250 | chickenpox |
| 'flu virus | 80–120 | influenza |
| poliovirus | 10–20 | poliomyelitis |

# 1 Structure

Structurally, viruses are very simple. An individual virus is called a **virion**. At the centre is a strand of nucleic acid, DNA or RNA or a hybrid of the two. Enclosing the nucleic acid is a coat called a **capsid**, made of protein globules known as **capsomeres**. The capsomeres are often arranged in an orderly manner, giving the virion some form of geometric symmetry. Enveloped viruses have an outer covering of lipoprotein. Naked viruses lack the envelope (Fig 25.1).

**Fig 25.1** (a) A range of viruses: (i) Smallpox virus, (ii) mumps virus, (iii) influenza virus, (iv) poliomyelitis virus, (v) AIDS virus

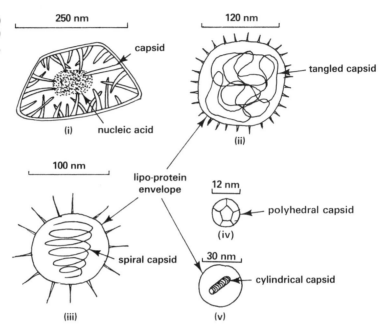

**Fig 25.1** (b) (i) Electronmicrograph of an influenza virus, × 3600

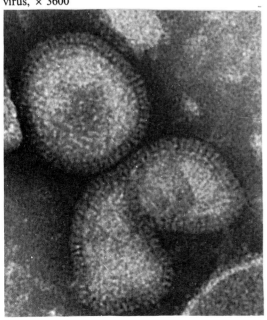

**Fig 25.1** (b) (ii) Electronmicrograph of a bacteriophage, × 135 000

473

## 2 Multiplication

The only feature viruses have in common with other living organisms is their ability to multiply. To do this they must have access to the living cells of a host. They are **obligate parasites**. Viruses multiply by a process called **replication**. First a virion becomes attached to the surface of a suitable cell. Each kind of virus usually parasitises only one kind of host and often a specific tissue of the host. For example, the poliomyelitis virus parasitises neurons whereas hepatitis virus attacks liver cells. A lock and key mechanism, comparable to that involved in enzyme action (Chapter 3) enables a virion to recognise the cell in which it can multiply. The virion becomes adsorbed onto the cell membrane much as an enzyme molecule binds to its substrate.

After **adsorption** the virion penetrates the host cell. Virions may be taken wholesale into the cytoplasm when the cell membrane invaginates during pinocytosis and phagocytosis (Chapter 6). Once inside the host cell, the coat of the virion is stripped off by enzyme action. Alternatively, the nucleic acid only may enter, as happens with bacteriophages, viruses which parasitise bacteria.

Once in the host cell, the virus loses its ability to infect other cells and is in the **eclipse stage**. Simultaneously the infected host cell's ability to synthesise nucleic acids and proteins is inhibited. If the viral nucleic acid is DNA, it acts as a template on which its base sequence is transcribed as viral mRNA. The base sequence of the viral mRNA is then translated by host tRNA, and viral proteins are assembled at the host's ribosomes.

The first proteins to be made are enzymes which catalyse the synthesis of large numbers of replicas of the viral DNA. Next to be made are viral protein molecules which coat the replicated DNA to form hundreds of complete new virions. They can now be seen under the electron microscope as **inclusion bodies** in the nucleus or cytoplasm of the host cell. Finally the virions are released. This can happen by reverse pinocytosis, but often the host cell bursts open (lysis) soon after the virions have matured. For this reason the sequence of events involved in viral multiplication is called the **lytic cycle** (Fig 25.2).

**Fig 25.2** Lytic cycle of the influenza virus

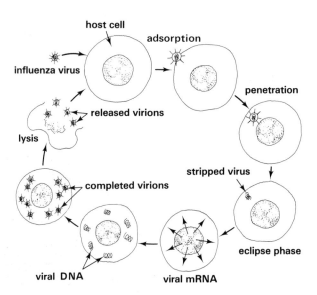

Viruses containing RNA produce an enzyme called reverse transcriptase which allows the viral nucleic acid to act as a template on which DNA is assembled. Otherwise the cycle is similar.

### 3 Cultivation

Two ways in which viruses are cultivated are to provide tissue cultures or live chick embryos in which they can multiply.

**Fig 25.3** Cytopathic effect of a virus

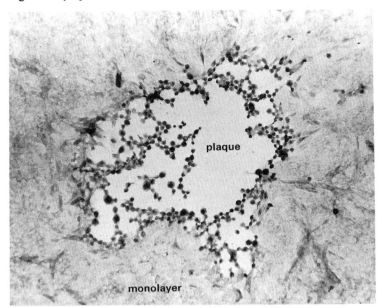

**i. Tissue culture** Most cells have first to be grown in a nutrient solution. Human tissue such as amnion, placenta and foetal lung can be used for such purposes. The tubes or flasks in which the tissue is placed are rotated during incubation, causing the tissue to grow as a single layer of cells called a **monolayer**. The culture is then inoculated with the virus and incubated further. If the virus is able to parasitise the tissue it causes **plaques** to develop in the monolayer where the cells are lysed. This is called the **cytopathic effect** (Fig 25.3) and can be of help in identifying viruses.

Another form of tissue culture uses minced kidneys of freshly killed monkeys as the source of live cells. Once more the cells are cultivated in a nutrient solution before the virus is introduced. Poliovirus is cultivated in this way. On lysis of the host cells, virions are released into the surrounding medium from where they can be harvested. After attenuation (Chapter 11) the live poliovirus is used to prepare the Sabin oral vaccine which can protect us against poliomyelitis.

**ii. Chick embryos** As a chick embryo grows inside its shell it becomes enclosed in embryonic membranes as it uses up the yolk enclosed in the yolk sac (Fig 25.4). Once the membranes and yolk sac have developed, the egg can be inoculated with viruses. A small hole is cut in the shell with a dentist's drill and a suspension of virions is then injected into a suitable part of the egg. Vaccinia virus, the cause of smallpox, multiplies well in the

**Fig 25.4** Chick embryo culture of viruses

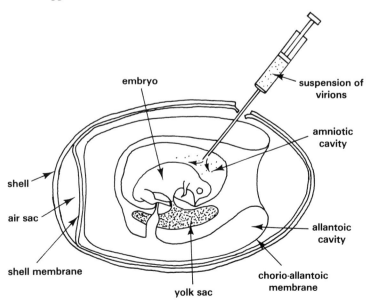

chorio-allantoic membrane, while myxovirus the cause of influenza and mumps does better in the amniotic and allantoic cavities. On incubation, the viruses attack the egg membrane and its cells then lyse. Large numbers of virions are released into the enclosed fluid which can be withdrawn from the various cavities of the egg. Vaccines which have eradicated smallpox and which can protect us from 'flu can be prepared from the harvested virions after they have been attenuated.

## 4 Viral diseases

Viruses can enter our bodies in several ways. One of the most important is the **respiratory route**. People infected with the influenza and common cold viruses release tiny infective droplets when they sneeze. The droplets remain suspended for long periods of time in the air and may be breathed in by other people. Hence the adage 'coughs and sneezes spread diseases'! Dust from infected clothing and bedding may also be inhaled. Smallpox was spread in this way. Another route of entry is the **alimentary tract** when infected food or water is consumed. Poliomyelitis has frequently been contracted by people swallowing infected river water in which they were bathing. Viruses cannot penetrate unbroken **skin**, only skin which has been punctured. Rabies is acquired when people are bitten by a rabid dog, fox or cat. Yellow fever is transmitted in the bite of infected mosquitoes. Yet another route of entry is the **placenta**, causing infection of the developing foetus. Perhaps the best known viral disease transmitted in this way is German measles, rubella. A detailed study of influenza and rubella will provide models for appreciating the threat posed to mankind by some viral diseases.

**i. Influenza**   Four strains of **influenza** virus are known: A, B, C and D. Strains C and D are stable, whereas variants of strain A commonly occur; strain B variants are less frequent. Influenza is always with us somewhere in the world. **Epidemics** of it break out regularly, infecting large numbers of people in a particular place. Occasionally the disease becomes **pandemic**, spreading quickly across continents. The worst pandemic of influenza happened in 1918–1919 when the disease killed more than 20 million people worldwide. About 150 thousand died in Britain alone. The strain of the virus was identified as $A_1$. Another pandemic, caused by the $A_2$ strain, occurred in 1957–58. It began in China and quickly swept across south-east Asia, followed by the other continents within months. This variety of the disease was called Asian 'flu. Yet another strain of the virus appeared in 1968, starting a pandemic in south east Asia again. Known as Hong Kong 'flu the disease lasted almost the entire winter of 1968–69 in Britain, killing at least 1000 people and infecting about a quarter of the entire population. Nationwide epidemics of influenza caused by the A strains of the virus recur every 2–3 years, while B strains have a 4–6 year cycle.

On entering the respiratory system in infected droplets, the virus attacks the epithelial lining of the air passages over an incubation period of 2–3 days. The symptoms characteristic of the disease quickly appear. In the first day the body temperature goes up rapidly and may reach as high as 40 °C. It is accompanied by shivering, headache, sore throat and nasal congestion. Adults often have aches in the back and limbs. After 3–4 days the temperature normally begins to fall, but a cough develops because of the damage to the trachea and bronchi. Perhaps the most characteristic symptom is the marked general weakness of the patient who can often do little more than lie down until the disease subsides. An unwelcome complication of influenza is invasion of the damaged air passages by pathogenic bacteria leading to bronchitis and pneumonia.

**ii. Rubella** Rubella is **endemic** in nearly all countries, small numbers of the population having the disease at any time. Occasionally, epidemics flare up. Rubella is not a notifiable disease in Britain so no accurate figures are available of the number of cases. However, the virus can pass from mother to child across the placenta, causing abnormalities in the developing foetus. Hence, the number of children born with **congential rubella syndrome** is an index of the threat posed by the disease. About 200 babies with rubella syndrome are born each year in Britain. During the 1964–65 epidemic in the USA, more than 20 000 children were born with the syndrome. It was also estimated that at least 30 000 foetuses were aborted or stillborn as a consequence of rubella infection. The greatest risk of foetal infection occurs in the first few weeks of pregnancy. If the mother contracts the disease at this stage there is more than 50 % probability that her child will develop the rubella syndrome. As the gestation period advances the risk diminishes, but there is still a 20 % probability of foetal infection at four months of pregnancy. Infants born with the syndrome excrete the virus for several months after they are born. Typical effects of congenital rubella are retarded mental and physical development, defects of vision and hearing, and heart disease. These come about because the virus parasitises the tissues of the foetus as it develops.

Rubella is much less of a threat after birth. The disease is most common in older children and young adults. The virus is spread by droplet infection and breathed in. After a usual incubation period of 17–21 days, a rash appears on the face and begins to spread to other parts of the body. Swelling of the lymph glands below the ears and in the nape of the neck is another characteristic symptom. The patient is often listless and may develop a mild fever. Within 2–3 days the rash disappears and after about a week the patient is normally free of infection and fully recovered.

## 25.1.2 Bacteria

**Bacteria** are visible with the aid of a light microscope and were first seen by van Leeuwenhoek in 1683. Very little was known about them until the work of Louis Pasteur and Robert Koch in the last half of the nineteenth century. They were the first to prove the link between bacteria and some of the common diseases of man. Relatively few species are pathogenic. The majority are harmless saprophytes which play an important role in the cycling of matter (Chapter 23). Ironically, some saprophytic bacteria are the producers of antibiotics which can be used to treat and cure bacterial diseases.

### 1 Structure

Most bacteria exist as single cells or as clusters of similar cells. Individual cells are shaped as rods (**bacilli**), bent rods (**vibrio**), spirals (**spirilla**) or as spheres (**cocci**). The regular shape is maintained by a **cell wall**. Outside the cell wall some bacteria have a thick gelatinous **capsule** which reduces the risk of dehydration and prolongs survival in air. It may also protect against the antibodies of a host's immune system. Some bacteria have **flagella** which can be seen with a light microscope if they are thickened and then stained. Flagella provide the propulsive force for movement. Other bacteria move by bending back and forth (flexing) or by undulating movements. Several kinds of inclusion bodies can be seen inside the cell wall, especially if they are stained. The most prominent is the **nucleoid**, in which DNA is concentrated. In adverse conditions a few species of bacteria form **endospores** inside each of their cells. Endospores have very thick

Fig 25.5 (a) Features of bacteria visible
with a light microscope

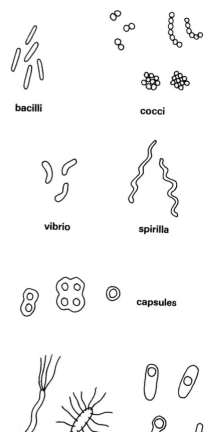

bacilli

cocci

vibrio

spirilla

capsules

flagella

endospores

walls enclosing a copy of the cell's DNA. The thick wall protects its
contents from desiccation, chemicals such as disinfectants, extremes of
temperature and from radiation. When favourable conditions return, an
endospore germinates to release a young bacterium (Fig 25.5).

Fig 25.5 (b) Photomicrographs of cocci and bacilli, × 1000

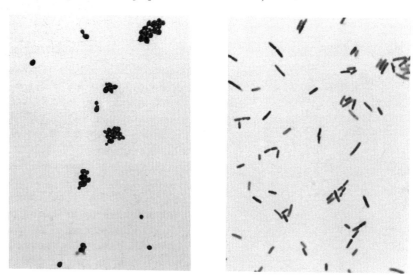

One of the most important techniques of staining and categorising
bacteria was devised by Christian Gram in 1880. A thin smear of the
bacteria is first stained with a purple dye, crystal violet. An attempt is then
made to decolourise the smear using an organic solvent such as ethanol.
The smear is finally counterstained with a red dye, carbol fuchsin. Some
bacteria retain the purple colour of the first dye. They are **Gram positive.**
Others are decolourised and take up the red colour of the counterstain.
They are **Gram negative.**

With the electron microscope much more detail of the internal structure
can be made out (Fig 25.6). The selectively permeable plasma membrane is
folded to form **mesosomes** in the outer cytoplasm. Mesosomes contain
enzymes used in respiration, and in cell wall and toxin production.

Fig 25.6 (a) Ultrastructure of a bacterium based on electronmicrographs (from J I
Williams and M Shaw 1982)

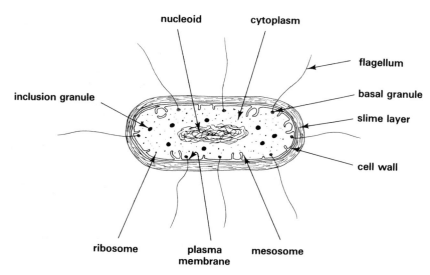

nucleoid

cytoplasm

flagellum

basal granule

slime layer

inclusion granule

cell wall

ribosome

plasma
membrane

mesosome

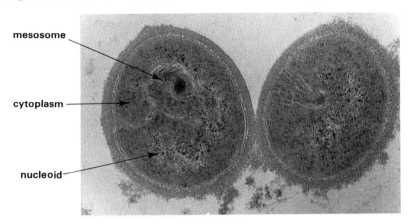

Mitochondria are not found in bacteria but the cytoplasm does contain **ribosomes**. They are suspended freely in the cytoplasm, not attached to membranes as in eukaryotic animal and plant cells (Chapter 6). The nucleoid does not have an external membrane and when it divides chromosomes do not appear. Because they lack a true nucleus, bacteria are described as **prokaryotic**. Some species of bacteria have fine adhesive threads called **fimbriae**.

Biochemical studies have revealed quite a lot of the chemical nature of the cell wall of bacteria. A cell wall component found in all bacteria is **mucopeptide**. It is a mesh-like polymer made of two monomers, glucosamine and muramic acid. Embedded in the mucopeptide mesh are proteins and lipids. In Gram negative bacteria the mesh is covered externally with a thick layer of lipid. No one is sure why different species of bacteria react differently to the Gram staining technique. One explanation is that the thick lipid layer of Gram negative species allows the organic decolouriser to enter the cell and dissolve out the crystal violet dye.

## 2 Multiplication

Bacteria multiply asexually by **fission**. Prior to fission the cell's DNA replicates. The nucleoid becomes pinched into two more or less equal parts, each containing a copy of the DNA. Simultaneously the cell wall grows inwards, cleaving the cell into two identical progeny, each half the size of the parent cell (Fig 25.7). Bacteria with one or more characteristics different from their parents occasionally appear. Gene mutations (Chapter 4) may cause this to happen. Mutations can be of survival advantage to bacteria. They have enabled some strains of pathogenic bacteria to develop resistance to antibiotics.

Given ideal conditions, some species of bacteria take as little as 12 minutes to multiply. Such a rate of reproduction creates millions of progeny in a matter of hours. It explains why many kinds of bacterial diseases have very short incubation periods.

**Fig 25.7** Fission of a bacterium

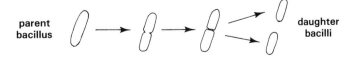

## 3 Cultivation

Most kinds of bacteria can be cultivated relatively easily. Unlike viruses they do not require live cells as a source of raw materials. Bacteria can be provided with non-living nutrients in the form of liquid or solid **media**.

Liquid media are called **broths**; one of the commonest is **nutrient broth**. It contains peptone (a source of amino acids), meat and yeast extract, sources of additional amino acids, vitamins and minerals, all dissolved in water. The amounts of nutrients are small, giving a very dilute solution. Its water potential is made suitable for bacterial growth by adding sodium chloride. Broths can be made into solid media by adding **agar**, a seaweed extract which causes media to set like jelly at room temperature.

After the ingredients have been dissolved in water, the acidity of the media is measured and, if necessary, adjusted to pH 7.0–7.5. It is within this narrow range of pH that most species of bacteria grow best. It is then necessary for the media to be sterilised to kill any contaminant organisms that may have entered. Sterilisation is normally achieved using an **autoclave** (Fig 25.8) in which bottles of the media are steamed under pressure so that a temperature of about 120 °C is reached. This temperature has to be held for 15 minutes to ensure all contaminants including bacterial endospores are destroyed. After sterilisation the media are allowed to cool.

Fig 25.8 (a) An autoclave

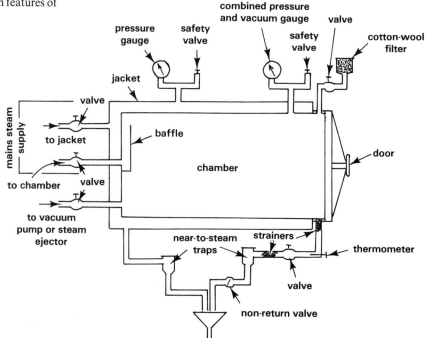

Fig 25.8 (b) Diagram showing the main features of an autoclave

480

Bacteria which cause human diseases are often more exacting in their nutritional requirements. These can be provided by adding blood to nutrient broth and nutrient agar, making **blood broth** and **blood agar** respectively. This must be done after the nutrient media are autoclaved because many of the nutrients in blood are destroyed at the high temperature required for sterilisation.

Media can be dispensed in various ways for use in the laboratory (Fig 25.9).

Fig 25.9 (a) Various ways of dispensing bacteriological media

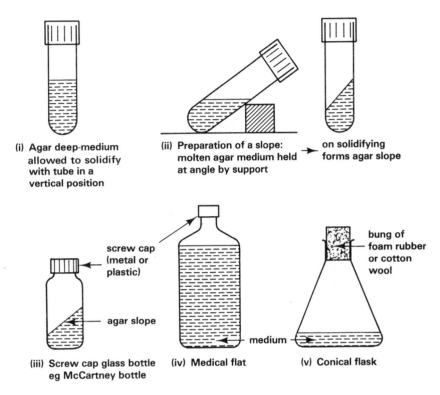

(i) Agar deep-medium allowed to solidify with tube in a vertical position

(ii) Preparation of a slope: molten agar medium held at angle by support → on solidifying forms agar slope

screw cap (metal or plastic)

agar slope

(iii) Screw cap glass bottle eg McCartney bottle

(iv) Medical flat

medium

bung of foam rubber or cotton wool

(v) Conical flask

Fig 25.9 (b) Pouring an agar plate (from J I Williams and M Shaw 1982)

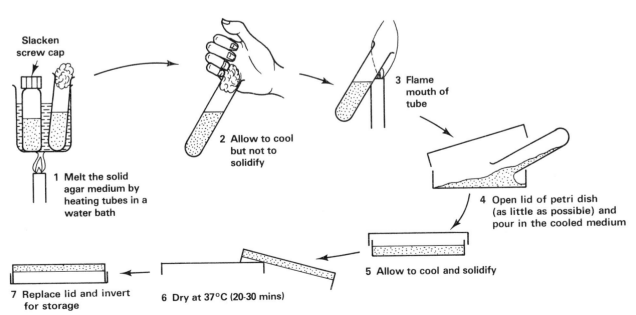

Slacken screw cap

1 Melt the solid agar medium by heating tubes in a water bath

2 Allow to cool but not to solidify

3 Flame mouth of tube

4 Open lid of petri dish (as little as possible) and pour in the cooled medium

5 Allow to cool and solidify

6 Dry at 37°C (20-30 mins)

7 Replace lid and invert for storage

**Fig 25.10** (a) An anaerobic jar

Mixed colonies of bacteria normally appear within 2–3 days when a sample of almost any liquid or solid is inoculated into bacteriological media and incubated at a suitable temperature. Strict anaerobes grow only if the preparations are incubated in an **anaerobic jar**, sometimes for several weeks (Fig 25.10). Pure cultures, containing a single species, can be prepared from the mixed colonies in various ways. One of the methods most often used is the **streak-plate** technique (Fig 25.11). A battery of tests, including staining, biochemical, physiological and serological tests, may then be performed to identify the species.

**Fig 25.10** (b) Diagram showing the main features of an anaerobic jar (from J I Williams and M Shaw 1982)

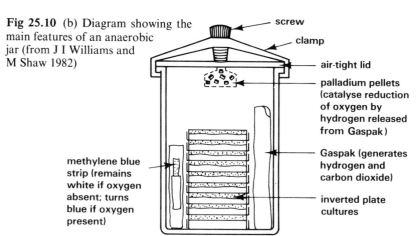

### 4 Bacterial diseases

Bacteria are everywhere, in the air, soil, water, food, on and in humans and livestock. Table 25.2 summarises some of the more common bacterial diseases of man. Note that the routes by which they enter our bodies are the same as those used by viruses. Again, a detailed study of two bacterial diseases, typhoid fever and gonorrhoea will illustrate the problems caused by pathogenic bacteria.

**Table 25.2  Some bacterial diseases of man**

| Route of entry | Diseases |
| --- | --- |
| respiratory | diphtheria, tuberculosis, scarlet fever, lobar pneumonia, whooping cough |
| gastro-intestinal | typhoid and paratyphoid fevers, bacillary dysentery, cholera, undulant fever, gastro-enteritis |
| urino-genital | syphilis, gonorrhoea |
| broken skin | tetanus, anthrax, gangrene |

**i. Typhoid fever**   This disease is caused by a Gram-negative, non-sporing motile bacillus of the species *Salmonella typhosa*. Typhoid occurs in all parts of the world but is most frequent where sanitation is poor and water purification inadequate. The infective cycle is straightforward. Faeces of infected individuals contain the pathogenic bacilli. If there are inadequate means of sewage disposal, water supplies become contaminated. Healthy individuals drinking the water or eating food washed in it, acquire the disease. Flies may also carry the bacilli from faeces to food.

People who have had typhoid fever sometimes continue to egest the bacilli, often for years. They are healthy **carriers** of the disease and are a danger to others, especially if their standards of personal hygiene are poor.

**Fig 25.11** (a) A streak plate technique (from J I Williams and M Shaw 1982)

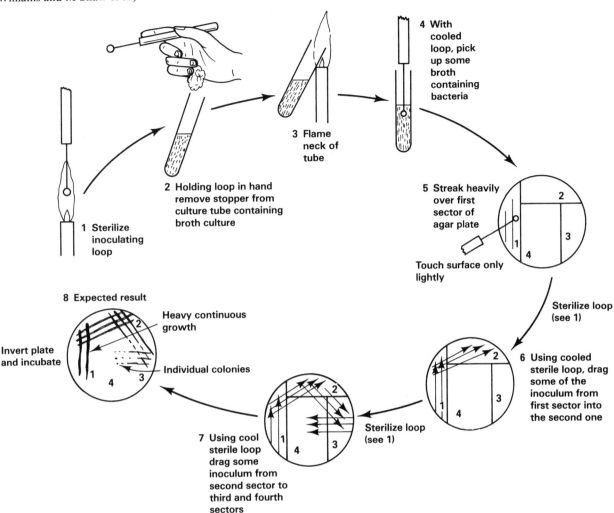

1 Sterilize inoculating loop

2 Holding loop in hand remove stopper from culture tube containing broth culture

3 Flame neck of tube

4 With cooled loop, pick up some broth containing bacteria

5 Streak heavily over first sector of agar plate

Touch surface only lightly

Sterilize loop (see 1)

6 Using cooled sterile loop, drag some of the inoculum from first sector into the second one

Sterilize loop (see 1)

7 Using cool sterile loop drag some inoculum from second sector to third and fourth sectors

8 Expected result

Heavy continuous growth

Individual colonies

Invert plate and incubate

**Fig 25.11** (b) A streak plate after incubation

A number of cases have been identified where carriers have been employed in food preparation and have passed on typhoid fever to people eating the prepared food. A notorious carrier was an American named Mary Mallon. In the 1930s she is known to have caused at least ten outbreaks of the disease in which three people died and over fifty acquired typhoid fever. For this reason she was nicknamed Typhoid Mary.

Epidemics of the disease have occasionally been started by carriers in Britain. In 1937 there were 43 deaths and over 300 cases of typhoid fever in Croydon. People got the disease by drinking contaminated water. An employee at the town's waterworks was later found to be a carrier. Food is less often a cause of typhoid fever in Britain. The last serious outbreak of typhoid begun in this way was in Aberdeen in 1964. Corned beef was identified as the source. It had become contaminated by the water used to cool the cans after they had been heat-sterilised but inadequately sealed. Shellfish

483

are a likely source of the disease. They are filter-feeders, straining small particles from the water in which they live. If the water is contaminated with sewage they may concentrate typhoid bacilli in their bodies. Swallowing sewage-contaminated water while bathing is another way in which people acquire typhoid fever.

In recent years the number of cases of typhoid in Britain has been between 100 and 200 annually. About half of these people are thought to have contracted the disease abroad. Typhoid fever is still a major problem in many of the countries bordering the Mediterranean Sea and also in tropical and sub-tropical areas of the world.

The incubation period for the disease is usually 10–14 days. During this time the ingested bacilli penetrate the bowel wall and migrate to the lymph glands and blood. A fever which increases in intensity develops and usually lasts for about a week. The abdomen is tender and may have a faint rash. In the second week the patient has toxaemia and becomes listless and mentally confused. Death may follow in the third week after further progressive deterioration in the patient's condition. Over the course of the disease ulcers appear in the bowel. They may burst, causing severe haemorrhage into the bowel. At the same time some of the bowel contents may enter the abdominal cavity through the perforated ulcers. Up to 20 % of typhoid deaths are probably due to bowel perforation. The condition of some patients improves after the third week. Their appetite returns, they become mentally alert and gradually recover.

**ii. Gonorrhoea**  Gonorrhoea is caused by bacteria of the species *Neisseria gonorrhoea*, a Gram-negative, non-motile diplococcus which cannot form spores. It is a **venereal disease**, usually spread directly from one person to another during sexual intercourse. At one time syphilis was the most common venereal disease but these days gonorrhoea is more frequent and has reached epidemic proportions in many countries. A large proportion of gonorrhoea cases are people between 17 and 35 years of age. Fig 25.12 shows the increase in incidence of gonorrhoea in Britain in recent years. From 1957 to 1967 the number of diagnosed cases rose from about 216 500 to 375 000. It has been estimated that more than 60 million new infections occur worldwide each year.

The incubation period of gonorrhoea is usually about five days. Infected males then begin to experience a burning feeling in the penis when passing urine. A yellow discharge comes later on from the urethra. If the disease is untreated the gonococcus infects the epididymis and prostate gland, causing them to swell and become abscessed. The urethra may become so constricted that it is impossible to pass urine.

The most common sites of infection in females are the urethra and the cervix. In the early stages of infection there are often no symptoms. From the cervix, infection may spread into the uterus and up to the oviducts. At this stage there is abdominal pain and in time the oviducts become blocked, making the woman sterile. Infection of the urethra may cause a burning feeling when urine is passed. A yellow discharge usually appears when the uterus is infected. This is often accompanied by irregular or heavy menstruation.

Infants born to diseased mothers may become infected at birth. While inside the amnion the foetus is protected from the gonococcus. Children sharing a bed with an infected parent may also acquire gonorrhoea.

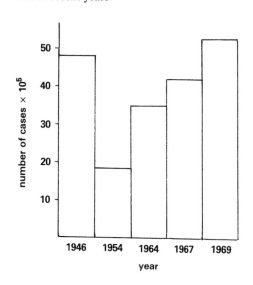

**Fig 25.12** Incidence of gonorrhoea in the UK in recent years

**Fig 25.13** (a) Spore production by two moulds

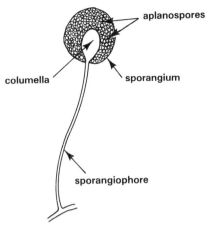

*Mucor*

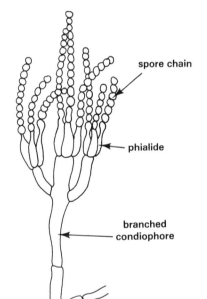

*Penicillium*

## 25.1.3 Fungi

At least 100 000 species of **fungi** are known to man. Many, such as yeasts and moulds, are microscopic. Others, such as mushrooms, toadstools and puffballs, are big enough to be seen with the unaided eye. As with bacteria, the majority of species are harmless saprophytes, some of them producing life-saving antibiotics. However, a few species are pathogenic.

### 1 Structure

Some fungi such as baker's yeast are **unicellular**. Most are **filamentous**, existing as a **mycelium** made up of a network of branching threads called **hyphae**. They have a cell wall which is usually made of chitin, a fibrous polymer of glucosamine. The protoplast inside contains one or more nuclei, each bounded by a nuclear membrane. Chromosomes appear during nuclear division. Also present are mitochondria and endoplasmic reticulum. Septate hyphae have cross walls called **septa**, which divide the protoplast into cells. Septa are absent from aseptate hyphae. All these structures distinguish fungi from bacteria.

### 2 Multiplication

Many species reproduce sexually and asexually. **Spores** are formed by most fungi as a means of multiplication. The asexual spores formed by many filamentous fungi are formed directly on the mycelium and are called conidiospores. Others produce sporangia containing sporangiospores. The manner in which spores are borne on the mycelium, their shape, size and colour, are characteristic of a species and an important feature in

**Fig 25.13** (b) Photomicrograph of conidiospore production by *Penicillium*

identification (Fig 25.13). Yeasts multiply by **budding** (Fig 25.14). In ideal conditions spores and buds are formed in enormous numbers and are often dispersed by air currents. When they land on a suitable substrate they

**Fig 25.14** Budding of a yeast

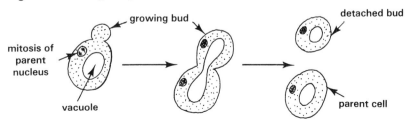

485

germinate to form a new mycelium. In this way fungi multiply rapidly. The nuclei of spores and buds are formed by mitosis and thus normally contain the same number of chromosomes as in the parent. Unless there has been a mutation, the offspring are genetically identical to the parent.

When sexual reproduction occurs, sexual organs are formed. They vary greatly among fungi and need not be studied here. The sexual organs come together and nuclei from both eventually fuse. Later the nuclei undergo meiosis which restores the original chromosome number. Crossing over and random segregation of alleles occur in meiosis (Chapter 6) and create new genotypes among the progeny.

## 3 Cultivation

Agar-based media are often used to cultivate colonies of moulds and yeasts. Nutrients can be provided in the form of malt extract or yeast extract in **malt extract–agar** (MEA) and **yeast extract–agar** (YEA) respectively. Moulds usually grow well on MEA and yeasts on YEA. **Sabouraud's medium**, which contains maltose and peptone, is often used for pathogenic fungi. The acidity of the media is tested and adjusted if necessary to pH 5.5–6.0. Sterilisation by autoclaving then follows and plates are poured when the media have cooled but not yet set. When the agar has set, the plates are inoculated with hyphae or spores of a mould or with a few yeast cells (Fig 25.15). On incubation at an appropriate temperature moulds grow as fluffy or velvety colonies. Yeast colonies are slimy and are smaller.

**Fig 25.15** Inoculation technique for a freely sporulating mould

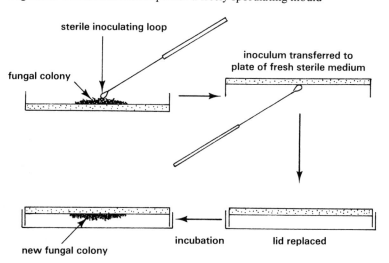

## 4 Fungal diseases

Compared with viral and bacterial diseases, there are few fungal diseases of man. They fall into two categories, **superficial** and **systemic mycoses**. Systemic mycoses are infections of internal tissues and organs. They are rarely seen in Great Britain but are quite frequent in localised areas of other countries. **Cryptococcosis** is an example of this category of fungal disease in people living in the San Joaquin Valley of California. It is caused by a capsulated yeast *Cryptococcus neoformans* which grows on pigeon droppings. When the droppings are disturbed the yeast is dispersed into the air. If breathed in, it infects the lungs and spreads to the brain, causing symptoms similar to meningitis.

Superficial mycoses are infections of the skin and the mucous membrane of the mouth and vagina. **Athlete's foot** and **ringworm** are two of the superficial mycoses which occur often in Britain. They are caused by fungi called **dermatophytes** which attack the skin. **Thrush** is another disease in this category. The pathogen infects the mucous membrane of the mouth (oral thrush) or vagina (vaginal thrush).

**i. Athlete's foot and ringworm** The dermatophytes which cause these diseases are several species of the genera *Epidermophyton*, *Microsporum* and *Trichophyton*. They are filamentous moulds with septate hyphae on which conidia develop. *Microsporum* and *Trichophyton* produce large, septate macroconidia as well as small unicellular microconidia. *Epidermophyton* produces macroconidia only (Fig 25.16). Several of the species have alternative hosts, such as cats, dogs and cattle. Athlete's foot is common among people who use communal changing and bathing facilities. The skin between the third and fourth toes is most often infected, although the infection may spread between other toes. The infected areas are red, and may peel or crack (Fig 25.17), becoming intensely itchy. Unless treated, they often become secondarily infected with bacteria.

**Fig 25.16** Spore-production by dermatophytes: (a) macroconidia of (i) *Microsporum*, (ii) *Trichophyton*, (iii) *Epidermophyton*; (b) microconidia of *Microsporum* and *Trichophyton* (after M P English 1980)

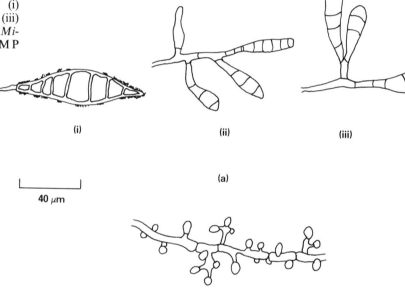

(i)　　　　　　　　(ii)　　　　　　　　(iii)

(a)

40 μm

(b)

**Fig 25.17** Athlete's foot

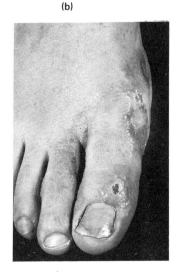

Ringworm of the scalp, *Tinea capitis*, is usually an ailment of childhood. Epidemics of it often break out in schools. Other humans or occasionally cats and dogs are the source of the infection. The fungus grows in the shafts of hairs causing them to break at the base. A small scaly spot with broken hairs is the first symptom of the infection. The spot enlarges as the infection spreads into surrounding hairs and becomes covered with thick scales at its edge. Several infected areas may later appear on the scalp. Sometimes they are inflamed. Farm workers may have similar symptoms in their beard. The source of infection here is usually infected cattle. Ringworm of the skin produces the classical symptoms of the disease. It often accompanies ringworm of the scalp in children. The lack of body hair makes it possible to see the infected sites more readily. The circular lesions are scaly and raised at the edges where they often become blistered, red and swollen (Fig 25.18).

**Fig 25.18** Ringworm

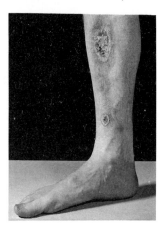

**ii. Thrush** *Candida albicans* causes thrush. It is yeast-like, with branched elongate cells from which buds arise. The term **pseudomycelium** is used to describe its structural organisation. When grown on a medium containing low concentrations of nutrients and in an atmosphere containing little oxygen, *C. albicans* form thick-walled **chlamydospores** (Fig 25.19).

**Fig 25.19** Pseudomycelium of *Candida* (after M P English 1980)

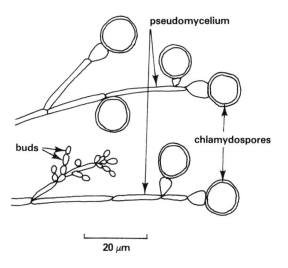

*Candida* is a commensal organism of the human gut. When the resistance of its host is lowered it becomes an **opportunistic pathogen**. Several factors may bring about a lowering of resistance. If broad-spectrum antibiotics (section 25.5.5) are taken for a long time, much of the normal commensal microflora of the digestive system is destroyed. Micro-organisms such as *Candida* which are resistant to the antibiotics then multiply to take their place. Immunosuppressive drugs and steroids inhibit the body's immune system with comparable results. Consequently, patients treated with such drugs frequently contract **oral thrush** (Fig 25.20). The fungus forms white fluffy patches on the lining of the mouth, rather like the speckles on the breast of the thrush. Beneath each patch is inflamed, irritating oral tissue.

**Fig 25.20** Oral thrush

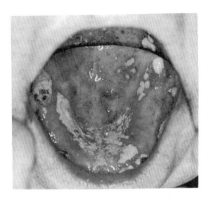

**Vaginal thrush** is relatively common among women who are pregnant. During pregnancy the epithelium of the vagina is stimulated by the large output of oestrogen from the placenta to accumulate excess quantities of glycogen. When the epithelial cells are sloughed off the glycogen is converted to glucose which is fermented by commensal bacteria to form organic acids. Diabetic females are also prone to vaginal thrush because of the high sugar content of their urine (Chapter 16). The fermenting activity of the bacteria causes the vagina to become very acid and encourages growth of *Candida*. The high oestrogen content of the birth pill has a similar effect. Many babies with oral thrush probably became infected when passing through their mother's vagina at birth.

### 25.1.4 Protozoans

These are unicellular animals which belong to the phylum *Protozoa*. It is divided into four classes, all of which have parasitic representatives (Fig 25.21). The Sarcodina are amoeboid protozoans including *Entamoeba histolytica*, the cause of amoebic dysentery in man. It parasitises the lining of the bowel, forming ulcers. Transmission occurs when the parasite forms resistant cysts which are passed in the faeces and ingested by another host.

**Fig 25.21** Some protozoan parasites (a) *Entamoeba histolytica*, (b) *Trypanosoma gambiense*, (c) *Trichomonas vaginalis*, (d) *Balantidium coli* (after J D Smith 1962)

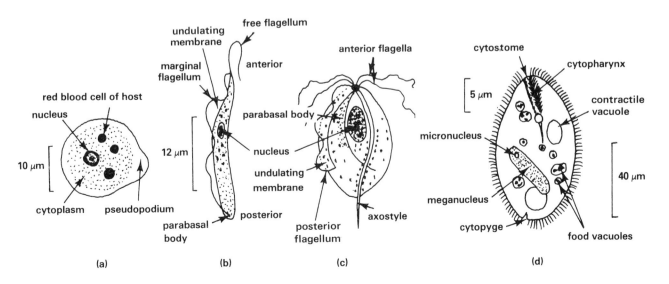

Flagellate protozoans are placed in the class Flagellata. A parasite in this class is *Trypanosoma*, the cause of sleeping sickness in man. It inhabits the bloodstream into which it is introduced by the blood-sucking tsetse fly. At a later stage the parasite invades the central nervous system. It is then that the symptoms of the disease appear, the patient shivering and wanting to do little other than sleep. Other flagellates are parasites of the gut and genital tract. *Trichomonas vaginalis* causes vaginitis, inflammation of the vagina in women.

Ciliated protozoans belong to the class Ciliata which includes one parasite of man, *Balantidium coli*. Similar in its habits to *Entamoeba*, it causes balantidial dysentery.

The class Sporozoa includes the malarial parasite *Plasmodium*. The incidence of malaria is highest in tropical and subtropical countries inhabited by the anopheline mosquito which transmits the disease from one person to another. About 2000 million people, about half the world's population, live within this area, and some 400 million have no protection against the disease. The majority of people with malaria live in Africa where over a million children die of the disease each year. In other areas much progress has been made in controlling the spread of malaria (section 25.5), but there is a great deal to be done before malaria is eradicated. About 2 million people in the world die of malaria each year. It takes more lives than any other disease.

At the present time some of the world's best cell biologists, biochemists, epidemiologists and immunologists are contracted to the World Health Organisation in an attempt to resolve the many problems which the disease presents. A detailed study of the life cycle and the characteristics of *Plasmodium* illustrates some of the difficulties involved.

## 1 The malarial parasite, *Plasmodium*

Four species of *Plasmodium* are parasitic in man, *P. vivax*, *P. ovale*, *P. falciparum* and *P. malariae*. The first three species bring about fever every 48 hours, the fourth every 72 hours. Hence various forms of malaria exist, according to the species of *Plasmodium* infecting the host (Table 25.3). As the disease progresses, fever is accompanied by severe anaemia, with enlargement of the liver and spleen. There may be kidney damage in malignant and quartan malaria. The parasite of malignant malaria may also lodge in blood capillaries, causing the flow of blood to stop locally. This is especially dangerous if it occurs in the brain.

**Table 25.3 Some forms of malaria**

| Species | Form |
| --- | --- |
| *P. vivax* | benign tertian |
| *P. ovale* | ovale tertian |
| *P. falciparum* | malignant tertian |
| *P. malariae* | quartan |

**Fig 25.22** (a) An anopheline mosquito

**i. Transmission**  Female mosquitoes of the genus *Anopheles* are vectors of the disease (Fig 25.22(a)). Unlike the nectar-consuming males they are blood-suckers. During daylight they rest in shaded places such as the interior of houses or in vegetation. At night they are active, searching for a host from which they can take a meal of blood. They are small insects with long, slender legs and land on the skin of a human without their presence being felt. The mosquito then pierces the skin with sharp stylets which penetrate as far as a capillary in the outer dermis. Before sucking blood the mosquito injects into the wound a drop of saliva. It contains an anticoagulant which prevents the blood clotting in the insect's mouthparts. The mosquito's saliva also contains the malarial parasite. A sleeping person would be unaware that blood was being taken in this way.

**Fig 25.22** (b) (i) Life cycle of *Plasmodium vivax*

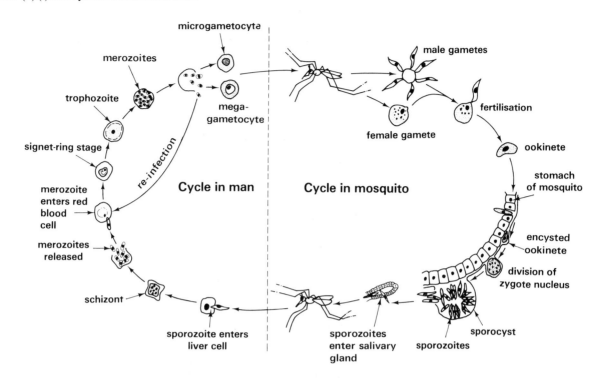

**ii. Life cycle** The stage of the life cycle injected by the mosquito is called the **sporozoite** (Fig 25.22(b)). Sporozoites are tiny spindle-shaped cells with a prominent nucleus. They circulate in the blood for about half an hour and are later found as spherical bodies called **schizonts** inside the cells of the host's liver. Schizonts then undergo rapid mitosis over the next two weeks or so, depending on the species. They produce thousands of rounded, uninucleate **merozoites** which leave the liver cells and enter the bloodstream.

**Fig 25.22** (b) (ii) Stages in life cycle of *Plasmodium* (from J I Williams and M Shaw 1982)

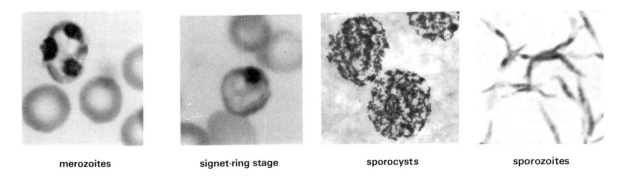

merozoites     signet-ring stage     sporocysts     sporozoites

The exoerythrocytic stage (the stage outside red blood cells) now ends for some species of *Plasmodium*. In others some of the sporozoites remain in the liver as dormant **hypnozoites**. They mature slowly, undergo **schizogony** (multiplication), and intermittently release further crops of merozoites into the blood later on. The activity of hypnozoites causes relapses in recovery from malaria.

491

In the blood, merozoites become elongate and identify receptors on the surface of red blood cells. A merozoite attaches itself to the receptors and is quickly taken into the red cell's cytoplasm. Here it feeds on haemoglobin and absorbed glucose and amino acids from the blood plasma. As it grows the parasite develops a large food vacuole which pushes its nucleus to one side. This is the 'signet-ring' stage, typical of all malarial parasites. The feeding stage is called the **trophozoite**. A red cell usually contains only one trophozoite. When fully grown, it undergoes schizogony, producing 8, 16 or 24 merozoites which resemble those formed in the liver. Division takes 24, 48 or 72 hours according to the species. The red cells burst open, releasing batches of merozoites which may invade fresh red cells and repeat the cycle. Several million merozoites appear within a few days of infection by a single sporozoite. The cyclical release of merozoites coincides with repeated bouts of fever.

After several cycles of schizogony, the sexual stage of the life cycle appears. Instead of feeding and growing after entering the red cells, some merozoites develop into **micro-** and **macrogametocytes**. Others continue to multiply in red cells. The numbers of gametocytes in the blood are at their highest at night when the mosquito feeds. The remainder of the life cycle occurs in the vector, and gametocytes which are not quickly taken by the mosquito soon die in human blood.

In the mosquito's gut the membrane of the red cell is digested and the gametocytes escape. Each microgametocyte gives rise to several uninucleate spindle-shaped **male gametes**, whereas a single spherical **female gamete** develops from each megagametocyte. Male gametes swim about until they contact a female gamete. The nucleus of one male gamete enters and fuses with that of the female to produce a diploid motile zygote called an **ookinete**. It wriggles its way through the epithelium of the mosquito's gut wall and comes to lie under its basement membrane. Here it rounds up and its nucleus divides by meiosis many times. Within 10–20 days it matures into a **sporocyst** containing thousands of sporozoites. They are released into the haemocoel of the mosquito and migrate to its salivary glands ready to be injected into a new host.

## 25.2 Macrobial parasites

Two important categories of parasites big enough to be seen without a microscope are the **helminths** and the **arthropods**. Helminths include flatworms and roundworms. Among the arthropod parasites are fleas, lice, ticks and mites.

### 25.2.1 Helminths

Two important phyla of helminths are the Platyhelminthes to which flatworms belong, and the Nematoda which are roundworms. Flatworms which parasitise man include the blood fluke *Schistosoma* which causes **schistosomiasis** (also known as **bilharzia**). The condition is debilitating but not usually fatal and is prevalent in tropical and subtropical Africa, America and Asia. Also in this category are the **tapeworms** which parasitise the gut. Among the nematode parasites of man are **hookworms**, **pinworms** and large **roundworms**, all gut parasites. **Filarial worms**, the cause of **elephantiasis** (Fig 10.46) are also nematodes. A detailed study of one parasite from each group will illustrate the ways in which they have adapted to a parasitic way of life.

## 1 The beef tapeworm, *Taenia saginata*

Man is the **definitive host** for two species of *Taenia*. At one time the pork tapeworm *T. solium* was frequent. However, these days the beef tapeworm, *T. saginata*, is much more common. Over 60 million people are thought to be parasitised by it. *T. saginata* occurs in all countries where beef is part of the diet, although it is rare in countries where standards of hygiene are high.

Ruminants, cattle especially, are the **intermediate host** in which the larval stages of the life cycle develop.

**i. Structure** A mature beef tapeworm can be up to 17 metres long. Specimens 8 metres in length are common. Man is called the definitive host because the adult tapeworm lives in his small intestine. At one end of its body the tapeworm has a **scolex** (head) on which are four suckers by which it clings to the intestinal wall of its host (Fig 25.23). Behind the scolex the tape-like body consists of 700–1000 **proglottids**. New proglottids are

Fig 25.23 (a) Scolex, gravid proglottid and egg of *Taenia saginata* (after R A Avery 1972)

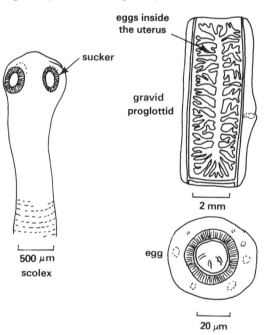

Fig 25.23 (b) Entire tapeworm (Dr A R D Adams)

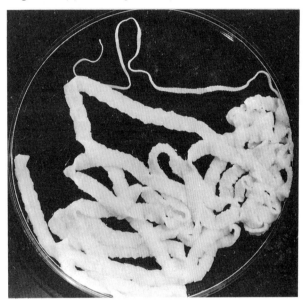

Fig 25.24 Mature proglottid of *Taenia*

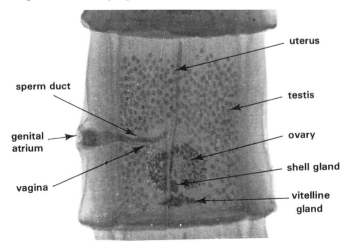

constantly budded off in the neck region and replace older ones shed from the rear of the worm. As each proglottid matures it develops male and female reproductive systems (Fig 25.24). Eggs are fertilised by sperm from the same proglottid and then receive yolk and a shell. Fertilised eggs collect in the uterus which grows as it fills up with between 30 000 and 40 000 eggs. Within each egg a six-hooked embryo called an **onchosphere** develops. Every day 3–10 ripe proglottids are shed and pass out with the faeces of the definitive host.

493

The beef tapeworm is well adapted to living as an endoparasite in the gut in various ways. The suckers on its head enable it to grip onto the intestinal wall so that it is not moved by peristalsis and other gut movements. Although it does not have a gut of its own, the flatness of the tapeworm's body gives it a large surface area relative to its volume. No part of its interior is far from its body surface through which nutrients are absorbed. The actively growing anterior end of the tapeworm lies in the duodenum where nutrients consumed by its host are plentiful as amino acids, sugars and vitamins. Covering the tapeworm's body is a **cuticle** bearing numerous microvilli which further increase the surface-area-to-volume ratio, thus enhancing absorption. The cuticle resists attack by its host's digestive enzymes. Tapeworms generate the little energy they require by anaerobic respiration. They are thus suited to the gut environment where oxygen is scarce.

**ii. Life cycle**   The embryo will not develop unless eaten by the intermediate host. Cattle are the usual intermediate host in Britain and the USA. Other ruminants such as goats and sheep may serve the same purpose elsewhere. Ripe proglottids passed in the faeces of an infected human wriggle about, causing the eggs to be squeezed out. The vast number of eggs produced by the adult tapeworm increases the probability that some will be consumed by the intermediate host.

The majority of people with tapeworms live in urban places so there is little chance of the eggs being transferred directly to pasture where cattle graze. However, there are other ways in which pasture can become contaminated. The eggs survive sewage treatment and may be dispersed onto land when sewage sludge is used as a fertiliser. Where sewage is discharged near the coast, seabirds such as gulls may eat tapeworm eggs and egest them onto farmland. Eggs of the beef tapeworm can survive outside for up to a year.

Following ingestion by a cow, the egg shell is digested by pepsin in the ruminant stomach. Breakdown of the egg shell is completed in the duodenum where it is further attacked by pancreatic juice which releases the onchosphere. The embryo secretes histolytic enzymes which assist its hooks to penetrate the mucosa of the intermediate host's gut. It enters a blood vessel and is carried around the body. Eventually it settles, usually in muscle but occasionally in fat deposits. Almost any part of the body may be the site where the embryo develops into a bladderworm or **cysticercus**.

At this stage the tapeworm is a small spherical bladder containing an inverted head. The cysticercus is fully formed within 7–10 weeks and remains viable for 4–9 months, when the head is everted (Fig 25.25). If within this time the intermediate host is slaughtered for human consumption, the cysticercus may be eaten unnoticed with the meat. When this

**Fig 25.25** Bladderworm

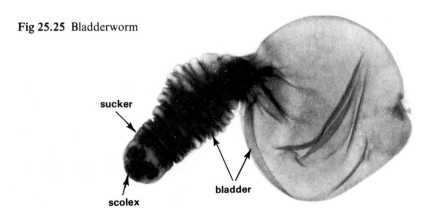

sucker

scolex

bladder

happens the bladder is digested in the person's stomach and the head fastens itself to the wall of the duodenum where it starts to bud proglottids.

The life cycle of *T. saginata* is summarised in Fig 25.26. One of the measures which can be taken to break the life cycle and hence to minimise the spread of the parasite is to provide adequate means of sewage disposal (section 25.5.1). Also the carcasses of animals killed for humans to eat can be inspected for bladderworms which are visible to the naked eye. Thorough cooking of the meat usually kills any bladderworms that have been missed. Eating 'rare' beef is probably the reason why so many people are infected with *T. saginata*.

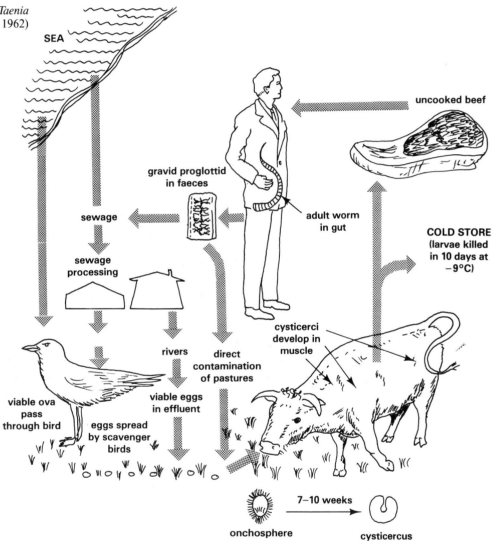

**Fig 25.26** Life cycle of *Taenia saginata* (after J D Smyth 1962)

Individuals infected with tapeworms often have no symptoms at all. Others experience nausea and mild abdominal pain. It is possible for man to act as intermediate as well as definitive host for *T. saginata* but this happens rarely. A poor standard of personal hygiene is the usual cause. The danger is that the bladderworm may develop in a vital organ such as the heart, and impair its function.

Niclosamide is an effective drug against tapeworms. It is taken orally and inhibits anaerobic respiration by the worms. The parasite is then digested by the host's peptidase enzymes.

## 2 The large roundworm, *Ascaris lumbricoides*

About 1000 million people are hosts to the large roundworm. It parasitises the small intestine of man, apes and pigs. The variety found in pigs does not parasitise man.

**i. Structure**   The worms are of separate sexes (Fig 25.27). Female worms are larger than males and up to 35 cm in length. The body is **unsegmented** and covered with a tough **cuticle** which protects the worm from its host's digestive enzymes. At the anterior end is a mouth into which the semi-liquid gut contents of its host are drawn by the sucking action of the muscular pharynx. The intestine of the worm opens at its anus at the posterior end of the body which is curved in the male. The female has a pair of thread-like ovaries. They discharge eggs into narrow oviducts which lead into wider uteri. The uteri open about half way along the length of the body. The male has a single testis, leading to a seminal vesicle which opens into the cloaca at the rear of the body. Also opening into the cloaca is a pouch containing copulatory spicules.

**Fig 25.27** (a) Male and female nematodes (after R A Wilson 1967)

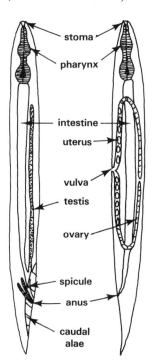

**Fig 25.27** (b) *Ascaris lumbricoides*, entire male (smaller) and female (Wellcome Tropical Institute Museum)

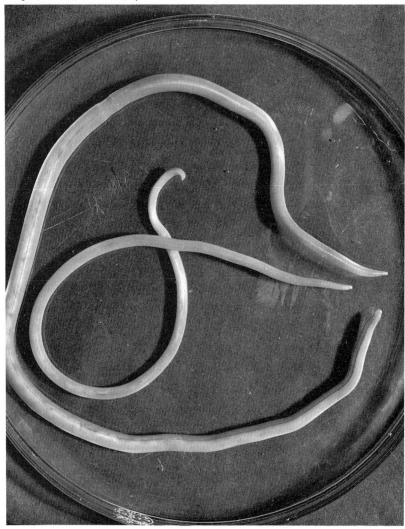

**ii. Life cycle**   There may be many roundworms in the host's intestine. Here worms of opposite sex copulate and sperm is transferred into the female's reproductive tracts. Soon after, the females begin to lay fertilised eggs which are passed out of the host's body in its faeces. Roundworm eggs have a three-layered thick coat and can remain viable outside a fresh host for up to six years. In cool moist conditions a coiled **larva** develops inside each egg. The larva then moults but cannot develop further unless eaten by a human.

The stimuli for the egg to hatch are the relatively high internal temperature of its host's body and the high tension of carbon dioxide in the host's gut. Stimulated by these conditions the infective larva secretes a hatching fluid containing enzymes which digest the egg coat. The larva wriggles out, penetrates the intestinal mucosa and enters a blood vessel. It is transported first to the liver where it may stay for a few days before continuing to the heart and lungs. Here the larva moults twice before breaking through a capillary wall into an alveolus. It then crawls into the air passages and up to the pharynx from where it is swallowed. In the intestine it moults for the last time and matures within about two months. During its life cycle the large roundworm moults four times, growing at each stage (Fig 25.28).

**Fig 25.28** Life cycle of *Ascaris lumbricoides* (after J D Smyth 1962)

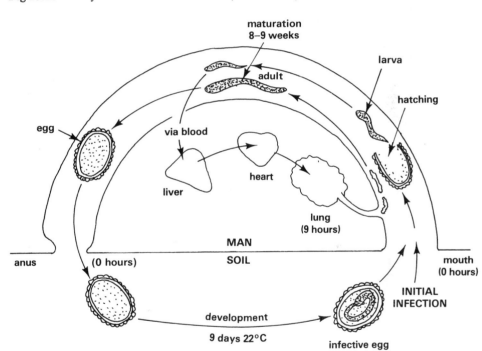

When a host is infected with many larvae, as often happens, they cause inflammation of the lung tissue they have damaged. Breathlessness on exertion may occur because of the reduced alveolar surface area through which oxygen is absorbed. The wriggling larvae in the air passages also stimulate the coughing reflex. Large numbers of adult worms in the gut may block the small intestine, preventing the passage of food. This, together with the worm's habit of feeding on its host's food, often results in the host being malnourished. Ineffective attempts to eradicate the parasite with medication may cause roundworms to migrate up the bile duct and damage the liver. The drug piperazine is used to treat patients who harbour roundworms. Taken orally, it paralyses the worms which are moved along the gut by peristalsis and egested.

Compare the adaptations to a parasitic mode of life shown by roundworms and tapeworms.

497

### 25.2.2 Arthropods

Animals of this group have **jointed legs**. The phylum Arthropoda includes a number of common ectoparasites of man. Some of them spend their entire life on their host, such as mites which cause scabies. Others such as bedbugs only visit their host at intervals to feed. The disorders caused by arthropods are often mild and are mainly irritation of the skin where the parasite has bitten the host, and secondary infection of the broken skin. However, many arthropods are alternative hosts or vectors of a wide variety of microbial diseases.

### 1 Insects

**Insects** include lice, fleas and bedbugs. All insects have three pairs of jointed legs. **Lice** are of two kinds, biting and sucking. The latter are parasitic on man. Two species occur, the **human louse** of which there are two varieties (the head and body louse), and the **pubic louse** or crab louse (Fig 25.29). Head and pubic lice lay eggs which are cemented to the hairs of the host as nits. Body lice lay their eggs among the fibres of the host's clothes. Within a few days the eggs hatch into nymphs which moult three times before becoming sexually nature adults. A single louse can soon give rise to a severe infestation.

**Fig 25.29** (a) A sucking louse

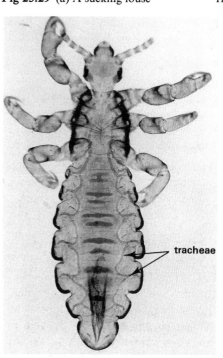

tracheae

**Fig 25.29** (b) A crab louse

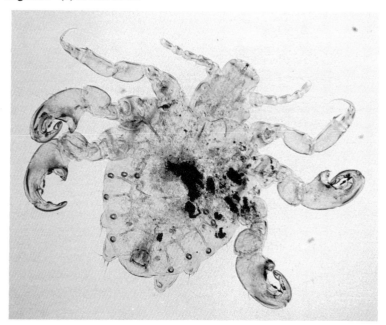

All three kinds of louse are spread by direct contact. Head lice are frequent among children of school age and are passed from one child to another when their heads come into contact. Crab lice are usually passed to a fresh host during sexual intercourse. Body lice crawl onto a new host from infested clothing or bedding. The main symptom of infestation with lice is intense itching. Scratching the skin can result in **impetigo**, blisters infected with staphylococci. Lice can carry the bacterial disease **typhus**. The organism which causes typhus is often present in the faeces of lice. If the faeces contaminate the skin when the host scratches, the disease usually follows. Dried faeces may also be inhaled. Hospital staff are therefore at risk when dealing with louse-ridden patients. A face mask and goggles reduce the risk of breathing in the infected faeces and contamination of the eyes.

**Fleas** are wingless insects whose bodies are flattened laterally (Fig 25.30). This enables them to move easily between the hairs of a host. They spend most of their time on the host but they leave the host to breed. The eggs are laid in crevices between floorboards or in carpets and furniture. Within a few days they hatch into white, grub-like larvae which pupate after feeding for about two weeks. The pupae later metamorphose to adult fleas. Both male and female fleas suck blood. Their bites usually become red and swollen and may become infected with bacteria. The infection is sometimes carried by the flea's mouthparts. Rat fleas which sometimes infest humans are vectors of the bacteria which cause **bubonic plague** (black death) and **endemic typhus**. It is estimated that in the Middle Ages about 25 million of the population of Europe died from bubonic plague.

Fig 25.30 A flea

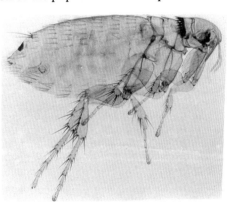

Fig 25.31 A bed bug

**Bedbugs** are flattened dorso-ventrally (Fig 25.31). During the day they hide in crevices in masonry, under loose wallpaper and in furniture. They emerge at night to feed on human blood. Their bites have a central red area around which is a weal. The bites itch intensely. Female bedbugs lay clusters of eggs in bedding and furniture. Nymphs hatch from the eggs and moult several times before reaching maturity. Between each moult the nymphs take a blood meal. Bedbugs are not known to transmit diseases.

## 2 Arachnids

**Arachnids** have four pairs of jointed legs and include mites (Fig 25.32). The **itch mite** is the cause of **scabies**, an infestation of the skin. The mite is transmitted by direct contact. Fertile female mites bore into the epidermis where they lay their eggs in burrows. The burrows can be seen on the surface of the skin as greyish, wavy lines 1–10 mm long. At the far end of each burrow is a red spot where the mite is feeding. From here it can be extracted with a needle. Within three or four days of being laid the eggs hatch into nymphs which crawl into the hair follicles. After two moults they reach adult size.

At first the host is unaware of the infestation but later the burrows itch intensely. Scratching of the skin may cause the burrows to become infected with bacteria giving rise to **impetigo** and **boils**.

Fig 25.32 (a) The itch mite, *Sarcoptes scabiei* (after R A Avery 1972)

Fig 25.32 (b) Photomicrograph of a mite

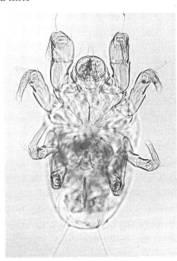

200 µm

## 25.3 Response of the host to parasites

The immune response to microbial infection is described in Chapter 11. The same principles apply to all microscopic parasites and also to some of the larger parasites.

### 25.3.1 Tissue reactions

Tissues damaged by parasites become inflamed as circulation of blood to the site increases. Phagocytic white blood cells may engulf and destroy small parasites. Alternatively, fibroblasts may secrete a capsule to surround and contain larger parasites. This sometimes happens with the larval stages of parasitic nematodes. Tissue reactions are non-specific, many different kinds of parasites being dealt with in the same way.

### 25.3.2 Immune responses

Antigens on the surface of a parasite may stimulate the host to produce specific antibodies which may destroy the parasite. In some instances the antibodies give protection against subsequent infestations by the same parasite. Antibodies formed in response to one species of parasite are ineffective against another. Variations of immune response occur.

**1 Long-lived protection**

Antibodies produced in an initial attack by a strain of influenza virus usually provide life-long protection against the variety of 'flu caused by the strain.

**2 Short-lived protection**

Antibodies formed in response to a first attack of typhoid fever give resistance for a short time against a subsequent infection. This is why travellers to countries where typhoid is endemic are advised to be vaccinated against the disease every three years.

**3 No protection**

Here the host produces antibodies but they give no protection against future infestations such as sleeping sickness and amoebic dysentery.

None of these mechanisms can cope with the larger parasites which are too big to be phagocytosed and against which antibodies are ineffective. Parasites can also evade the host's immune mechanisms in various ways. Each of the stages of the life cycle of the malarial parasite is antigenically different from the next. By the time the host develops immunity to one stage, another stage appears. The influenza virus and malarial parasite are genetically unstable, constantly producing new antigens. The host then constantly lags behind in forming appropriate antibodies. Parts of the body where antibodies cannot reach may be occupied by parasites. Gut parasites avoid our immune response in this way.

Attempts are being made to prepare vaccines which can give us active immunity against important parasites such as *Plasmodium* and *Trypanosoma*. They involve transferring the genes for antigen production from the parasites to bacteria by genetic engineering (Chapter 4). When grown in culture, the bacteria then produce the antigen. Such bacteria can be attenuated and used to produce vaccines.

## 25.4 Occupational diseases

The respiratory system is a route through which several occupational diseases are contracted. In 1979 almost 15 % of the deaths in England and Wales were attributed to **respiratory ailments**. Whereas a high proportion of the deaths were caused by pneumonia and influenza, a significant number was due to industrial diseases such as pneumoconiosis and asbestosis. A smaller number died of an agricultural disease called farmer's lung.

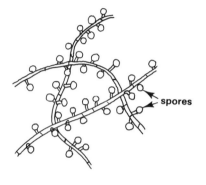

Fig 25.33 *Micromonospora*, a cause of farmer's lung disease

### 25.4.1 Farmer's lung

This is an allergy caused by inhaling spores of thermophilic bacteria of the order Actinomycetales such as *Micromonospora* and *Thermopolyspora*. They are filamentous bacteria which produce spores on their thread-like bodies (Fig 25.33). Thermophiles grow well at relatively high temperatures (50–65 °C). A temperature of this order occurs in self-heated hay. Self-heating occurs when energy is released by saprophytic moulds and bacteria living on damp hay. When the damp, mouldy hay is moved, millions of spores are released into the air and may be breathed in. Within hours the symptoms of farmer's lung appear. The heart begins to race, fever develops, the patient's breathing is shallow and sputum is repeatedly coughed up. The symptoms quickly reappear on subsequent exposures. Eventually extensive lung damage and heart failure may occur. Early diagnosis is necessary so that future risk can be avoided. If mouldy hay has to be moved it is sensible to wear a respirator which filters out the spores and to work in a well ventilated area. Avoiding contact with self-heated hay is the only certain way of preventing the disease. Hay which is properly dried before it is collected and stored in dry accommodation should not self-heat.

Fig 25.34 (a) X-ray of lungs showing shadowing caused by pneumoconiosis (St Bartholomew's Hospital Department of Medical Photography)

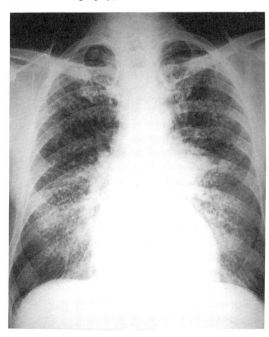

### 25.4.2 Silicosis

**Silicosis** may occur following breathing in particles of silica over a long period of time. Many kinds of rock and stone are made of silica. Hence the disease is common in those occupations where dust from rock or stone is generated. Quarrymen, miners and sandblasters are especially at risk. The inhaled particles become trapped in nodules formed in the alveoli from fibres secreted by the lung tissue. As they enlarge, adjacent nodules join together causing the lungs to become fibrous and distorted. The surface area for gas exchange is reduced, as are lung volumes and capacities (Chapter 12).

The first symptom of silicosis is breathlessness on exertion. The heart has to work harder to compensate for poor lung function. The condition gets progressively worse and the patient is often reduced to the life of an invalid. Patients with silicosis are prone to tuberculosis of the lungs and if they are smokers they frequently suffer from chronic bronchitis too. Radiographs prepared from chest X-rays show shadows on the lungs of diseased persons (Fig 25.34).

The amount of dust in mines and quarries can be greatly reduced by sprinkling the working area with water. Respirators which filter dust from inhaled air can also be worn.

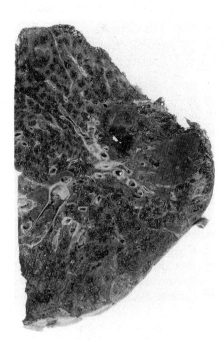

**Fig 25.34** (b) TS of part of the lung showing accumulation of coal dust (St Bartholomew's Hospital Department of Medical Photography)

### 25.4.3 Pneumoconiosis

This is a disease common among coalminers. It is caused by inhalation of stone and coal dusts over a long period of time. The dust is engulfed and held by macrophages in fibrous nodules which develop in or near the bronchioles. The nodules grow to block the flow of air to the alveoli. The alveoli become distorted and adjacent air sacs join to each other through breakdown of adjoining walls, a condition called **emphysema**. Continued inhalation of coal dust causes wholesale fibrosis of both lungs. Breathlessness on exertion, and a persistent cough which produces black sputum, are the early symptoms of **pneumoconiosis**. As with silicosis, the patient develops heart strain and becomes an invalid, often at a relatively young age. Chronic bronchitis is a common side-effect.

The measures of control and prevention of pneumoconiosis are similar to those for silicosis.

### 25.4.4 Asbestosis

**Asbestosis** is caused by breathing in particles of asbestos. Crocidolite and chrysotile asbestos are more likely to cause the disease than others. People at risk are those employed in mining asbestos and in the manufacture of asbestos products. Asbestosis has long been identified as an occupational disease of asbestos miners in South Africa. However, only since the 1960s has there been much concern about asbestosis in the manufacturing industry.

Asbestos is widely used as a heat insulator, in making ceiling tiles, roofing, brake linings, fireproof suits, fireblankets and fireproof curtains. In the production of these items asbestos fibres are released into the air and may be inhaled. They become trapped in and thicken the alveolar walls. The lower parts of the lungs in particular become diffusely fibrous over many years. Shortage of breath on exertion and a sputum-producing cough are the early symptoms of asbestosis. The sputum contains fibrous microscopic **asbestos bodies**. Heartstrain may follow. Even worse, the risk of lung cancer developing in workers employed in the asbestos industry for more than twenty years is ten times that of the general population. Smoking increases the risk.

**Mesothelioma**, cancer of the pleural membrane around the lungs, is also linked with inhalation of asbestos. The condition arises many years after only small amounts of asbestos are taken in. In a 1965 study of 83 Londoners with mesothelioma, half had a history of industrial or domestic contact with asbestos. Of those with no such history, a third lived within half a mile of an asbestos factory. Asbestos is obviously a potentially dangerous material to be handled with great care. The **Asbestos Regulations** in Britain specify the working practices which minimise the amount of dust generated and released in manufacture of asbestos products. Respirators are available for employees who are likely to come into contact with asbestos dust.

### 24.4.5 Byssinosis

This is a respiratory allergy caused by inhaling threads of cotton (*byssus* Latin = thread). The cotton threads stimulate constriction of the bronchi causing asthma-like symptoms. The symptoms disappear at weekends and during holidays when the patient is away from the working environment. Adequate ventilation of the workplace minimises the hazard. Face masks may also be worn to filter out cotton fibres from inhaled air.

## 25.5 Prevention and control of parasitic diseases

The only certain way of preventing parasitic diseases is to stop the parasites entering the host. One of the most important routes by which parasites enter our bodies is the gastro-intestinal system. **Waterborne diseases** have plagued humanity ever since we gave up our way of life as hunter-gatherers. As soon as we began to live in settlements and towns we were confronted with the problem of what to do with domestic waste, especially faeces.

The earliest attempt to deal with the problem was to bury human excrement in the soil. In time saprophytic bacteria in the soil broke it down into harmless products. Later came the earth closet, a bucket containing soil into which faeces were egested. When the closet was full its contents were emptied into a pit dug into the earth and then covered over. These practices are still used today in remote parts of Britain. However, most houses and workplaces now have a water closet which carries human faeces and urine to a local sewage treatment works.

### 25.5.1 Sewage disposal

**Sewage** is the water-borne waste of a community. It has three main components: domestic waste, industrial waste and surface water. Domestic waste includes tea leaves from our kitchen sinks, detergents from washing machines, soaps in bathwater, faeces and urine from lavatories. Industrial waste contains animal waste such as blood from slaughterhouses, plant waste such as cellulose from paper mills, pharmaceuticals, clay from potteries, oil from garages and engineering works, heavy metals such as lead, zinc and mercury. Surface water includes rain water from paved areas, roads and roofs. Many of the older towns in Britain are sewered on the combined system in which surface water drains are connected to foul sewers carrying domestic and industrial waste. To prevent the sewers from flooding when there is heavy rainfall, overflows release excess sewage into natural watercourses such as streams, rivers and the sea. Because this arrangement results in pollution of watercourses, newly installed sewers have a separate system for collecting surface water and discharging it into natural watercourses. Domestic and industrial wastes are channelled to sewage treatment plants.

### 1 Composition of sewage

The composition of sewage varies from place to place and from time to time. An approximate analysis of 'typical' sewage shows it to contain 99.9 % water, 0.02–0.03 % suspended solids, and 0.07–0.08 % dissolved organic and inorganic compounds. The amount of organic material determines the **Biochemical Oxygen Demand, BOD**, of the sewage. The larger the amount of organic material, the higher is the BOD. The BOD is the quantity of oxygen required, mainly by decomposer organisms, to break down the organic matter into simple, stable products. Sewage also contains micro-organisms. Among the bacteria present are relatively harmless species, for example *Escherichia coli* which comes from the intestines of healthy people, and pathogens which come from the bowels of people suffering from enteric diseases such as typhoid fever, cholera and dysentery. Viruses in sewage include harmless bacteriophages as well as pathogens such as the viruses which cause poliomyelitis and infectious hepatitis.

## 2 Aims of sewage treatment

There are two main aims of sewage treatment.

**i.** To lower the BOD sufficiently so that the effluent from the sewage works can be discharged into a natural watercourse without grossly upsetting its ecological stability (Chapter 23).

**ii.** As far as possible to destroy or eliminate pathogens which may endanger wildlife such as shellfish and birds living in the water. Humans may also eat shellfish from, or bathe in natural watercourses into which treated sewage is discharged.

Both these aims can be achieved by a combination of physical, chemical and microbiological processes.

## 3 Stages of sewage treatment

There are many different ways in which sewage can be treated. The choice depends on the amount of sewage to be dealt with, the fate of the effluent and its possible ecological effects and the cost of treatment. In general there are three main stages (Fig 25.35).

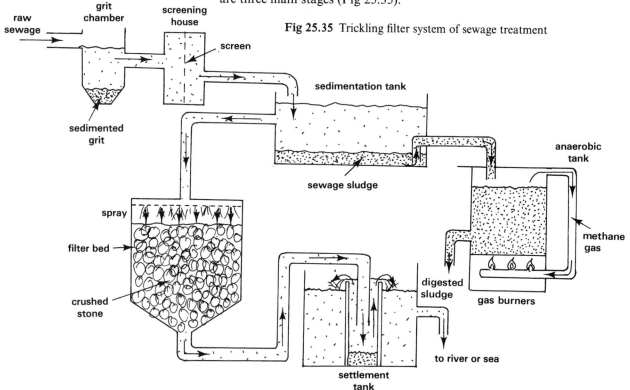

Fig 25.35 Trickling filter system of sewage treatment

**i. Primary treatment**   physically removes the larger suspended and floating solids by **sedimentation** and **screening**. Raw sewage is first channelled into small detritus tanks where soil and grit form a sediment. Grit is removed to prevent excessive wear on the pumps which move the sewage from one part of the treatment plant to another. The supernatant liquid then passes through a screen house where floating debris such as paper, rags and wood are filtered out using coarse metal screens. Screening is carried out to minimise the risk of pump and pipe blockages in the later stages of treatment. The screened sewage enters large sedimentation tanks where after several weeks many suspended solids are precipitated as crude **sewage sludge**. It contains about 4 % of the total sewage solids. The supernatant liquid and sludge are then subjected separately to secondary treatment which is entirely microbiological.

**ii. Secondary treatment**    There are two main ways of dealing with the supernatant liquid from the sedimentation tanks the trickling filter system or the activated sludge process.

*Trickling Filter System.* The filter is a bed of crushed stone onto which the liquid sewage is sprayed through rotating pipes (Fig 25.36). Spraying enriches the liquid with oxygen and is carried out intermittently to maintain aerobic conditions in the filter. The stone becomes coated with a thick film

**Fig 25.36** Trickling filter system: circular tanks in foreground (Water Authorities Association)

of slime containing bacteria, fungi, protozoans and algae. Aerobic saprophytic bacteria such as *Zooglea ramigera* and fungi such as *Fusarium* and *Geotrichum* inhabit the surface of the slime where they decompose organic substances in the sewage as follows:

Carbohydrates     $\rightarrow CO_2 + H_2O$
Proteins     $\rightarrow NH_3 + CO_2 + H_2S$
Lipids and soaps     $\rightarrow CO_2 + H_2O$
Urea     $\rightarrow NH_3 + CO_2$

Aerobic chemo-autotrophic bacteria such as *Nitrosomonas* and *Nitrobacter* also live near the surface of the slime. They carry out important oxidations such as the conversion of ammonia to nitrate (nitrification).

$$2NH_4^+ + 3O_2 \rightarrow 2NO_2^- + 4H^+ + 2H_2O$$
$$2NO_2^- + O_2 \rightarrow 2NO_3^-$$

Others such as *Thiobacillus* oxidise hydrogen sulphide to sulphate:

$$2H_2S + 5O_2 \rightarrow 2SO_4^{2-} + 2H_2O$$

As a result of these activities the BOD of the sewage is greatly reduced and the products are simple, relatively harmless substances. With little further treatment the liquid leaving the filter can be released into a nearby watercourse. Many motile protozoans are found in the upper part of the filter, stalked forms lower down. They consume the bacteria and maintain a suitable balance of organisms in the filter.

505

*Activated Sludge Process.* In recent years the trickling filter system has been superseded by the activated sludge process. About half of the sewage in Britain is now treated using the activated method. The supernatant liquid from the primary treatment is pumped into **aeration tanks** (Fig 25.37). The liquid is constantly stirred and vigorously aerated, resulting in the formation of small aggregates of finely suspended colloidal organic matter called **floc**. The formation of floc is not fully understood. It has a gelatinous texture and contains numerous slime-forming bacteria, notably *Zooglea ramigera*. Nitrifying bacteria are also present as well as protozoans. Stalked protozoans are more common than motile species. Fungi are normally absent. The biochemical changes which take place in the floc are similar to those which occur in the bed of a trickling filter system, but because the sewage is better aerated the breakdown of organic matter is much more rapid. Within 5 to 15 hours the BOD of the liquid is lowered by 85–95 %. The contents of the tank are then pumped into settlement tanks for tertiary treatment.

**Fig 25.37** Activated sludge process of sewage treatment

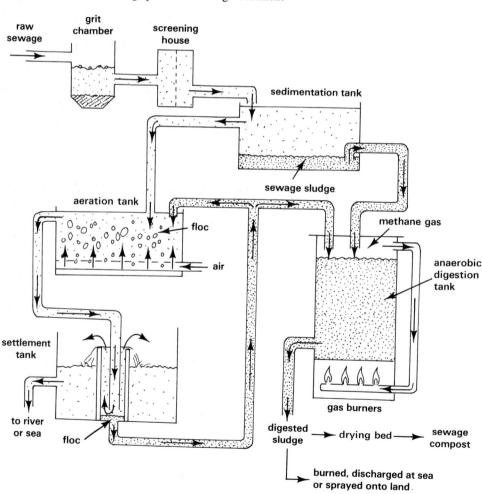

Several methods are used in Britain for the treatment of raw sewage sludge. One of the most important is **anaerobic digestion**. About half of the sewage sludge in the country is now treated in this way. The digester is a large closed tank which is almost filled with sludge so little air is present and thus very little oxygen (Fig 25.38). Facultative bacteria such as *E. coli* multiply and break down carbohydrates, lipids and proteins into a variety of products including fatty acids, alcohols, hydrogen sulphide, ammonia and carbon dioxide. Methane-producing bacteria such as *Methanobacillus* then metabolise some of the fatty acids and carbon dioxide as follows:

$$CH_3CH_2CH_2COOH + H_2O + CO_2 \rightarrow CH_4 + CH_3COOH$$

butanoic      + water + carbon    → methane +   ethanoic
acid                    dioxide                 acid

**Fig 25.38** (a) Activated sludge aeration tanks (Water Authorities Association)

**Fig 25.38** (b) Anaerobic digestion tank (Water Authorities Association)

The digestion process is speeded up by constantly stirring the sludge and heating it to 50–55 °C. At this temperature many pathogens in the sludge are killed. Within 2–4 weeks about 50 % of the organic matter is broken down, 70 % to methane and 30 % to carbon dioxide. The methane is burned and the heat generated is used to warm the digestion tanks or converted to electrical energy which drives the pumps in the sewage works. About two thirds of the works' energy requirements comes from sewage in this way.

In Britain the digested sludge is dealt with in several ways. Roughly one fifth of the treated sludge produced at inland sewage works is dumped at sea, two fifths applied to agricultural land and two fifths disposed of in other ways such as burning or burying. Wet sludge is sometimes applied to land and ploughed in after it has dried sufficiently. Alternatively, the sludge is dried at the sewage treatment plant and sold as 'compost'. These methods make use of the limited mineral constituents in the sludge and its organic matter contributes to humus formation in the soil. However, land-disposal is not a feasible long-term proposition because the sludge is of variable and limited quality and may contain high concentrations of pesticides and heavy metals.

Another process which has recently been introduced into this country is to incinerate the sludge. In 1976 the largest British sludge incineration plant was opened at Coleshill, near Birmingham. The plant occupies 2.9 hectares compared with 54.6 hectares which would have been required for drying the sludge in the conventional way. An additional important benefit is that the mass of ash to be disposed of is only one eighth of air-dried sludge prepared from the same volume of sewage. The sewage works at many coastal towns discharge untreated sludge directly into the sea. The fate of sludge dumped at sea and its effects on sea-life including fisheries is not fully known.

The time taken for the microbiologically-based secondary treatment to occur depends mainly on how much organic matter there is to be broken down. Effluent from abattoirs, dairies and food-processing industries often pose problems to sewage treatment works because of their high organic content. For similar reasons Water Authorities are becoming concerned at the increased use of kitchen sink homogenisers. They grind up most forms of soft waste such as vegetable peelings which hitherto were disposed of in domestic refuse. Another difficulty which may arise in the secondary stage of treatment is the presence in sewage of wastes such as phenols which are toxic to the decomposer micro-organisms. Phenol is a byproduct of the coal tar industry which produces antiseptics and disinfectants. This problem can be overcome by introducing phenol-oxidising bacteria to activated sludge tanks.

**iii. Tertiary treatment**  Where a high standard of liquid effluent is required, the liquid from the secondary stage is treated further. At many sewage works the liquid is simply pumped into settlement tanks where the floc forms a sediment. About 10 % of the floc is used to inoculate fresh batches of sewage entering the aeration tanks. The remainder is pumped to the anaerobic tanks for digestion. The supernatant liquid is then run into a watercourse. Liquid from the settlement tanks may be disinfected with chlorine or with ozone to kill pathogenic bacteria. However, this is only done if the water is required for immediate re-use. Chlorine is toxic to fish and other aquatic organisms and would have a catastrophic effect on the ecological balance of rivers and streams. Other components in the liquid effluent such as organic matter, phosphate, nitrate and other inorganic ions, can be removed by filtration through activated carbon and ion-exchange resins. These methods are however comparatively expensive and at the moment are not widely practised.

## 25.5.2 Water supply

The incidence of waterborne diseases has been greatly reduced in many developed countries by the provision of a domestic supply of water fit for drinking, food preparation, bathing and washing. Many forms of industry can function efficiently only if they have access to clean water. Early man relied on springs, streams and natural lakes for a water supply. This is still the case in many parts of the world. Springs and streams can be intermittent, drying up in the hotter months of the year. Alternative supplies of water may be obtained by digging wells. As the human population grew and the Industrial Revolution occurred, more people began to live in towns and cities. The volume of water required for domestic and industrial use could no longer be met by traditional means. It was then that municipal authorities embarked on making use of natural lakes and the construction of artificial lakes, reservoirs, as sources of water. In this way we have modified the natural cycling of water (Chapter 23). It has also enabled us to help nature in purifying water so that it is suitable for human consumption.

### 1 Sources of water

Water precipitated as rain, hail or snow percolates through the ground or may drain as surface water into streams which feed **lakes, reservoirs** and

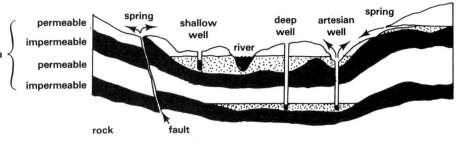

**Fig 25.39** Sources of drinking water (after A G Clegg and P C Clegg 1973)

rock strata

permeable

impermeable

permeable

impermeable

spring  shallow  deep  artesian  spring
well  well  well

river

rock  fault

rivers (Fig 25.39). Alternatively, it may penetrate deeper into the earth until it reaches an impervious layer of rock. If there is a geological fault nearby, the water may emerge as a **spring**. Above the impervious rock, the permeable upper rock strata retain water which can be obtained by sinking a shallow **well**. Deep wells can be sunk to extract water from natural underground reservoirs where water has collected in permeable lower rock strata. If the opening of a deep well is below the water table, water rises continually without pumping. This is an **Artesian well**. Water obtained from springs and wells is called **ground water**.

### 2  Impurities in water

As it falls through the air, rain collects dust and dissolves gases. Some of the suspended and dissolved impurities of the atmosphere are harmless. Others are a threat to health (Chapter 24). When flowing over the ground and into streams and reservoirs, ground water collects particles of soil and humus, bacteria, manure, dead leaves and twigs, pesticides and fertilisers. River water contains all these, and possibly sewage and industrial effluent too. Spring water may become contaminated with heavy metals from lead, zinc and copper mines. Excrement buried in nearby ground may drain into shallow wells, carrying with it pathogenic bacteria. Contamination of this kind is less likely to be found in water from deep wells where the thick layers of rock above act as a filter. However, deep well water in limestone areas contains excessive quantities of salts of calcium and magnesium.

At this stage, whatever its source, water is probably unacceptable for domestic or industrial use on account of one or more of the following undesirable properties:

**i. Colour**  This is caused by dissolved organic matter, often present in water which flows through peaty ground. The colour is not thought to be detrimental to health.

**ii. Taste and odour**  Decaying vegetation and sewage may make the water stagnant, giving it an unpleasant taste and smell.

**iii. Micro-organisms**  All natural waters contain small numbers of bacteria. However, the numbers are high and likely to contain pathogenic species if the water is contaminated with human or animal excrement.

**iv. Suspended matter**  Particles of soil and humus may cause water to be murky.

**v. Excessive hardness**  Calcium and magnesium salts, carbonate and sulphate especially, make it difficult to form a lather with soap and cause the build-up of scale in water piping.

**vi. Toxic materials**  They include pesticides and heavy metals.

**vii. Excess nitrogen and phosphorus**  Water draining from arable land may contain relatively high concentrations of ammonia, nitrate and phosphate which were applied as fertiliser. They enrich the water and encourage the growth of algae in static water such as lakes and reservoirs. Excess nitrate may also be a health hazard, especially to infants (Chapter 24).

509

### 3 Purification of water

The three basic processes in purifying water are **sedimentation**, **filtration** and **disinfection**. Additional steps are required to purify water of poor quality.

**i. Sedimentation**  When water stands in lakes and reservoirs, suspended particles settle out. Dissolved organic matter in stored water is bleached by sunlight and hardness is reduced as carbonates are used in photosynthesis by algae in the water. Anything floating in the water is removed by passage through coarse screens. River water is usually **screened** before it is stored.

**ii. Aeration**  If the water is still contaminated with organic matter it is **aerated** by cascading it down a tower or, more effectively, by allowing the water to pass down through a bed of metallurgical coke through which air is forced upwards. The effect is to eliminate odours by oxidising objectionable dissolved gases such as hydrogen sulphide, and to clear the water of much dissolved organic matter.

**iii. Flocculation**  Tiny suspended particles such as bacteria and some dissolved organic matter are still present. They are precipitated by the addition of aluminium sulphate (alum). It reacts with carbonates and hydrogencarbonates in the water to form a sticky **floc** which slowly settles. As it does so it removes micro-organisms and absorbs colour from the water. Sedimented floc forms a **sludge blanket** which is bled off regularly to keep it at a constant level. The supernatant water is then filtered.

**iv. Filtration**  The purpose of **filtration** is to remove suspended particles still in the water. They may have been present in the stored water or produced by flocculation. Filtration is normally achieved by passing the water downwards through a bed of fine inert material such as sand or anthracite.

In **slow filters**, algae, bacteria and protozoans develop as a layer of slime on the particles of the filter bed. They are similar to those which appear in the trickling filters of sewage treatment works. Their activities are similar too. The algae removes carbonate and hydrogencarbonate ions for photosynthesis. The oxygen they release helps to oxidise residual organic matter. Among the bacteria are nitrifiers which oxidise ammonia to nitrate, and oxidisers of hydrogen sulphide. The protozoans engulf bacteria which become trapped in the slime layer. In this way 99 % of the bacteria in the water are removed. Slow filters are used mainly for water which has not been flocculated.

Water which has been subjected to flocculation is more often passed through **rapid filters**. They have a filtration rate 30–50 times higher than slow filters. However, fast filtration is not as effective in removing bacteria, taste or odour from the water.

**v. Disinfection**  Most of the remaining bacteria are destroyed by **chlorination**. Disinfection is achieved by adding sufficient calcium hypochlorite to provide $0.2–1.0\,\mathrm{mg\,dm^{-3}}$ of available chlorine. This ensures that all pathogenic bacteria are killed. The chlorine is undetectable to the consumer. After treatment, the water enters trunk mains which feed small service reservoirs or towers near points of water consumption. They make it possible to store water locally and meet fluctuations in demand without affecting the treatment process. Outlet mains carry the water near to where consumers live. Smaller pipes convey the water to houses and workplaces.

Fig 25.40 summarises the main stages in preparing water for domestic and industrial use.

**Fig 25.40** Main stages of water purification

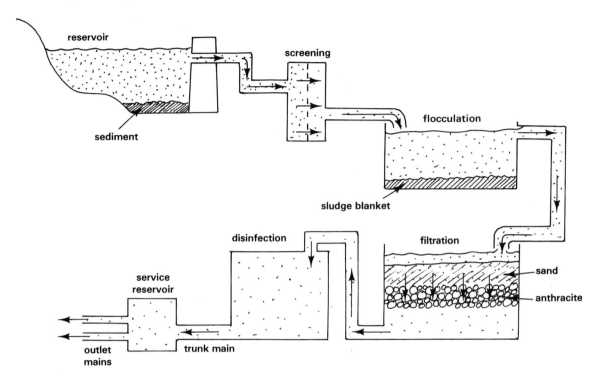

## 4 Water quality

Chemical, physical and microbiological analyses are regularly carried out on samples of water taken from service reservoirs. The pH is checked and, if too acidic or alkaline, is usually adjusted to pH 7.0. This is because acid water will attack metal fittings in the distribution network. If too alkaline, it encourages scale to form inside water pipes. Hard water may be **softened**. The fluoride content is monitored if the water is to be **fluoridated** to help reduce dental caries. High concentrations of ammonia indicate contamination by sewage.

The **coliform test** is the most frequently performed microbiological investigation of water quality. It requires the water to be analysed for the presence of *Escherichia coli*, an inhabitant of the bowel which is present in large numbers in our faeces. Water containing few *E. coli* is unlikely to contain pathogenic gut bacteria which cause waterborne diseases such as typhoid. However, high numbers of coliforms indicate faecal contamination and the probability that pathogens are present.

The World Health Organisation recommends an upper limit of 50 coliforms per 100 cm³ of raw water if it is only to be disinfected before use. If the water is to be flocculated and filtered before disinfection it may contain 50–5000 coliforms per 100 cm³. After treatment, 95 % of samples should be free of coliforms and coliform bacteria should never be present in numbers higher than 10 per 100 cm³ for any two consecutive samples.

### 25.5.3 Immunisation

Diseases such as typhoid fever, cholera, bubonic plague and smallpox were rife before the 1900s. Smallpox has since been eradicated and the others are now much less common.

Several eminent scientists have contributed to the **germ theory** of disease. Notable among them was **Louis Pasteur** and **Robert Koch** (Fig 25.41). In 1877, Pasteur used a microscope to examine the blood of sheep

Fig 25.41 Louis Pasteur and Robert Koch (Ann Ronan Picture Library)

and cattle dying of anthrax. He saw the cause of the disease, a bacillus. He later succeeded in growing the organism in the laboratory. Also in the 1870s Koch isolated anthrax bacilli from diseased sheep, injected them into mice and noticed that they developed the disease. He re-isolated the bacilli from the mice and saw that they were similar to those he had isolated from the sheep. It was the first time that anyone had proved that a bacterium caused a disease. The procedure led to the development of **Koch's postulates** by which the link is established between the symptoms and cause of disease. The years between 1875 and 1900 are sometimes called the Golden Age of bacteriology in which the causes of many microbial diseases of man and livestock were discovered by Pasteur and others.

Once the cause of diseases had been determined, attention was focused on methods of prevention and treatment. Three main lines of attack were developed, **immunisation, antisepsis** and **chemotherapy**.

The way in which Edward Jenner successfully immunised a boy against smallpox in 1797 is described in Chapter 11. Nearly 100 years later Pasteur developed a vaccine against rabies (Fig 25.42). First he inoculated rabbits with the saliva of rabid dogs. The rabbits soon died of rabies. Pasteur removed their brains and spinal cords. These he pulverised and suspended in sterile saline. Dogs inoculated with the suspension were found to be immune to rabies. Later a boy named Joseph Meister who had been bitten by a rabid wolf was brought to Pasteur who promptly inoculated the youngster with his crude vaccine. Meister lived.

Fig 25.42 Development of rabies vaccine (after M J Pelczar and R D Reid 1972)

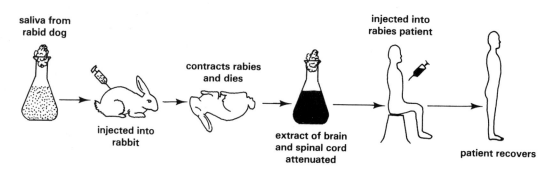

saliva from rabid dog

injected into rabbit

contracts rabies and dies

extract of brain and spinal cord attenuated

injected into rabies patient

patient recovers

Subsequently vaccines have been developed against many of the major microbial diseases. The ways in which they are prepared and give us immunity are described in Chapter 11 and in section 25.1.1.

### 25.5.4 Antisepsis and disinfection

Antiseptic measures are those taken to combat sepsis, infection of wounds. At one time puerperal (childbed) fever was a common infection in maternity wards. When Ignaz Semmelweis joined the staff of the Vienna General Hospital in 1844 the maternity unit was divided into two departments. Medical students were instructed in one of them. Between 1840 and 1846 the overall death rate in the unit was about 10 %, but in the department without students the rate was about 3 %. The reason for the differences was a mystery until the death of a Professor of Medicine at Vienna University. Professor Kolletschka died of puerperal fever within a few days of carrying out a post-mortem on a woman who had died of the disease. During the post-mortem a student's scalpel had slipped and cut the professor's finger. Semmelweis was convinced that the scalpel was infected with minute pieces of flesh from the corpse.

It was the practice after post-mortems for students of obstetrics at the university to attend the maternity unit at the hospital. Here they washed their hands in soap and water before examining women in labour. Semmelweis deduced that despite washing, particles of decayed flesh were transferred to the sexual organs of the women from the hands of students and their tutors. From then on he insisted that students wash their hands in chlorinated lime water before they attended expectant mothers. In 1848 the mortality rate in the teaching unit fell below 3 % for the first time.

By the 1850s the administration of anaesthetics had taken away the most feared aspect of surgery, pain. However, almost as many people died as lived after surgery, mainly because surgical wounds became infected. In the 1860s **Joseph Lister** (Fig 25.43), a surgeon at Glasgow University, developed a procedure which greatly reduced the incidence of wound infection. Lister was familiar with the **germ theory** of disease from Pasteur's writings. He had also read that carbolic acid could be used to disinfect sewage. When an 11-year-old boy, James Greenless, was admitted to the university hospital in 1865 with a compound fracture of the leg, Lister smeared the wound with carbolic acid before dressing the injury. When he later removed the dressing, the wound was not septic. Lister later used carbolic acid as an antiseptic on surgical incisions and noted a drastic reduction in the number of infections. His observations were published in the *Lancet* in 1867 and opened the era of **antiseptic surgery**. Today, iodine, chlorhexidine and alcohols such as ethanol and isopropanol are the antiseptics most commonly used in surgery.

**Disinfection** is the destruction of infective organisms on or in non-living objects. Hypochlorite disinfectants are used in purifying water so that it is fit to drink, and in the water in public swimming baths. They are also used in the home to disinfect babies' feeding bottles (Milton) and toilets (Domestos). Disinfectants work best in warm solution. This is because heat increases the kinetic energy of disinfectant molecules causing them to move more quickly. It increases the chance of them contacting a target organism in a given time. At low concentrations there may be insufficient disinfectant molecules in contact with the target organisms to kill them. However, there may be enough to stop them multiplying. The term **bacteriostatic** is used to describe this effect on bacteria. At a high concentration the same disinfectant may be **bacteriocidal**, killing the target bacteria.

**Fig 25.43** Joseph Lister (Ann Ronan Picture Library)

Some disinfectants are most effective within a narrow pH range. Hypochlorites are best at pH 5. At this pH they form the highest concentration of hypochlorous acid HOCl, the active ingredient. The term 'available chorine' refers to the amount of HOCl in solution. The acid kills by releasing its oxygen to oxidise microbial proteins. The same process is the reason why hypochlorites are good bleaching agents. If organic matter is present it may be oxidised too, thus lowering the number of HOCl molecules reaching the target organisms. This difficulty can be overcome by increasing the concentration of hypochlorite in solution. Hypochlorites kill most kinds of microbes. Other disinfectants have a much narrower killing spectrum. Spores are more difficult to kill than vegetative cells, a problem overcome by increasing the concentration of the disinfectant and giving it more time to work.

The properties of other disinfectants are well documented elsewhere. Some of those on sale to the general public are much less effective than hypochlorites and do little more than provide a pleasant odour.

### 25.5.5 Chemotherapy

**Chemotherapy** is the use of drugs to treat diseases. Mercury was used as long ago as the 1400s to treat patients suffering from syphilis. Many of the patients eventually died of mercury poisoning. It was not until the early part of this century that modern chemotherapy began.

In 1906 **Paul Ehrlich**, a German clinician, began the search for what he called '**magic bullets**', substances that would destroy the cause of diseases without harming the patient. He believed that a chemical would be effective in curing a disease if it would bond to the infective microbe in much the same way as an enzyme–substrate complex is formed. One of the earliest compounds he tested was an arsenic compound called **atoxyl** (Fig 25.44). Ehrlich produced hundreds of derivatives of atoxyl and each was tested against a known disease in laboratory animals. In 1910 one of the derivatives called Salvarsan was found to be effective in treating rabbits infected with syphilis. Salvarsan was then used successfully to treat syphilitic humans. In Ehrlich's time 6 % of deaths in Germany were due to syphilis.

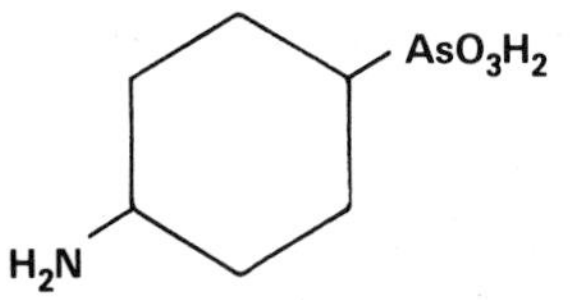

For the next 20 years or so there were no important advances in chemotherapy. However, in the early 1930s a German chemist **Gerhard Domagk** obtained encouraging results when investigating the effect of various dyes on streptococcal infections of mice. One of the dyes was prontosil which he subsequently used to cure his own daughter of a similar infection she acquired after pricking her finger with a knitting needle. Prontosil is a derivative of **sulphanilamide** (Fig 25.45). By introducing new chemical groups in its side chains, as Ehrlich had done with atoxyl, a wide range of **sulphonamide drugs** was made. They are still used to treat several important bacterial diseases of man.

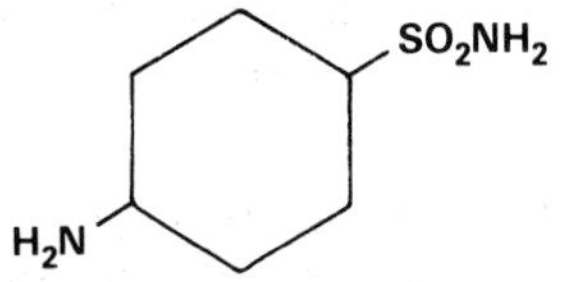

The first **antibiotic** was discovered almost by accident. **Alexander Fleming** was a bacteriologist working at St Mary's Hospital, Paddington, London. In 1928 he had prepared a plate culture of staphylococci for experimentation which had become contaminated with a mould, *Penicillium notatum*. Fleming observed that the bacteria were unable to grow adjacent to the mould colony. He published his findings the following year but there was little interest in them at the time. Ten years later the antibacterial substance involved, called **penicillin**, was isolated by Howard Florey and Ernst Chain. By the end of the Second World War penicillin was manufactured on a large scale in the USA and proved extremely successful in treating a wide range of bacterial diseases including gonorrhoea,

**Fig 25.44** Atoxyl

**Fig 25.45** Sulphanilamide

514

bacterial meningitis, pneumonia and scarlet fever. Soon penicillin replaced sulphonamides in the treatment of puerperal fever and Salvarsan for syphilis.

Several derivatives of penicillin have since been manufactured including ampicillin, benzylpenicillin and methicillin. They all have a $\beta$-lactam ring as part of their structure (Fig 25.46). From the mid 1940s a number of other antibiotics have been discovered. They include chloramphenicol, griseofulvin, nystatin and streptomycin. All are produced by moulds or actinomycete bacteria.

**Fig 25.46** The $\beta$-lactam ring

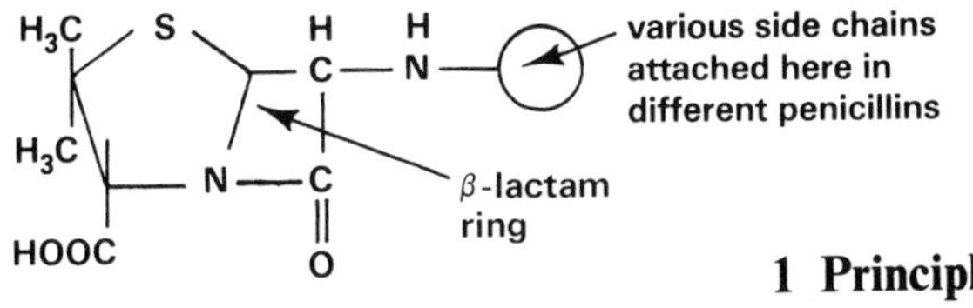

various side chains attached here in different penicillins

$\beta$-lactam ring

## 1 Principles of chemotherapy

Chemotherapeutic substances are toxic to pathogenic microbes but are non-toxic or only slightly toxic to the cells of the host. The aim is to take advantage of biochemical differences between pathogens and the host and to exploit the differences to the disadvantage of the pathogen. Some chemotherapeutic substances such as griseofulvin are **narrow spectrum** because they act against a small range of pathogens. Others such as streptomycin are **broad spectrum** and are effective against a wider range. Antimicrobial substances usually act in one of the ways described below.

**i. Inhibition of DNA synthesis** Sulphonamides are competitive inhibitors for the enzyme used by some pathogenic bacteria to convert *p*-aminobenzoic acid (PABA) into folic acid, a co-enzyme required for the synthesis of nucleotides. At first the bacteria continue to multiply, using folic acid they have stored. However, when the stores have been used up, the pathogen stops multiplying but is not killed. Sulphonamides are thus bacteriostatic. *Plasmodium falciparum* is affected in the same way. The host cells are unaffected because they, unlike the pathogens, can absorb ready-made folic acid from the blood stream.

**ii. Inhibition of cell wall synthesis** Bacteria differ from mammalian cells in having a cell wall. When grown in the presence of penicillin they become bloated and burst. This happens because penicillin prevents the coupling of mucopeptide to polypeptide chains in the cell wall. Hence the bacteria fail to develop a cell wall. They then absorb water and lyse.

**iii. Lysis of cell membranes** Some chemotherapeutic substances cause the breakdown of membranes containing cholesterol. Membranes of bacteria contain little cholesterol, whereas fungal membranes have a lot. Mammalian cell membranes also contain cholesterol. Hence this category of substances can only be used to treat superficial fungal infections such as ringworm and thrush. Nystatin is an example of a chemotherapeutic substance of this kind.

**iv. Inhibition of protein synthesis** Some of the antimalarial drugs such as primaquine and chloroquine bind to the DNA of the parasite. Consequently it is unable to code the synthesis of messenger RNA and protein synthesis is inhibited. Chloramphenicol and the tetracyclines bind to the ribosomes of bacteria, preventing messenger RNA from becoming attached. Translation of the genetic code cannot then take place so bacterial proteins are not assembled. The effect is reversible, so both drugs are bacteriostatic. Streptomycin distorts the messenger RNA of bacteria, so that mistakes are made in translation and grossly abnormal proteins are synthesised. Hence streptomycin is bacteriocidal.

515

## 2 Spectra of chemomotherapeutic substances

The following is a brief review of some of the many kinds of chemotherapeutic drugs commonly in use.

**i. Penicillins**   The original penicillin was a mixture in which benzylpenicillin, also called penicillin G, was the most effective component. It has an intermediate spectrum of activity and is effective against Gram-negative rods as well as Gram-positive cocci. It is the antibiotic of choice in the treatment of syphilis. Staphylococci are resistant to penicillin G. They secrete an enzyme called $\beta$-lactamase which destroys the lactam ring in the penicillin molecule. Erythromycin is used to treat infections caused by resistant staphylococci. Some people are hypersensitive to penicillin G. The allergic symptoms include a skin rash and nausea.

**ii. Cephalosporins**   These are broad-spectrum antibiotics. Though related to penicillins, they are less likely to be broken down by bacteria which secrete $\beta$-lactamase. Consequently they are often used as alternatives to penicillins where resistance is encountered.

**iii. Chloramphenicol**   This broad-spectrum antibiotic was widely used at one time. Unfortunately it can stop the production of red cells in the bone marrow. Hence its use is now confined to the treatment of a few bacterial diseases such as meningitis and typhoid fever.

**iv. Griseofulvin and nystatin**   Griseofulvin is effective against fungi which cause ringworm and athlete's foot. Nystatin is used to treat oral and vaginal thrush. Both antibiotics are narrow-spectrum in their activity.

**v. Streptomycin**   This broad-spectrum antibiotic is used mainly to treat tuberculosis. The disease does not always respond to treatment with other drugs because the pathogen soon develops resistance to them. Streptomycin has to be given in doses of low concentration because it can cause degeneration of the auditory nerve.

**vi. Tetracyclines**   Tetracyclines are broad-spectrum antibiotics used in the treatment of gonorrhoea.

**vii. Sulphonamides**   These were the first of the modern antibacterial drugs. When antibiotics were introduced in the 1940s sulphonamides became less popular. They are especially useful in treating infections of the urinary tract. Cases of malaria which do not respond to traditional anti-malarial drugs can also be successfully treated with sulphonamides.

## 3 Selection and testing of antibiotics

The advantage of broad-spectrum antibiotics is that they eliminate the need to find the cause of an infection. However, long-term treatment with them can grossly alter the commensal flora of the body, which may lead to infections with opportunistic pathogens such as *Candida*. For this reason, except where treatment is urgently necessary, it is better to isolate the cause of the infection, test its sensitivity to a range of antimicrobial drugs and select the most appropriate for chemotherapy even though it may have a narrow spectrum of activity.

The **agar diffusion test** is used to investigate the sensitivity of the pathogen to a range of anti-microbial drugs. Small paper discs impregnated with known concentrations of the drugs are placed on the surface of an agar-based medium on which a suspension of the pathogen has been evenly spread (Fig 25.47). Control organisms whose sensitivity to the drugs are known in advance are tested in the same way at the same time. On incubation, the drugs diffuse radially from the discs into the agar and may prevent growth of the organisms under investigation. If so, a halo of growth inhibition is seen around one or more of the discs. The diameter of the halo is a measure of the effectiveness of the drug. The control plates are required to ensure that the result is reliable and not affected by the experimental procedure, media and equipment.

**Fig 25.47**   The agar diffusion test
(Oxoid Limited)

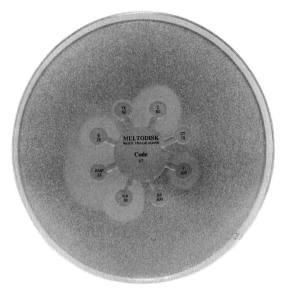

### 4 Resistance to chemotherapeutic drugs

In a population of pathogens, many may be sensitive to a drug and a few may be unaffected. When the drug is administered to a patient infected with the pathogen the sensitive cells are killed or prevented from multiplying, whereas the resistant cells are not. The resistant cells multiply and a resistant population of the pathogen is established. It is a form of artificial selection (Chapter 21) induced by man. **Resistance** among bacteria may be due to their ability to destroy the drug. Penicillin-resistant bacteria for example, produce the enzyme $\beta$-lactamase which cleaves the lactam ring of penicillins. Alternatively, a metabolic pathway may be used which bypasses the reaction normally inhibited by the drug. The ability to resist can come about by mutation or by the transfer of genes from other bacteria.

When sulphonamides and antibiotics were first used, resistance to them was infrequent. However, with chemotherapy so widely available today, resistance is common. Development of resistance can be minimised by avoiding the use of antibiotics for non-clinical purposes. The Swann Report (1969) drew attention to the dramatic increase in the number of strains of resistant gut bacteria in farm animals. Pigs, cattle and poultry frequently have antibiotics added to their food to stimulate their rate of growth. Some of the resistant bacteria are eventually ingested by humans. Resistance may also arise where human infections are treated with too low a concentration of the drug to kill the pathogen quickly. In some countries it is possible to buy antibiotics across the counter without a doctor's prescription. There is concern that careless use of unprescribed drugs may accelerate the appearance of resistant pathogens.

Many clinicians now recommend the initial administration of a single antibiotic for the treatment of serious bacterial infections compared with combinations of antibiotics recommended previously. Explain in terms of resistance.

**Table 25.4 Diseases transmitted by arthropods**

| Disease | Vector |
| --- | --- |
| poliomyelitis | housefly |
| intestinal infections | housefly |
| malaria | mosquito |
| sleeping sickness | tsetse fly |
| typhus | louse |
| bubonic plague | flea |

## 25.5.6 Vector control

Many of the vectors of infectious diseases are arthropods (Table 25.4). One way of controlling outbreaks of such diseases is to attack the vector. Many vectors are insects and can be killed using insecticides such as DDT. Where insects have developed resistance to DDT, alternative insecticides such as benzene hexachloride (BHC) and dieldrin may be used. The same substances can be used to kill other arthropod parasites of man such as lice, fleas and bedbugs. A knowledge of the habits and life history of the target arthropod is essential if control is to be effective. A study of the **anopheline mosquito**, the vector of the malarial parasite, illustrates why such knowledge is essential.

The mosquito lays its eggs in still water, such as a pond, lake or swamp. The eggs hatch into larvae, and thence pupae which breathe and feed just below the surface of the water to which they cling by surface tension (Fig 25.48). One way of eliminating the vector is therefore to eliminate wet spots where the larvae and pupae live. In some parts of the world such as China this policy has been quite effective. In others the bodies of water are too extensive and often too remote for drainage to be feasible. Biological control, introducing fish that eat the larvae, has also been tried but with little success. Better results are obtained by pouring oil on the water to suffocate the larvae, but again this is feasible only on a small scale.

Fig 25.48 Mosquito (a) larva and (b) pupa

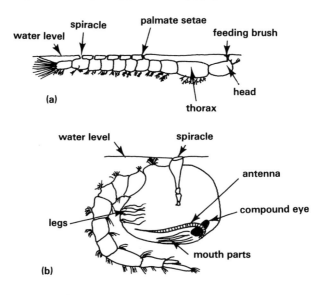

Eventually the pupae metamorphose into adult mosquitoes. Females act as vectors and they have the habit of resting in the shade of trees and in houses during the daytime. This is where the attack on the vector has been mainly focused. Spraying houses with insecticides kills any mosquitoes there, and leaves a residue which deters others from entering. To break the life cycle, spraying for at least three years is required. Even then, setbacks occur where mosquitoes develop resistance to the insecticide of choice, DDT. Alternative insecticides such as dieldrin must then be used. At the same time a programme of treating human cases of the disease must be set up. Considerable planning and organisation is required. It is an enormous task which the WHO has so far failed to solve.

One or more of the measures described above can eradicate infectious diseases. One of the greatest triumphs in this respect has been the recent worldwide eradication of smallpox. It is a viral disease with man its only host, so vaccination is the most important single measure. To get rid of the disease a very large proportion of mankind had to be vaccinated shortly after birth. Vaccination of contacts of smallpox cases was carried out as soon as any fresh outbreaks occurred. The programme was embarked on by the WHO in 1967. By 1968 the number of cases of smallpox was down by 40 % (Fig 25.49). So successful was the strategy that the last recorded case was in Somalia in 1977.

Fig 25.49 Deaths from smallpox this century (from M J Pelczar and R D Reid 1972)

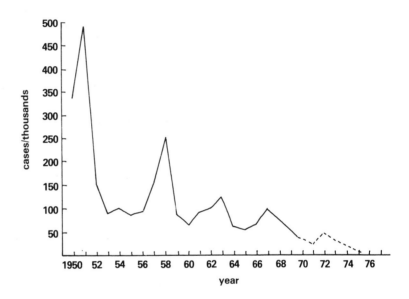

# Appendix 1:
# Chemical nomenclature

In recent years the International Union of Physics and Chemistry (IUPAC) and the International Union of Biochemistry (IUB) have reached agreement on the rules of nomenclature for many compounds of biological interest. The systematic names of most biochemicals are extremely cumbersome and require a knowledge of chemistry beyond what is expected of GCE Advanced Level Human Biology students. We have used systematic names where students should reasonably be able to cope with them, or where they present little or no more difficulty than the trivial names. Trivial names have been given where it is not anticipated that the systematic names will be used at this level of study in the near future. Table A1 caters for students who may wish to know the systematic names of some of the biochemicals given trivial names in the text. The list should also be useful for those who are not familiar with the systematic names used in the text.

## Table A1

| Trivial name | Systematic name |
|---|---|
| acetaldehyde | ethanal |
| acetic acid | ethanoic acid |
| acetone | propanone |
| adenine | 6-aminopurine |
| benzoic acid | benzenecarboxylic acid |
| carbon tetrachloride | tetrachloromethane |
| chloroform | trichloromethane |
| cytosine | 2-oxy-4-aminopyrimidine |
| ethyl alcohol | ethanol |
| formaldehyde | methanal |
| glycollic acid | hydroxyethanoic acid |
| guanine | 2-amino-6-oxypurine |
| hippuric acid | *N*-benzoylglycine |
| oleic acid | *cis*-9-octadecanoic acid |
| palmitic acid | hexadecanoic acid |
| phenylthiourea | phenylthiocarbamate |
| thymine | 5-methyl-2,4-dioxypyrimidine |
| stearic acid | octadecanoic acid |
| uracil | 2,4-dioxypyrimidine |
| urea | carbamide |
| xylene | dimethylbenzene |

The names of mammalian hormones have recently been revised by the IUB and the revised names have been used in the text. Abbreviations for the old names have also been included as these are in such wide use. Table A2 on page 522 lists the recommended and other names of such hormones.

**Table A2**

| Recommended name | Other names |
|---|---|
| calcitonin | thyrocalcitonin |
| choriogonadotropin | chorionic gonadotrophin (CG) |
| corticotropin | adrenocorticotrophic hormone (ACTH) |
| follitropin | follicle-stimulating hormone (FSH) |
| glucagon | hypoglycaemic factor |
| gonadotropin | gonadotrophic hormone |
| lutropin | luteinizing hormone (LH) |
| lutropin | interstitial cell-stimulating hormone (ICSH) |
| ocytocin | oxytocin |
| pancreozymin | cholecystokinin |
| pancreozymin | chloecystokinin–pancreozymin (CK–PZ) |
| parathyrin | parathormone |
| prolactin | mammotrophic hormone |
| somatomedin | sulphation factor |
| somatotropin | somatotrophic hormone (STH) |
| thyroliberin | thyrotrophin-releasing factor (TRF) |
| thyrotropin | thyroid-stimulating hormone (TSH) |
| vasopressin | antidiuretic hormone (ADH) |

# Appendix 2: Units, symbols and quantities

## SI and SI-derived units, symbols and quantities

Système International (SI) and SI-derived units, symbols and quantities have been used throughout the text. Non-SI units have been given in addition, where it is anticipated that the SI system may not be adopted in the near future. Non-SI units and their conversion to SI units are given in brackets under each section.

### Length

$$1 \text{ metre (m)} = 1000 \text{ millimetres (mm)}$$
$$1 \text{ mm} (10^{-3} \text{ m}) = 1000 \text{ micrometres } (\mu\text{m})$$
$$1 \text{ } \mu\text{m} (10^{-6}) = 1000 \text{ nanometres (nm)}$$
$$(1 \text{ in} = 25.4 \text{ mm})$$

### Energy

$$1 \text{ kilojoule (kJ)} = 1000 \text{ joules (J)}$$
$$(1 \text{ calorie} = 4.187 \text{ J})$$

### Pressure

$$1 \text{ kilopascal (kPa)} = 1000 \text{ pascals (Pa)}$$
$$(1 \text{ atm} = 760 \text{ mmHg} = 101.325 \text{ kPa})$$

### Mass

$$1 \text{ kilogram (kg)} = 1000 \text{ grams (g)}$$
$$1 \text{ gram} = 1000 \text{ milligrams (mg)}$$
$$1 \text{ mg} (10^{-3} \text{ g}) = 1000 \text{ micrograms } (\mu\text{g})$$
$$1 \text{ } \mu\text{g} (10^{-6} \text{ g}) = 1000 \text{ nanograms (ng)}$$
$$1 \text{ ng} (10^{-9} \text{ g}) = 1000 \text{ picograms (pg)}$$
$$(1 \text{ lb} = 0.454 \text{ kg})$$

### Volume

$$1 \text{ cubic decimetre (dm}^3) = 1000 \text{ cubic centimetres (cm}^3)$$
$$1 \text{ cubic centimetre} (10^{-3} \text{ dm}^3) = 1000 \text{ cubic millimetres (mm}^3)$$
$$(1 \text{ litre} = 1 \text{ dm}^3)$$
$$(1 \text{ ml} = 1 \text{ cm}^3)$$

### Electrical potential difference

$$1 \text{ volt (V)} = 1000 \text{ millivolts (mV)}$$

### Time

$$1 \text{ second (s)} = 1000 \text{ milliseconds (ms)}$$

### Temperature

$$\text{thermodynamic temperature} = \text{kelvin (K)}$$
$$\text{degree Celsius} = {}^\circ\text{C}$$
$$(t \text{ } {}^\circ\text{Fahrenheit} = \tfrac{5}{9}(t - 32){}^\circ\text{C})$$

# Questions

The questions have been arranged so that they can be answered after the appropriate chapter has been studied. In some instances it may be necessary to use information from more than one chapter. All questions which are not acknowledged are written in the style of BTEC.

London = University of London Examinations Council
AEB = Associated Examining Board

## Chapter 1

1 Discuss those properties of water that are of importance to living organisms.    (London: June 1981, special paper)

2 The graph shows erythrocyte osmotic fragility curves for two breeds of pig. 0.02 ml of freshly collected blood was mixed with 10 ml of each concentration of saline. After 30 minutes the tubes were centrifuged at 2 000 rpm for 10 minutes. 3.5 ml of supernatant liquid were transferred from each tube to a colorimeter and the absorbance measured at wavelength 540 nm. The percentage lysis for each sample was calculated assuming 100% haemolysis in the tube containing distilled water.

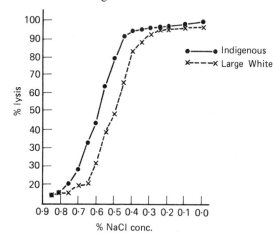

Erythrocyte osmotic fragility curves in the two breeds of piglets (from M O Makinde, *Animal Technology* 1986, p. 75).

(a) The saline solutions were prepared by diluting a 1.0% stock solution of sodium chloride with distilled water. Calculate the volumes (in ml) of stock solution and distilled water used to prepare the 0.3% NaCl solution.
(b) Explain why erythrocytes lyse when immersed in dilute solutions.
(c) Discuss the probable reason why only a proportion of cells lysed at each NaCl concentration.
(d) Why was each sample centrifuged?
(e) Name the substance in the supernatant liquid, the absorbance of which was measured.
(f) Suggest a hypothesis to explain the difference in osmotic fragility shown by the two breeds.
(g) Outline another way in which comparable data could have been obtained by microscopy.
(h) What is the potential clinical significance of the results of osmotic fragility tests?

3 Mammalian red blood cells are sensitive to a change in salt concentration of the external solution. If they are transferred from plasma to a less concentrated solution they swell and, if they swell sufficiently, may burst, in which case they are said to be haemolysed. In an experiment to find the percentage of human red cells haemolysed at different concentrations of salt solution, the following results were obtained:

| Percentage salt concentration, g $100\,cm^{-3}$ | 0.33 | 0.36 | 0.38 | 0.39 | 0.42 | 0.44 | 0.48 |
|---|---|---|---|---|---|---|---|
| Percentage red cells haemolysed | 100 | 90 | 80 | 68 | 30 | 16 | 0 |

(a) Plot the results on graph paper, using the horizontal axis for the percentages of salt concentration.
(b) Explain why red cells swell and burst when placed in a dilute salt solution.
(c) At what percentage salt solution is the proportion of haemolysed to nonhaemolysed cells equal?
(d) Suggest a hypothesis to account for the red cells haemolysing over a range of salt concentrations rather than at one particular salt concentration.
(e) What is the significance of these observations in relation to the functioning of the human body?

## Chapter 2

1 **EITHER**   Discuss the structural and metabolic roles of lipids in the body.
  **OR**   Compare and contrast the structures of proteins and nucleic acids.
    (AEB: June 1983, paper 3)

2 (a) Using chemical formulae, show how TWO amino acids may be chemically linked into a dipeptide.
(b) State the name for this type of reaction.
(c) Describe what is meant by:
   (i) the secondary structure of a protein,
   (ii) protein denaturation,
   (iii) a conjugated protein,
   (iv) an essential amino acid.
    (AEB: June 1984, paper 1)

3 (a) The diagrams show structural formulae of a compound.

Write down the name commonly given to this compound.
(b) Using the abbreviated form of the formulae show how molecules of this compound are linked together in chains in glycogen.
(c) In which cells in the body would you expect to find stores of appreciable amounts of glycogen?
    (AEB: June 1985, paper 1)

# Chapter 3

1 (a) Explain systematically how an experiment could be performed to determine the rate of a NAMED enzyme-catalysed reaction.

(b) (i) Name TWO factors which affect the rate of enzyme activity.

(ii) Describe how you would investigate experimentally the effect of one of these factors.

(c) With reference to the structure and mode of action of enzymes, explain the terms;

(i) specificity,

(ii) inhibition.

(d) Suggest ways in which enzyme inhibitors may be of use in human biology. (AEB: June 1982, paper 2)

2 The diagram summarises four stages (**A**, **B**, **C** and **D**) in the "lock-and key" theory by which an enzyme brings about the digestion of sucrose to monosaccharides.

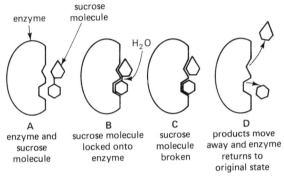

A
enzyme and sucrose molecule

B
sucrose molecule locked onto enzyme

C
sucrose molecule broken

D
products move away and enzyme returns to original state

(a) To which one of the following categories of enzymes does this one belong?

Isomerases, transferases, oxido-reductases, hydrolases, or lyases.

(b) Suggest an appropriate name for the enzyme.

(c) What are the names of the products?

(d) Where in the body would you expect to find this enzyme?

(e) What is the name often used to describe the total combination shown at **B**? (AEB: June 1986, paper 1)

3 (a) Define the term *enzyme specificity*.

(b)

| Substrate concentration millimols dm$^{-3}$ | 0.25 | 0.5 | 0.75 | 1.0 | 1.5 | 2.0 | 2.5 | 3.0 | 4.0 |
|---|---|---|---|---|---|---|---|---|---|
| Rate of reaction $\mu$g product min$^{-1}$ mg$^{-1}$ enzyme | 100 | 160 | 190 | 250 | 280 | 290 | 290 | 290 | 290 |

(i) Plot the data on a graph.

(ii) Give the name for the shape of this curve.

(iii) Give the factor which might have limited the rate of this enzyme-catalysed reaction above 2.0 millimols dm$^{-3}$.

(iv) Give details of the experiment you would do to test your hypothesis.

(c) Using a diagram to assist you explain the relationship between enzyme activity and temperature.

(d) Name THREE enzymes in the body which each have a different pH optimum. State the pH optimum for each enzyme.

(e) Define *coenzyme* giving ONE named example.

(f) State the single type of chemical reaction which is catalysed by lipases, peptidases and phosphatases.

(g)

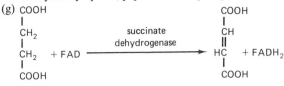

succinic acid

fumaric acid

The reaction shown occurs during the Krebs tricarboxylic acid cycle. An effective inhibitor of this reaction is malonic acid shown below.

COOH
|
CH$_2$
|
COOH

malonic acid

(i) Suggest how malonic acid may inhibit the enzyme.

(ii) Explain how the effects of the inhibitor could be overcome.

(AEB: June 1983, paper 2)

# Chapter 4

1 (a) The units from which DNA is constructed are called nucleotides. Draw a simple diagram to illustrate the nature of the link between two consecutive nucleotides.

(b) Lysozyme is a protein made up of 129 amino acids.

(i) How many DNA nucleotides are needed to encode for this chain of amino acids?

(ii) A complete turn of the DNA double helix contains 10 pairs of bases and is 3.4 nm long. What length of the DNA molecule is occupied by the gene for lysozyme?

(iii) How many turns of the DNA double helix does this represent?

(c) Explain concisely how nucleic acids are involved in the synthesis of proteins.

(d) Mutations can alter the structure of the DNA molecule. Describe FOUR changes in DNA which may result from mutation.

(e) (i) Explain the cause of sickle-cell anaemia in Man.

(ii) Explain why there is an association between the frequency of the sickle-cell–haemoglobin allele and the incidence of malaria in East Africa.

(AEB: June 1984, paper 2)

2 The following table shows some of the RNA (ribonucleic acid) codons (triplets of bases) together with the amino acids for which they code. U, C, A and G, are the initial letters of the bases that occur in ribonucleic acid:

| | | |
|---|---|---|
| **UUU** Phenylalanine | **UUA** Leucine | **UCU** Serine |
| **UAU** Tyrosine | **UGU** Cysteine | **CCU** Proline |
| **CAU** Histidine | **CAA** Glutamine | **CGU** Arginine |
| **AUU** Isoleucine | **AUG** Methionine | **ACU** Threonine |
| **AAU** Asparagine | **AAA** Lysine | **GAU** Aspartic |
| **GUU** Valine | **GCU** Alanine | acid |
| **GAA** Glutamic acid | **GGU** Glycine | |

**AUG** codes for the start of the peptide molecule ("capital letter").

**T** is the initial letter of the base **Thymine** which occurs in DNA.

(a) What sequence of amino acids could the following code in the DNA (deoxyribonucleic acid) give rise to?

—|GCA|GGA|CCA|—

(b) The following amino acid sequence is from the beginning of a human peptide.

Lysine—glutamine—threonine—alanine—

Give a possible DNA (deoxyribonucleic acid) code in the chromosome for this fragment of protein.
(AEB: June 1985, paper 1)

3 (a) Include in any order the following keywords in a concise account of protein synthesis in a cell:

ribosomes; nuclear pore; anticodon; amino acids; triplet; messenger RNA; transfer RNA; peptide; peptide bonds; protein.

(b) State ONE reason for the physical stability of the double helical structure of DNA.
(AEB: June 1983, paper 1)

# Chapter 5

1 The simplified diagram below outlines the principal stages in Man for the metabolism of glucose to yield energy.

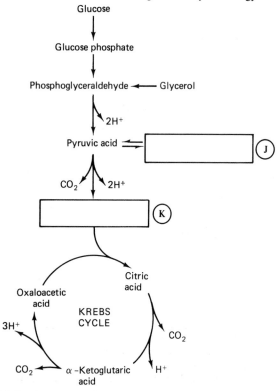

(a) In the rectangular boxes labelled **J** and **K**, write the names of the appropriate substances formed from pyruvic acid.
(b) (i) State under what conditions substance **K** would be formed,
    (ii) State where, and under what conditions, substance **J** is commonly formed in relatively **large** amounts.
(c) Show on the diagram, by means of arrows and by writing clearly, points at which "**fatty acids**" and "**amino-acids**" respectively, enter metabolism.
(d) Similarly, show on the diagram by arrows and by writing clearly the words "**ATP in**" and "**ATP out**", ONE point where ATP is consumed and ONE point where ATP is produced.

(e) Describe briefly what happens to the hydrogen atoms leaving the Krebs cycle between the time of their attachment to hydrogen acceptors and their final appearance as water. (AEB: June 1982, paper 1)

2 (a) The diagram shows the effect of different light intensities on the rate of photosynthesis of a green plant at 20°C and carbon dioxide concentration at 0.03%.

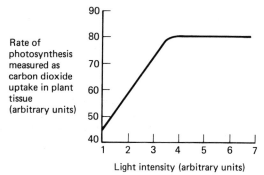

(i) Suggest a possible hypothesis for the levelling off of the curve.
(ii) How could the conditions be altered to test your hypothesis?
(b) (i) Define the term *absorption spectrum of chlorophyll.*
    (ii) State briefly how the absorption spectrum of chlorophyll corresponds with its action spectrum.
(c) (i) Name the technique used to separate the constituent pigments extracted from a green leaf.
    (ii) Describe how you would carry out this technique to achieve the separation.
(AEB: June 1983, paper 1)

3 (a) Show how the data in the following equation can be used to calculate the respiratory quotient:

$$C_{17}H_{35}COOH + 26O_2 \rightarrow 18CO_2 + 18H_2O + energy$$

(b) At a certain stage in their germination sunflower seeds have a respiratory quotient (RQ) of 0.7 and broad bean seeds have an RQ of 1.0.
    Infer from these RQ values the nature of the food reserves of each of sunflower and broad bean seeds.
(c) State TWO advantages of fats over carbohydrates as energy storage molecules. (AEB: June 1984, paper 1)

# Chapter 6

1 (a) Make a labelled diagram to represent the way in which molecules are arranged in the cell membrane.
(b) Name TWO sites inside a cell where you would expect to find such a membrane.
(c) Give THREE functions of the external cell membrane.
(AEB: 1985, paper 1)

2 (a) For a diploid cell with 3 pairs of chromosomes, each pair being of a different length, show, by means of clearly labelled diagrams *only* , the appearance of the chromosomes at;
    (i) metaphase of mitosis,
    (ii) metaphase I of meiosis,
    (iii) metaphase II of meiosis.

(b) Copy and complete the following table:

| Condition | Sex chromosomes | No. of Barr bodies |
|---|---|---|
| Normal woman | ................................ ................................ | |
| ................................ ................................ | | 0 |
| Turner's syndrome | ................................ ................................ | |
| Klinefelter's syndrome | XXY | ........................ |
| Down's syndrome | ........................ | 1 or 0 |

(AEB: June 1982, paper 1)

3 Using the symbols below make a diagram to illustrate the latest view of the structure of a section through the plasma membrane of an animal cell. Show on your diagram a membrane pore and label the inside and outside of the membrane.

(b) (i) What name is given to the layer to which glycolipids and glycoproteins both contribute?
(ii) State TWO functions of this layer.
(c) Name ONE other compound which is normally incorporated into the plasma membrane of higher animals cells such as those of Man.
(d) State TWO ways in which a relatively large, positively charged, water-soluble particle may be;
(i) hindered and
(ii) accelerated in its passage through the plasma membrane of a cell. (AEB: June 1982, paper 1)

# Chapter 7

1 (a) Give TWO differences between ways in which specimens are prepared for light microscopy and electron microscopy.
(b) (i) Define *resolution* of a microscope.
(ii) Give the resolution of the electron microscope.
(c) The diagram shows the thick and thin filaments of a skeletal muscle fibril as they appear in the electron microscope.

(i) Draw the muscle in a state of contraction.
(ii) Explain briefly how the change from the relaxed to the contracted state is thought to occur.
(AEB: June 1983, paper 1)

2 (a) For each type of exocrine gland shown diagrammatically below and in the next column, state ONE example of where it may be found:
(i) Simple tubular   (ii) Coiled tubular

(iii) Compound tubulo-saccular

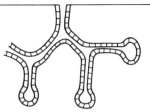

(b) For each of the following glands, state its location and one of its main functions:
(i) parotid glands,
(ii) Brunner's glands,
(iii) lachrymal glands,
(iv) sebaceous glands. (AEB: June 1982, paper 1)

3 Discuss the structure of (a) bone, and (b) cartilage in relation to their functions.

# Chapter 8

1 The diagram shows a section through the hip joint.

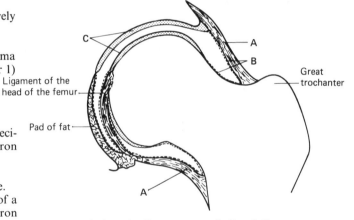

(a) Label on the diagram parts **A**, **B** and **C**.
(b) Suggest ONE important function of the following parts in the diagram.
(i) ligament of the head of the femur,
(ii) pad of fat,
(iii) great trochanter. (AEB: June 1986, paper 1)

2 (a) (i) Draw a fully labelled diagram of a named synovial joint.
(ii) State the movement permitted by the joint you have drawn and suggest which structures impose limitations on its movement.
(b) By reference to named examples in each case, state TWO other types of joint which differ in their degree of movement from that you have drawn.
(AEB: June 1983, paper 1)

3 (a) State TWO functions of the skeletal system OTHER than blood cell formation.
(b) Sketch the **anterior** aspect of a typical vertebra and label all those features possessed by all **unfused** vertebrae. (AEB: June 1981, paper 1)

527

4 Describe the macroscopic and microscopic structure of a long bone and discuss the ways in which this structure is related to its functioning. (AEB: June 1985, paper 3)

# Chapter 9

1 (a) Draw a labelled diagram to show the layers of the stomach wall in vertical section.
 (b) Describe how peristalsis occurs in the stomach to aid digestion.
 (c) (i) Define the term *sphincter*.
  (ii) Explain how sphincters control the movements of food contents through the stomach.
(AEB: June 1983, paper 1)

2 (a) Construct a table to show THREE differences in structure and function between incisors and premolars.
 (b) Bone is formed in a similar way to tooth dentine. Name the cells involved in
  (i) dentine formation,
  (ii) bone formation.
 (c) Name TWO dietary requirements which are essential to ensure normal tooth and bone growth.
 (d) By means of a flow diagram show the normal relations between vitamin D, parathyroid hormone (PTH) and plasma calcium levels.
 (e) Name ONE other hormone which has a direct effect on plasma calcium levels. (AEB: June 1982, paper 1)

3 The diagram shows a stomach with part of the wall cut away.

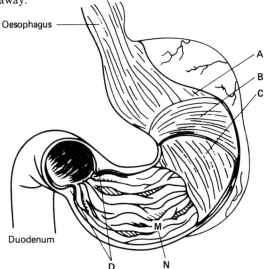

 (a) Identify:
  (i) the parts of the wall **A**, **B** and **C**,
  (ii) the valve **D**.
 (b) Make a diagram of a transverse section of the stomach wall through **M—N** to show one of the folds and label the layers of tissue you show (detail of cells is not required).
 (c) Explain how the stomach is not damaged by its own secretions.
 (d) Suggest a method whereby the fluid contents of the stomach may be obtained and the presence or absence of proteases in it established.
 (e) Briefly explain the part played by the nervous system in bringing about gastric secretion.
(AEB: June 1986, paper 2)

4 **EITHER** Give a reasoned account of the ways in which the stomach and the small intestine are adapted to their functions.
 **OR** Give a detailed account of the absorption of the end products of digestion by the ileum.
(AEB: June 1985, paper 3)

# Chapter 10

1 The diagram shows the relative distribution of the blood to the various organs at rest (lower scale) and during exercise (upper scale). During exercise the circulating blood is primarily diverted to the muscles. The area of the black squares is roughly proportional to the minute volume of blood flow. Not included is an estimated blood flow of 5 to 10% to fatty tissues at rest, and about 1% during exercise.

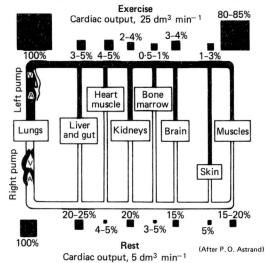

 (a) From the information given on the diagram calculate
  (i) the blood volume travelling through the lungs per minute when the body is at rest and during exercise;
  (ii) the increase in blood volume flowing through the heart per minute when the body engages in exercise from rest;
  (iii) the approximate amount of blood flowing through the muscles per minute during exercise.
 (b) Explain why changes in the blood supply are necessary to the heart muscle when the resting body becomes engaged in exercise.
 (c) What changes would occur in the heart activity to increase the cardiac output?
 (d) Explain the changes in the proportion of blood flowing through the gut and liver when the body is at rest and when it is engaged in exercise.
 (e) Outline the mechanisms by which cardiac output and blood flow through the muscles would be increased when exercise is undertaken.
(AEB: June 1986, paper 2)

2 (a) Construct a table to show as many ways as you can in which arteries differ from veins.
 (b) State THREE different advantages which are to be gained from the fact that blood contains mainly water.
 (c) (i) By means of a simple diagram and using the following data, describe the way in which tissue fluid is formed and replaced.

Blood pressure at the arterial end of capillaries
= 4.27 kPa
Blood pressure at the venous end of capillaries
= 1.60 kPa
Plasma osmotic pressure at the arterial end of capillaries = 3.33 kPa
Plasma osmotic pressure at the venous end of capillaries = 3.33 kPa

(ii) Which substance is largely responsible for the generation of the plasma osmotic pressure?

(iii) Explain why the plasma osmotic pressure at the arterial end may exceed that at the venous end of the capillaries.

(d) (i) If blood vessels are cut, blood loss is normally stopped quite quickly by clot formation. Re-arrange the components listed below to show how clot formation normally proceeds:
prothrombin, calcium ions, thromboplastin, fibrin, fibrinogen, thrombin.

(ii) Name ONE other component which is required for normal clot formation.

(iii) Blood flow from a wound often slows down before a clot forms. How may this observation be explained?            (AEB: June 1982, paper 1)

3 (a) State the site of production of each of the following substances and for each substance give ONE function.
   (i) heparin
   (ii) histamine
   (iii) prothrombin
   (iv) erythropoietin

(b) Describe how the Erythrocyte Sedimentation Rate (ESR) is measured and explain how this may vary with the health of the individual.

(c) Briefly outline the fate of worn-out red blood cells and their components.            (AEB: June 1983, paper 1)

# Chapter 11

1 (a) Describe the circumstances in which haemolytic disease of the newborn (HDNB) might arise.

(b) How is the problem usually averted?

(c) Why does HDNB only arise in about 10% of possible cases?

(d) Why is HDNB usually not a problem when the foetus and the mother are of different ABO groups?

2 Write short notes on EACH of the following:
(a) Immunoglobulins.
(b) The cellular immune response.
(c) Vaccines.
(d) The ABO blood group system.

3 Write brief notes on the following:
(a) Immunoglobulins,
(b) Rhesus blood groups,
(c) allergy, and
(d) tissue types.

# Chapter 12

1 A student measured her breathing volumes. When at rest and sitting down she found that her inhalation volume was $450 \, cm^3$ and she was inhaling every 6 seconds. When she

forcibly exhaled to a maximum having just breathed out, she found she blew out another $1750 \, cm^3$. When she filled her lungs to capacity and breathed out forcibly as far as she could she exhaled altogether $4100 \, cm^3$.

(a) What was her
   (i) vital capacity?
   (ii) inspiratory reserve volume?
   (iii) expiratory reserve volume?
   (iv) respiratory minute volume?

(b) If her maximum lung volume was $5000 \, cm^3$, calculate her residual volume.            (AEB: June 1986, paper 1)

2 (a) Give a description of how the anatomical features of the thorax allow inspiration to take place. In your description state clearly how the changes in volume and pressure in the thorax are brought about.

(b) A spirometer is a piece of apparatus that can be used to help measure lung volume. The tracing was made by a student who used a spirometer. He was asked to breathe steadily at rest, then to breathe in and out as deeply as possible and finally to breathe steadily when exercising.

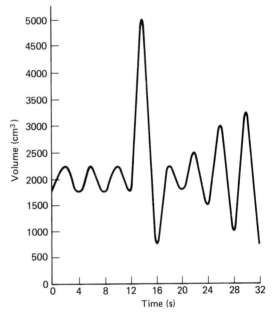

(i) How long after the start did exercise begin?

(ii) Use the spirometer tracing to calculate: a. tidal volume at rest, b. vital capacity, c. inspiratory capacity, d. expiratory reserve volume.

(c) The ventilation rate can be defined as the volume of air breathed per unit time. The resting ventilation rate of a person was found to be $8 \, dm^3$ per minute. Two experiments were carried out. In one experiment the person inhaled pure oxygen and in the other, a mixture of 92% oxygen and 8% carbon dioxide. The following results were obtained.

| Time after start of experiments (minutes) | Average ventilation rate ($dm^3 \, min^{-1}$) | |
|---|---|---|
| | 92% $O_2$:8% $CO_2$ | Pure $O_2$ |
| 0.5 | 11 | 7.5 |
| 1.0 | 24 | 7.2 |
| 1.5 | 35 | 6.8 |
| 2.0 | 42 | 6.5 |
| 2.5 | 46 | 6.5 |

(i) Explain the difference in ventilation rate with reference to the nervous system and the blood system.

(ii) In the interval between 1.0 and 1.5 minutes on pure oxygen the person breathed in and out nine times. Calculate the tidal volume under these conditions and show the method of your calculation.

(d) From the table of data on p. 525, suggest ONE reason for mouth-to-mouth resuscitation being preferred to other methods of artificial respiration.

(AEB: June 1985, paper 2)

**3** Write an account of the physiological changes which occur in the body during exercise. Discuss the possible long-term effects caused by taking regular strenuous exercise.

(AEB: June 1984, paper 3)

**4** Plot the following data as oxygen dissociation curves:

| Oxygen tension (kPa) | Percentage saturation of haemoglobin with oxygen | | Volume of oxygen in blood (cm³ per 100 cm³) |
|---|---|---|---|
| | Carbon dioxide tension (5.3 kPa) | Carbon dioxide tension (9.3 kPa) | |
| 1.3 | 7 | 4 | 2 |
| 2.6 | 27 | 15 | 4 |
| 3.9 | 53 | 35 | 6 |
| 5.3 | 70 | 58 | 8 |
| 6.6 | 79 | 71 | 10 |
| 7.9 | 85 | 82 | 12 |
| 9.3 | 90 | 88 | 14 |
| 10.5 | 95 | 94 | 16 |
| 11.8 | 98 | 98 | 18 |
| 13.2 | 100 | 100 | 20 |

In the tissues, as a result of cellular respiration, the carbon dioxide tension rises from 5.3 kPa to 9.3 kPa. The oxygen tension in the same tissues is 4.2 kPa.

(a) What proportion of the haemoglobin in the blood in the tissues capillaries is induced to dissociate?

(b) By how much would the oxygen tension in the tissues have to drop to cause the same amount of dissociation if the carbon dioxide tension remained constant?

(c) What happens to the oxygen released from the haemoglobin?

(d) What volume of oxygen would be released from oxyhaemoglobin if the carbon dioxide tension were to rise as described above AND the oxygen tension were to drop as in (b) above?

(e) The atmosphere contains about 21% by volume of oxygen. However, because of the mechanics of breathing, the air in the lungs contains about 14% oxygen. If the atmospheric pressure is 100 kPa, what proportion of the haemoglobin in the pulmonary capillaries should become oxygenated?

(f) If 1 g of haemoglobin combines with 1.36 cm³ oxygen at STP, calculate the haemoglobin concentration for the individual whose blood was used to obtain the data for the dissociation curves above. Assume standard conditions and show your working clearly.

# Chapter 13

**1** (a) A healthy young man was given 1 dm³ of water to drink on each of two occasions. On the first occasion only water was given; on the second occasion a subcutaneous injection of 0.5 units of posterior pituitary extract was given at the same time as he drank the water. The table shows the volume of urine produced on each occasion.

| Time from start of experiment (min) | Volume of urine produced (cm³ min⁻¹) | |
|---|---|---|
| | 1 dm³ water only taken | 1 dm³ water taken and posterior pituitary extract injected |
| 0 | 0.5 | 0.5 |
| 15 | 2.0 | 0.6 |
| 30 | 4.0 | 0.75 |
| 45 | 16.0 | 0.6 |
| 60 | 11.0 | 0.5 |
| 75 | 9.0 | 0.4 |
| 90 | 5.0 | 0.3 |
| 105 | 3.0 | 0.3 |
| 120 | 2.0 | 0.3 |
| 135 | 0.5 | 0.3 |
| 150 | 0.5 | 0.3 |

(i) Plot these data on ONE graph.

(ii) Discuss the results of this experiment.

(b) An artificial kidney consists of blood flowing directly from a patient's body through cellophane tubing immersed in a fluid as illustrated in the sample diagram below.

**Artificial kidney**

(i) For what structure does the cellophane tubing substitute?

(ii) Explain how a salt, e.g. sodium chloride, is removed from the blood using an artificial kidney.

(iii) Explain how the artificial kidney is used to remove water.

(iv) A person undergoing dialysis must avoid certain foods. Name three substances which should be restricted giving a reason in each case.

(AEB: June 1984, paper 2)

**2** Give an account of water balance and acid–base balance in Man and discuss their control.

(AEB: June 1982, paper 3)

**3** (a) Diagrams (**A**) and (**B**) represent sections through a Malpighian body. Name the parts indicated by the letters **D**, **E**, **F** and **G**.

(b) Ultrafiltration takes place in the Malpighian bodies.
  (i) State concisely what provides the pressure to form the glomerular transudate.
  (ii) Explain why the Malpighian body would not function efficiently if it looked like diagram **B**.

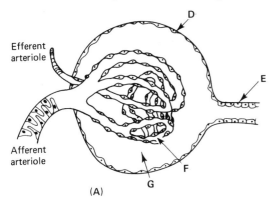

Efferent arteriole

E

Afferent arteriole

F

G

(A)

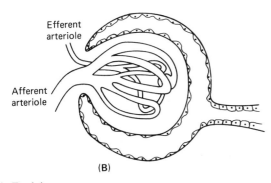

Efferent arteriole

Afferent arteriole

(B)

(c) Explain
  (i) why the OVERALL diameters of the afferent and efferent arterioles are different.
  (ii) how this difference enables the afferent arteriole to control the volume of blood filtered in a given time.

(d) Name the enzyme produced by the kidney which increases blood pressure in the Malpighian bodies.

(e) (i) Calculate the rate of renal clearance of a substance **X**, given that the concentration of substance **X** in plasma is $0.2\,\mathrm{mg\,cm^{-3}}$, but it is secreted in urine at a rate of $0.2\,\mathrm{mg\,min^{-1}}$. Show your method of calculation and indicate the units in your answer.
  (ii) How long does it take $1\,\mathrm{dm^3}$ of plasma to be cleared of substance **X**?

(AEB: June 1985, paper 2)

# Chapter 14

**1** (a) (i) Define a *simple reflex action*. Give a suitable NAMED example.
  (ii) Draw a **TRANSVERSE SECTION** through the spinal cord to show all the components involved in your NAMED simple spinal reflex.
  (iii) Explain briefly how the brain may become aware of, and able to exercise control over, a simple spinal reflex.

(b) Define the term *conditioned reflex*. Give a suitable example.

(c) State TWO features of the axons of neurons which may result in an increased rate of nerve impulse transmission. (AEB: June 1983, paper 1)

**2** The diagram below shows a transverse section of spinal cord.

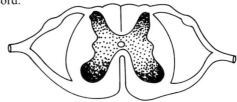

(a) Label the following structures on the diagram;
  (i) spinal ganglion (sensory nerve ganglion),
  (ii) ventral root,
  (iii) posterior horn of grey matter,
  (iv) lateral column of white matter,
  (v) anterior (ventral) fissure,
  (vi) somatic sensory part of the grey matter,
  (vii) somatic motor part of the grey matter.

(b) State ONE major difference between the appearance of a section through the cerebral cortex and that of the spinal cord shown above.

(c) The central canal of the spinal cord is continuous with the ventricles of the brain.
  (i) What do they both contain?
  (ii) Name ONE other place where this substance is found.
  (iii) State where this substance is produced.
  (iv) State TWO functions of this substance.

(AEB: June 1982, paper 1)

**3** The diagram shows a vertical section through the longitudinal fissure of the brain.

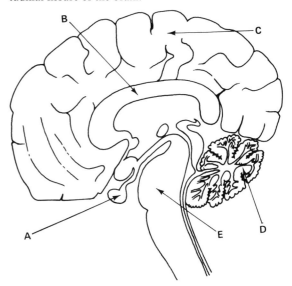

(a) Name the features **A**, **B**, **C**, **D** and **E**.

(b) State ONE function for each labelled feature.

(c) State precisely in which part of the brain the following body functions are regulated.
  (i) speech    (iii) temperature control
  (ii) water balance    (iv) ventilation.

(d) Briefly explain the role of the sympathetic nervous system in the regulation of body temperature.

(AEB: June 1984, paper 1)

**4** The graph below shows an intracellular recording of the potential within a nerve cell and shows changes resulting from its excitation by other nerves which synapse with it.

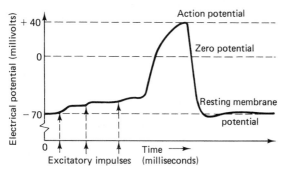

(a) Explain the following terms;
  (i) resting membrane potential,
  (ii) action potential.
(b) Explain why, individually, the excitatory impulses shown did not generate an action potential, but collectively they did.
(c) Name TWO transmitter substances which would depolarise the post-synaptic membrane.
(d) Explain why the release of transmitter substance into a neuro-muscular junction does not stimulate the muscle to go into a **prolonged** contraction.
(AEB: June 1982, paper 1)

# Chapter 15

**1** The diagram represents the retina of the human eye as seen in section.

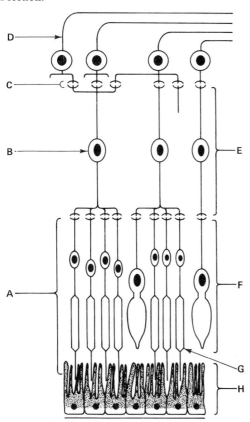

(a) Name the parts **A** to **H**.

(b) Account for the following:
  (i) Rod cells produce an indistinct image.
  (ii) Rod cells are concerned with night vision.
  (iii) Cone cells are capable of colour perception.
  (iv) Visual acuteness is greatest at the centre of the fovea.
  (v) When a person enters a dimly lit room from bright sunlight, the room at first seems dark but gradually objects become visible.
(London: Adv. Biology, June 1983)

**2** (a) (i) The diagram shows a section through a part of the human ear. Identify the parts labelled **I**, **II** and **III**.

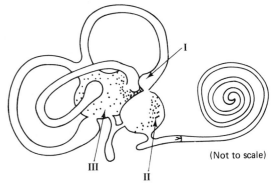

(Not to scale)

  (ii) Draw an enlarged, labelled view of the interior of structure **I**, giving a brief description of how its structure is related to its function.
  (iii) Which parts of the brain receive stimuli from this region of the ear?
  (iv) Explain briefly how the muscles are involved in the maintenance of the body's position.
(AEB: June 1983, paper 1)

**3** (a) Make a large, clearly labelled diagram of the inner ear.
(b) Describe how each of the following is achieved:
  (i) hearing,    (ii) balance.

**4** (a) What are the main features of receptors in the human body?
(b) Described how light is channelled onto the receptors in the retina in such a way as to produce a sharp image.
(c) What is (i) myopia, and
      (ii) hypermetropia?
(d) How may myopia and hypermetropia be corrected?

# Chapter 16

**1** (a) Explain the effects of parathyroid hormone (PTH) on plasma calcium levels.
(b) State TWO functions of Vitamin D and its derivatives on calcium metabolism.
(c) Give THREE possible effects of a **low** plasma calcium level on body functions.   (AEB: June 1984, paper 1)

**2** (a) The diagram shows interaction between the hypothalamus, pituitary and thyroid glands. Arrows indicate probable pathways of direct influence.

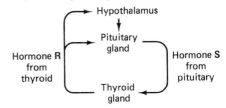

(i) Give the names of the hormones **R** and **S**.
(ii) State the effect of hormone **S** on the thyroid.
(iii) State the effect of hormone **R** on the hypothalamus.
(iv) Describe the control of the pituitary by the hypothalamus in this situation.
(v) What is this kind of control of the pituitary by the thyroid called?

(b) Read the following passage and then answer the questions below. Water soluble hormones and steroid hormones both alter cell function by activation of genes but by somewhat different mechanisms. Since steroid hormones are lipid-soluble, they easily pass through the plasma membrane of the target cell. Upon entering the cell, a steroid hormone binds to a protein receptor site in the cytoplasm and the hormonal-receptor complex is translocated into the nucleus of the cell. The complex interacts with enzymes necessary to alter cell function in a specific way.
(i) Explain what is meant by a target cell.
(ii) Why should lipid-soluble hormones pass through plasma membranes easily?
(iii) Suggest a mechanism and a pathway by which the hormonal-receptor complex may be translocated.
(iv) Describe a way in which the complex may alter cell function.

(c) Explain clearly the part played by hormones in regulating the concentration of calcium in the blood.
(AEB: June 1986, paper 2)

**3** Describe the feedback loops which control the release of hormones making particular reference to the thyroid gland and the ovary. (AEB: June 1984, paper 3)

**4** (a) What general features characterise endocrine co-ordination? Illustrate your answer with reference to the islets of Langerhans and their secretions.
(b) Briefly compare and contrast the mode of action of hormones and the autonomic nervous system.

# Chapter 17

**1** (a) (i) Define *basal metabolic rate* and state the conditions under which it should be measured.
(ii) Explain why a very young baby has a slightly higher basal metabolic rate than an adult.
(iii) Suggest why the basal metabolic rate in females is slightly lower than in males.

(b) The body cannot store oxygen but its six cubic decimetres of blood can hold about one cubic decimetre of oxygen. Every time a cubic decimetre of oxygen is consumed approximately 20 kilojoules of heat energy are produced.

The table shows the total oxygen consumption for different body activities.

| Body activity | Speed km $h^{-1}$ | Heart stroke volume $dm^3 min^{-1}$ | Total oxygen consumed by body $dm^3 min^{-1}$ |
|---|---|---|---|
| Rest | 0 | 5 | 0.25 |
| Walking | 3.2 | 10 | 0.8 |
| Running | 12.0 | 25 | 3.0 |

(i) A. If a person walked 16 kilometres as a speed of 3.2 km $h^{-1}$ calculate how much total oxygen he would need to consume.
B. Calculate the number of cubic decimetres of blood which would be pumped through the heart during this walk.
(ii) Calculate the energy output when running for 15 minutes at 12.0 km $h^{-1}$.
(iii) A. Calculate how long one cubic decimetre of oxygen in blood would last if a person held his breath while resting.
B. Describe the effect on the resting oxygen consumption in a person who develops hyperthyroidism.

(c) There are TWO different methods of determining basal metabolic rate
(i) human calorimetry
(ii) indirect calorimetry.
Briefly discuss EACH method. State any advantages or disadvantages associated with its use.

(d) (i) State all the different ways in which heat losses occur from the body surface.
(ii) Describe how the temperature regulating mechanism of the body normally operates to conserve heat when the internal temperature of the blood is falling below normal. (AEB: June 1983, paper 2)

**2** The White, Eskimo, Alacaluf (from Tierra del Fuego) and Australian Aborigine are four ethnic groups. In a study of response to cold in these groups, individuals were subjected to a cold environment and their skin temperatures recorded. By measuring their oxygen uptake over the period of the experiment, their heat production could be estimated. The results of studies on four individuals, one from each ethnic group, are summarised in the graph.

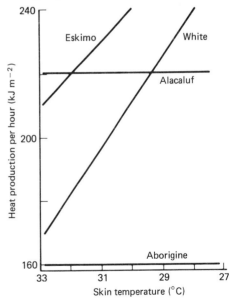

(a) What are the differences in metabolic activities between the individuals in this study?
(b) How is heat normally generated in the human body?
(c) How might the differences shown in the graph be explained physiologically?
(d) What further experimental evidence is necessary to enable the experimenter to draw conclusions concerning human physiological patterns of response to cold?
(AEB: June 1986, paper 2)

**3** (a) Define the term endothermy, regarding body temperature regulation.

The diagram below shows the relation between resting metabolic rate and ambient (environmental) temperature in a mammal.

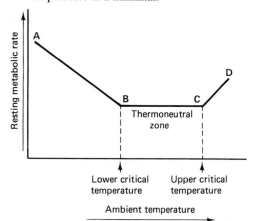

(b) (i) Explain the significance of the thermoneutral zone between points **B** and **C**.
(ii) What initial effect does a rise in metabolic rate have on internal body temperature?
(c) Suggest why the resting metabolic rate rises when a mammal is subjected to environmental temperatures between (i) points **B** and **A** and (ii) points **C** and **D**.
(d) Describe and explain one type of physiological mechanism by which mammals dissipate body heat.
(e) Explain why a small person is liable to lose heat to the environment more rapidly than a larger person.

# Chapter 18

**1** (a) Describe the hormonal and structural changes that constitute the ovarian cycle.
(b) How is the cycle altered during pregnancy?
(c) How is the cycle re-established after pregnancy?

**2** (a) Identify the structures labelled **L**, **M**, **N** and **O** on the diagram of the female urinogenital system shown below.

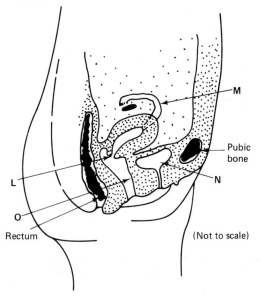

(b) Mark on the diagrams by writing the letters **P**, **Q** and **R** respectively, the points at which:
(i) sperm is normally deposited during coitus (**P**),
(ii) fertilization of the ovum normally occurs (**Q**),
(iii) implantation normally takes place (**R**).
(AEB: June 1982, paper 1)

**3** Give an account of the changes in maternal physiology that occur from the moment an ovum is fertilised until the onset of parturition. Suggest any special nutritional requirements that are necessary to maintain the mother in good health.
(AEB: June 1983, paper 3)

# Chapter 19

**1** The graph below shows typical examples of annual growth velocity curves for the height of boys and girls.

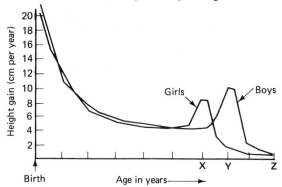

(a) (i) State the approximate ages indicated by the points **X**, **Y** and **Z** respectively.
(ii) Name the period during which the increases in height shown at **X** and **Y** occur.
(b) Using information from the graph ONLY, describe the pattern of growth in boys and girls.
(c) Explain how growth is controlled by hormones during the period shown and especially why the growth rate increases at **X** and **Y**.

The second graph shows the level of two hormones during the menstrual cycle of a woman.

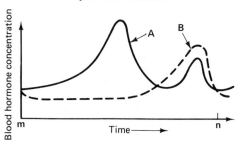

(d) (i) What is the average length of the menstrual cycle (i.e. time **m**—**n**) in women?
(ii) Identify the hormones **A** and **B**.
(iii) a. How many days before the onset of the next menstrual cycle does ovulation normally occur?
b. On which day of the menstrual cycle does menstruation commence and for how many days does it usually continue?
(e) Describe FULLY the changes which occur in reproductive hormone levels during one menstrual cycle if fertilisation does not occur.

(AEB: June 1982, paper 2)

**2** "Growth is not a simple increase in size". Discuss this statement with reference to the growth of a human foetus.

**3** Write an essay on ageing.

**4** Give an illustrated account of the foetal development of the central nervous system.

# Chapter 20

**1** (a) Man has an extended juvenile phase compared with other Primates. Give **three** examples of this and suggest an evolutionary importance for each one.

(b) List **four** of the conditions which are necessary for the Hardy–Weinberg equilibrium to occur.

(c) Sickle-cell anaemia is a condition in which erythrocytes containing abnormal haemoglobin break down if they are subjected to low oxygen tensions. It is common in West Africa where malaria is endemic.

Sickle-cell anaemia is controlled by a single pair of alleles. If the allele for normal haemoglobin formation is **H** and the allele determining abnormal haemoglobin is **h**, then individuals with normal haemoglobin have genotype **HH**; those with the fatal sickle-cell condition have genotype **hh**; and heterozygotes with genotype **Hh** suffer a milder form of the anaemia called **sickle-cell trait**.

In some areas of West Africa 1 in 4 of the population at birth showed the fatal sickle-cell anaemia. Assuming that the Hardy–Weinberg principles apply:

(i) Calculate the frequency of occurrence of the genotypes for
  1. **sickle-cell trait** and
  2. normal haemoglobin
  using the Hardy–Weinberg formula where $p$ and $q$ are the frequencies of the two alleles.

(ii) Explain the following observation:
  Among the descendants of African slaves transported to the Americas, the sickle-cell trait frequency is about 10%.

(d) (i) List **four** different types of change in DNA which may result in mutation of a gene.

(ii) Explain how a gene mutation results in sickle-cell anaemia. (AEB: June 1986, paper 2)

**2** (a) The ability to taste low concentrations of phenylthiourea carbamide (PTC) is only possessed by seventy-five per cent of the population and is due to a dominant autosomal allele.

One form of migraine is the result of an independently-segregating dominant autosomal allele.

Using appropriate symbols, with full explanation of the genetic basis of your reasoning, work out the genotypes and phenotypes of the possible offspring of a mating between:

(i) two individuals each heterozygous for both PTC tasting and migraine.

(ii) a double heterozygote for PTC tasting and migraine and a double homozygote for non-tasting and non-migraine.

(b) For a diploid cell with TWO pairs of chromosomes, each pair being of a different length, show by means of simple, clearly labelled diagrams the appearance of the chromosomes during:

(i) prophase I of meiosis,
(ii) metaphase II of meiosis,
(iii) anaphase of mitosis,
(iv) anaphase I of meiosis.

(c) (i) State TWO ways in which genetic variation can arise as a consequence of meiosis.

(ii) Briefly outline the process of spermatogenesis using labelled diagrams.

(iii) Give ONE difference between spermatogenesis and oogenesis during meiosis. (AEB: June 1983, paper 2)

**3** (a) Distinguish between an *autosomal dominant allele* and a *sex-linked dominant allele*.

(b) Haemophilia is inherited as a sex-linked recessive trait. By means of a diagram, using appropriate symbols, illustrate the possible offspring of a mating between a phenotypically normal woman, whose father was a haemophiliac, and a phenotypically normal man. Identify any symbols you use.

(c) With reference to the above mating:

(i) if the couple's first child is a son, what is the probability of this son being a haemophiliac?

(ii) what proportion of their daughters will be haemophiliacs?

(iii) what proportion of their daughters will be carriers? (AEB: June 1984, paper 1)

# Chapter 21

**1** Critically discuss the statement that, "The course of evolution is determined by mutation and selection". (AEB: 1982, paper 3)

**2** What is meant by the term *species*?

Discuss the genetic processes whereby new species could have arisen in the course of evolution. (AEB: June 1985, paper 2)

**3** The diagram illustrates the "mosaic" character of hominoid evolution. This means that different features have evolved at different rates so that primitive and advanced characteristics are found side by side.

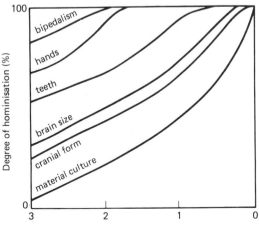

Discuss how development in this mosaic pattern may have enabled *Homo sapiens* to emerge as the dominant animal species. (AEB: June 1983, paper 3)

# Chapter 22

**1** (a) The human population of the earth is now increasing exponentially.
   (i) Explain the difference between exponential and arithmetic growth increases.
   (ii) Identify the two most important changes in Man's way of life which have been instrumental in bringing about this exponential growth.
(b) Fundamentally, the present explosive rise in the world's population is due to the fact that birth rate exceeds death rate.

The graph shows an example of the demographic transition which has taken place in an industrial country as a result of these trends.

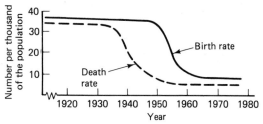

Describe the changes which would have taken place in the population of such a country during the years:
   (i) 1920–1930,
   (ii) 1940–1950,
   (iii) 1950–1960,
   (iv) 1970–1980.
(c) The crude birth rate is the ratio of the number of births in a given population to the total population during a specified period (usually 1 year).

State one reason why this ratio is rather uninformative and suggest how a more informative ratio could be calculated.
(d) It is expected that there will be about 6000 million humans on the earth by the year 2000 AD.

Discuss the changes which will have to take place in Man's way of life to allow the continued survival of this number of people.     (AEB: June 1982, paper 2)

**2** The growth rate of an aerobic bacterium was measured by inoculating some cells into a sterile nutrient broth maintained at a constant temperature. At regular intervals 1 cm$^3$ samples were removed and the number of living cells in the sample was determined. The results are represented below in graphical form.

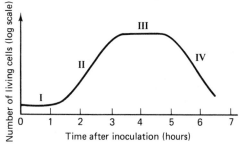

(a) The graph shows four phases of growth **I, II, III** and **IV**.

Explain fully the changes in the growth rate of the culture which occurred in each phase.
(b) Describe briefly an experiment to investigate further any explanation you have given in (a).
(c) Explain FOUR activities of bacteria beneficial to Man.

(d) Many bacteria can be grown on nutrient agar but viruses cannot. Explain this statement.
(e) Penicillin is effective against bacteria, but not against viruses. Explain why this is so.
(f) Some bacterial infections can be treated using antibiotics. What are the problems associated with the use of:
   (i) narrow spectrum antibiotics, e.g. penicillin?
   (ii) wide spectrum antibiotics, e.g. streptomycin?
                              (AEB: June 1984, paper 2)

**3** (a) What is meant by
   (i) crude birth rate?     (ii) infant mortality rate?
(b) With reference to populations that have been
   (i) increasing
   (ii) stable
   (iii) declining,
   what is the effect on the numbers if the birth rate stays about the same but death rate falls? (Assume no significant emigration or immigration)
(c) Giving suitable examples, distinguish between *birth control* and *contraception*.
(d) In a culture of bacteria, set up in a test tube under controlled conditions of temperature and light, the rate of increase in numbers may fall to zero within 72 hours. Suggest THREE factors that could be responsible for this.
(e) Give THREE methods that can be employed to control the numbers of internal parasites, such as roundworms, infecting human beings.
(f) Although food production could be sufficient for the whole of the human population on this planet Earth, malnutrition may not be eradicated. Suggest FOUR reasons for this.     (AEB: June 1985, paper 2)

# Chapter 23

**1** (a) Distinguish between:
   (i) a *community* and an *ecosystem*;
   (ii) a *food chain* and a *food web*.
(b) Explain the following observations:
   (i) the number of links in any food chain is usually less than five;
   (ii) organisms usually decrease in number but increase in mass towards the end of a food chain.
(c) The diagram shows three ecological pyramids of number, **A, B** and **C**.

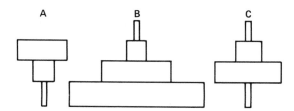

   (i) Deduce which pyramid or pyramids include(s) a parasitic relationship.
   (ii) Deduce which pyramid or pyramids correspond(s) to a single large primary producer which is able to support decreasing numbers of consumers at each trophic level.
   (iii) Give ONE example of a food chain of organisms for EACH of the pyramids **A, B** and **C**.

536

(d) The diagram and table show a pyramid of energy.

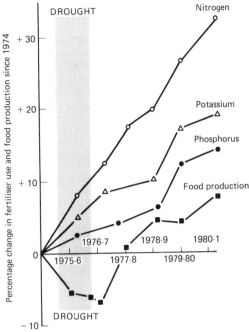

| Total energy assimilated as new biomass (kJ m$^{-2}$ y$^{-1}$) | Total energy flow through each trophic level (kJ m$^{-2}$ y$^{-1}$) |
|---|---|
| 25 | 87 |
| 280 | 1603 |
| 6186 | 14 098 |
| 36 974 | 87 110 |

(after Bushell)

(i) Suggest TWO reasons for the energy losses which occur at each trophic level.
(ii) Ecological efficiency can be defined as *the net production of new biomass at each trophic level as a percentage of the total energy flowing through that trophic level.*
   Calculate the ecological efficiency of the **secondary** consumers in this community.
(iii) List TWO ways in which the ecological efficiency could be improved between plant production and the biomass of beef cattle as primary consumers.
(e) Fish-eating birds by a Canadian lake were found to be poisoned by applications of DDT which had been sprayed onto the water to kill the midge larvae in the mud at the bottom of the lake.
   Suggest ways in which the fish in the lake may remain unharmed by the insecticide.
(f) Proliferation of algae often occurs in lakes in warm weather.
   (i) Suggest TWO reasons for this occurrence.
   (ii) Briefly describe any problems which may result.
(AEB: June 1986, paper 2)

**2** "It is ironic that seven tenths of the earth's surface is covered with water and yet there is a world wide demand for water far in excess of supply. Water holds a key position in Man's economy and its control is altering the face of the earth." Discuss this statement.
(AEB: June 1985, paper 3)

**3** Discuss the ways in which man modifies the natural cycling of carbon and nitrogen in the biosphere.

# Chapter 24

**1** The graph shows the use of fossil fuels by Man. Discuss these observations and comment upon their implications.
(AEB: June 1982, paper 3)

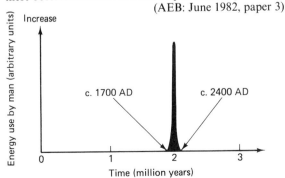

**2** The diagram shows the percentage change in the use of fertilisers and food production since 1974.

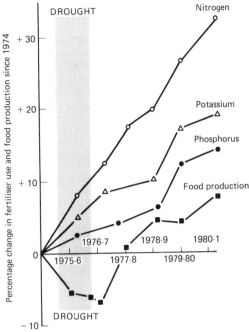

(a) Comment on the data above in a general discussion outlining the advantages and disadvantages of modern farming methods of food production.
(b) Suggest ways in which present food productivity might be improved. (AEB: 1984, paper 3)

**3** The data below indicates death and pollution levels in London in December 1952. Pollution levels were measured at 12 different stations throughout the city.

| Date in December 1952 | Mean atmospheric sulphur dioxide (µg m$^{-3}$) | Mean atmospheric smoke (µg m$^{-3}$) | Total number of deaths per day |
|---|---|---|---|
| 1 | 300 | 180 | 250 |
| 3 | 450 | 300 | 290 |
| 5 | 950 | 900 | 490 |
| 7 | 2300 | 1600 | 875 |
| 9 | 2250 | 1100 | 825 |
| 11 | 1500 | 230 | 530 |
| 13 | 1350 | 175 | 490 |
| 15 | 1350 | 175 | 400 |

(Pollution levels from the Royal College of Physicians 1970; Deaths from the Registrar General's Office)

(a) (i) Plot the three sets of data on a single graph.
   (ii) Critically discuss any causal relations that may be inferred from these data.
   (iii) How would you test your conclusions, either by experiment or by the use of further data?
(b) (i) Name FOUR major sources of sulphur dioxide in the atmosphere.
   (ii) Explain THREE methods that could be used to limit the amount of sulphur dioxide in the atmosphere.
(c) Name THREE other common atmospheric pollutants, and for each name a possible source of the pollutant and suggest a method by which its level could be reduced. (AEB: June 1984, paper 2)

# Chapter 25

**1** (a) The bacterium *Gonococcus* is the organism responsible for the venereal disease gonorrhoea.

(i) An agar surface on a Petri dish was seeded with infected pus containing the *Gonococcus*. Filter paper discs impregnated with four different concentrations of penicillin were then laid on top before incubating the dish for 48 hours at 30 °C.

This procedure was carried out on pus from different patients in 1956 and 1976 and the results are shown after incubation.

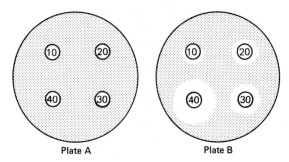

Plate A          Plate B

(The relative concentration of penicillin in each disc is shown in arbitrary units.)

1 Suggest which agar plate, **A** or **B** showed the result obtained in 1976.
2 Explain the reasons for your answer in 1.
3 Describe THREE aseptic procedures that would be necessary in carrying out an antibiotic sensitivity test similar to the one described.

(ii) Explain why gonorrhoea is easier to detect in males than in females.
(iii) Gonorrhoea has reached epidemic proportions in many countries. Suggest FOUR different factors which have influenced the incidence of the disease.

(b) Many bacterial and viral diseases can be prevented by vaccination. The graph shows the effects of antigen injections and the amount of antibody in plasma.

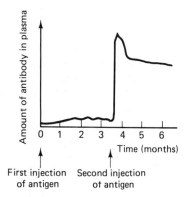

(i) Using data from the graph and your own knowledge, explain why a vaccination programme includes spaced intervals of injections.
(ii) Distinguish between the terms *active* and *passive* immunity.
(iii) State which type, active or passive immunity, is achieved by vaccination.
(iv) Suggest why a vaccine for the common cold has not yet been successful.     (AEB: 1986, paper 2)

**2** Give an illustrated account of the large-scale disposal of sewage at a modern urban treatment works. Suggest any problems which might occur if the efficiency of this process is low.                    (AEB: June 1983, paper 3)

**3** (a) Give FOUR important "portals of entry" by which micro-organisms obtain access to deep inside the human body. In each case name ONE organism or disease entering by that route.
(b) What is an *antigen*?
(c) What are the main phases of attack and destruction employed by neutrophils in the elimination of bacteria from the human body?
(d) On a nutrient agar plate seeded with a suspension of the bacterium *Staphylococcus aureus* small plugs of agar containing three different species of *Streptomyces* (species **X**, **Y** and **Z**) were placed as shown in **A** in the diagram.

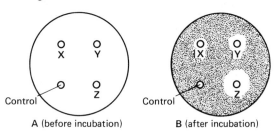

A (before incubation)          B (after incubation)

The plate was then incubated for 48 hours at 30 °C and the result is shown in **B**. A clear area is evident around each of the plugs containing the *Streptomyces*. How may the results be interpreted?
(e) Smallpox has been eliminated as a major disease in the world. What important measures have been necessary to effect this?                    (AEB: June 1985, paper 2)

# Index

Figure references are set in bold type.